PETERSEN'S CHEVROLET HIGH-PERFORMANCE

Introduction

"Baseball, Hot Dogs, Apple Pie and Chevrolet" is the title that C.J. Baker, HOT ROD Magazine's Executive Editor, gave to a story of his that concerned the development of the Chevy small-block V8. And I can think of nothing more appropriate; Chevy's ubiquitous engines—not only the small-blocks, but all of Chevrolet's powerplants, especially in high-performance form—are indeed a part of Americana. Ever since that memorable day in late 1954 when the Chevrolet Division announced the release of their amazing 265-cubic-inch V8, hot rodders around the world have revered the lightweight, high-revving screamers for use in virtually every vehicle that rolls, flies or swims.

On the heels of the Chevy V8's introduction came a cascade of special high-performance parts that continues to this day, as hot rodders, specialty parts manufacturers and even factory engineers themselves sought to improve the horsepower and torque of every member of this engine family. And along with the trick parts came countless thousands of hours spent in developing engine-modifying techniques; tips and clues on not only urging extra horses out of the Chevy, but permitting all this exotica to "live" under the heat of prolonged battle.

Perhaps no one will ever boil down all this development and experimentation to a set of standards and place it between a single set of covers. But we've done our best to bring at least the current state-of-the-Chevy-art to a sharp focus in this **Chevrolet High-Performance** book, a part of our new HOT ROD SHOP SERIES. The result is a careful gleaning of only the best, most authoritative Chevrolet material from HOT ROD Magazine's files. What's more, we've also put together for you the most complete listing of Chevrolet factory-engineered high-performance parts and suggested where and how to get them.

So sit back, flip through the pages that follow, and get ready for a slice of real Americana: the realization of even greater performance from the world's best-known engines.

Spence Murray

CHEVROLET HIGH-PERFORMANCE

Edited by Spence Murray and the staffs of HOT ROD Magazine and the Book Division. Coordinated by Marlan Davis, associate editor of HOT ROD Magazine.

COVER: Legendary Keith Black took time off from his, er, "other" engines to assemble this exotic-looking, impressively strong-running, aluminum-block Chevrolet 350. Its 366 horsepower wouldn't have raised an eyebrow a decade ago, but with today's gas and required low compression, this is a hi-po yet streetable, modern state-of-the-art engine. Photo by Pat Brollier, PPC Photographic. Cover design by Dick Fischer.

ISBN 0-8227-6005-3

Library of Congress Catalog Card No. 80-80176

Contents

ENGINE BUILDUPS

4 **NUMBER ONE** Why Chevy leads the pack
10 **SMALL-BLOCK MAGIC** Inside our cover engine
14 **ALL IN THE FAMILY** Small-block racing comparison
16 **611-HP SMALL-BLOCK** Mario Rossi does his thing
20 **MAXI-MOUSE** Getting 454 hp from the small-block
23 **ACCEL'S 350 SHOWCASE** For street and strip
26 **501/302** Through an Edelbrock buildup
28 **TRIPLE THREAT** Detailing a short-block small-block
36 **SMOKEY'S SMALL-BLOCK** The man himself shows how
53 **BUILDING A BLOWN 350** On huffin' and puffin'
58 **ENGINES YOU'VE PROBABLY NEVER HEARD ABOUT** Really!
60 **PROVEN PERFORMANCE: THE BIG-BLOCK CHEVROLET** Improving on Chevy
66 **SMOKEY'S 600-HP ROAD RAT** On Chevy's largest

BOLT-ON POWER

72 **STREET POWER** The ways to achieve it
76 **TURBOCHARGING** Performance key to the '80s
82 **IMMACULATE DECEPTION** Don't tell 'em we told you
84 **BLOWIN' IN THE WIND** On huffin' and puffin'
91 **FAST AND FLASHY** Something for everyone
94 **GREATER THAN AN EIGHT** Ever heard of the Six?
98 **500-HP SMALL-BLOCK BUILD-UP** Somethin's in the wind
101 **DOUBLE TROUBLE** And twice as nasty

RODS, PISTONS AND CRANKS

104 **LIGHT TOUCH** Milodon's aluminum "mouse motor"
106 **PANNING FOR GOLD** Sumps from Moroso for Jenkins
108 **THE K-ROD** Somethin' stout between piston and crank

CAMS, VALVES AND EXHAUST

110 **TICKET TO RIDE** Big-block rockers, small-block head
113 **THE LITTLEST PORCUPINE** "Secret" heads
114 **THE OFFSET STUD** Little-motor tricks
115 **BROWNFIELD HEADS** Watchin' your diet
116 **HIGH C** Trick factory big-block heads
118 **CLAMP DOWN** Sealing in high compression
120 **FINE TUNING** On degreeing a cam

FUEL SYSTEM

122 **HOLLEY CARB MODS** On doin' it yourself
127 **KITCHEN TABLE CARB REBUILD** Honest!
130 **THE PERFECT MATCH** Intake manifold alignment
133 **WATER POWER** Water injection—it *does* work
134 **SUPER TUNING** Mild mods for street Holleys
138 **Q-JET QUICKIE** A how-to on Rochesters

TRANSMISSION

142 **TAKING 5** The Doug Nash street 5-speed
145 **MIX AND MATCH** A converter treatise
146 **EARLY DEPARTURE** Updating the venerable T-10
150 **TORQUE CONVERTER TIPS** An automatic how-to
151 **'GLIDE GUIDE** An extensive look-see

REAR ENDS

162 **12-BOLT MAGIC** Doin' it to it
165 **TAXING THE I.R.S.** More beef for the Vette

CHASSIS

168 **CHEVERRA: POOR MAN'S PORSCHE** Makin' it handle
172 **GETTING IT HANDLED** Makin' it handle even better
176 **THE NAKED LAUNCH** Bracket car suspension
178 **V8 VEGA SWAP KITS** When and where to get 'em
180 **GET DOWN AND BOOGIE** The best in roadability
183 **SWAP MEAT** What to look for and where

FEATURE CARS

186 **BRACKET ATTACK** Approaching it properly
198 **HOW TO BUILD A STREET MACHINE** The right way
206 **NOVAMOTION** A Nova in motion—get it?
212 **DOUBLE VISION** The same machine two ways
218 **STRIP SHOW** Jenkin's Pro Camaro
221 **WINNING WAGON** Yep, that's what we said
224 **TURBO DIES EL-CAMINO** Matching the new with the old
228 **A FACTORY FUELIE** Dream about it, but don't touch
232 **LO 'MAD** Ground-hugging Nomad stripper
234 **CHEVY'S ULTRA Z** Trickest Camaro of all
236 **CLONING CAMAROS** Look-alikes, or are they?
240 **BIRTH OF AN $1100 BRACKET RACER** For real

SPECIAL 242 **THE COMPLETE HIGH-PERFORMANCE PARTS GUIDE**

Number One

After some 25 years of production, the small-block Chevrolet V8 is still the standard by which others are judged.

(Special thanks to Chevrolet Public Relations, Buick Public Relations, Herb Fishel, Vince Piggins and Paul Prior)

The year 1952 was an exciting point in American automotive history. New overhead-valve V8 engines were being developed by various manufacturers: Oldsmobile and Cadillac had debuted their new engines in '49, the Chrysler Hemi had appeared in '51, and Lincoln had introduced their new OHV V8 for the '52 model year. On the horizon, the Dodge Hemi was to appear in '53, Ford and Mercury would get their versions of the OHV V8 in '54, and Pontiac, Plymouth and Chevrolet V8s were scheduled for introduction in '55.

It was the spring of '52 when Ed Cole was named chief engineer at Chevrolet. At the time of his appointment, Cole was given a mandate by GM—upgrade Chevrolet's image from that of a stodgy old six-cylinder manufacturer who made cars for the lower-priced market segment, to that of a contemporary manufacturer offering exciting products to the American motoring public. His first effort was to upgrade Chevrolet's image from the standpoint of performance. What followed became one of the greatest automotive success stories of all time: the small-block Chevrolet V8.

To understand why the small-block Chevy V8 emerged as a technological revolution in engine design, a little must be known about the people who were responsible for the philosophies and concepts behind the engine. We've already mentioned the late Ed Cole. Cole was a dynamic, flamboyant individual who wouldn't take "no" for an answer. He was always pushing his people beyond the limits at which they felt comfortable, extending the existing technology. Many see that attitude as a key to Cole's success—he eventually became president of General Motors.

Of course, there's always an element of risk involved in trying new engineering concepts. In the early '70s, Ed Cole allegedly said, "If you wanted to take the same sort of risks that I proposed at Chevrolet today, you'd promptly be fired." However, Cole had surrounded himself with a group of brilliant engineers to do the job. In 1946, Cole was chief engineer at Cadillac, where he and Harry Barr were the driving forces behind the design of the 1949 Cadillac OHV V8. Barr transferred to Chevrolet in '52, at Cole's request, as assistant chief engineer to help with the design of a new V8.

"We had a very talented engine designer already working at Chevrolet by the name of Al Kolbe. Al had been a drafting room supervisor and had management experience, so he was delighted to become chief designer of the new engine."—Harry Barr

---

The late Al Kolbe was a meticulous engineer and a super engine designer. He took on Ed Cole's ideas for a new V8 engine as a personal challenge. A 230-cubic-inch V8 was already on the drawing boards under the auspices of the previous chief engineer, Ed Kelley, but it was a very conservative, heavy engine that would have been expensive to produce. Cole scrapped that design, specifying a new lightweight, high-revving, high-power-output, free-breathing V8 utilizing thin-wall precision-casting techniques (previously unheard of in automotive foundries) and a minimum of casting cores to reduce the likelihood of core shifts and casting alignment problems. Cole also wanted to incorporate inexpensive green sand cores (cores made of packed moist sand, as opposed to dry sand cores, which are hardened by baking) in an effort to make this the most inexpensive V8 in the industry.

At the time, John Dolza was heading an advanced engineering group at GM Central Engineering. With Dolza's help and green-sand, inverted-casting techniques, it became possible to cast the

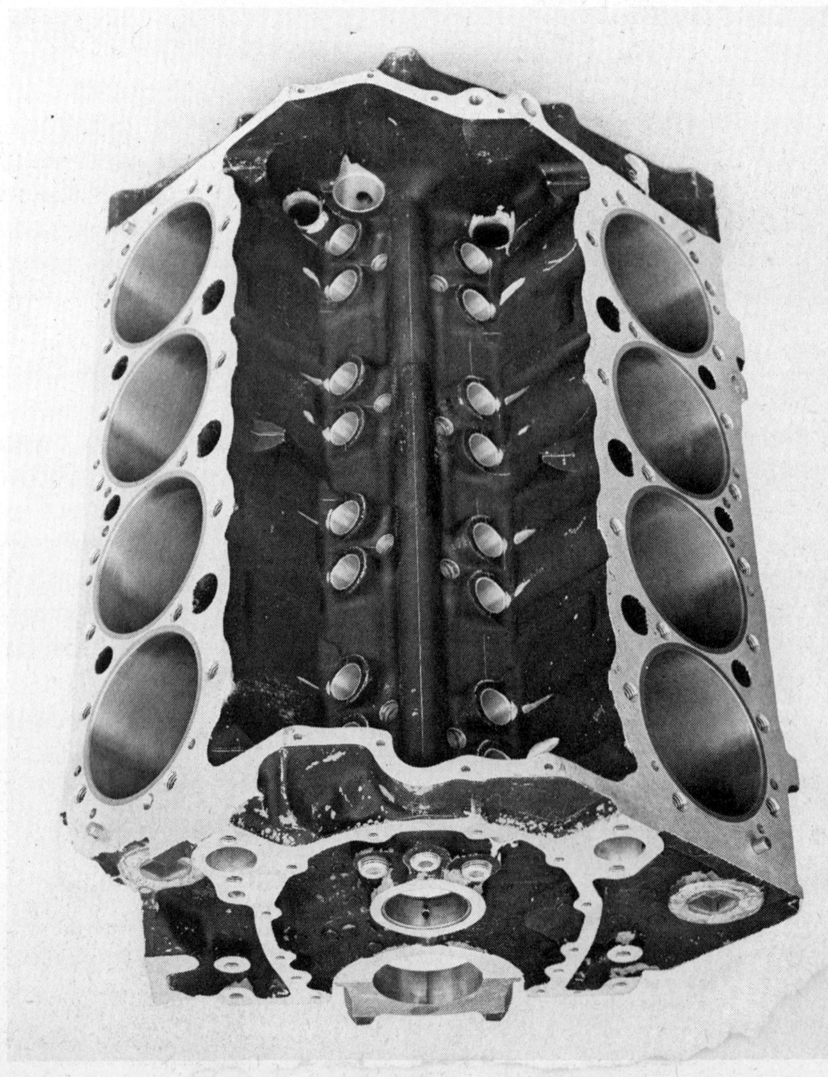

block with only nine major and three minor cores. Other manufacturers were using more than 20 cores.

The basic dimensions of the block were determined by the crankshaft. After totaling the widths of the rod and main bearings, crank arms, bulkheads, etc., the total crank length dictated a block 21¾ inches long. Wherever metal or water could be removed from the block design, it was, in an effort to make the block as light, compact and economical as possible. Bore centers were established at 4.4 inches.

The original design called for a displacement of 245 cubic inches with a 3¾-inch bore, but as the engine began to take shape on the drafting boards, it was concluded that it could displace 260 cubic inches just as easily, which would have called for a 2.93-inch stroke. At the time, Ed Cole was enthusiastic about hydroplane boat racing, and he suggested rounding off the bore to an even 3 inches, producing a displacement of 265 cubic inches, which just happened to coincide with a 266-cubic-inch class limit for hydroplanes!

"When we initially designed that block at 265, we thought we were stretching the foundry to the absolute limit from the

2

3

1. Considered almost miniscule in the world of contemporary V8s, the new Chevy bent-eight had a block only 21¾-ins. long. The drawing board had called for 245 cu. ins., but in the end, it was taken to 265 cu. ins.

2. The original design for '55 was so conservative that many bore/stroke ratios could be utilized and output grew as the years passed. By 1970, for example, this 350 cu.-in. version put out 300 horses.

3. Two of the men under Ed Cole's direction who helped immeasurably in making the small-block the envy of the automotive world. At left, Don McPherson, drafting room supervisor and head of the cylinder head design group. Right, Loren Papenguth, project engineer who solved the valvetrain oiling problems.

standpoint of technology—wall thickness, water between the bores and bolt pattern . . . we thought that was pretty well as big as the engine would ever get."—Don McPherson

---

There's more to the story, though, than minimal block dimensions and casting techniques. Al Kolbe's assistant, Don McPherson (who went on to be general manager of Buick) was the cylinder head group leader in the drafting room, and as any racer will tell you, the cylinder head is the key to the performance of any engine. McPherson prepared numerous cylinder head sketches based on what his group thought the head should have. The sketches deviated from conventional designs that were basically rectangular in cross section. Again, cast iron and water were eliminated wherever possible to make the head light, compact and inexpensive. Since it was to be a high-output engine, provisions for large valves were included, along with free-breathing ports. Eventually, they arrived at a sketch that is basically what the head still is today. Cole approved the design and told Kolbe to make a workable head from the sketch. The result was a head with cast-to-shape, wedge-type combustion chambers for high turbulence and combustion control, high exhaust ports (for smooth flow and to elevate the exhaust manifolds for steering box clearance) and five bolts around each cylinder.

"The engine was designed to be free-breathing to start with, and we designed the intake and exhaust ports so that they were as clean as possible. Then we went back and put the head bolts in. Because of the configuration of the ports, we couldn't get a standard four-bolt pattern in. That's the reason we went to five bolts around each cylinder. It was expedient, and Ed Cole agreed to it, after much browbeating."—Don McPherson

---

Five bolts around each cylinder—it was unconventional, but it turned out to be one of the main features that added versatility to the design. Because the five bolts could achieve adequate clamping force with less torque loading than a four-bolt arrangement, internal block ribbing was not required and there was minimal block distortion. Gasket sealing was also enhanced, which helped permit the eventual growth of the small-block to 400 cubic inches.

Another consideration was connecting rod length and piston compression height. Since the original criterion was to make a high-revving engine of oversquare design (bore diameter larger than stroke), a very short connecting rod length was deemed desirable, along with a short, flat-top piston. This approach reduced the reciprocating weight and the block height, once again compacting the engine.

Probably the most controversial feature of the small-block Chevy V8 was the valvetrain, featuring stamped steel rocker arms pivoting on independent stud-mounted balls. Equally unique was the through-the-pushrod oiling system for the valvetrain. Our own *Rod & Custom* reported in March 1955, "The rocker arm system of the [Chevy] V8 is receiving ridicule from automotive enthusiasts. Whether such ridicule is warranted or not can only be determined by the test of time." The reliability of the system is now a matter of fact.

Actually the stamped steel rocker arm was not a Chevrolet invention—it was a Pontiac development. Cole liked the idea and received permission from GM to use it on the Chevrolet V8, so both Pontiac and Chevy engines introduced them in 1955. The man who designed the stamped steel rocker arm for Pontiac was Clayton B. Leach, a motor development engineer who later became an assistant chief engineer at Pontiac.

"I developed the stamped rocker arm at home in my basement shop back around 1947. I worked on it at both places—Pontiac and at home, too. The project occupied all of my time. I had a keen interest in the stamped rocker arm concept. When you get a radical idea, it stays with you. The finished piece looked like a flimsy piece of tin, but it worked."—Clayton B. Leach

---

There was more to the system, however. Instead of utilizing conventional rocker arm shafts for the rocker arms, each rocker pivoted on an independent

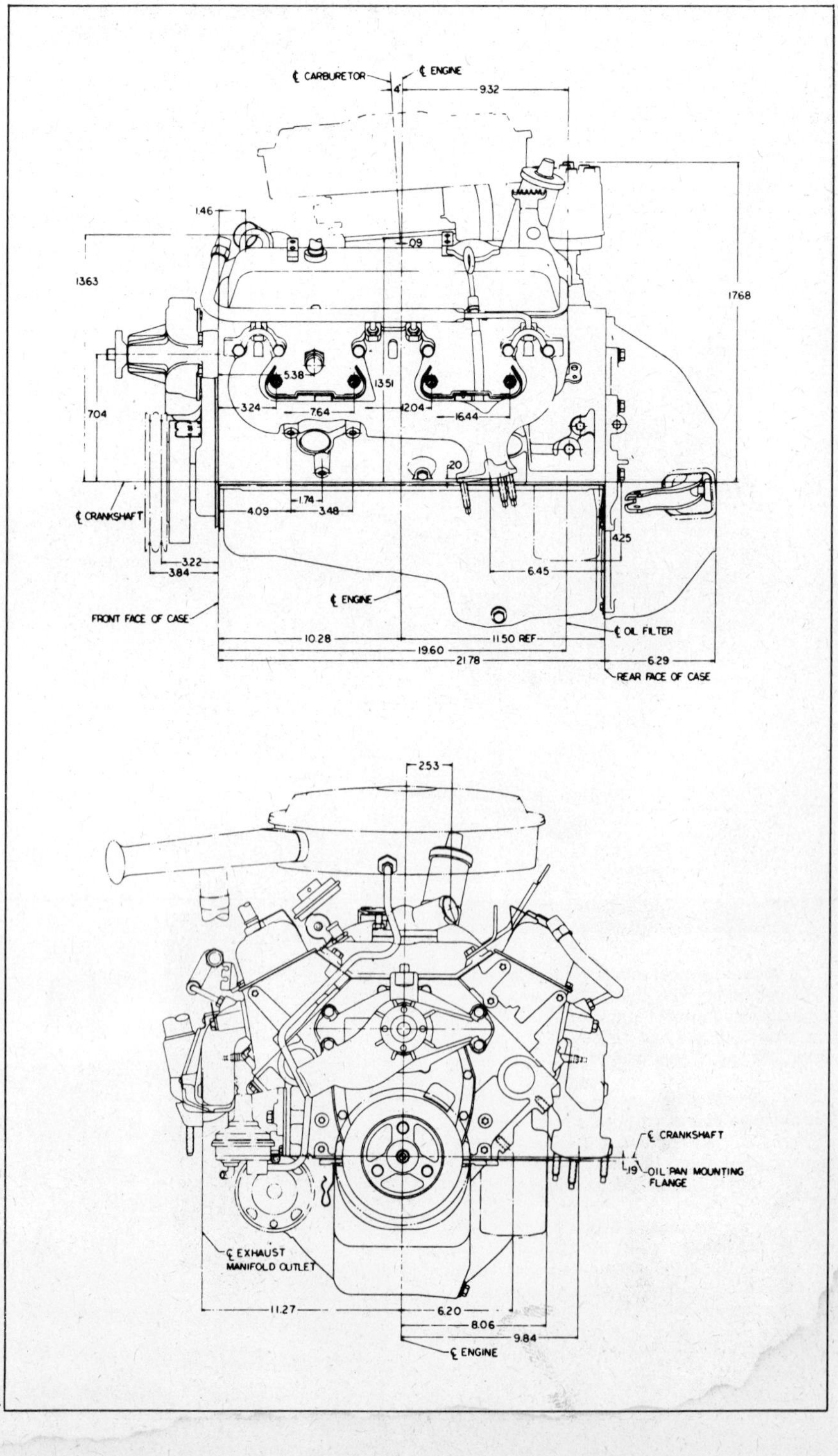

4

ball located on rocker studs that were pressed into the head, thus simplifying cylinder head machining and assembly for reduced weight and cost. The result was a very lightweight valvetrain, but here the similarity to the Pontiac system ended. Pontiac had chosen to lubricate their pushrod tip, rocker balls and valve tips by pumping oil through drilled passages in the rocker studs. Chevrolet took a different approach, oiling the valvetrain through the tappets and hollow pushrods, thus keeping all pressurized oil passages within the block. It was one of the biggest problems confronting the engineers as the engine was designed, because using the pushrods as oil galleries tended to flood the top of the engine with oil.

The problem was finally solved by a Chevrolet project engineer, Loren Papenguth. He developed the piddle valve, a tiny metal wafer, in the tappet. It was inexpensive and it worked. Once the oil flow was controlled and delivered to the rocker arm, the rocker arm stamping served as a trough to carry the oil to the rocker ball and valve tip. Oil splash provided the valve spring and valve stem oiling. The system worked so well that Pontiac eventually copied it. The final combination also provided the breakthrough for increased crankshaft speeds due to the minimal valvetrain weight.

"The valvetrain itself, as it finally evolved, was lighter than any other OHV V8 valvetrain in use at the time. I think it's still one of the lightest. We were able to get extremely high speeds compared with other engines of the day without losing control of the valvetrain—up to about 5600 rpm. When we added more sophisticated dampening mechanisms, we were able to exceed 6000 rpm. Of course, that inspired a lot of interest in the racing fraternity."—Loren Papenguth

---

The basic oiling system of the engine was every bit as good as the valvetrain lubrication. With the single addition of full-flow oil filtration, the oiling system remains unchanged after 25 years of production.

Buttoning up the top of the engine was yet another innovation—an intake manifold that served a variety of purposes. It sealed the top of the engine without a separate valley cover, provided for distributor mounting, provided a take-off for heater water, included an exhaust heat crossover for carburetor heat, had an oil filler opening, a thermostat housing, a single water outlet for the radiator and a temperature gauge sender pad—all in one casting.

The final engine design that was put into production late in '54 for the 1955 model year fulfilled all Ed Cole's requirements. It was lightweight—about 40 pounds lighter than the standard Chevrolet straight six. It was high-revving, free-breathing, and it was powerful. A 2-barrel carbureted version rated at 162 hp, and a Power Pack 4-barrel, dual-exhaust version rated at 180 hp were both offered in '55.

"It was a damn good performing engine. I had one of the Power Pack versions in a new '55 Chevrolet with a manual transmission. The first year I got three tickets within three months' time. Everywhere I went, I got caught. I damn near lost my license."—Don McPherson

---

Technologically, the 265 was decades ahead of its time, but even so, its potential was far from being realized in 1955. What was to happen in the next 10 years was a chain reaction of events that captured the spirit of the automotive enthusiast and motoring public alike. It also established the small-block Chevrolet as the most popular performance engine of all time.

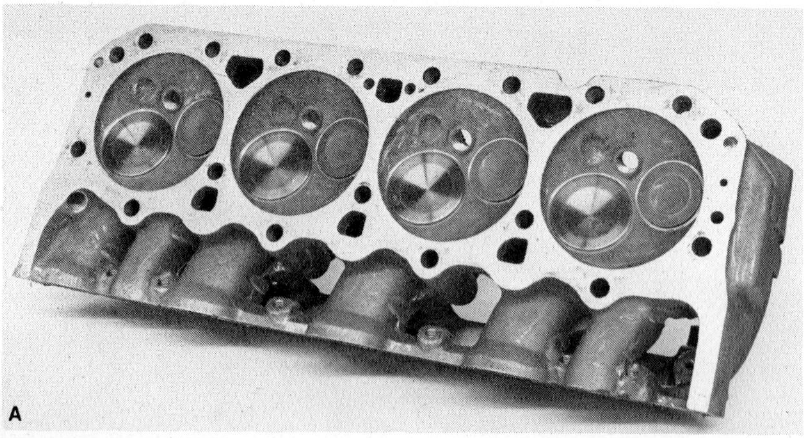

A

4. Throughout its production, the external dimensions of the small-block remained the same, even though the displacement grew from 265 to 400 cu. ins.

5. Over the years, a number of exotic pieces were tested. Shown here are several that never made it to production, such as the (A) semi-hemi aluminum porcupine head that featured identical intake and exhaust ports for each cylinder; an unusual overhead-cam aluminum head (B) with true hemispherical chambers; a very exotic overhead-cam head (C) with square intake ports, a hemi chamber and spark plug access ports above the exhaust ports; and a couple of aluminum manifolds (D) for Weber carburetors.

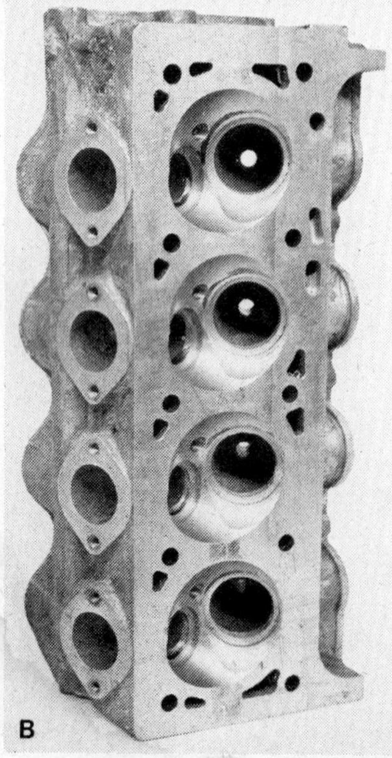

B

C

D

5

The year 1955 was a time the likes of which we will never see again. Prices were low, speed limits were high, auto insurance rates weren't prohibitive, and gasoline was both plentiful and cheap. NASCAR stock car racing was beginning to mature, and the good ol' boys were racing with essentially *stock* automobiles. There were no thoroughbred racing cars blistering around the tracks at 200 mph thinly disguised as something vaguely resembling a production body, as we have today. Even more amazing, the factories were directly sponsoring cars in competitive events, and the horsepower race was on in a very big way. Before 1955, hot rodding had been the private domain of Ford, and Hudson Hornets and Oldsmobile 88s were dominating stock car racing. Selling the motoring public the same kind of performance in the dealer showroom on Monday that had won at Darlington on Sunday was a way of life, and then suddenly, leading the pack was the small-block Chevrolet.

"Herb Thomas had been racing Hudsons back in the late '40s and early '50s. I had built most of the cars for him. He came to me in '55 and wanted to run a Chevy. I was semi-retired in those days with other interests—those were my drinking days. At the time, I had been doing some work on Buicks—funny how things come full-circle. Ed Cole wanted some NASCAR involvement, and at that point Chevy race cars were unheard of. Well, one thing led to another, and Chevrolet supplied a car and $10,000 to get it ready for Darlington. Cole wanted to win Darlington the first time out, but I was just concerned with going the distance to prove the car and engine. Herb Thomas was going to drive.

"We got the car ready. I found a loophole in the rules—they had a rim width rule, but no diameter restriction. At the time, everybody was running 15-inch wheels. I put 16-inchers on and found some Firestone tires that had been specially made for another application—hell, they even had whitewalls, but they did have a big footprint.

"We won the race by three or four laps, running on one set of tires, stopping only for fuel. It worked 'cause Herb had been 'shook' by a previous crash, so he was kind of scared and held back, staying out of trouble. Of course, after Darlington, they outlawed the 16-inch wheels."—Smokey Yunick

---

These were the formative years when the small-block began to be developed to release some of the potential in its inherently good design. Of particular significance were the contributions of two men—Zora Arkus-Duntov and Henry "Smokey" Yunick. Duntov was in the process of establishing the performance image of the Corvette for Chevrolet, and Smokey was heading up the factory's highly successful racing efforts in '55.

"Zora Arkus-Duntov came into the operation with the Corvette program in 1953. The first Corvette had a six-cylinder with three sidewinders [sidecraft carburetors] on it. Zora then went to V8s in the Corvette in '55. His input into the Power Pack program was very instrumental in what we got out of the engine. He had the public relations ability and charisma to sell such programs to higher management. Some of the stuff he developed went into the aftermarket and heavy-duty-parts program. It was readily available, low-priced, well-engineered and easy for everybody to use."—Don McPherson

---

The availability of low-priced, thoroughly tested, durable performance parts directly from Chevrolet certainly was a key factor in the small-block's popularity. No other engine has ever benefited from such an array of aftermarket performance parts either, thanks in part to the engine's acceptance, helping to perpetuate that popularity. In truth, it was the racing community, and Smokey Yunick in particular, that developed most of the significant performance improvements on the small-block Chevy V8. Perhaps it's best to let Smokey tell you about it in his own words:

"I was doing all sorts of heavy-duty-

6

7

6. Much of the factory development of the small-block must be credited to this man, Zora Arkus-Duntov, the father of the Corvette. He designed many pieces for the small-block that were to enhance the performance image of the Vette.

7. Frequently contracted by Chevrolet to develop high-performance pieces for the small-block was Smokey Yunick. The list of items originated or refined by Yunick reads like the Chevy heavy-duty parts catalog.

8. Simple and compact, all extra cast iron had been removed from the design to save weight and costs. Clean port design was a criterion for performance in the small-block. Note that a full third of the engine's total height consisted of the carb and oil-bath air cleaner.

parts development on rods, heads, pistons, etc., for Chevrolet. I even wanted to move the spark plug location to where it is now in the angle-plug heads clear back in '56, but Cole thought it was already correct. Much of what was going on then is similar to what I'm doing for the Buick V6 today. We had to modify *everything* for power and durability on the Chevy.

"The pistons were a disaster. We needed a piston that was round in the bore when it was hot and running, regardless of what it looked like cold. We developed the barrel-faced piston. All of the performance-piston dome configurations came out of my shop, too.

"Some of the other things that were developed here or originated here for the Chevy V8 are: virtually all of the factory aluminum-intake manifolds; the four-bolt block; blind head bolt holes in the block; the angle-plug head; heat riser elimination in the heavy-duty head; the 2.05-inch intake valve; the titanium intake valve; the dual 4-barrel carburetion setup; the first mushroom tappet cam for a Chevy; gear drives, screw-in rocker studs; pushrod guide plates; the full-floating pin piston/rod setup; double Tru-Arc piston pin retention; the moly-faced compression ring; the nodular iron flywheel and harmonic balancer; and a bunch of other stuff that still hasn't seen the light of day, like the tall deck block and the spiral combustion chamber head.

"I also did some work on aluminum blocks, aluminum heads, overhead-cam heads, air-gap manifolds, special head gaskets and internal dry-sump oil systems. There's been so much stuff, I get tired just thinking about it.

"In fact, it was Ralph Johnson and I who worked out the fuel injection system for 1957. That's when we painted all the windows in the shop black and put up chain-link fences around the place 'cause Cole was afraid somebody would find out what we were up to."

Of course, the racers weren't the only ones improving the small-block. Chevrolet was doing its part, implementing the innovations of people like Smokey and Duntov, along with constantly refining existing components. They were also busy enlarging the mighty mouse, as it had come to be called, to compete in the horsepower and displacement wars. In 1957 the 265 grew to 283 cubic inches. It grew again in '62 to 327 cubic inches, and finally to 350 cubic inches in '67, with nothing more than bore-and-stroke enlargements of the original design. Compression ratios had also risen from the original 8:1 to over 11:1 on the highest output versions."

"The 283 turned out to be, to my mind, kind of a happy marriage. It was probably the best combination of bore and stroke we'd had up to that time. It was a lot better performing than the 265. We stayed with the 3-inch stroke and just bored it out from 3¾ to 3⅞ inches, but that little increase in bore seemed to help quite a bit. We just kept expanding the bore until we went out to 4 inches for the 327.

"The 327 was a major challenge to the foundry. They'd never done reliable sand cores that were that thin. We were going way beyond anything else in the industry at that time. We thought you couldn't go any further with bore increases, so we started stroking the engine. It was then that we realized that we really hadn't stretched that bugger at all. We went up to a 3¼-inch stroke [also for the 327], and before we were done, we'd increased the stroke to 3.48 inches [for the 350] . So you can see how conservative we were with the original design in '55."—Don McPherson

The final step was siamesing of the cylinder bores to permit a 4⅛-inch bore and stretching the stroke to 3¾ inches for a displacement of 400 cubic inches, which is, in all probability, as large as the small-block will ever get in production. Since its introduction 25 years ago, roughly 35,000,000 small-blocks have been built by Chevrolet. Various intermediate displacements have appeared over the years, such as the 302 with a 4-inch bore and a 3-inch stroke, which was probably the closest the factory ever came to building a true racing engine in production. It was fitted in the '67 Z28 Camaro. Other displacements include the 307 (3.875x3.25), the 305 (3.736x3.48), the 262 (3.671x3.10) and the 267 (3.500x3.48). This last group of engines was produced primarily for emissions and mileage applications.

What is the future of the small-block Chevy V8? Production of engines larger than 305 cubic inches for passenger car use will end with the 1980 model year. The only likely exception will be the Corvette, which may continue to receive the 350. Also, 350- and 400-cubic-inch versions for truck applications will continue for some time, but small-block production for any application beyond 1985 is, at best, vague. It is almost certain that aluminum will replace cast iron for some major components, such as cylinder heads, intake manifold, water pump and possibly even the block, for production engines, in an attempt to reduce weight for energy conservation before the plug is finally pulled on the small-block. Similarly, factory turbocharger development will continue, again with an eye toward mileage rather than performance. The costs of the final stages of development will be high, and although the small-block will eventually disappear from production, it is a tremendous tribute to its original designers—Ed Cole, Harry Barr, Al Kolbe, Don McPherson and all the others—that it not only survived but prospered for so long. As Harry Barr commented, "A good engine design ought to last a manufacturer 12 to 15 years." The small-block has now been around for 25 years, with production anticipated for at least another five years. Perhaps Don McPherson summed it up best:

"The small-block was exceedingly well done, and it set the stage for everybody else to copy, which they did, but in all honesty, nobody has improved on it. They've copied it, and they've picked up some of the features, but no manufacturer has put the total package together any better than that engine."

C.J. Baker

Small-Block Magic

Our cover engine combines the latest Chevrolet aluminum block, a pair of rare factory aluminum heads, a host of aluminum aftermarket performance components, and the engine-building skills of Keith Black.

mag-ic (maj'ik), n.—1. the art of producing a desired effect or result through the use of various techniques that presumably assure human control of supernatural agencies or the forces of nature. 2. the exercise of this art. 3. any extraordinary or irresistible influence, charm, power, etc. 4. mysteriously enchanting.

Random House Dictionary

mag-ic (maj'ik), n.—1. an all-aluminum small-block Chevrolet V8 built by Keith Black Racing Engines, Inc.

C.J. Baker

Although my definition of magic is a bit different from that of the folks at Random House, it's really the same thing. Just mention a professionally built all-aluminum Chevy small-block in the midst of any group of car nuts and all the conditions of the first definition are met, especially if the sorcerer in question is none other than the legendary Keith Black. Begin describing the engine and the crowd becomes quiet; place the finished engine in front of them and they speak in whispers and hesitate to reach out and touch the dull alloy castings or polished and anodized covers. You could almost classify it as reverence—or at least a respect for something that is valued at over $13,000.

In truth, there is nothing supernatural about the engine at all. Yes, it is exotic and expensive, but it is really nothing more than the product of technological evolution carefully assembled by a very talented and experienced group of people at Keith Black's shop in South Gate, California. Even the material, aluminum, isn't especially exciting. Yes, aluminum is light, strong, a good conductor of heat and electricity, but it's also the same stuff beer cans are made of.

So now that we've stripped away some of the wondrous nature of this powerplant, let's get down to the specifics, because as availability of Chevrolet's aluminum block increases, and with the likelihood of new aluminum heads looming on the horizon, the feasibility of duplicating such an engine is a definite reality. We might add that Keith Black contributed his highly respected talents to the assembly of this engine because he wants the automotive world reminded that Keith Black's Racing Engines (KBRE) builds all kinds of racing engines for a full spectrum of applications.

The accompanying photos and spec charts give a detailed look at the components and techniques used. The result was a very potent 366 hp at 6000 rpm, which was achieved with a smooth idle, clean, crisp throttle response and, oh yes, 9:1 compression, because from the very start this was intended to be a mild-mannered street engine, suitable for a street rod. And a street rod engine is just what it will be. So beer can metal or not, look for some very magic power in a '32 Ford in the near future.

—C.J. Baker

1. The heart of this exotic engine is the rare aluminum block that features siamesed cylinder bulkheads similar to the cast-iron 400 block, but steel cylinder liners have a 4-inch bore for 350-cubic-inch displacement. Main cap studs are included from the factory, but 7/16-inch head studs are from Milodon. The block requires align boring/honing. Cylinder honing was done on a Sunnen CK-10 with torque plates in place—a mandatory step on the aluminum block. Decks were square, as delivered, so no deck cut was made.

2. The pistons used are 9:1 Chevrolet aluminum forgings (part No. 336744) with 5/64-inch moly-filled top and second rings from Chevrolet (part No. 3995665), which includes a 3/16-inch chrome/stainless oil ring. Note the use of the small-block-truck double-roller timing chain and sprockets.

3. Keith Black's very talented engine assembly specialist and dyno operator, Larry Ingram, handled the blueprinting, assembly and testing of the aluminum small-block. Here Larry installs one of the rare Chevrolet aluminum heads that were developed for Trans-Am racing in '63 and '64. The heads were cc'd to 69.5cc, and the ports were cleaned up and polished. The valves are 2.02-inch intakes and 1.6-inch exhausts. Fel-Pro Teflon-coated head gaskets, with a compressed thickness of .040-inch, are used.

1

AFTERMARKET CONTRIBUTORS

ACCEL—Ignition system, coil, plug wires
JOHN BUTTERA—Valve cover adapters
PHIL ENDERLE—Carb linkage, stainless steel fuel lines
HOLLEY—Manifold, carburetors, fuel pump
HOOKER—Headers
MILODON—Cylinder head studs, valve cover hold-downs
MOROSO—Valve covers, oil pan, oil pump pickup
NGK—Spark plugs

2

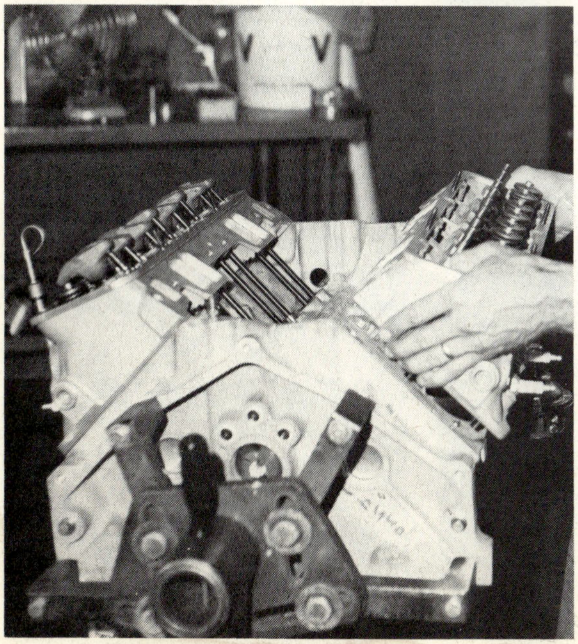

3

ALUMINUM SMALL-BLOCK CHEVROLET RECOMMENDED CLEARANCES

Piston to Bore:	Iron-lined engines: .0035-.0045-inch measured at centerline of wrist pin hole, perpendicular to pin. Finish bores with 280-320 grit stones or equivalent (smooth). (Above clearance for LT-1 pistons).
Piston Ring:	Minimum end clearance: Top—.018, second—.016, oil—.016.
Wrist Pin:	.0004-.0008-inch in piston.
Rod Bearing:	.002-.003-pinch, side clearance .015-.025-inch (minimum preferred).
Main Bearing:	.002-.003-inch minimum preferred, .005-.007-inch endplay.
Piston to Top of Block: (Deck Height)	.0-.005-inch average above deck, with piston centered in bore. Deck height specified is for a .040-inch (compressed) Victor composition head gasket. If thinner head gasket is used, deck may be reduced accordingly. For best results, piston deck to cylinder head clearance should be established at .035-.040-inch.
Valve Lash:	.024 intake and .030 exhaust. Cold lash all-aluminum engine .008 tighter than hot lash.

4

4. The crankshaft is the forged high-performance version. It was straightened, Magnafluxed, nitrited and polished at Black's. The oil holes were also chamfered. The Chevy neoprene rear seal is used. The connecting rods are the 3/8-inch bolt high-performance versions. They, too, were sized and Magnafluxed at Black's. Rod length is a standard 5.7 inches. All bearings are GM Morraine 400s. The oil pump had end clearance set at .0025-inch and the relief spring shimmed. The Moroso pickup was added to fit the Moroso aluminum oil pan.

5. The guide plate arrangement on the aluminum heads is unlike that used on cast-iron heads. Instead of going under the rocker studs, these guide plates are installed under the head stud nuts.

6. The rocker studs on these aluminum heads screw into place and a jam nut locks them. The valve springs are installed at 1.650 inches, which yields a seat pressure of 125 pounds. The solid-lifter cam used in this engine has .434- and .455-inch intake and exhaust lifts, respectively, which gives an open spring pressure of 295 to 300 pounds. Standard 1.5:1 rocker arms are used with 5/16-inch pushrods and part No. 5231585 lifters. Note the canted bolt pattern on the end exhaust port pads, which readily distinguish these heads.

7. The cam used is the street Z28 version (part No. 3972178), which has 300 and 314 degrees duration and 72 degrees overlap. The cam was installed 2 degrees advanced to compensate for dynamic loading and chain stretch.

8. The aluminum heads were designed for unique cast-aluminum valve covers with eight hold-down bolts. Unfortunately, Chevrolet was unable to supply such covers, since they are rarer than the cylinder heads! Consequently, we used a pair of Moroso's lightweight aluminum valve covers with the help of a pair of special aluminum adapters fashioned by John Buttera. Milodon hold-downs attach the covers to the adapters.

9. The unique bolt pattern on the aluminum heads required special header flanges on the 1 7/8-inch-diameter Hooker headers. The primaries are 33 inches long, feeding into a 3½-x12-inch collector. Special header gaskets also had to be made.

10. Completing the engine is an aluminum water pump, a Holley 2x4 dual 4-barrel manifold with a pair of 600-cfm carbs, an Accel BEI II ignition system with a Super Coil and 8mm Fat-Stuff plug wire firing NGK No. B8H plugs on the dyno. On the street, NGK No. B7HZ plugs are used. The headers are installed backward for dyno-testing only.

5

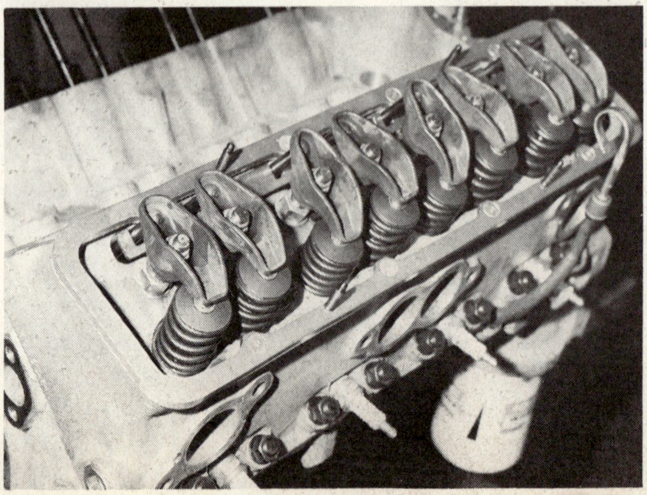

6

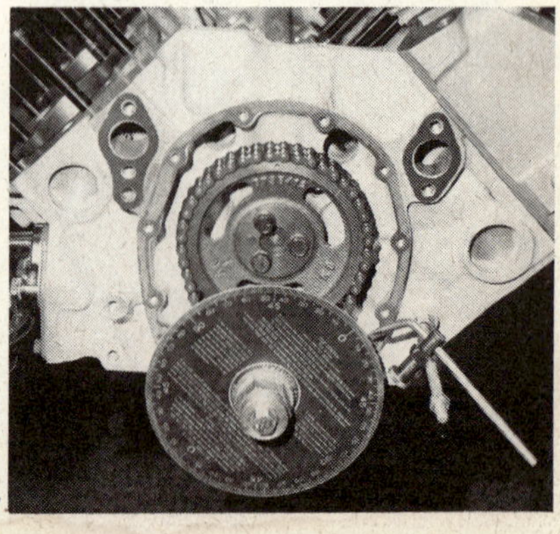

7

RECOMMENDED BOLT TORQUE AND LUBRICANT FOR THE ALUMINUM SMALL-BLOCK

		Torque	Lubricant
Main Bearing Stud Nut	Inner:	70 Ft.-Lb.	Oil
	Outer:	65 Ft.-Lb.	Oil
Connecting Rod Bolt (⅜-inch)		45-50 Ft.-Lb. (.006-inch stretch preferred)	Oil
Cylinder Head Stud in Block		10 Ft.-Lb.	Loctite Sealant
Cylinder Head Stud Nut	Long:	70 Ft.-Lb.	Oil
with Hardened Washer	Short:	60 Ft.-Lb.	Oil
Rocker Arm Stud		50 Ft.-Lb.	Oil
Camshaft Sprocket		20 Ft.-Lb.	Oil
Intake Manifold		25 Ft.-Lb.	Antiseize
Flywheel		60 Ft.-Lb.	Oil
Bellhousing		25 Ft.-Lb.	
Spark Plugs		15 Ft.-Lb.	Antiseize
Exhaust Manifold		20 Ft.-Lb.	Antiseize
Oil Pan		165 in.-lb.	Antiseize
Front Cover		75 in.-lb.	Antiseize
Valve Cover & Adapter		25 in.-lb.	Antiseize

DYNO TEST RESULTS

RPM	Observed Torque (Ft.-Lbs.)	Corrected HP
3500	307	211
4000	322	285
4500	334	295
5000	335	329
5500	326	352
6000	311	366
6500	275	351

All tests run on Keith Black's Heenan & Froude dyno. Data obtained after 45-minute break-in and warm-up. Water temp: 180 degrees. Oil pressure (20W-50 Pennzoil): 35 psi (idle) and 65 psi (max.). Ignition timing (total): 38 degrees @ 4000 rpm. Spark plugs: NGK No. B8H (.040-inch gap). Fuel: 110-130 aviation gasoline.
Note: Start-up oil pressure can exceed 100 psi, so engine should be allowed to warm up before revving it to prevent possible oil filter failure.

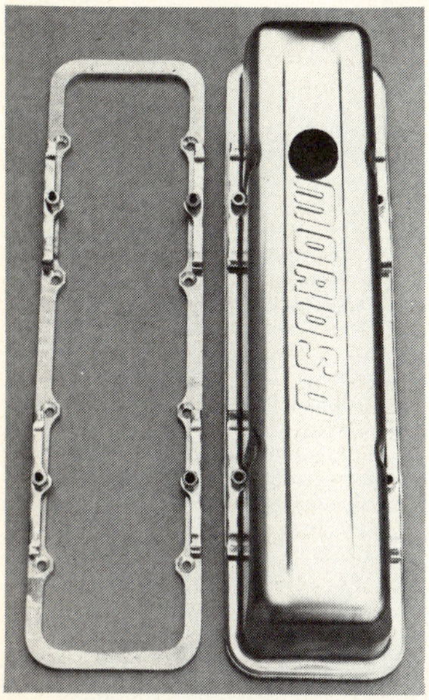

8

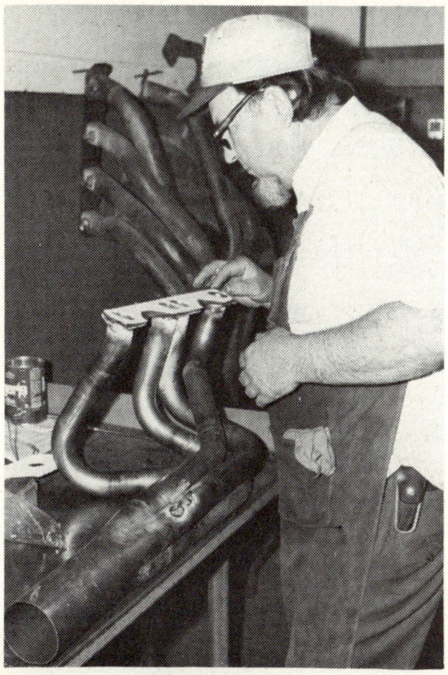

9

Editor's note: Before we get any hot letters wanting to know why our aluminum small-block Chevy 350 didn't pump out 600+ horsepower, let's qualify the parameters by which this engine was built. First of all, our intent was to build a totally streetable engine using all-Chevy internal parts. We decided to go with the 9.0:1 compression because the quality of today's gasoline effectively prohibits the higher compression ratios synonymous with high-horsepower production. Our goal was to produce an engine that would put out better than one horsepower per cubic inch while working under the handicaps of pump gasoline and lower compression. Strictly as a concession to the visual aspect of our cover engine, we chose the Holley Pro Dominator 2x4 race manifold and 600-cfm carbs, knowing that the engine would ultimately end up in someone's street rod engine compartment, where one look is worth a thousand words. Both Keith Black and Holley engineers warned us that this setup was not the ultimate street package, stating that a good 4-barrel system would be more responsive in the lower rpm ranges, but we opted for the 2x4 arrangement on the strength of its strong performance appeal. Although the resulting 366 hp that we obtained is nothing to sneeze at, especially in today's low-performance world, Keith Black could easily have assembled this engine with more radical parts to produce almost double that horsepower rating. But we got what we wanted: a strong, reliable, one-of-a-kind, high-performance, all-aluminum Chevy street engine that will produce neck-snapping acceleration and turn a few heads, too!

10

All in the Family

A quick look at the small-block Chevy as it's run in three of the most sophisticated areas of auto racing.

No American production engine has ever enjoyed greater popularity in virtually every aspect of motor racing than the small-block Chevrolet. That in itself is a tribute to the basically sound engineering design of this engine.

Of course, these engines are modified to suit the racers' needs for each application. For instance, a drag racing engine need operate for only a matter of seconds, but it must do so at very high crankshaft speeds. NASCAR engines must provide rugged durability to withstand 500-mile flat-out races, but at least the builders need concern themselves only with producing peak power in a very narrow rpm range. Formula 5000 engines must work in a fairly broad rpm range and offer durability, too. Consequently, the components used in these forms of racing vary considerably.

This chart provides a quick reference to the combinations currently being used in the more successful cars in each of the three categories mentioned above. You will also note that the price of racing is never cheap! *—C. J. Baker*

NASCAR SUPERSPEEDWAY

Displacement:	355 to 356 cubic inches
Operating Range:	6500 to 7200 rpm
Power:	575 hp @ 7000 rpm
Torque:	445 ft.-lbs @ 5500 rpm
Induction:	Single 4-barrel carburetor
Manifold:	Single-plane or box design
Heads:	Ported (little or no welding)
Valve Size: Intake	2.05 inches
Exhaust	1.60 inches
Valve Spring Pressure: Seat	165 pounds
Open	460 pounds
Camshaft: Type	Flat-tappet (mushroom lifters)
Lift	.600-inch
Duration @ .050	282 degrees
Lobe Centers	109 degrees
Rocker Arms: Material	Aluminum or steel
Ratio	1.5:1 to 1.6:1

Rocker Studs:	7/16-inch
Rocker Stud Girdle:	Used by some engine builders
Pistons: Material	Forged aluminum
Compression Ratio	12.5:1
Rings: Top	Either .043- or 1/16-inch Moly
Second	1/16-inch cast iron
Oil	3/16-inch
Piston Pins:	Full-floating
Rods: Material	Steel (Carillo or stock)
Length	5.7 to 6.0 inches (3 1/2-inch stroke)
Crankshaft:	2.45-inch mains, 2.10-inch rods
Main Caps:	4-bolt
Ignition:	Chevy transistorized
Headers:	2-inch tube, 32-34 inches long
Fuel:	High-octane gasoline
Oil System:	Dry sump
Cost:	$7500

PRO STOCK DRAG RACING

Displacement	322 to 340 cubic inches
Operating Range:	7500 to 9500 rpm
Power:	700 + hp @ 8500 rpm
Torque:	450 ft.-lbs. @ 7500 rpm
Induction:	Dual 4-barrel carburetors
Manifold	Tunnel-Ram
Head Work:	Extensive welding and porting
Valve Size: Intake	2.07 inches
Exhaust	1.62 inches
Valve Spring Pressure: Seat	140 to 180 pounds
Open	540 to 600 pounds
Camshaft: Type	Roller
Lift	.700-inch
Duration @ .050	286/296 degrees intake/exhaust
Lobe Centers	109 degrees
Rocker Arms: Material	Aluminum (roller design)
Ratio	1.6:1
Rocker Studs:	7/16-inch
Rocker Stud Girdle:	Yes
Pistons: Material	Forged aluminum
Compression Ratio	13.3:1
Rings: Top	.043-inch moly
Second	1/16-inch cast iron
Oil	3/16-inch
Piston Pins:	Full-floating, lightweight
Rods: Material	Aluminum
Length	5.9-inch (3 1/4-inch stroke)
Crankshaft:	2.30-inch mains, 2.00-inch rods
Main Caps:	4-bolt
Ignition:	Battery-powered electronic
Headers:	2 1/8-inch tube, 28-30 inches long
Fuel:	High-octane gasoline
Oil System:	Dry sump
Cost:	$9000

FORMULA 5000

Displacement:	305 cubic inches
Operating Range:	5000 to 8500 rpm
Power:	550 hp @ 5500 rpm
Torque:	415 ft.-lbs. @ 5500 rpm
Induction:	Lucas timed fuel injection
Manifold:	MacKay fuel injection manifold
Head Work:	Ported (little or no welding)
Valve Size: Intake	2.05 inches
Exhaust	1.60 inches
Valve Spring Pressure: Seat	165 pounds
Open	460 pounds
Camshaft: Type	Roller
Lift	.550-inch
Duration @ .050	280 degrees
Lobe Centers	108 degrees
Rocker Arms: Material	Aluminum (roller design)
Ratio	1.5:1 or 1.6:1
Rocker Studs:	7/16-inch
Rocker Stud Girdle:	Rarely used
Pistons: Material	Forged aluminum
Compression Ratio	11.5:1 to 11.7:1
Rings: Top	1/16-inch moly
Second	1/16-inch cast iron
Oil	3/16-inch
Piston Pins:	Full-floating
Rods: Material	Steel (Carillo)
Length	6.0 inches
Crankshaft:	2.45-inch mains, 2.00-inch rods
Main Caps:	4-bolt
Ignition:	Vertex magneto
Headers:	1 3/4-inch tube, 24-30 inches long
Fuel:	High-octane gasoline
Oil System:	Dry sump
Cost:	$10,000

611-HP Small-Block

Mario Rossi reveals the "state of the art" combination that produces a reliable 1.7 hp per cubic inch in today's NASCAR V8.

When viewing a NASCAR Grand National race, you can't help being impressed by the incredible power and durability of the engines that propel those heavyweight stockers at awesome velocities, lap after lap, mile after mile and hour after hour. Building such engines is truly an outstanding achievement. One such well-known engine builder, Mario Rossi, working in conjunction with Reed Cams, Inc., in Decatur, Georgia, has consented to show us the combination of readily available components that will pull the dyno needle past the 600-hp mark with today's mushroom-cam, single 4-barrel, 358-cubic-inch NASCAR small-block Chevy engine.

Although performance figures in excess of 600 hp are something that have just come about within the past year, there is no real magic involved—just careful preparation and, most important, the right combination of components that complement each other. Drag engines of similar displacement, using roller cams, multiple carburetion and tunnel ram manifolds, have been able to produce power peaks of over 600 hp for some time but not with the broad power range and durability that a NASCAR engine must achieve.

Even though this feature details the specifics of a NASCAR engine, the basics apply to any performance engine. A change in cam profile, manifolding and carburetion would produce an equally impressive drag engine, street engine, sprint car engine, etc. Each application has its own peculiarities, but the fundamentals are the same, so follow along as Mario Rossi reveals the current state of the art.

BUILDING BLOCKS

Something new is now available to the engine builder—a fully prepped, sonic-tested, machined and seasoned block. These blocks, which are available from either Mario Rossi Racing Engines, (5859 New Peachtree Rd., Bldg. 128, Doraville, GA 30340, 404-451-8750), or Roth Automotive (Rte. 1, P.O. Box 6, Oswego, IL 60543, 312-859-2360), are hand-picked from the latest high-strength Chevrolet castings (part No. 366246), using the best known criteria for spotting and eliminating blocks with unacceptable core shift. The blocks are then sonic tested to assure adequate wall thickness in the high-stress areas. At this point the blocks are fully deburred, inside and out, and generally cleaned up. The blocks are then stress relieved, using a thoroughly researched procedure to add durability and reduce final assembly clearance changes during actual engine operation. After the stress relief process, the blocks are chemically descaled (pickled).

With the basic inspection and seasoning completed, machining begins. Deck plugs are installed for added strength. All pressed-in oil plugs and freeze plugs are replaced with screwed-in cast-iron plugs. Four-bolt Milodon main caps are properly fitted to replace the stock cast iron four-bolt caps. Main cap studs are also used. The blocks are align bored and align honed to Chevrolet low specs for maximum bearing crush. A unique rear main support is added that does not require machining of the stock rear cap,

thus allowing the option of a wet sump, if so desired. The decks are machined with a minimum cut parallel to the crankshaft centerline and perpendicular to the bore centerlines. The bores are similarly trued to be perpendicular to the crankshaft centerline while simultaneously being enlarged .025-inch for an eventual .030-inch overbore. Final honing for piston-to-wall clearance and final decking are left for the individual engine builder, although these operations are also available upon special request. All honing is done with a Sunnen CK-10, using torque plates. After boring, .004-inch diameter is honed with Sunnen JHU-625 stones and the final .001-inch is honed with Sunnen JHU-820 stones, using light honing pressure on the final few machine strokes, but being careful to avoid burnishing. And last, Milodon head studs are included.

Although these fully prepped blocks are expensive, the average individual doesn't have the ability to hand-pick a good block, nor can he return a block if it doesn't pass sonic testing. More important, no individual or heat-treating shop has the scientifically designed procedure for *properly* stress relieving the blocks. This is a very exact science and it is extremely critical. So in the end, maybe one of these blocks is quite a bargain after all.

ENTER THE SCAVENGERS

The lubrication system is an important part of this engine in terms of both durability *and* power. Rossi used the complete Moroso dry sump system, including the lastest low-windage dry-sump pan. This special pan features "kick-outs" to allow oil to readily escape from the rotating assembly, internal scrapers and windage screen as it drains back to the scavenge pickups. The actual dry-sump pump is a three-stage assembly with two scavenge sections and one pressure section. Oil pressure is regulated to a maximum of 70 psi, utilizing Union 76 40W racing oil on the dyno and 50W oil on the race track. Oil temperature usually runs about 220 to 230 degrees in either situation.

Anything that can be done to reduce the amount of oil around the crankshaft and connecting rods reduces windage, oil foaming and, most important, the amount of oil thrown on the cylinder walls. If too much oil is thrown on the walls, the rings will be overwhelmed and oil will begin to contaminate the combustion chambers, reducing power and creating detonation. The Moroso system is so clean that when engine speed is increased from 3000 rpm to 7600 rpm, the level of oil in the dry-sump tank only drops one quart. The scavenge pumps are so effective that they actually form a pan evacuation system that pulls air *into* the valve cover breathers.

Oil is restricted to the lifter galleys in this engine using .090-inch restrictors, thus reducing the accumulation of oil in the lifter valley and valve cover areas. The oil-drainback holes in the valley are not plugged, as the relatively quick profiles used on the mushroom cams create a high load between the camshaft and lifter and the extra lubrication is essential to durability.

DIMINISHING DOMES

Along with cylinder head and manifolding improvements, piston dome science has also contributed to power gains by increasing flow and flame propagation within the combustion chamber. The pistons selected by Rossi are the Hank the Crank/BRC silicone-aluminum versions. These pistons feature a semi-full skirt, double .043-inch Spirolok pin retention (optional), a .150-inch down top ring placement (optional), 14 oil ring groove drainback holes, pin oiling holes and a very low (.250-inch) dome.

Dome height is the key. Ideally, a flat-top piston would be the best, since any dome impedes intake and exhaust flow and flame propagation. Unfortunately, since the heads cannot be milled without sacrificing durability and inviting cracking, some dome is still required to achieve a suitable compression ratio for racing, such as 12:1 or better. Using the HTC/BRC piston with a minimal 6 or 7cc dome, a 358-cubic-inch engine can be built with 12.4:1 to 12.5:1 compression. This combination assumes a block deck height .010-inch above the piston decks at TDC and using *two* .017- to .018-inch-thick Chevrolet stainless steel head gaskets (part No. 3916336) under each head. Such an assembly, using GM gasket sealer on both sides of each gasket, provides a quench clearance (piston-to-head) of .045- to .050-inch. This is a common configuration on the top-running cars.

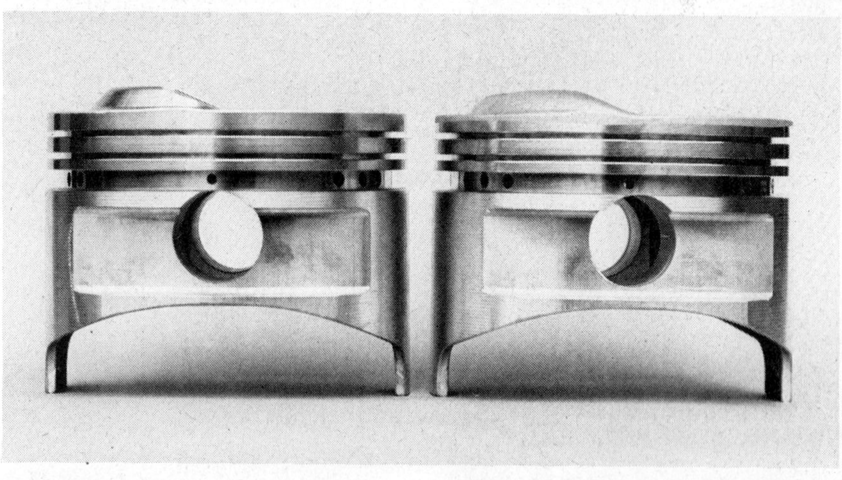

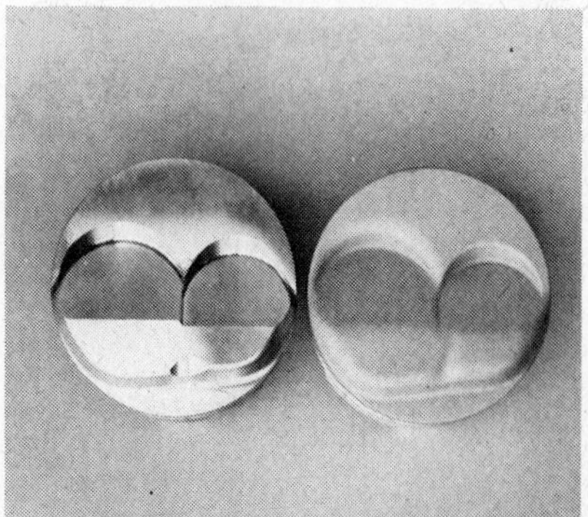

Flame propagation is further enhanced by careful massaging of the piston top, as shown in the photos. Note that the intake valve relief has been laid back and radiused while a relatively abrupt edge has been left on the exhaust valve relief. Note, too, that the sides of the dome are laid back to allow flame travel around the sides of the dome. And incidentally, to get the flame started in the first place, Rossi uses the potent Autotronics MSD-7A multiple-discharge ignition system and Champion BL-60Y plugs gapped at .060-inch. Lead is held to a maximum of 37 degrees.

Completing the rotating assembly are Speed Pro rings. A BR-18PF 1/16-inch Plasma Moly top ring is used with an .018-inch end gap. The second ring is the RBT-10 reverse-taper cast-iron version with a .015-inch end gap. The SS-50U 3/16-inch standard tension (not low tension) oil ring is used. BRC tapered pins hold the pistons to the lightweight Carrillo 6.0-inch steel rods. The crankshaft is a fully prepped, 3.5-inch-stroke, knife-edged forging from Hank the Crank with 2.1-inch and 2.45-inch rod and main journal diameters, respectively. Clevite 77 bearings are used, and the Chevrolet malleable iron harmonic balancer (part No. 364709) helps keep it all alive.

A MATTER OF FLOW

Moving air and fuel into the combustion chambers and exiting the exhaust gases is of critical importance to any racing engine, and this is one of the areas where continued development has helped contribute to the 600+ horsepower figures. Since flow is so important, the cylinder heads, intake manifold, carburetor and headers must precisely complement each other. To guarantee this, the Edelbrock Scorpion Victor intake manifold, Holley R4779-2 carburetor and the latest version of the Chevrolet angle-plug heads (with internal reinforcing ribs and blocked heat riser, part No. 3965784) used on Rossi's engine are all prepared by Reed Cams, Inc. The intake manifold and heads are ported, polished, flowed and matched for optimum flow of 270 cfm on the intakes and 211 cfm on the exhausts at a depression of 28 inches (H_2O). The carburetor is fully reworked for peak performance and Stahl headers, mated to the cylinder heads with carefully matched adapter plates, handle the exhaust.

Preparation of the cylinder heads includes Chevrolet 2.050-inch-diameter stainless steel intake valves (part No. 366285) and Manley 1.600-inch-diameter stainless steel exhaust valves. Winona spiral bronze inserts are used in the guides with .001-inch and .0015-inch stem clearances on the intakes and exhausts, respectively. The valve stems are deburred with 320-grit paper to provide a light crosshatch pattern prior to installation, and one-piece Teflon valve seals are used. A careful three-angle valve job is done to provide a .070- to .080-inch-wide intake seat and an .080- to .090-inch-wide exhaust seat.

MUSHROOM MANEUVERING

Of prime importance to this combination is the Reed M-314-20F-6A2 mushroom tappet camshaft and all of the remaining valvetrain components. Chevrolet .960-inch-diameter mushroom tappets (part No. 366253) are used in conjunction with Reed 5/16-inch pushrods, Manley No. 42151 guide plates, Reed 7/16-inch rocker arm studs, Iskenderian rocker arms (1.6:1 intake and 1.5:1 exhaust), Reed springs and retainers and the Reed valve lash retention fixture. Camshaft drive is accomplished with a Cloyes Tru-Roller chain and sprocket assembly. A cam button and thrust washer are essential to prevent cam end play with a mushroom tappet cam.

Installation and initial break-in of a mushroom cam is critical to its durability. All lifters and lifter bores must be deburred to permit free rotation. Upon initial installation with light oil, each lifter should revolve freely when the cam is rotated—this is essential. When free lifter rotation is achieved, the cam and lifters must be removed, cleaned and dried. Then each lobe and lifter face only must be coated with MoS_2 (or other suitable break-in lubricant) and assembled. Using *only* the outer valve springs and a relatively tight valve lash (.016-inch), the cam must be run-in for one hour at 2500-3000 rpm. Mushroom cam failure almost always results from inattention to free lifter rotation, too much initial spring pressure, loose initial lash or valvetrain binding.

FINAL EXAMS

Measuring the state of development in racing engines is a much discussed and cussed subject. No two dynamometers ever seem to read exactly the same, and in reality such variations don't matter a

great deal, for a dyno is actually little more than a measuring device to indicate relative gains or losses. Still, everyone is looking for those big horsepower numbers, so we've gone along with the game—just to prove a point. What we've done is to run this engine on two well-known dynos, of different manufacture and, just to make it more meaningful, on different sides of the country.

The dyno in question are the Go Power unit at Reed Cams, Inc., in Decatur, Georgia, which is typical of the dynos used by the majority of NASCAR engine builders across the country, and the ultra-sophisticated Heenan & Froude Ltd. dyno at Edelbrock Equipment Co. in El Segundo, California.

Initial tests at Reed's produced power readings of 611 hp at 7600 rpm. With a little additional tuning at a later date, a maximum of 616 hp was achieved. The engine was then crated and shipped to California, where it produced a maximum of 580 hp at 7400 rpm. Why the big difference? Some experts contend that there are differences in the dampening and sensing characteristics of the two makes of dynamometers that show up only at high engine speeds (7000 + rpm). Such an explanation may be true, or it may simply be a matter of calibration. However, it should be noted that at 6000 rpm, the conservative Edelbrock dyno showed the engine to be producing *more* torque than it did at 6000 rpm on the Reed dyno, and it was only in the upper rpm ranges that the Edelbrock dyno indicated less peak power output. Additionally, the folks at Edelbrock indicated that this was the strongest 358-cubic-inch NASCAR engine they'd ever seen, so you be the judge. In the end, the real test of power is on the race track, and that's where engines like this one have equaled or exceeded the performance of some of the best engines in racing today. *—C.J. Baker*

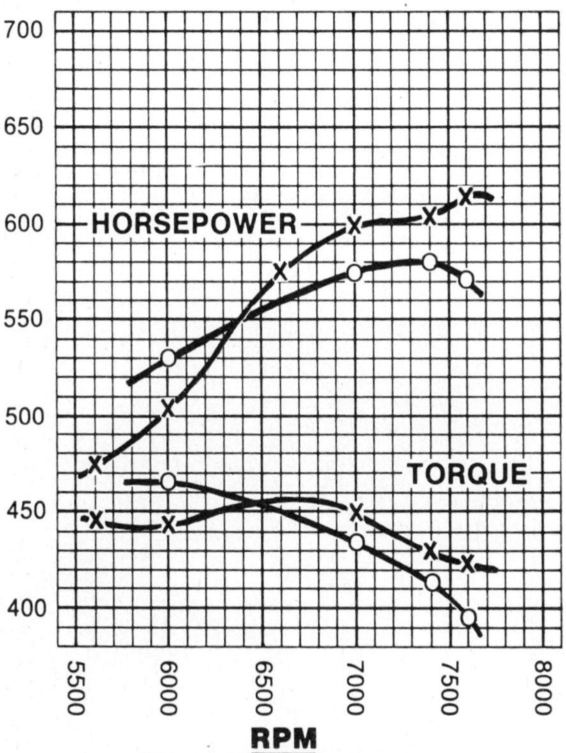

358-CUBIC-INCH NASCAR V8 CHEVROLET
X= REED CAMS INC.—GO POWER DYNO
O= EDELBROCK EQUIPMENT CO.—HEENAN & FROUDE LTD. DYNO

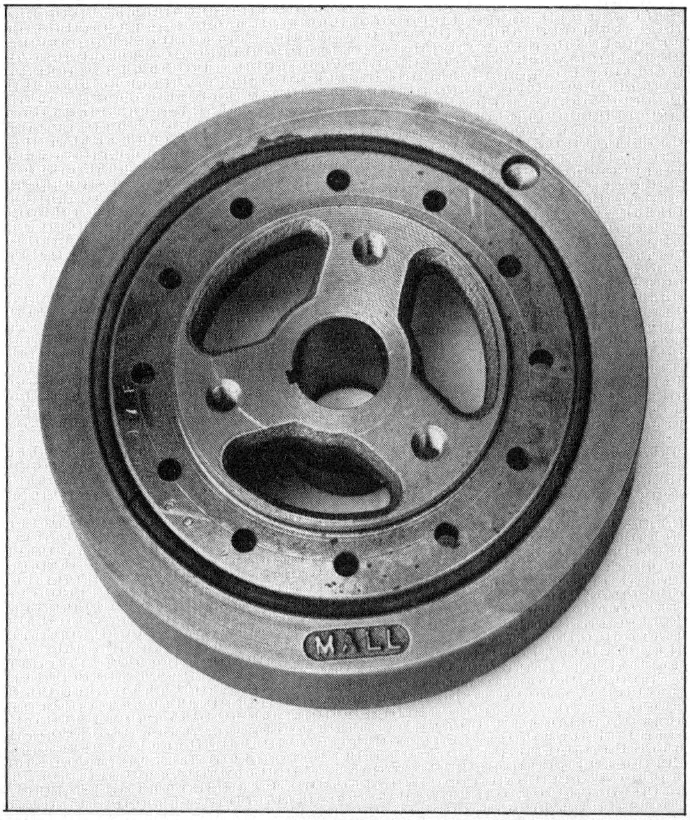

CLEARANCES	INCHES
Main Bearings	.00275 to .003
Rod Bearings	.0025 to .00275
Crank End Play	.004 to .007
Rod Side Clearance	.018 to .020
Piston Pin	.0008
Pin End Play	0 to .005
Piston-to-Wall	.008
Cam End Play	0 to .010
Piston-to-Valve—Intake	.070
—Exhaust	.090
Valve Lash—Intake	.020 (hot)
—Exhaust	.022 (hot)
Valve Spring Installed Height	1.87

Maxi-Mouse

Building the small-block Chevy into a fire-breathing monster of 440, 454 or even 482 cubic inches for any type of racing you want to do.

If you are a Chevrolet racer—bracket, roundy-round, or otherwise—who could use the power potential of 450 or more cubic inches, but your racer can accommodate only the small-block engine because of its physical size, you may be able to have your cake and eat it, too, with a maxi-mouse powerplant. At Conely Speed & Machine Shop (807 E. Grand River Avenue, Brighton, MI 48203). 454-inch small-blocks are commonplace and 480-inchers abound.

The guy who makes these monstrous motors happen is Jack Conely, a man who has had a close working relationship with Chevrolet Motor Division research and development programs since 1957. Jack built and drove brutal unlimited race cars on a Michigan circuit for many years, developing the big-inch small-block engines, at first for his own use as injected nitro-burners in the unlimited car, and later figuring out combinations for alcohol, gasoline and carburetors. The things Jack learned on the big engines he has put to use in refining the construction of his NASCAR Grand National 355-inch engines for many of the biggest names in the Chevrolet camp.

The heart of a Conely maxi-mouse—in 440, 454, or 482-inch form—is the cylinder block. The standard blocks simply won't work for these sleeveless engines, so Conely uses the limited production 0382318 or 3970010 Chevrolet blocks, the former a high-nickel-content alloy steel and the latter a high-tin-content alloy block. Both were developed for endurance racing applications by Chevrolet, based on their 400-inch passenger car block with its two center cylinders on each bank siamesed together with no water circulation between them. These blocks are less than 10 pounds heavier than standard 350 passenger car blocks, but the increased weight is in all the right places and combines with the alloy properties themselves to afford incredible strength and durability, even in the 480-inch versions where walls are thinnest. The 440 version measures 4.1875 inches bore and 4.000 inches on stroke; the 454 measures 4.1875 inches in the bore and 4.125 inches on the stroke; and the giant 482-inch version measures 4.250 inches on both bore and stroke.

Conely and his staff of machinists and assemblers follow proven preparation

1

2

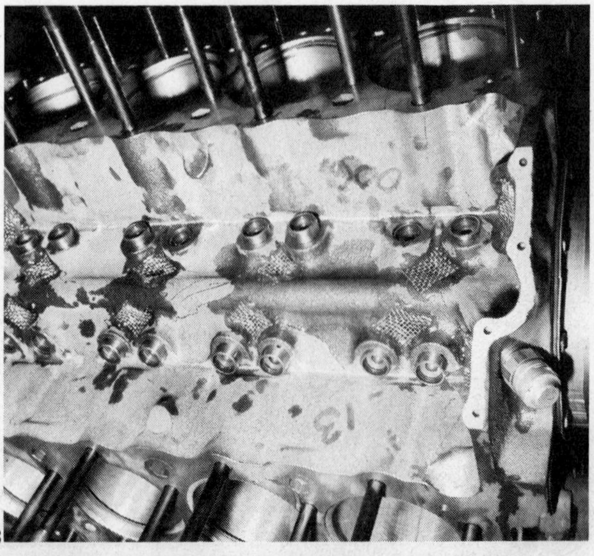

3

techniques in doing the big-inch small-blocks with the following extra steps. All freeze plugs are replaced by new plugs, and the plugs are physically restrained from ever coming out by installation of capscrewed crossbars. All blocks are relieved on the underside of the lifter bores to accept mushroom tappets and are thoroughly ground around the bottoms of all eight bores for rod cap rotating clearance. Lifter valleys are ground free of all casting flash and troughed out to aid drainback of oil. *All* oil drain holes are enlarged slightly in the valley area and then are fitted with epoxied-on mesh screens which will allow passage of oil but will prevent metal chips or valvetrain components from getting down inside the crankcase. On the deck surfaces of each block, Jack installs steel screw-in plugs in all water passages to create dry decks, mills the tops of the plugs off when the blocks are decked, and installs .062-inch copper O-rings in the finished decks, with the Chevrolet .017-inch stainless steel head gasket providing a trouble-free seal everywhere else. The upper plugs are then drilled out to maintain water circulation.

The crankshafts for all of Jack Conely's big-inch mouse motors are sourced by Moldex at about $600 a pop. The 4340 steel forgings are made with 2.450-inch main bearing journals, like the standard 350 engine, but have 2.200-inch journals for the rods, to accept the 6.135-inch big-block L88 rods (PN 3969804). The crankshafts use no center counterweights, but rather up to 15 slugs of Mallory heavy metal for balancing, since crankcase space is limited.

The rods themselves are crack-checked, beam-polished, lightened and shortened on the big ends to clear the block and the camshaft. Conely uses the best available Chevrolet 7/16-inch rod bolts with machined nuts to provide additional rotating clearance, but before the job is done, Jack routinely machines at least .060-inch of material from both rod and cap faces to shorten the whole assembly, touches up the beams and balancing pads with grinders, and then rebores the big end. The small end of the rod houses a full-floating, .990-inch 4340 steel wrist pin, locked in place by a .072-inch Spirolock ring. The short 2.920-inch pin is used, because in large-displacement engines, Conely uses only two rings: a 1/16-inch ductile moly Sealed Power top ring and a 3/16-inch oil ring that almost touches the pin in that portion of the land that passes *across* the pin hole.

The pin location for Conely's stroker small-blocks is set at .912-inch below the deck of the flat-top JE piston, and the thick-section but lightweight two-ringer pistons have held up beautifully in racing service despite the somewhat odd two-ring layout and high pin location with static compression ratios approaching 12.7:1. Not bad for a 475-gram piston, either.

An interesting sidelight to the 454 and 482 small-blocks is the meticulous boring and honing operation Conely uses for maximum ring and piston life in his engines. All operations are performed with a honing plate in place atop the block deck, and Jack uses a Sunnen 300 hone down to the last .005-inch. For the next .002-inch of metal removal, Conely uses a Sunnen AN301 stone. The next .002-inch is honed away with an AN500 stone, and the last .001-inch of material is taken off with an NN40-J85 or NN40-J87 stone setup using manual Sunnen equipment as opposed to the automatic CK-10 Cylinder King machine. The process may sound like overkill and nit-picking, but if Conely didn't believe it contributes to the life of his engines, he wouldn't be doing it to each and every one of them.

With these kinds of displacement figures and high compression ratios, Conely maxi-mouse engines don't need really radical cam and valvetrain components, just reliable ones. For that reason, Jack

1. Conely maxi-mouse is housed in the 0382318 high-nickel alloy block, which is made like the 400 passenger car block, with two center cylinders on each bank siamesed and with no water in between bores for cooling. This block has stock-type fuel pump mount pad, full-width part number pad on right front of block, to resemble ordinary production pieces. Unfinished bore comes to 3.900 inches.

2. The high-nickel block is made with the 400-type siamesing, but it doesn't use the big 2.650-inch mains. Rather, this racing piece uses the 2.450-inch main sizes from the 350 passenger car castings. Though it easily takes bores and strokes up to 4¼ inches without liners or sleeves, this block weighs only about 8 pounds more than the standard Chevrolet passenger car 350.

3. Valley of block has been ground in troughs and corners to aid oil return, with screens epoxied over drain holes to keep pieces out. Finished bore for 454 small-block is done in five-step cuts.

4. Freeze plugs in all of Conely's NASCAR 355 engines (as well as the monster motors) are standard plugs, but they are mechanically retained in the block by a cross-strap that's bolted into the block with capscrews.

5. This big-inch block has been fitted with studs for head retention, and threaded socket-head plugs have been placed in the water passage entries to block them off, creating dry-deck block, using O-ring seals.

6. Jack uses L-88 big-block rods for strokers, shortening the caps and using machined nuts for retention and block clearance. Rods are beam-polished, ground and then lightened before final bottom-end balance.

4

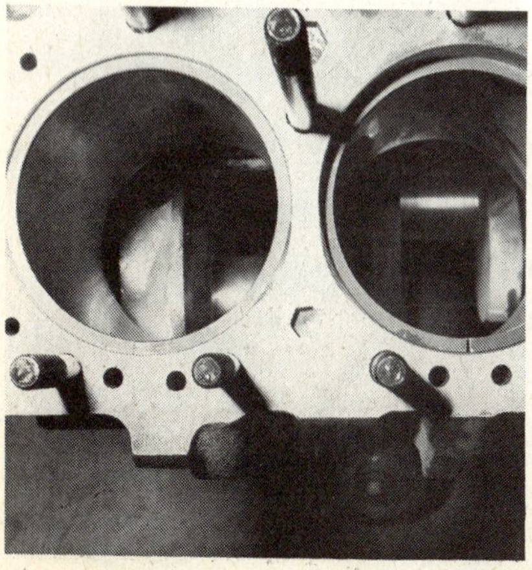

5

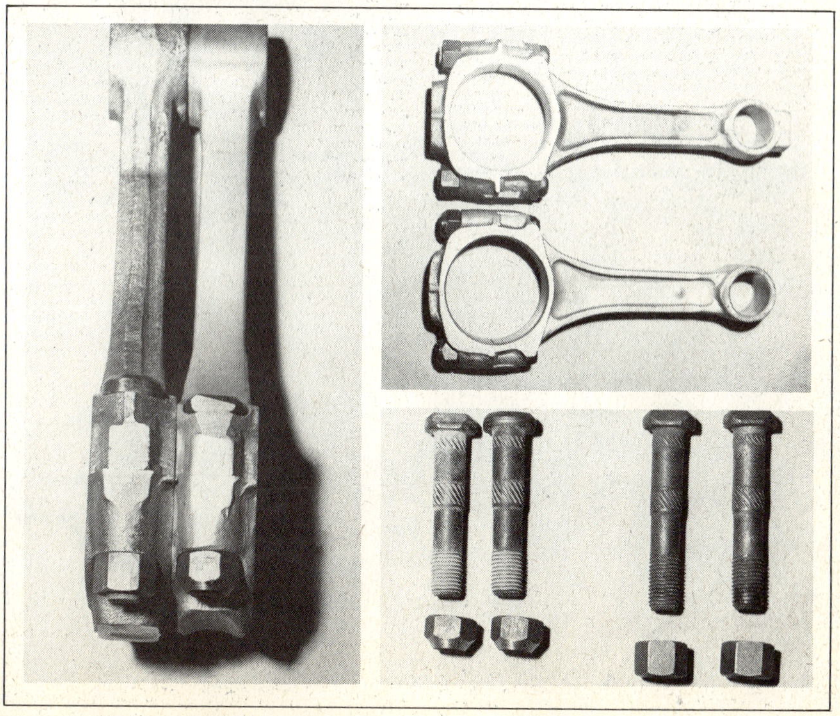

6

has worked exclusively with Bruce Crower for many seasons perfecting the equipment, including reduced-base-circle cams to clear the big ends of the rods, ranging in duration from 272 to 280 degrees at .050-inch lift; Crower rollers and 7/16-inch pushrods; Crower steel rocker arms in a variety of ratios; and Chevrolet valves, including a 2.060-inch intake and a 1.625-inch exhaust. It's all installed around Crower Vasco-Jet springs using 165 pounds of seat pressure with Crower and Chevrolet hardware and guide plates.

Standard cylinder head selection for the big-inch motors is the 340292 "turbo" casting from Chevrolet using 72cc combustion chambers. The heads have lots of hours of port grinding in them, based on the sum total of Jack Conely's experience with this type of small-block engine, which is considerable. All we can say is that the ports are enormous and very shiny from initial entry to final exit, and they are ported according to the fit of cast-rubber plugs that Jack keeps handy in the "flow room" of his unassuming but extremely busy shop. More detail than this on the cylinder heads we can't supply, except to say that Jack has run these engines using every type of experimental head Chevrolet has cast since 1963, including tunnel-ports, canted-valve aluminum, straight-valve aluminum, and several variations of the basic cast-iron "turbo" head; but he uses the standard 292 casting because it makes excellent horsepower and is readily available. Besides standard head modifications, Conely drills and taps both ends of each cylinder head to accept a ½-inch-pipe dry-sump scavenging pickup. Jack has a limited quantity of Chevrolet's 3824059 aluminum water pumps which he uses on the big engines as a further weight saving, since they pump about the same as the standard cast-iron water pumps.

To light off the highly compressed mixture in those tiny chambers, Conely uses the complete Delco electronic ignition system (part No. 3997782), which includes tach-drive distributor, control box and wiring harness. To get the message to the plugs, Conely uses 7mm solid-silicone-insulated stainless steel-wire ignition cables. Jack has tried to destroy this wire with heat, chemicals, abrasives, corrosives and solvents and has found it unkillable, so he'll be marketing it with molded straight and 90-degree ends in sets.

To exhaust the chambers, Jack recommends that his stroker customers use 1⅞-inch headers with 32-inch-long primaries and collectors from 14 inches long and longer, depending of course on the type of car and its application. We know it sounds strange, but those are Conely's considered opinions for his big-inch engines.

And if you're not using nitromethane and/or alcohol with Crower port injectors, which Jack considers a standard setup for many of his roundy-round customers, he will fit the engine instead with an 830 Holley double-pumper, a Holley Strip Dominator small-block manifold, and a 2-inch-thick cast-aluminum spacer plate. This combination will yield in the neighborhood of 600 horsepower on good gasoline, lubricated and cooled as they all are when Conely builds them by a custom-made oil pan and Weaver dry-sump pump and tank system for competition applications. Of course, such a 440, 454 or 482 could also be lubed by the more conventional wet-sump, deep-pan and heavy-duty pump system, if the customer was interested in a big-inch small-block for street, drag or bracket racing use. The advantages of a lightweight, compact powerplant that takes up small-block space and puts out big-block power are not hard to see, and the life-giving tricks that Jack Conely has devised after years of building, tuning and racing these maxi-mouse engines make them as nearly bulletproof as a combination like this can be. GM may be phasing out the big engines, but the small-block will be with us for a while at least, and Jack Conely will be building the big ones as long as there is a demand for them. *—Jim McGraw*

7
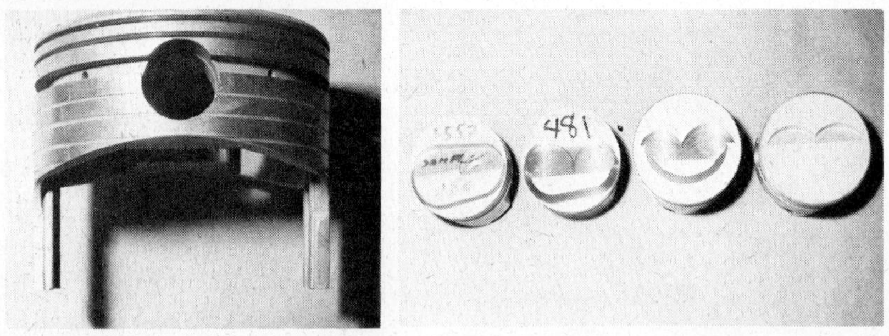

8
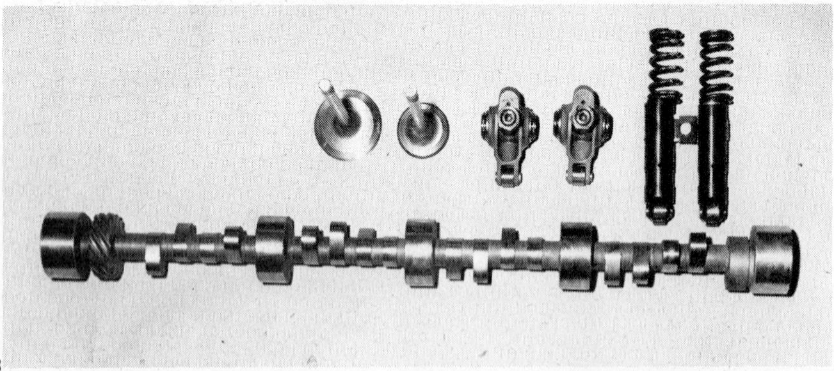

9

7. Conely prefers JE forgings for his big-inch mouse motors, and they are made to use 4340 .990-inch-diameter pins 2.920 inches long to fit behind the 3/16-inch oil ring. Piston array below shows 355, 440, 454 and 482-inch JE pistons designed by Conely.

8. Valvetrain in a typical maxi-mouse uses a Crower reduced-base-circle cam for clearance; Crower rollers, springs and rocker arms; and Chevrolet 2.050-inch and 1.625-inch valves.

9. Moldex forgings are used for all Conely strokers, at $600 apiece, made with 2.450-inch mains and 2.200-inch throws for least friction problems. Each crank uses 15 slugs of Mallory heavy metal for balancing, with short pistons and big-block rods.

Accel's 350 Showcase

HOT ROD Magazine and Accel join forces with Sid Waterman on a street/strip 350 build-up.

Regular readers of HOT ROD should be aware by now that the magazine from time to time involves itself in project cars. These have run the full gamut of auto-involvement; drag machines, kit cars, street rods, bracket racers, cruisers—just about anything you care to name. Not long ago, we joined with the good folks at Accel to put together a Camaro combo that would serve cross purposes; shows, street driving, and drag duty. The heart of this beauty is, of course, its 350 Chevy engine, and this is how that old master Sid Waterman put another winner together.

The first step, in the powerplant end of things, was pulling out the original equipment (OE) mouse motor (a dead player, anyway) and tired 2-speed automatic; in went a four-bolt 350 by Sid Waterman, the noted nitro-burning engine builder, and a Borg-Warner 4-speed. Virtually every phase of the operation was performed by the staff and facilities of Waterman Racing Engines (WRE) (3410 W. Fordham, Santa Ana, CA 92704), which now offers blueprinting service to pro racers and "little guys" alike.

For the business end of Accel's Camaro WRE selected a powertrain combination just wild enough for respectable showings at the dial-in drags, yet mild enough for occasional cruising through the local Jack in the Box. A thorough blueprint brought the 350 block out to 357 cubic inches (350 + .030), and WRE filled the space with eight of their own aluminum rods and a set of cast-aluminum Arias pistons. (The cost of alloy rods, which are custom-designed for bolt-on installation, is comparable to buying and properly race-prepping today's state-of-the-art steel rods, explains Mike Rizuto of Waterman's rod shop. Mike further prefers aluminum for its superior heat dissipation and shock ab-

Waterman Racing Engines has the kind of machinery that the school board never buys for your auto shop teacher. Basic WRE engine-block blueprint includes trips to both the table-model boring bar, which grips the mains rather than the top of the block (per conventional boring bars), and later the Storm-Vulcan milling machine for deck surfacing. (Felt-tip numerals indicate deck-height measurements gleaned from early engine mock-up with crank, pistons and rods in place.) Boring bar takes each cylinder to within .002-inch of the finish bore; the rest is done by hand.

sorption qualities, among other things.) At their recommended final compression ratio of 10.5 to 11.0:1, the Arias Speedmaster slugs team up with angle-plug heads to create a package compatible with today's so-called "premium" fuel.

The valvetrain is almost exclusively Crane's, activated by the dual-pattern HT-284-2-NC hydraulic camshaft. With its lift of .430-inch (intake)/.448-inch inch (exhaust), and an advertised duration of 284°/294° (intake/exhaust), the bumpstick is especially well suited to low- and mid-range acceleration. The incoming fuel is handled by Holley, from electric pump to Street Dominator manifold and No. 6619 carburetor, the 600-cfm 4-barrel with vacuum-operated secondaries and provisions for OE smog equipment. Milodon made the oil pan, tray, pickup and anodized (gold) rocker covers. Waterman's own race-bred goodies include screw-in aluminum freeze and oil galley plugs, O-ringed for perfect sealing and anodized red.

Passing the horsepower along is an

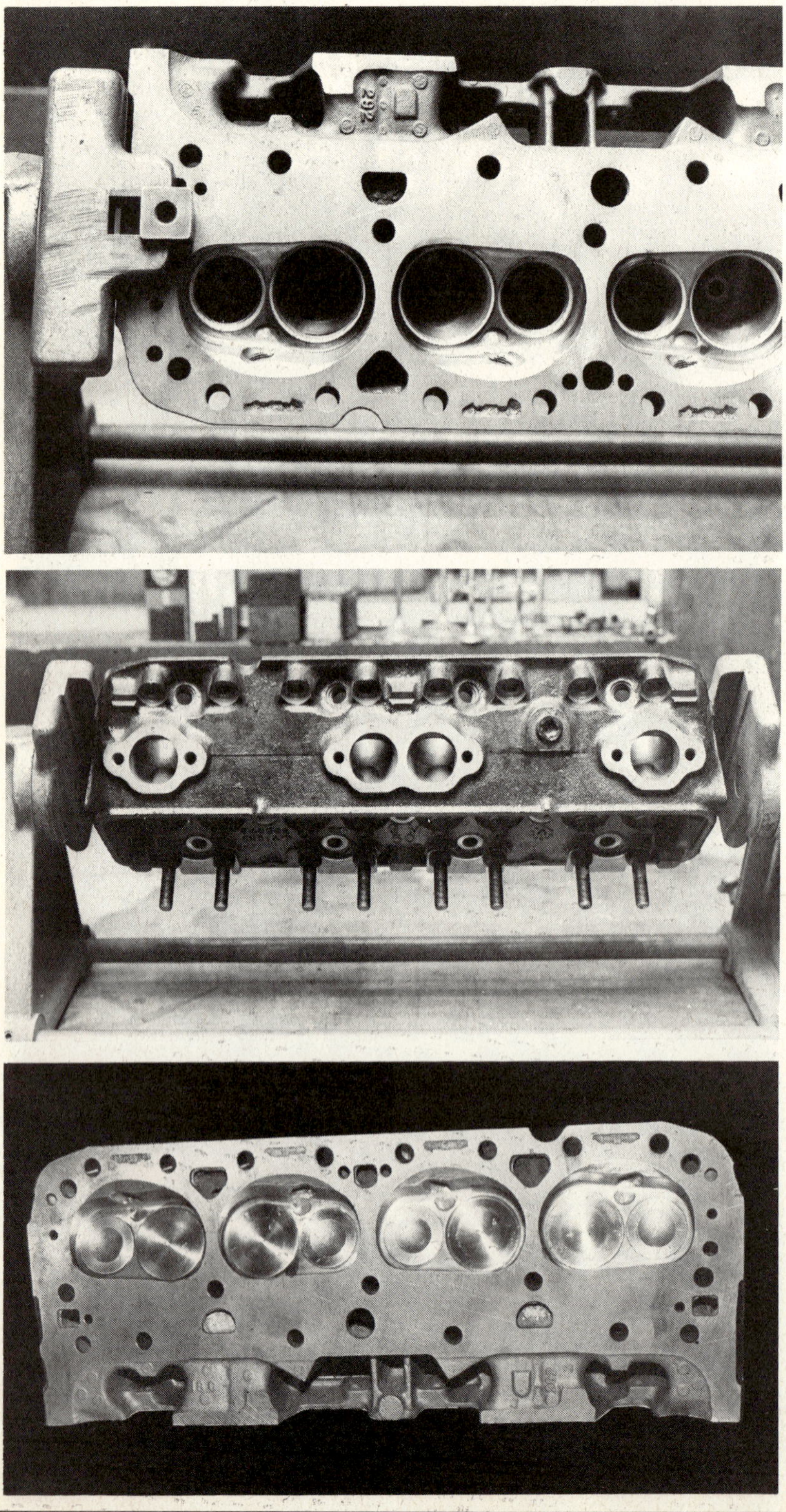

2

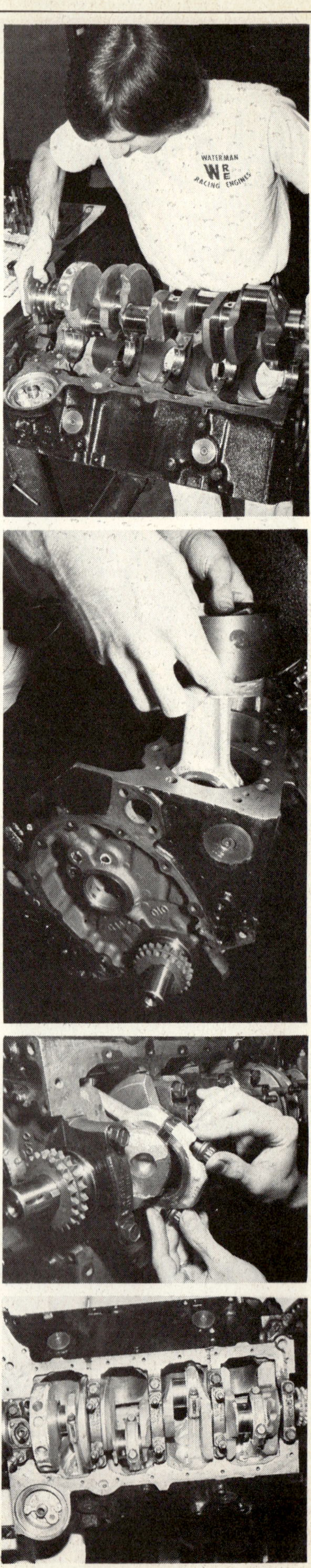

3

all-Weber setup, including steel flywheel, 11-inch disc and diaphragm pressure plate. A Hurst Street Super Shifter sends directions to Borg-Warner's Super T-10 gearbox.

—Dave Wallace

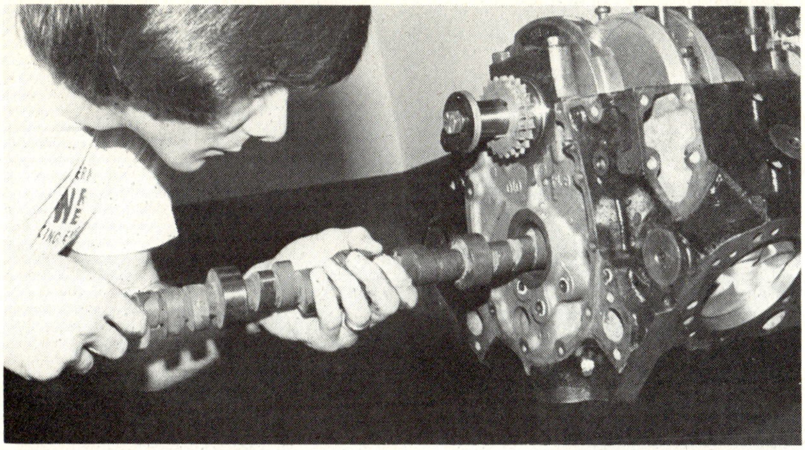

2. Bare angle-plug cylinder heads received three-angle valve job, three-angle seats (intake and exhaust), .001-inch valve guide overbore, surfacing and angle-milling from WRE's Chuck Cerny. Intake ports are matched to Holley Street Dominator manifold, exhaust ports to Hooker headers. Steel valves (2.02-inch intake, 1⅝-inch exhaust) are from Speed-Pro.

3. General Manager Paul Cavallero personally guided our tour through the WRE assembly procedure. (Note Waterman freeze plugs, just like the pros use.) Velasco Crankshaft gets credit for crank preparation, one of the few services not performed in-house by WRE. Trick WRE ring compressor leaves one hand free to guide connecting rods toward crank, minimizes risk of "nicking" a journal in the process. Pistons are Arias Speedmasters, cut for stock-type ring grooves. Completed bottom-end photo shows .060-inch clearance grinding required for aluminum small-block rods, which "grow" at high rpm.

4

4. Once the four-bolt mains are buttoned up, Paul inverts the engine to install Crane camshaft, then Speed-Pro timing gears and double-roller chain. Victor Core gaskets separate the block from the heads. Finished cylinder heads include Crane double valve springs; 1.5:1 Crane aluminum, roller-tip, needle-bearing rocker arms; Perfect Circle seals; and WRE keepers. The Waterman crew prides itself on clean, orderly, straightforward engine assembly.

5

5. With the high-performance Chevy balancer, tach-drive Accel distributor, Holley manifold, clutch and gearbox in place, brothers Mike (left) and John Rizuto slide in the small-block. Note quick-release radiator mounts, frame and suspension detailing. Underhood beauty will not be hidden, thanks to Kolor Me Kustom's trick tilting front end.

501/302

A dynamometer exercise turns into an exciting package for econo-bracket race applications with a single 4-barrel.

It was an exercise, a planned integration of parts and pieces. It was founded on an extremely popular engine and the results were predictable and uncompromising. They succeeded in producing 501 horsepower from a single 4-barrel, 302-cubic-inch Chevy engine.

Jim McFarland and Murray Jensen do this kind of thing for a living, and they do it with pride and no uncertain amount of zeal in Edelbrock's white and gray research and development center. This remarkable pair had but one goal and planned on receiving an answer that would satisfy many questions. Although every part they used is a catalog item, their limitations and value for this particular project were subject to scrutiny.

The exercise was intended to produce a strong, reliable single-carbureted powerplant that could motivate any number of today's econo-race cars. And as a basis for construction, they made full inquiry about the engine's time of peak efficiency. They followed the engine cycle which has to do with powermaking and found (1) that it's all a matter of cylinder pressure and (2) just precisely *where* it's the greatest during the engine's power stroke. But this happens in microseconds, of course, and they are apt to contain this process by extreme slow motion, such as talking about it.

The key, as it were, is getting "a hold on the piston"—"stopping" it at a place where the crankshaft has the piston/rod in the same position for the longest period of time. That is the place or time period when the intake velocity is at its greatest. When that is ascertained, then the compression ratio and camshaft designs could be drawn to maximize that effect. Herewith are the items used for the conversion.

BLOCK: A four-bolt main bearing "service" block which was bored .030-inch, decked and align-bored.

CRANKSHAFT: Standard 302 high-impact steel, indexed, balanced and checked for straightness. The stroke remains stock.

RODS: Service connecting rods with beam flash removed, glass-beaded, sized on both ends. The bearings were Federal-Mogul 2555-AP.

PISTONS: Arias No. 12-19 hollow-dome, slipper-skirt, with 4.030-inch bore. Sealed Power rings, 1/16, 1/16, 1/8, moly top, cast-iron second with a clearance of .010-inch end gap.

CAMSHAFT: Lunati roller tappet, No. RRA-675/669, with the following specs:

.448-inch lift (intake)
320 degrees @ .020-inch
282 degrees @ .050-inch
.444-inch lift (exhaust)
332 degrees @ .018-inch
290 degrees @ .050-inch

HEADS: Cylinder heads were Cylinder Heads America slant-plug "turbo" castings with 2.02-inch intake, 1.60-inch exhaust valves; no welding employed; "Super Eliminator" design for single 4-barrel applications. Valve spring pressure, 145-150 pounds seat, 420 pounds open.

ROCKER ARMS: Crane aluminum with a 1.5:1 ratio on both intake and exhaust valves.

INTAKE MANIFOLD: Edelbrock Scorpion II with ports matched to cylin-

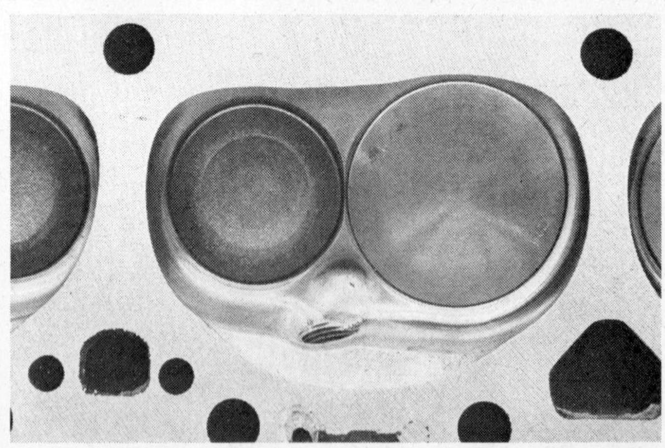

1

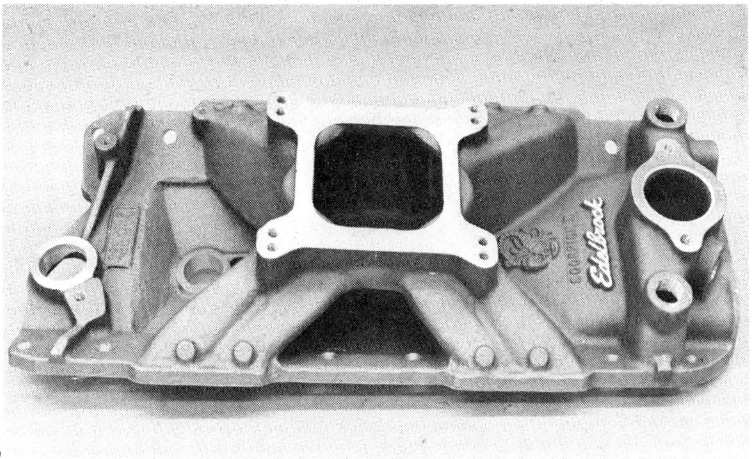

2

3

der heads (1.3 inches wide/2.25 inches high) used with a 1-inch-thick, open-throat carburetor spacer. Carb spacer in this instance stops fuel reversion @ 5500-6000 rpm and serves to broaden that rpm range.

CARBURETOR: Holley 830-cfm (part No. 4788) completely flogged by Murray Jensen in his secret lab. Murray says if you want to know what's inside, contact him at Edelbrock so you can buy one. Known as the "Herderator" or "Turkey Herder," Murray has had same on the fastest NASCAR Chevys.

IGNITION: Stock Accel BEI.

EXHAUST: Hooker NASCAR-type, no available number, but use 2-inch o.d. x 34-inch primary with 10x3.5-inch collector. Also had 3-inch crossover pipe between collectors. *—R. McGonegal*

1. What with the influx of bracket racing and econo classes, Cylinder Heads America has developed a different version of the "turbo" head expressly for single 4-barrel carburetor applications. Called the "Super Eliminator", it uses stock-size valve diameters and employs no welding. Spring pressure was 145-150 on the seat and 450 pounds wide open.

2. The intake was most naturally an Edelbrock Scorpion II. The ports were matched to those of the cylinder head (1.3x2.25). An open-throat carb spacer was also used to combat a reversion problem at 5000-6000 rpm.

3. Pistons were Arias, and the domes were treated and hand-finished at Edelbrock to help produce a compression ratio of 12.1:1.

4. Connecting rods (as well as the crankshaft) were Chevrolet "service" items. The stock-length forgings were cleaned of beam flash, sized and glass-beaded. Murray Jensen's prize, the "Turkey Herder," is for sale by its father. It possesses some interesting peculiarities, which have appeared on NASCAR's fastest Chevys and are worth at least a tenth of a second in e.t. *over* the already trick stuff.

5. Hooker headers used for the dyno thrash were in-car NASCAR type (Jack Davis can fix you up) with 2-inch-o.d. primaries and a 34-inch length. Note the crossover between the collectors.

4

5

COMPRESSION RATIO:

57.00cc combustion chamber
8.54cc gasket (.036 thick)
66.54cc
-9.20cc piston dome
56.34cc total

627.07cc cylinder volume
8.54cc gasket volume
57.00cc cylinder head volume
692.61cc
-9.20cc piston dome
683.41cc total

$$\text{Comp. ratio} = \frac{683.41}{56.34} = 12.13:1$$

CLEARANCES:

Ring end gap (top)	.010-inch
Main bearings	.002-.003-inch
Rod bearings	.0025-inch
Rod side	.016-inch
Crank endplay	.005-inch

ENGINE DYNO DATA

RPM	Cor. BHP	Cor. Torque	Fuel Flow	Cor. BSF
5500	364	348 lbs./ft.	153 lbs./hr.	0.455 lbs./hp-hr.
6000	412	361 lbs./ft.	176 lbs./hr.	0.463 lbs./hp-hr.
6500	447	362 lbs./ft.	200 lbs./hr.	0.485 lbs./hp-hr.
7000	475	356 lbs./ft.	222 lbs./hr.	0.506 lbs./hp-hr.
7500	483	338 lbs./ft.	232 lbs./hr.	0.521 lbs./hp-hr.
8000	501	329 lbs./ft.	238 lbs./hr.	0.514 lbs./hp-hr.

$$\text{Cor. Brake Spec. Fuel} = \frac{\text{Obs. Fuel Flow}}{\text{Obs. BHP}} \times \text{Correction Factor}$$

Correcton factor based on SAE compensation for barometric pressure, air density and carb air temperature.

Triple Threat

Once upon a time, HOT ROD Magazine built a project car. Consider this jewel: The why's, wherefore's and short-block build-up.

Over the years, HOT ROD Magazine has built many a project car, for many different pruposes. Staffers have put together, or at least been deeply involved with, Bonneville cars, street rods and street machines and drag cars, some of which even doubled for two purposes. One of the more notable efforts was a quick Chevy II, designed to be versatile enough to run in more than one drag racing class. While the entire series covering the project gave detailed bodywork and chassis modifications, we thought it appropriate to bring you here just the engine mods.

As mentioned, our intent was to construct a vehicle that would be versatile from a competitive standpoint—a car that could run on a national level in assorted classes to take advantage of the imbalance in the factoring system. Your intention may be considerably simpler. A bracket racer approach would be far less expensive and require fewer exotic components and, if constructed properly, would not demand nearly as much midweek maintenance.

The accompanying photographs will give you some idea as to the nature of the components we'll be dealing with, along with some guides as to which components go best in particular combinations. Check them over, and if you haven't yet decided just where you'd like to compete, they may help you decide.

THE GAME PLAN

The most important question is: How much can you spend putting a race car together? There's little sense attempting to build a Modified car that could conceivably cost in excess of $14,000 if the budget is not there to support such a venture. Understandably, it's difficult, if not impossible, to accurately guess the cost of a new race car, but you can come close if the major expenditures are considered and the hundreds of seemingly unimportant items are considered.

There's also the consideration of what it'll cost to campaign the car of your choice. It's commonly known that a multi-carbureted, high-revving, megapower racer will eat the tires off the car, not to mention such expendable parts as rearends, transmissions, U-joints, light valves, thin-wall ported cylinder heads and all the rest of those high-buck parts that consistently seem to plague the serious racer. And we won't even go into those little gems called valve springs that last a run or two and eat pistons, like the Cookie Monster makes away with cookies. Yes, there's truth in the old saying that it costs money to go fast.

On the other side of the coin, nothing comes with a guarantee, not even a bracket racer. These cars can cost a guy numerous headaches and midweek repair tabs, but comparatively speaking, the luck of the draw usually runs in the bracket racer's favor.

Now we raise another question, equally important as the opening one: Just how fast do you want to go?

THE SKY'S THE LIMIT

Bracket racing is very rewarding to the type of guy who simply wants to mix it up on the weekends. There's no pressure from sponsors, national record holders or any of that. He simply brings

In a Super Stock application of either 327 or 350 variety, one is restricted to use of the OEM carb or suitable replacement. In this instance, that means running the Rochester Quadrajet. Now the Q-jet is a fine carburetor, but it was never designed for racing applications. Consequently, it takes considerably more work to get it operating properly for 1320-foot sprints. To top it off, exotic carb mods are illegal in S/S, so again you face a restriction. Here's where someone like Lingenfelter can save a guy money and hours of frustration.

By comparison, the 850 Holley is better suited to be a racing carburetor. It has many features to aid a tuner in adapting it to his engine combination and, for the most part, is highly adjustable from the standpoint of accelerator pump volume and capacity, idle circuitry, accelerator shot timing, fuel bowl capacity and pickup, etc. And although it's by no means as tunable as, say, a Weber carb, it's a good dollar investment for a bracket- or modified-type racer.

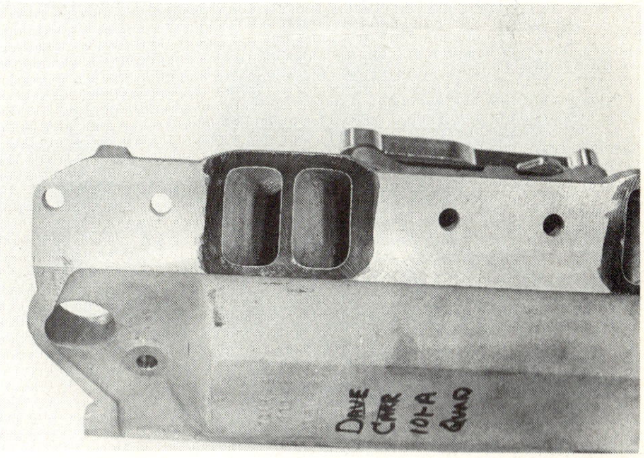

Although we'll be showing you more of this particular manifold modification, this is an example of the kinds of things a racer must take into consideration. For one, most people would be quite satisfied to buy a manifold and carb, go racing and, other than tuning, forget it. Not so with serious class racers. They work and massage to extract the very last bit of power within a given range.

Lingenfelter uses both a Holley and an Edelbrock manifold, as one gives him one thing and the other something else. So, depending on the track, conditions and the needs of his combination, he'll go with what best provides the results he needs. Here a modified Holley. manifold is fitted with an Edelbrock Q-jet adapter. The manifold plenum is ground in special areas to alter flow, and the adapter is blended accordingly. The ports are also matched or mismatched and given the dump angle John prefers. This setup racer-nets out for $195 and includes a manifold, the modifications and the adapter.

You can see how too many misjudgments could add up in a hurry, if a guy didn't know what he was doing. Of course, the above price includes the manifolds John destroyed getting to where he is now, not to mention the mid-week strip flogging that tends to wear down a race car quickly.

his racer to the strip, makes his time runs and dials his car in accordingly. If he breaks out, then you can usually assume that he underestimated the day's potential of his car, weather or whatever. But even at that, he is not out anything but his tow money and entry fee, and hopefully he had a good enough time to more than offset this expenditure.

Bracket cars are so varied and unique that an enthusiast's imagination and budget are the only restrictions that limit his attack. Our particular Chevy II would make a dynamite bracket car with a junkyard engine and an automatic. There's enough chassis and construction there that the car should run within thousandths of a second from run to run. This would be a very consistent car with a single four-barrel, mild cam and a good driver.

For someone who enjoys doing much of the work himself, the expected cost for a venture of this type would be considerably low, especially if the person leaned toward the frugal side. You know, a car of this type could pay for itself in a very short time. In all honesty, if the driver and equipment were good, this little bomb could not only pay its own way, but also its own development costs, because prize money gets fatter the quicker you go.

Super Stock is where the heartaches can begin. A perfectly good vehicle can be built, but there's little insurance on the longevity of its competitiveness. Also, this "professional" form of racing takes bigger bucks to continue to be competitive, as there's a constant flow of new cams, transmission ratios, trick gear sets, converters, manifold mods, etc. Of course, these kinds of developments are what makes racing the intense sport that it is; unfortunately, nowadays it costs a little more for the "trick stuff."

In highly competitive, organized class/eliminator racing, no racer worth his salt remains constant. There's a continuing drive to go that much faster than the next guy, and for some, to run at least quicker than the index, as it's the index that determines respective dial-ins for handicap racing. It's much the same as bracket racing, only the racer doesn't get to select the dial-in.

Although our car wasn't considered for Stock Eliminator, this can serve as a jumping-off point for future racers—only the allowable modifications are more restricted and policed.

Modified Eliminator is an extremely popular type of racing, especially

There's nothing funny about a fuel system. If not done correctly, it can cause fuel starvation or, worse, burn your racer to the ground. But even here, there are some questionable regulations that one must follow if he is to run Super Stock. For instance, you cannot run a fuel sump addition to the gas tank, even though it's the simplest and least expensive method of making sure the fuel lines are always completely covered. Of course, in other areas of racing, they're acceptable.

Note the arrow showing the sump on our modified tank. Dual Holley electric pumps and regulators have become the standard in all types of doorslammer racing, although some of the lower-class, low-horsepower cars would not require the delivery of two electric pumps. It's when velocity or G-force overcomes the fuel pressure to cause a pressure head (restriction) that additional pressure is necessary. A fuel filter is absolutely essential in a race car, as even the smallest particle of dirt can cause a needle valve to plug or stick. The Moroso filter is nice because it can be easily drained and the element replaced. For a more detailed explanation of a well-working fuel system, consult the accompanying drawing.

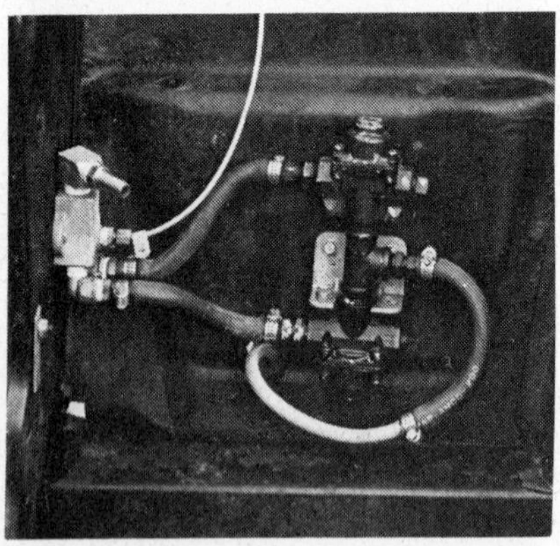

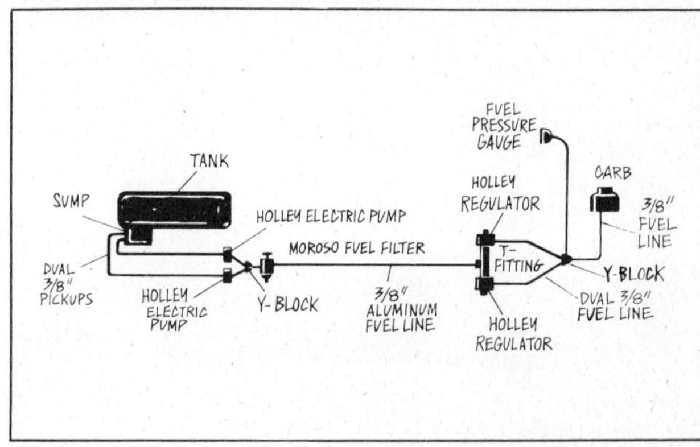

throughout the East and Midwest. This type of racing allows for varied engine modifications, which is usually the deciding factor between winners and also-rans. Modified racing is even more expensive than Super Stock. Cars cost more to construct, and such items as induction systems, complete with all the hand massaging, tend to deflate the old pocketbook by several hundred. But in spite of possible breakage and engine failure from these fast-breathing, high-winding machines, Mod Squad racing is a lot of fun and excitement. There's also no breakout as there is in Super Stock, so if you do get your car running that far below the index, then you can potentially swap a little of the brow sweat for some pretty heavy cash. These types of cars are really a kick to drive. All the popular modified-type, small-block-Chevy-dominated classes are at least in the high-ten-second range, and that makes for a good ride.

Our Chevy II will see a lot of Modified Eliminator activity as it makes assorted M/P (Modified Production) classes, along with B/ and C/Super Mod. And with very little changing, it can run the IHRA Heads-up Eliminator. Now if this very young eliminator catches on, we'll be seeing more and more of this form of heads-up racing on a regional basis. From our way of thinking, we feel these will be the real crowd-pleasing doorslammers of the future. They offer all the antics of a Pro Stocker, but at a third of the investment. At 10 pounds per cubic inch, there's little question that a whole herd of these cars will be running in the high nines. Some can do it now, but you can't build one for $5000.

INFLATED INVENTORY

When you're building your first race car, it's easy to make mistakes, to have to do things over and to simply buy some wrong parts. Buying unneeded race car parts is as easy as overcamming an engine, but it can be just as easily remedied if you do your homework out front. In a nutshell, that's what this story is all about.

If it's a 327 Super Stocker or single four-barrel modified combination that turns you on, then it really pays to at least speak with someone like a John Lingenfelter. People like John, who sell both components and complete racing engines, usually know a great deal, because they are appealing to a specialized market. The experience that Lingenfelter has compiled over the years can go a long way in saving newcomers a lot of cash, simply by supplying them with the right parts the first time. This eliminates having to buy two or three cams before finding the one that works well with a particular combination, or flogging a carburetor to rid it of a flat spot.

The same is pretty much true for all the specialized racer parts that are needed and for machine work. Those manufacturers who deal with doorslammers on a daily basis tend to know more of their needs and consequently are better equipped to solve their immediate problems. Consider this: You don't find many successful racers taking their engine blocks to an everyday engine-rebuilding shop to have their exacting machine work done. No, they take them to shops specializing in precision work. In the long run, it saves them money.

Now that we've discussed in some detail the many possibilities and potential costs involved in placing a similar Chevy II somewhere in the ranks of organized drag racing, let's take a serious look at building an engine for either brackets or Super Stock racing.

John Lingenfelter is directly responsible for programming our engine combinations, and that is where he really excels. John has so much testing and racing experience and what we call racer

It's been only in recent years that much has been changed on Chevy oiling systems, other than deepened oil sumps and a few seemingly trick pans. No one was having oiling problems, so why worry; but once Chrysler proved just how much an optimum oil system was worth, in terms of horsepower gains, a few Chevy racers began to take notice. Jenkins and Moroso were the first to devise an entire wet and dry Chevy system, and coupled with a little of the ol' Smokey Yunick savvy, things began to happen—then all of a sudden we found dynolike oil pans on Chevy drag cars. Outstanding, but as Dave Koffel and Tom Hoover would tell you, it's nothing new; they should have had them years ago.

In an effort to combat the ever-building problem of sloshing oil, attentive racers like John Lingenfelter resorted to such means as knife-edging cranks, trimming cylinder blocks, and grinding aluminum rods to eliminate edges, corners, etc. Crankshaft chamfering is a standard function of specialty crank shops, and even some resort to cross-drilling, but all this adds to the overall cost of that racer. Oil pans, for instance, are no longer a 20-buck item that you pick up at your local Chevy dealer. The new series of big-pan designs cost in the neighborhood of $200 and look something like the ones shown. The pan with the kicked-out rear sump is our initial Chevy II pan; however, it will be replaced with the (also shown) full flare-sided design. Also, there are more components to buy for oiling systems, like the Smokey Yunick/Moroso splash screen. Here it is shown with bolts for a two- and four-bolt Chevy block. The screen can be trimmed to fit a particular pan design, and can be complemented with scrapers and directing baffles. Ideally, if a good oiling system is assembled and the rotating mass is relieved of oil drag, a racer can expect to pick up 45 to 50 horsepower over a conventional stock system. Anything in between, of course, would reduce this figure.

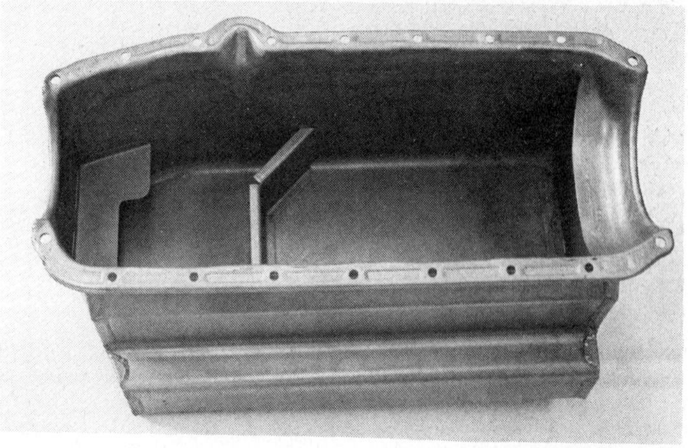

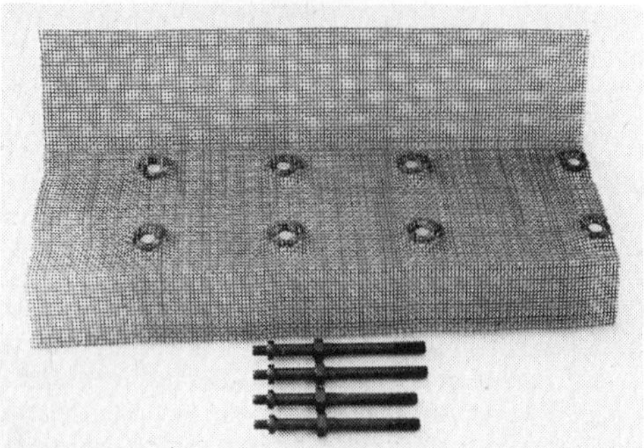

savvy, that his choices and combinations of engine components is what really is responsible for his outstanding performance levels. He doesn't subscribe to all the "trick-of-the-week" bunk; he simply figures out what's needed and does whatever is necessary to get it done and fitted to his present combo. And in this day of high-level sophistication, day-to-day maintenance and inflated costs, that's exactly what is required.

John doesn't waste time painting pretty blocks, polishing aluminum parts or anything else. He spends his valuable time cobbling up parts and fitting pieces in an effort to go quicker. He's constantly changing camshaft profiles, manifolds, flowing heads and induction systems, but leaves the rest pretty much straightforward. The lesson here is getting the most out of the time and parts you have.

The potential engine builder would be wise to follow this advice, as it'll save money in the long run, and if one were to follow John's advice and accompanying guidelines for either of these two engines, mid- to low-10-second times (de-

FLOWING CASH

Probably the single most expensive engine components today are cylinder heads. They're constantly being improved upon, and the more radical they become, the less dependable they are. Even a well-done set of S/S heads is pushing $500, as valve springs are currently bringing about $180 a set, for anything that lasts. Modified heads are getting higher, because more and more flow development is required to maintain any kind of performance edge. Lee Shepherd, who works in conjunction with Reher & Morrison Racing Engines, in Texas, does a great deal of the Modified and Super Modified heads around the country, and as competitive as their prices are, he still gets a tad over $600 for a bare set of ported S/M heads including new castings. But again, by going to someone like Shepherd, you're assured of getting the latest, most advanced head designs available from an aftermarket source. Lee spends most of his work week on the flow bench developing what he already does well, and this is something the average racer just can't afford to do. So again, you reach a monetary point of consideration. Super Stock heads are restricted to no porting modifications, brackets are wide open, and Modified Eliminator-type heads are under constant development. As you can see, it's a clear-cut choice.

PICKING PISTONS

There isn't too much difference in the purchase price of a set of pistons, between the stock and high-dome modified types. But once you get them, either considerable time or expense is involved in fitting them to particular cylinder heads, especially if a constant piston-to-head clearance is maintained. Shown are Duffy's new TRW forged Super Stock pistons, which take little or no work other than cutting deeper valve notches and sanding them a little. The comparison between an out-of-the-box and a hand-finished modified piston clearly shows the amount of hand and machine work involved in completing a set of domes. So remember, after spending a few hundred dollars on a set of pistons, leave a little extra in the budget to fit them to the engine.

pending on vehicle weight and remaining combination) could be expected.

Because there are no engine limitations, other than budget, in building a bracket-type engine, selections could range from mild to wild. With Lingenfelter's guidance, our selection for a bracket and Super Stock engine was of the 327 variety. The primary differences between the two engines are the valves, head work, pistons, cylinder block selection and induction. Both are assembled in the same fashion, only the Super Stock engine was built to pass the rigid teardown inspections of the NHRA tech crew. As you probably know, S/Sers are restricted to maximum overbores, stock crankshafts and rods, stock-type valves, and three-angle valve jobs and stock or OEM replacement carburetors.

CYLINDER BLOCK PREPARATION

A used two-bolt main 327 block works nicely for a Super Stock engine, as long as the finished bores will clean up within a .065-inch overbore. When overboring an engine, be sure to finalize the bore size so that there are no problems in ring selection. For our bracket engine we went with the high-tin-content, four-bolt main block, part number 366246.

The block was cleaned, visually checked for cracks and noticeable water damage, then align bored, squared, bored and honed. All head bolt holes were chamfered to eliminate pulling a thread, and the oil galley plugs were replaced with ⅛-inch pipe plugs. A word of caution when tapping the center front oil galley: Be sure not to cover the oil supply hole (see photo) with the plug, or the result is a wasted front cam bearing.

The oil drain holes in the lifter valley should also be tapped and plugged, and all casting flash ground from the rear two drain-back holes. A groove should also be cut above the right rear lifter boss, into the distributor shaft tower, to facilitate oil drain at the rear of the block. All sharp edges should be eliminated, especially in the main web area and at the bottom of the bores. This will relieve any possible stress risers which can lead to cracking. And before installing any of the new rings, the tops of the newly bored cylinders should be rounded slightly so as not to break the ring edge during fit-up and final assembly.

CRANKSHAFT PREPARATION

Our crank work was easy: We simply ordered a steel 327 high-performance crank from Lunati Cams & Equipment (3871 Watman Ave., Memphis, TN 38118). The crank was indexed, ground for the correct stroke, the rear flange ground perpendicular to the mains, and it was chromed and polished.

A blueprinted crank is essential but

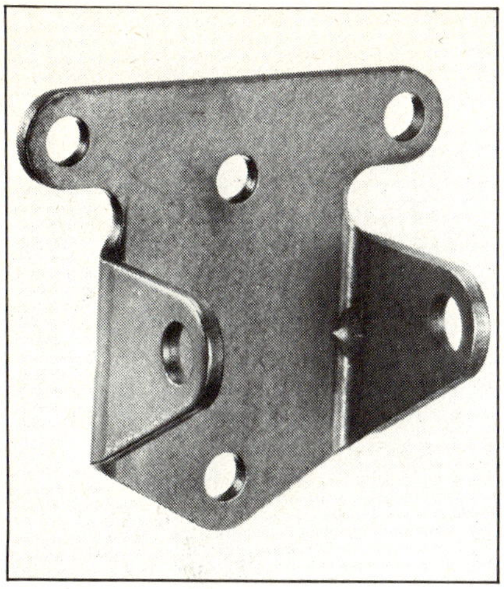

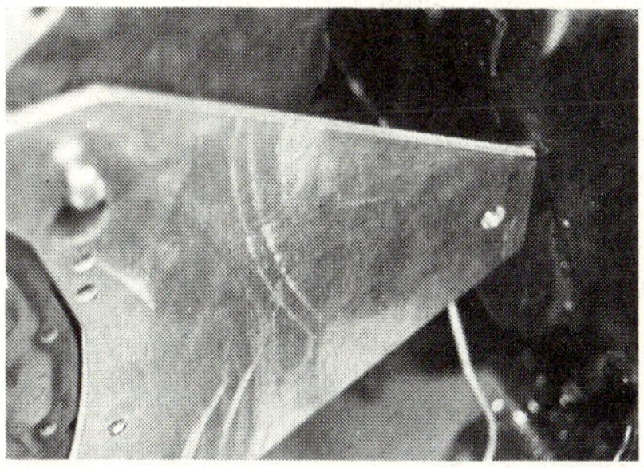

MOUNTED HIGH

So you think dropping the engine in your new wheels is a piece of cake. Not so, if you climb the ladder of fast class racing. The days of stock motor mounts for race cars are over. Even Super Stockers are using the stock frame connectors, but replacing the rubber insulators with steel motor mounts. This eliminates much of the engine movement, but is not quite as good as having a front engine mount supported with a mid-mount. Not only does this put the engine solidly in the chassis, but it also aids in transmission R&R, clutch replacing, etc.

Engine plates, as they're commonly called, are not legal in Super Sock, and are questionable, at the least, in Super Mod. But in Modified or the brackets, this is definitely the way to go. The mounts shown are stock Chevy frame connectors, Moroso steel mount and SRD's new Chevy II engine plate.

doesn't necessarily require all the gingerbread frills, as simple deburring of sharp edges will finish off a chromed Lunati just fine. Journal diameters were ground so that rod and main clearance could be held between .0025- and .003-inch, using TRW bearings.

PISTONS AND RODS

Here is where differences begin to come into play, depending on the type of engine you plan to build. For Super Stock, the original flat-top piston configuration must be retained, so we selected a set of the new Duffy/TRW pistons and fitted them with Childs & Albert chrome Dykes top rings, cast-iron second and chrome oil rails. Piston ring side clearance was held within .0025-inch and ring back clearance held as close to zero as possible, without the ring face protruding beyond the second ring land edge. End gaps can be critical if they're too large; we held ours to .012-inch for the top, .010-inch on the second and .022-inch on the oil rails. When checking ring gap, be sure to square the ring in the bore and mark the rings for the particular bore for which they were fitted, and don't fit the rings until the block has been final honed.

Lightweight pins and bushed heavy-duty Chevy rods complete the assembly. The rods are fitted with aluminum bronze bushings that are grooved 360 degrees and drilled with two oil holes at the bottom of the small-end boss.

A bracket engine could easily be equipped in the same fashion; however, pop-up pistons and aluminum rods might be a better selection, especially if the engine will be operated at high rpm (9000). Aluminum rods do require replacing more frequently, but they don't seem to break during their normal life span as easily as stock-type steel rods. Ring selection can vary, as it can in Super Stock, but the chrome Dykes will offer longer life than, say, the .043-inch moly rings. Also, due to the severity of usage, piston-to-wall clearance will more than likely need to be increased from .005-inch (as used in the Super Stock engine) to .007-.008-inch. Also, instead of holding something around .015-inch side clearance (steel connecting rods), aluminum rods require .030-inch, due to expansion.

SHORT-BLOCK ASSEMBLY

Final assembly is important, as you must make sure the pistons and rods go in their corresponding cylinders and that the crank and bearings are well-oiled before torquing and that the rear main seal

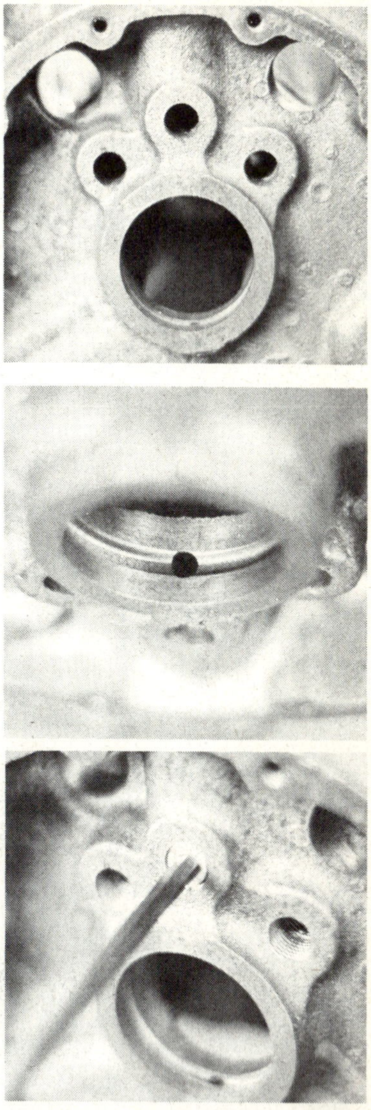

1

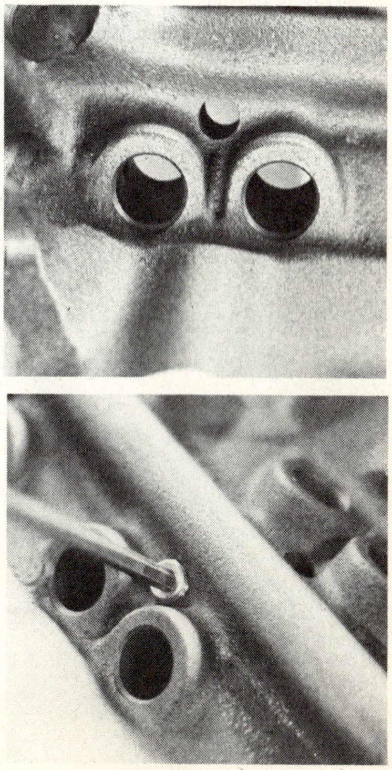

2

1. When threading the center main oil galley, make no mistakes here. Notice the oil hole in cam bearing journal; this must not be covered up. End your tapping procedure before the hole, and the plug will not eliminate or restrict oil to the camshaft.

2. The oil drain holes in the lifter valley should be tapped and plugged. This keeps excess oil away from the cam and off the rods and crank. Oil draining down the center of the engine only adds to the foaming problem, which is combated with a deep oil pan, screen, scrapers, etc.

is well-lubricated. Even more important though is making sure everything is clean. Once the crank is washed in hot soapy water, spray it well with WD-40 and make positively sure no trash is left in the oil holes. The cylinder bores should be wiped clean, using automatic transmission fluid and paper towels. Chopsticks can be used on the rod bolts so that the crank is not scratched while installing the piston/rod assemblies, and if these are not available, rubber fuel line or spark plug boots will work. Once the entire assembly is together, torque all the mains and rods, using some sequence so that there's no possibility that a bolt or nut is missed. We prefer using the taper bore ring installers, rather than the conventional ratchet type, as there's less chance of destroying a ring while tapping it in the bore.

Of course, before final assembly can be completed, all the fit-up steps should have taken place. Clearances such as deck heights and piston-to-valve clearance are critical and should have been done even before final honing. It's smart to leave an extra .001- to .002-inch extra material so that the engine can be assembled (without rings) and checked. Deck heights are measured at top dead center, and are judged by the distance between the top of the piston and the top of the block. In a modified engine this is critical so that the piston, flat or dome, does not come in contact with the cylinder head.

RECAP

That brings us up to completion of the short-block assembly, and since this book contains many stories on heads, induction systems, valve lash adjustment and so forth, suffice it to say you'll be able to button up your engine without further ado. The point of this exercise was to acquaint you with some of the *uses* of a go-fast small-block, as well as some handy yet often basic steps in short-block assembly, and that goal has been reached. Build for any class that satisfies both your whims and your pocketbook, but make sure you build it right.

—John Dianna

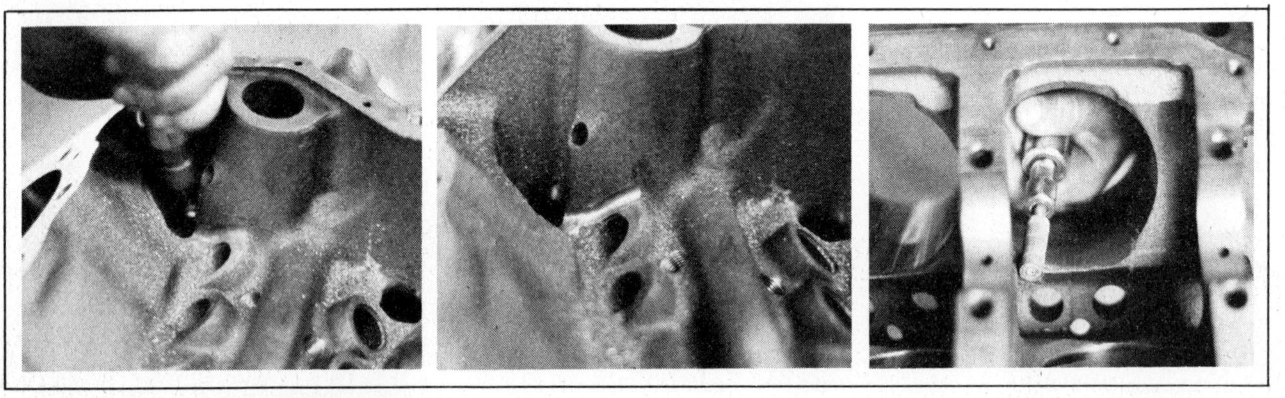

3

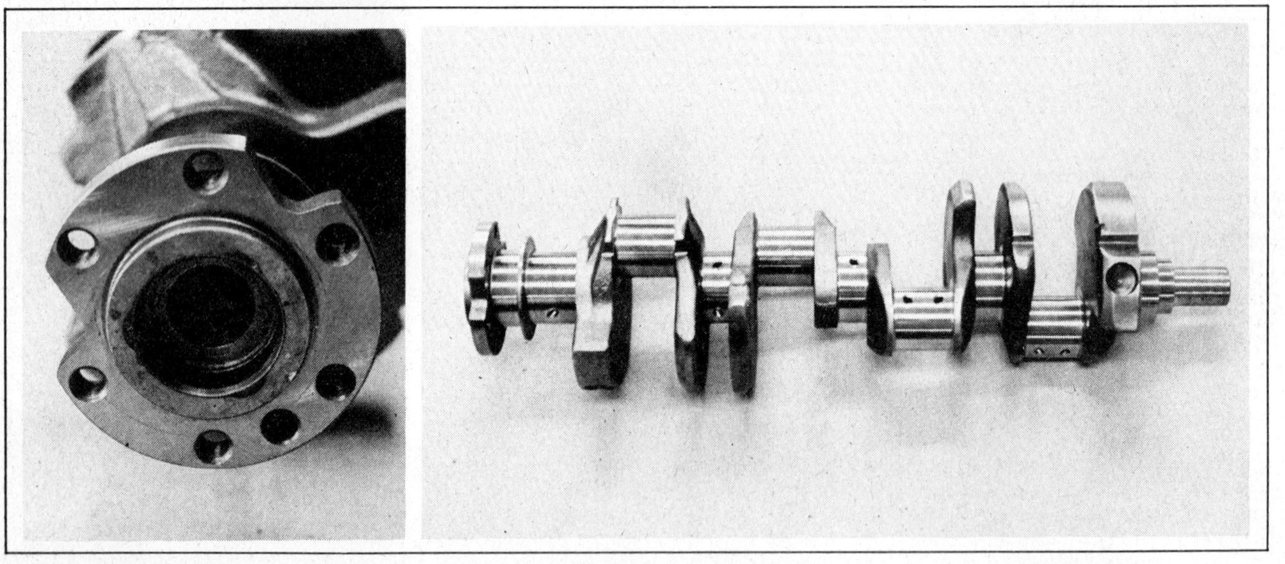

4

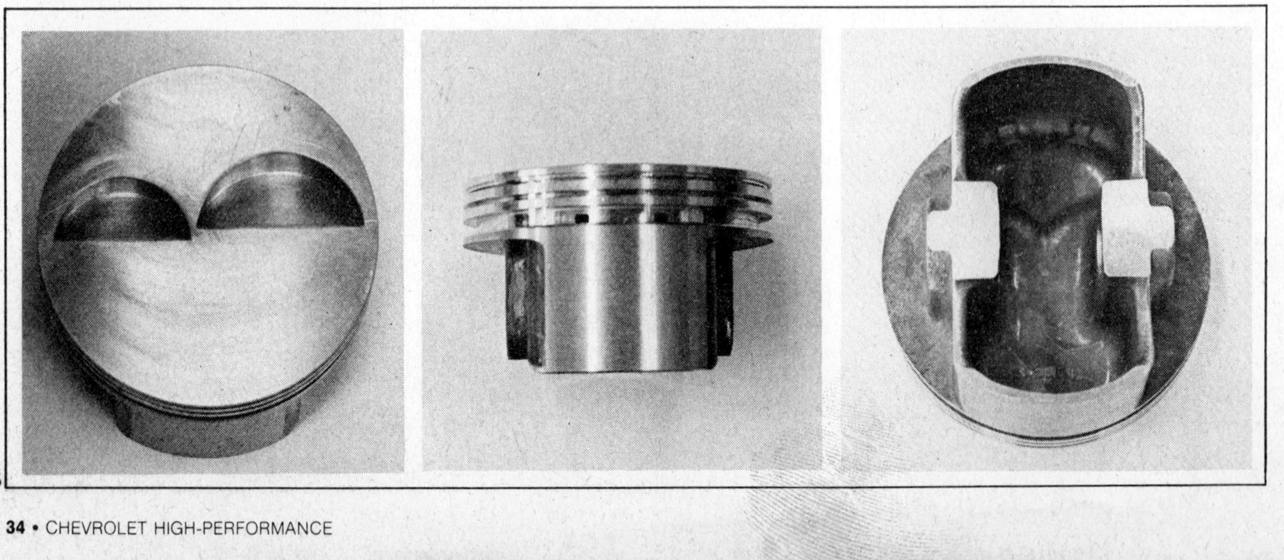

5

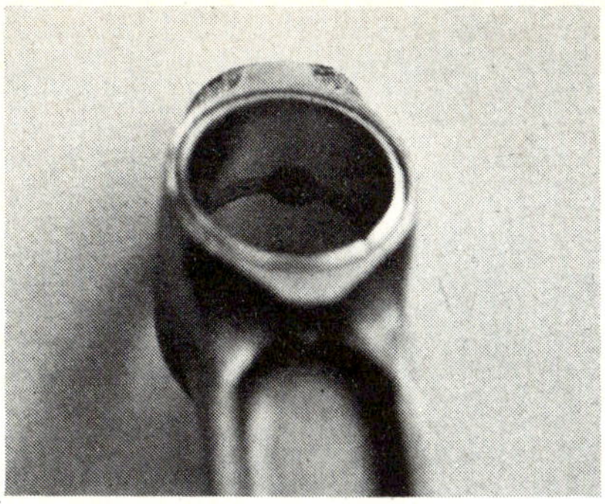

6

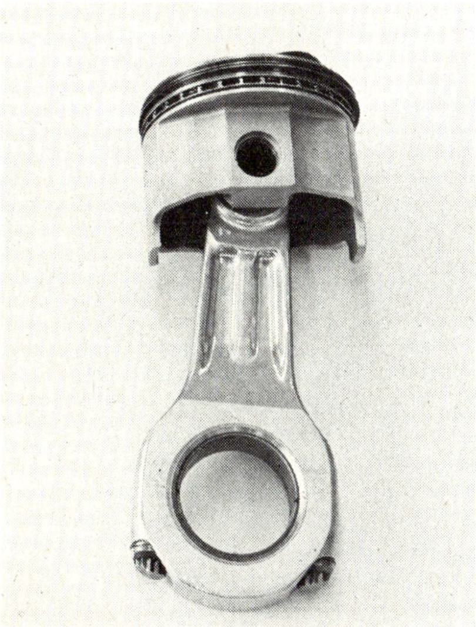

7

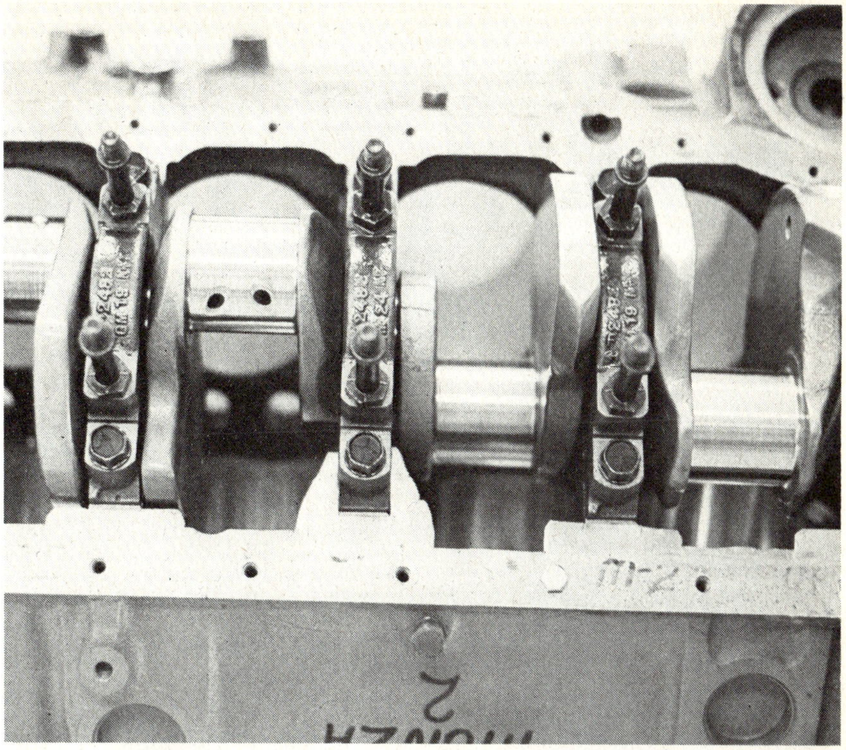

3. The two rear drain-back holes should be enlarged and cleaned of all casting flashing. Next a groove should be ground just above the rear lifter boss. Cut the slot with a ball-type grinding bit; then clean it up with a rosin roll. This groove allows oil to drain from the lower center portion of the valley. And don't skip the step of deburring the bottom end of the block. This operation will eliminate possible stress risers, which cause cracks.

4. A blueprinted crankshaft is necessary if all cylinders are to operate on the designed timing sequence. The crank should be indexed, checked for proper stroke, and polished or ground for adequate bearing clearance. We opted for a chrome crank and had the flywheel flange ground to ensure trueness to the block.

5. The Duffy/TRW Super Stock pistons are new and offer the desirable TRW barrel-face skirt design. The piston flats were cut to maintain the .015-inch deck height and the valve notches deepened to accommodate high-lift cam. Ring selection is the stainless steel, chrome-face Childs & Albert .031x4.063.

6. The rods used in our Super Stock engine are stock Chevy items, reconditioned for size and length. The small ends are fitted with aluminum bronze bushings with a 360-degree oiling groove and double oiling holes under the boss.

7. The BRC aluminum rod and high-dome piston assembly is for the bracket engine. Notice the piston domes before and after. This hand-blending takes considerable time, and you must know what you're doing to move the gases in the proper direction.

Smokey's Small-Block

None other than Smokey himself tells us in his own words how to get 580-plus horsepower out a 355-incher.

PHOTO CREDITS: C. J. BAKER, PPC PHOTOGRAPHIC

For the past 25 years Smokey Yunick has been known as one of America's foremost racing mechanics. His reputation has been built on his innovative (if not always within the letter of the rules) no-nonsense approach to building winning race cars and engines that produce a few more horsepower than those of the competition.

Although the happenings within Smokey's Daytona Beach shop, which he modestly calls "The Best Damn Garage in Town," are shrouded behind chain link fences, blacked-out windows and "KEEP OUT" signs, "the Smoke's" involvement with the small-block Chevrolet V8 is well known. He is recognized as one of the most knowledgeable authorities on this engine. He has been instrumental in the development of performance pieces for Chevrolet Engineering since the engine's introduction in 1955. Smokey is currently refining an injected small-block suitable for the grandest race of all, the Indianapolis 500.

So when we decided to do a Chevrolet performance book and needed an article on building a dynamite small-block Chevy, it was only natural that we turned to the man himself. One of Smokey's basic premises is that if it's worth doing, it's worth doing right. Sometimes doing it right involves more than a small outlay of cash, but we're sure that you'll find Smokey's straight-talk comments on small-block engine building both informative and revealing. —Editor

Let me start out by saying that I'm going to be positive as hell about what I'm going to tell you. This is because I've tried just about every combination known to man that'll pull both power and reliability out of the little Chevy. I'm not saying that there might not be other ways to put a small-block together, but if you follow what I say to the letter, you'll wind up with an engine that will run with the best of 'em *and live*.

The engine I'll be describing here will be a 355-cubic-incher, with a bore of 4.030 and a stroke of 3.48 inches. Since that displacement just happens to coincide with the current NASCAR cubic-inch limit, you might get the idea that what I'm saying would apply to an all-out race engine—and you'd be right. I'm gonna tell you the very best way, as I know it, to build one. Of course, a lot of what I'll mention won't be absolutely necessary, if you're building a street engine, and I'll try to point that out as we go along. The only big differences between a racing engine and a strong street engine are compression ratio, camshaft and the associated valvetrain pieces. Most everything else depends

2

3

1

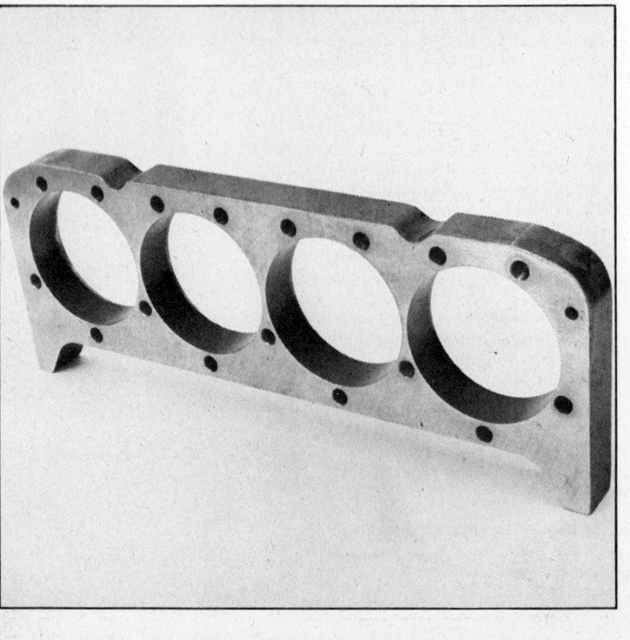

4

upon how much you can spend for power and reliability and just how serious you are about this in the first place. Do it my way, and you'll end up with 580+ horsepower in racing trim that'll last for 500 miles flat out.

THE BLOCK

Let's start at the beginning. For any high-performance engine, it's important to start with a good block. The most available heavy-duty block is the four-bolt-main-bearing-cap, 4-inch-bore standard high-performance production version (part No. 3970016). Also available as a service part is a stronger high tin content block (part No. 366246) and a special thickwall NASCAR block (part No. 366287), but you'll have to know somebody to get this last one.

There are some guys who like to use the pre-1968 blocks that have smaller main and rod journals, 2.30- and 2.00-inch as compared to the later 2.45- and 2.10-inch journals, respectively. The reason for using the small-journal block (part No. 3959588) is that the smaller journals create less internal friction than the larger journals, producing an extra 5 or 6 hp. But there's a trade-off here, since the small-journal block only came with two-bolt-mains. For racing any small-block over 350 cubic inches, I'd definitely use the later four-bolt block; but for smaller displacements and street engines of any displacement, the two-bolt setup will do just fine. I'll leave the choice up to you on this one.

Either way you go, a used block is generally more desirable than a new block, provided the used one isn't worn-out or cracked. The reason for picking the used block is that new blocks tend to shift around a little bit during the first few warm-ups and cool-downs until the cylinder bores and main webs take a set. This problem is common to everybody's blocks, whether they're Chrysler, Ford or GM. Consequently, with a used block you don't have to worry about this internal shift upsetting the precision machining that's necessary to properly set up an engine.

Naturally, you won't be likely to find the high tin content or NASCAR block used, since they were never installed in a production car. But the greater strength of these two blocks is critical in endurance or long-distance circle track racing. The new NASCAR, or "bowtie," block has blind deck stud holes. That's important since they tie the bores together. The deck is thicker on the bowtie block, too—at least .060-inch thicker, maybe twice that in some areas. This makes the block stronger, but it also provides less water jacketing for the top of the bore. This block is also 10-12 pounds heavier than standard blocks. The water core holes in the decks are still too big and must be restricted (I'll get to that later). At any rate, what I'm sayin' is that you probably will have to run these blocks green if you're going racing and need this extra strength.

WALL THICKNESS AND SONIC TESTING

The new NASCAR bowtie block has no water jackets between the cylinders, as on the 400 block, which permits thicker walls. On this block, wall thickness is *supposed* to be around .200-inch (compared to other blocks' minimum .135-inch). But there's a problem with cylinder block wall thickness nowadays, especially with some of the heavy-duty blocks where some of the first ones had a core shift problem with the No. 6 bore. So, ideally, every block ought to be sonic-tested. Sonic testing will give you a rough idea how thick the walls are, but let me say this: I haven't seen a sonic tester yet—and I've seen a lot of 'em—that's reliably accurate. Block cross section is just too difficult to measure. My experience is that you better allow .015- to .020-inch for error. If you don't believe it, the next time you junk a block that has been sonic-tested, saw that block up and check it and you'll get a big surprise.

I'm not saying there isn't a sonic tester made that does work on a block, but I haven't seen it yet—and I sure would like to. So, on sonic testing, I'd say it's a good approximation, and if it makes you feel better, do it.

You hear a lot about bad blocks today. All the big racers claim to be sonic testing blocks—and I believe they do. But they're still breaking blocks in race cars and then discovering they only had a .065- or .070-inch wall. I just offer that as further support about the lack of reliability of sonic testing. What I'd like to see these days is a minimum .125-inch wall, but I'd accept as little as .110-inch in a finished bore. It used to be you could get down as .095 on the wall thickness, but that was before we had all these detonation problems caused by poor gas—and before we were making close to 600 hp. Detonation is a mighty big factor in what will live and what won't.

BLOCK PREP

Wall thickness isn't the only problem. Once you've got the block, hot-tank it and check it for cracks, especially in the main bearing web area. If everything checks out all right, it's time to do some deburring and filing. You have to remember that when the engine is running, there's only one way that the oil gets back into the pan, and that's by gravity. For that reason, it's important to remove any flashings or roughness that might slow or hinder that oil return flow, such as in the lifter gallery area. But probably the most important reason for getting in there with a grinder or file is that certain areas of a block are highly stressed, such as the main webs and caps. That's where you really should be doing something, rounding off all the sharp edges and flashings that might act as stress risers and eventually cause cracks. Some time spent here can save a lot of grief later on.

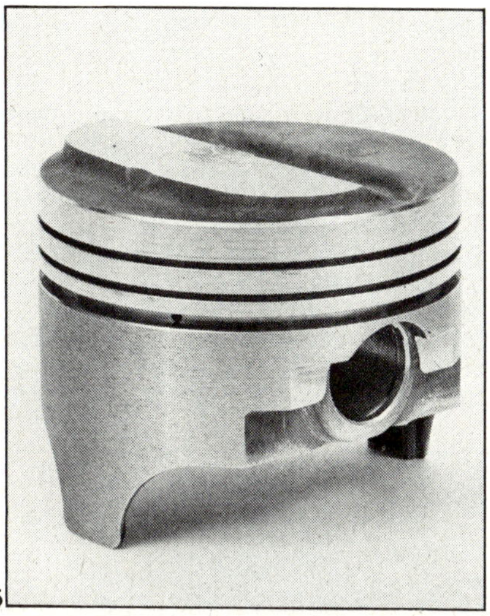
5

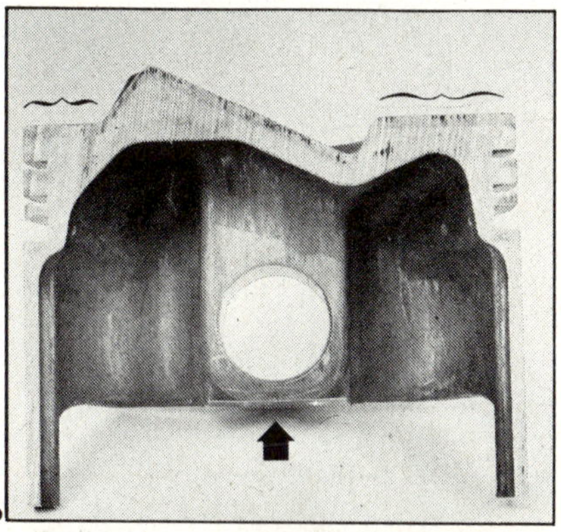
6

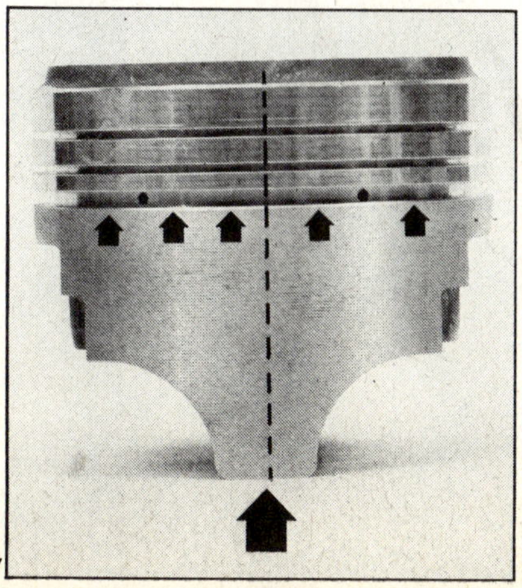
7

1. The latest heavy-duty block is readily identified by the Chevy "bow-tie" emblem on the side. The block weighs 12 pounds more than the standard four-bolt blocks.

2. The bow-tie block features thicker cylinder walls, thicker decks and blind cylinder head bolt holes for extra strength.

3. Smokey makes his own super-thick head plates for boring and honing his blocks.

4. Thinner head plates, such as this one from Hank the Crank, are readily available from a number of aftermarket manufacturers. Using one is a must.

5. The TRW 12.5:1 c.r. pistons offer power and reliability.

6. Mill the piston flats to get proper clearance. Sectioned view shows the thickness of the flats. Milling the pin bosses (arrow) for crankshaft clearance may also be required.

7. Drill five extra drain holes in the oil ring groove (arrows), but don't drill on the piston centerline (dotted).

CRANKSHAFT

Now we're ready for the actual machining operations. The first thing to consider is main journal bore alignment. Nowadays it's almost essential to alignhone the main bearing saddles in every block, but before you go to the trouble and expense, you should see if the crank turns freely in the block. Occasionally, it might not need to be alignbored. The way to do this is to install the crank in the block, using the proper inserts with the proper amount of clearance (.0025-inch). Since you need the crank and bearings for this check, we'll take time out to talk about these parts.

On this engine, we'll be using the Chevy high-performance 3.48-inch-stroke crank (part No. 3941184). A forging like this one is better than a billet, because the grain flow is not interrupted. You might think of a billet as being like a tree trunk with chunks sawed out of it. Obviously, it's weakened. But if you bend the tree to the right shape, it's still all in one piece. And while that's a little oversimplified, that's what in effect a forging accomplishes. Huge machines press and hammer the steel into the right shape. So a forged crank with the same cross sectional area as a billet will be considerably stronger.

Unfortunately, these Chevy cranks have two problems the way they're currently coming from the factory—they aren't straight and they have bum radii. Now I could go into a lot of discussion about how to straighten cranks, but the best bet is probably to just buy a good crank from Hank the Crank (7253 Lankershim Blvd., North Hollywood, CA 91605) or Moldex Tool (25249 W. Warren, Dearborn Heights, MI 48127) that's already been straightened and had the radii cleaned up. Naturally, the crank should be fully magnafluxed. If you want to have it cross-drilled, that'd be a good idea too, but it isn't absolutely necessary. If you want to spend some money, having the crank fully counterweighted is probably the best. This is worth more than knife-edging, but neither is worth it for anything but the super-speedways.

CHECKING CRANK STRAIGHTNESS

When you get the crank, it will be fully prepared and ready to run. All you have to do is give it a good cleaning and follow the crank shop's instructions to the letter. Of course, you should still check the crank to make sure it's straight when you receive it. This is so important that I'll outline how to check it. All you need is a block and a dial indicator. Lay the crank in the block, using bearing inserts on the No. 1 and No. 5 saddles only. Coat the inserts with a light oil and rotate the crank slowly, measuring runout on the center three open journals with a dial indicator. If there's less than .002-inch total indicator runout, you can stop right there; but if the readout is over .002-inch, then you'll want to proceed to find out just where the crank is bent. Next move the bearing insert from the No. 1 saddle to the No. 2 and recheck the open journals. This will tell you if the crank was bent in the middle or at the No. 1 journal. Then move the No. 5 insert to the No. 4 saddle and recheck, for the same reasons. If the crank has less than .001- to .0015-inch warpage (that's .002- to .003-inch total indicator runout), it's acceptable. Then, using some blueing, check to see that the edges of the bearing inserts aren't sitting up on the radii rather than the insert seating squarely on the actual journal.

BEARINGS

It used to be that the standard GM bearings were as good or better than anything else you could buy. That's no longer true. The Clevite 77 bearing is a better bearing than the GM Moraine bearing, in my opinion. The average guy who's just going to use the engine on the street is just fine with the Moraine 400, but if he's got a bearing problem, there's a better bearing available. Now I don't know about the other manufacturers because I've never used their bearings, but in our Indy engine we cannot use a GM bearing. The Clevite bearing is the only one that works. Clevite uses a heavier steel backing and they've built in a little more crush. Both of these are desirable. Their finished plating that touches the crankshaft is harder than the GM stuff too, but that isn't necessarily a desirable quality from the "embeddability" standpoint, so there's a little trade-off. Since there's not much difference in the price of the GM and Clevite bearings, you might as well use the Clevites to start with.

BLOCK ALIGNMENT

Now that we have a crank and the bearings, we can check the block alignment. The crank and inserts should be coated with a very light oil or automatic transmission fluid for this test. And of course the seals should be left out. If the block and crank are straight, then you should be able to rattle the crank from front to rear. If you've got the desired .004- to .006-inch endplay tolerance, you'll be able to hear it rattle. Being able to rattle the crank under these conditions means that it isn't binding. How-

8

9

10

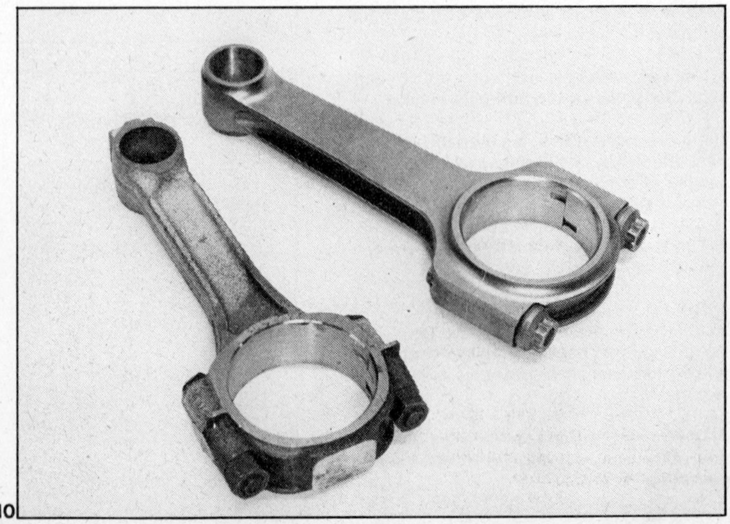

ever, you should rotate the crank about 10 degrees at a time and check it to see if it will rattle in each position. You might be able to rattle it in some positions and not in others. If you can rattle it all the way around, then I'd say the block doesn't need align-honing. On the other hand, if you can't rattle it in every position, you'd better start thinking about getting the block aligned. Naturally all this depends on the straightness of the crank, but you *did* check that when you bought the crank, didn't you?

Now, there are a lot of different machines around for align-boring cylinder blocks, and over the years I've tried all of 'em. For my money, you just can't beat the Sunnen align-hone. The Sunnen process involves milling about .002-inch off the mating face of each main cap. Then the caps are torqued back on, the hone makes several passes and the block is as straight as it's ever going to be. It's fast and it beats align-*boring* all to hell. And it doesn't take off enough metal to screw up the relationship between the timing gears or anything like that. So find a competent shop that has a Sunnen align-hone and have them resize and align those saddles to the minimum side of the diameter tolerance for a small-block. This will provide maximum bearing shell "crush," which is desirable in terms of heat transfer and to provide a better foundation for bearing inserts.

Lately a lot of guys have been making a big thing out of the side clearance of the caps where they index into the block. All that side clearance does is ensure proper side-to-side alignment of the cap. But while the engine is running, the sides of the cap do *not* try to spread out toward the indexing steps—they contract toward the crank journal. As the crank pushes down on the middle of the cap, the sides are pulled inward, pushing the bearing against the journal. This is one of the biggest causes of main bearing failure in the small-block Chevy. To fix this condition, machine the caps and the block for close-tolerance steel bushings around the main cap studs—yes, studs instead of bolts, and I'll get to that in a minute—to provide perfectly accurate alignment of the caps while simultaneously preventing any inward movement of the sides of the cap. Naturally this operation must be done *before* align-honing the block.

With the block properly aligned, the next step is to square the block to ensure that the cylinder head decks are exactly parallel to the crankshaft centerline. Any good machine shop should be able to handle this operation for you.

WATER CIRCULATION

Next we'll move up to the block decks. To gain additional strength in the deck area, tap and plug those big water holes, as shown, across the top of each deck. Then, using the head gasket as a template, drill ⅛-inch steam holes in each of the plugs.

On the bottom edge of each deck, just below each cylinder, you will see a small hole. This hole is intended to allow water to circulate up around the spark plug boss in the head. But, since the heads are designed to be used on either side of the engine, there is a slot on the deck of each head to allow the water to go around both sides of the plug boss, regardless of which end of the slot initially receives the water. Good water circulation in this area is critical, since it also helps cool the exhaust seat, so to increase the flow, drill a second small hole in the block corresponding to the second hole in the head. You'll also have to punch corresponding holes in the head gaskets. This provides about 5 gallons per minute of additional flow in this critical area.

OVERBORING

Now we're ready to open up those cylinder bores. You'll recall that at the beginning of this article I said we'd be building an engine with a 4.030-inch bore but that we'd start with a 4.000-inch bore block. Some of you might wonder why I recommend a .030-inch overbore instead of a .060-inch increase.

The reason for this is the wall thickness problem I already talked about. I've sectioned a lot of blocks, and nominal wall thickness, except for the NASCAR blocks, is in the neighborhood of .135- to .140-inch. Now one of the secrets of the Chevy block is that it's light. The Chevy engineers put the metal where it belongs, but they don't overdo it. I'm sure there's more to the story than just good engineering; it's also cheaper if a block weighs only 163 pounds instead of 173 pounds. So if you bore .060-inch oversize, which is a .030-inch cut all the way around, you're down to around .110-inch, assuming the cylinder core was dead on. But you have to believe that the cores don't usually go in exactly straight, that they do shift a little bit during many mass-production casting processes. Suppose the core had shifted only .015-inch (and .030-inch is not uncommon). Now you're down to a .095 wall thickness, and that, as I said, just won't do the job with 600 hp. With a .030-inch overbore, that same cylinder would have a .110-inch-thick wall, and that's okay.

Naturally, on those NASCAR blocks which supposedly have the .200-inch-thick walls, you could overbore the block more, provided the wall thickness checks out okay. But then you'd be de-

8. Racing engines not only get the 6-in. Carillo rod, but also the Diamond/TRW forged aluminum piston. Racing compression ratios may run as high as 13.0:1. Note the top ring's proximity to the piston deck.

9. Street engines should be fitted with the TRW turbo forged aluminum piston, which when used with the 67-68 cc combustion chamber of the turbo slant-plug head, will produce a compression ratio of approximately 8.6:1 for use with today's pump gasoline.

10. Shown here is the heavy-duty 5.7-in. Z28 stock rod and a 6-in. Carillo steel rod. The 6-in. rod is preferred, but the Z28 rod is acceptable for street use.

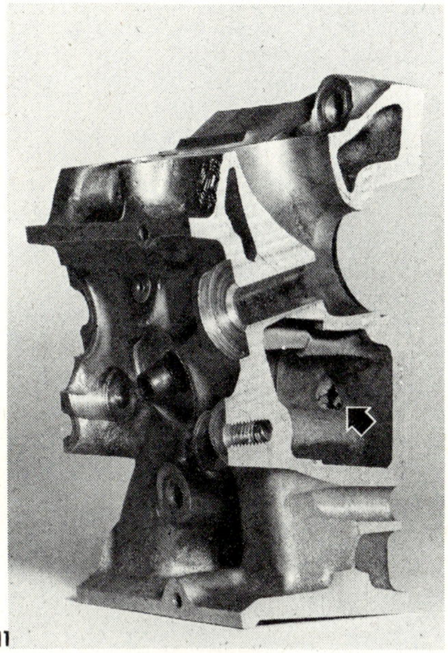
11

11. The importance of opening restricted water passages is shown in this sectioned (big-block) head. Note the crack adjacent to restriction (arrow).

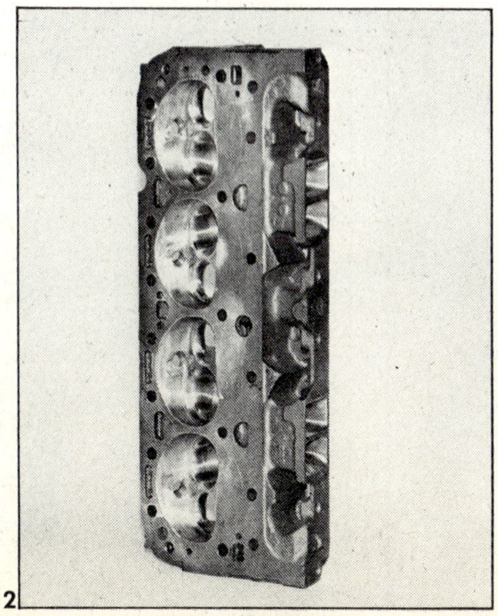
12

12. Although they are expensive, properly ported slant-plug heads, as this one from Valley Head Service, are mandatory for peak performance.

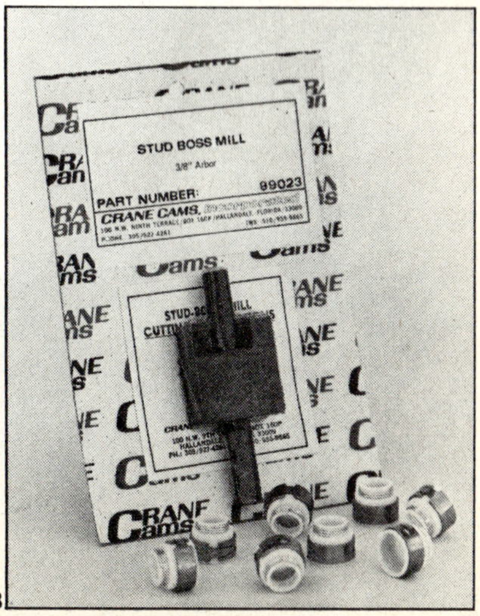

13

13. To machine the guides for Crane one-piece valve seals, which are similar to p.c. seals, a stud boss mill, which fits in any electric drill, is required.

feating the purpose of the block, which is to provide more wall strength. Anyway, at 355-cubic-inches, you'll have all the power you need.

Just remember that the standard production Chevy blocks aren't made out of very hard cast iron. It's fairly soft. A Mo-Par block, for example, has a much harder or tougher cast iron. Using a soft cast iron isn't necessarily a bad idea, though. In fact, for normal street use, which is what these engines were designed for, a soft cast iron lets the rings and everything get acquainted faster to seal better, and that's good. From the racer's standpoint, a harder cast iron would be more desirable, so when a guy overbores .060-inch, he's already made a mistake. Now, a lot of races have been won on .060-inch-over engines and a lot more will probably be won in the future. However, it's my guess that those same guys would not have gone faster with .030-inch-over engines. Any more, and you're asking for trouble.

This business of wall thickness has become more important because something else has happened down the line regarding boring with head plates. We've been talking about head plates for years and nobody disagrees that they're a good idea, but a lot of people would question just how much difference they make. There's also talk that at high rpm the cylinder walls vibrate enough to seriously detract from ring sealing anyway, and I don't doubt that a certain amount of that is true, but it would all be relative. A round bore has got to do a better job of sealing than a pentagonal one. There's just no substitute for a round hole; and unless you take a lot of precautions to get a round hole when you bore or hone these engines, that's not what you'll wind up with, especially at the top of the bore. I don't really believe that the head bolts in or out, affect the bore all the way down enough to be critical. We could afford a pentagonal-shaped hole down below, but it's in the upper 1½ or 2 inches of piston travel where your compression is made and unhappily that's where the head bolts are. That's why you should set up the ring gap about two inches down in the bore, below the bolt distortion area.

When you pull the head bolts up on any engine, they never pull straight up on the bore—they either pull the bore out at that point or force it in. After all, the head bolt holes aren't in a thick area of the block; they're simply hung on the sides of the bores internally. In the Chevy block, the bolts pull the bore in. If you check it, every place there's a bolt, the wall moves in there and twists a measurable amount (several thousandths) down about 1½ inches. This gives the bore a pentagonal shape, since the Chevy uses five head bolts per cylinder, and the compression rings simply don't seal.

A ring isn't pentagonal; it's round, and rings are designed to turn on the piston so they don't wear to a pentagonal configuration either. Even if they did, the bore is still round at the bottom, and the rings would leak down there like a son-of-a-gun. The proof of this is on a leak-down test.

You have to pay attention to things like these because there's horsepower to be gained by making a good seal. And if you use a head plate to get a round bore, then you can run a tighter end gap on the rings, down to as little as .012- or .014-inch, and they won't butt. Without the head plates we used to have to run .020- or .025-inch of end gap on the rings because of the wall vibration and the fact that the bore might have been 4.000 inches when we bored it, but squeezed down to 3.997 inches at the narrowest point when the heads were bolted down. Of course, this would change the end gap by a factor of 3.14 around the cylinder.

Boring and honing with a head plate just lets you build a better-sealing engine—it's as simple as that.

STUDS

From this discussion, you're probably starting to think that bolts put a helluva lot of stress on the cast iron—and you'd be right. Studs are absolutely necessary to derive adequate clamping force to contain 600 hp without overstressing the threads in the block. In addition, bolts simply impart too much stress into the cast iron during the torquing process. They twist the heck out of it.

So taking all that into consideration, studs are a better way to go—as long as they're installed correctly. Correct stud installation is important or you'll cancel out most of their benefits. The first thing to do is to chase both the hole and stud threads with a tap to clean them and make sure that everything is open. Then the top of each hole must be chamfered at an angle corresponding to the chamfer on the shoulder of the stud. Then screw each stud in finger-tight to be sure it will bottom on its shoulder rather than the threads bottoming out. If it doesn't bottom on the shoulder, grind off part of the threaded portion until it does. This is critical. After this clean the threads with solvent.

During final assembly, Loctite will be used to seal and lock the studs in place. When the time comes, the correct procedure is to coat the stud threads with Loctite red. Run the studs in and out several times, then hand tighten. While the Loctite is still wet, immediately install the heads and torque them down. This will insure the head studs are perfectly perpendicular to the deck surface.

Remember not to Loctite these studs until the engine is completely finished, has been trial-assembled at least once and is going together for the final time. If it ever becomes necessary to remove the studs after they've been Loctited, you'll have to heat the block, as per Loctite's instructions, to avoid thread damage.

The last operation on the head is to spot face the washer/nut face on the heads to assure the nuts seat squarely and don't distort the bore. This may sound nit-picky, but it's important because studs will twist that bore. The proof is this: Take a good engine that has been prepped like I described and do a leakdown test. If it's right, it'll be 3 to 4 percent. Tighten or loosen any one stud and the leakdown will increase.

LEAKDOWN TESTING

That brings up another point: leakdown testing. Any time you suspect an engine problem, that's the first test you ought to make. Around here, if an engine leaks over 7 percent, I pull the head to find out why. We check each cylinder before *and* after each dyno run. Otherwise, you don't know what you have.

GASKETS AND TORQUE PLATES

There's one more thing that I know for sure about boring or honing with a head

14

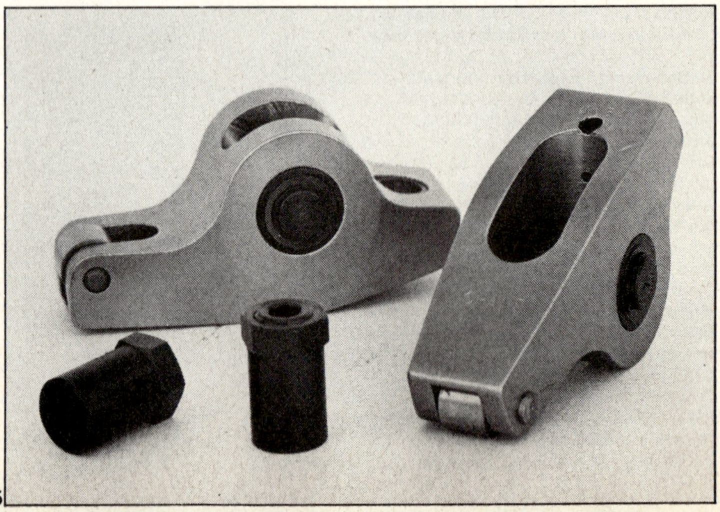

15

plate. When you put that plate on the block, if you don't use exactly the same kind of gasket between the plate and the block that you're going to run on the completed engine, you still won't wind up with a round bore when the heads are torqued down. This is because different types of gaskets load differently. I'm not saying that you have to use the same kind of gasket by brand name, but if you plan to run a steel gasket, use a steel gasket under the head plate. If you're going to run a composition gasket, that's the kind to use under the plate.

It's also important to torque the plates the same amount as you're going to run. And if you're going to run studs, use studs with the torque plate.

WHICH HEAD GASKET?

Getting back to head gaskets, to get a good seal, the steel head gasket is hands down the best when you're putting a cast-iron head on a cast-iron block. At one time, I recommended the Chevy stainless steel gaskets, but I don't use them anymore. Tests have shown it to be too stiff. Mild steel conforms better, so I now recommend you use two mild steel gaskets (part No. 3830711) on each head with a zero deck. This two-gasket stack will have a compressed thickness of .038-inch.

For putting an aluminum head on a cast-iron block, or with aluminum heads on an aluminum block, a composition-type head gasket is best because of the increased distortion of the aluminum between the head studs. Use the heavy-duty Victor-Core composition gasket (part No. 3965790) or the new Fel Pro-type O-ring gasket (part No. 14011041). These two gaskets both have compressed heights of .040-inch.

I haven't tried the new O-ring gasket on an iron engine, but I *know* two stacked mild steel gaskets work. That's also a safety valve. With iron on iron and studs, if they blow, that saves the piston or head.

Nowadays, most manufacturers are making gaskets for a 4⅛ bore, which is bad on a 4-inch bore. You can buy a mild steel gasket for the 4-inch bore, which is another good reason for using it. Fel Pro, or somebody, definitely ought to offer at least a 4-inch-bore gasket for the small-block. The guy with an old 3¾-inch bore small-block really has a problem these days. The big gasket leaves an area that doesn't get purged of contaminated gases. Also, it's so thin between the bores that it's more prone to blow between the bores. This problem is all because they want to make one gasket to fit all applications. Most gaskets are made to accept a .060-overbore, making the problem that much worse.

Regardless of what gasket you use, for extra sealing during the final engine assembly, I now suggest coating the gasket on both sides with Permatex Super 300 sealer. With this sealer it's no longer necessary to use high-temperature aluminum paint, as I've advocated in the past. This sealer acts as a filler for small imperfections in the mating surfaces. The sealer remains pliable in a semi-solid state—really it acts as a fluid and flows into any low areas when the heads are torqued down, making a better seal. I definitely don't recommend any of the copper sealers for head gaskets, since contrary to what most guys think, copper is very soft and has practically no strength. By the same token, I don't really care for those non-retorquing head gaskets, either.

PISTONS & COMPRESSION RATIO

Now that we've discussed the basics of head plates and gaskets, we'll get down to the actual boring. But before you can determine the final bore size, you have to get the pistons—and what kind of piston and compression ratio you use depends on whether you're going to be running on the street or racing, and

16

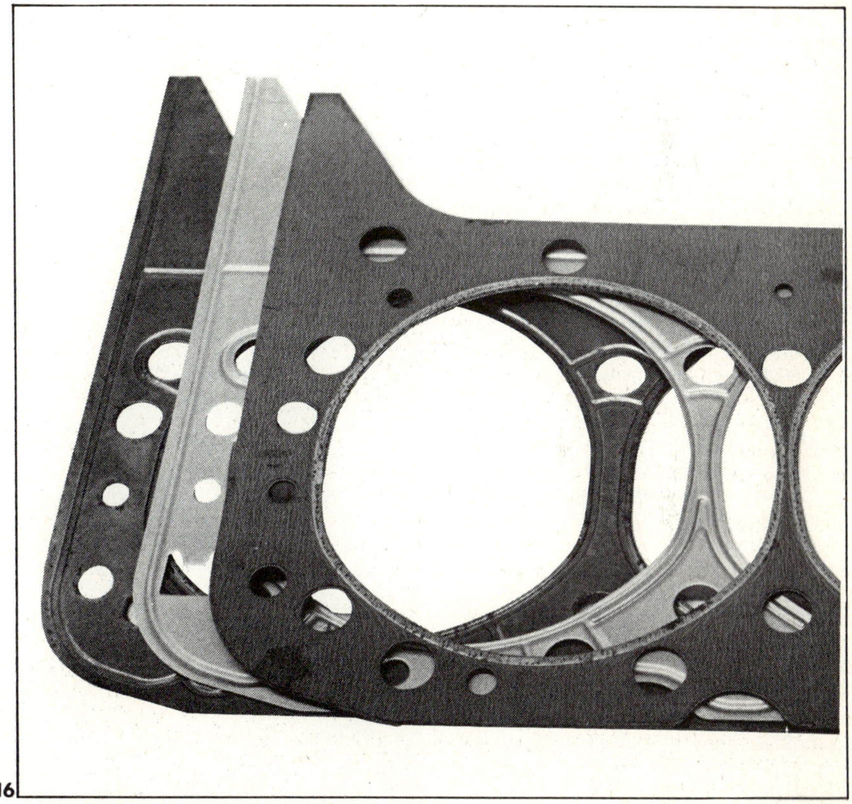

17

14. Stock Chevy rocker arms can be used, but a small sheetmetal cap should be welded over the pushrod end to prevent oil from squirting against the valve covers (arrow).

15. Since stock Chevy rockers have only a 1.5:1 ratio, extra valve lift can be attained by using Crane's 1.6:1 needle bearing rockers without moving the rocker studs. But check piston-to-valve clearance.

16. Shown here is a comparison of the between-the-bores width of the gaskets for the early 3-¾-in. bore, the 4-in. bore, and the 4-⅛-in. bore blocks.

17. All current gaskets now have the very narrow section width. Note how far from the edge of the bore the gasket edge is. The narrow gasket area between the bores also increases the likelihood of gasket failure.

just what kind of gas is available.

If you're going to race on a short track, I still recommend runnin' a 12.8 to 13.0:1 compression ratio, although on a long track you should drop it down to 12.5 to 12.6:1. I still think this is too high, but if you don't go this high, some other guy will and blow you away. It's expensive—you'll eventually destroy the engine—but that's what it's going to take to win: 13.0:1 at 8200 rpm. You've got to get the good racing gas to do this! It's a must: 100+ octane. You also have to have at least 3cc of lead per gallon, too, to keep the valves lubricated, even with methanol.

On the street, you have to drop to a *maximum* of 8.5:1 if you're using straight pump gas. Gas quality is so bad today and is likely to continue to get worse. There is no more 115/145 aviation gas to be had. Pure racing gas—Union 76, Daeco, Sunoco 280 or whatever—is illegal in many states. Some fuel additives are less effective than claimed, or it separates out, or is mixed incorrectly; you just can't achieve consistent results with an additive.

Now you can try mixing gas to raise the octane. My tests have shown that when one part of leaded regular (containing roughly 4cc of lead per gallon) is mixed with three parts of 91 octane unleaded premium, the result is a 96-octane fuel. However, if you add more lead, the effect tapers off, so 96 octane is about as high as you can get. Of course, the problem now gets to be one of finding out how much lead is in different brands of fuel, and that'll be tough. Anyway, even with mixing, you couldn't go much over 9.5:1. And with today's gas lines, how are you going to mix gas from two pumps? Forget compression.

Just remember, with straight pump gas, don't go over 8.5:1. With additives, gasohol, water injection or mixed gas, do not under any circumstances go over 10:1; 9.5 or 9.0 would be a lot better.

PISTON SELECTION

All that being said, it would appear that TRW (8001 E. Pleasant Valley Rd., Cleveland, OH 44131) has the technology and has gotten ahead of everybody else except those who are following TRW's lead. [*Smokey feels this way because TRW is the leading innovator in piston barrel skirt contours and taper.—Editor*] So it seems that using TRWs is the easiest, smartest and cheapest way out whenever possible. It's a well-built piston, very strong and the weight of it is generally lighter than anything else you might buy.

They recently came out with a new 12.5:1 high compression NASCAR piston with four valve reliefs (see chart), available for both a 5.7- and 6.0-inch rods (we'll get to rod lengths later). If you want to spend a little more money, the best bet for racing is to get a set of the special Diamond-Elkins pistons, which are forged for them by TRW without a pin hole or ring grooves. This means the piston has the premium alloy and the well-engineered TRW skirt, and Diamond-Elkins can put the pin hole and rings wherever you want them. This piston is the best on the market, and that includes the Cosworth Chevy piston.

For the street, there's a need now for good forged aluminum pistons in low compression. They would be flat-top or dished. TRW has a flat-top small-block forging, and a dished forging (see chart). Others will probably follow. [*TRW plans to introduce a series of 10:1 forgings this spring to fill in the gap between their high- and low-compression piston offerings.—Editor*]

As you can see from the chart, depending on how many cc's you run in the combustion chamber, compression ratios from 7.6 to 11.0:1 are available with TRW's current "low compression" forgings. Some have oil return holes, some have slots. Slots tend to weaken the piston, so I prefer holes. And incidentally, since the late Turbo heads clean up at 67 to 68cc, the TRW pistons will work with this head just fine on the street.

PISTON MODIFICATIONS

The first thing to do when you get the pistons from TRW is to have them machined for double Tru-arc pin retainers. I understand TRW is starting to make them this way, so you might not have to do this step. Drill about five extra oil drain holes in the oil ring groove on each side of the piston, being careful not to nick the ring lands or to drill any holes on a line exactly 90 degrees from the pin centerline. About the only other thing you might want to do to the pistons is to carefully remove any sharp edges from the piston top that might heat up and become hot spots in the combustion chamber. But don't round off the lower edges of the piston skirts, since they're designed to help scrub excess oil off the cylinder walls before it overloads the oil rings. Just as with the connecting rods (hold on, we're getting there!), it's best to consult with your crankshaft grinder on the pistons you plan to run.

When using the custom-tailored Diamond pistons, have them cut the top piston ring .150-inch down from the piston deck. Maintain normal spacing between the rest of the rings. Consult Diamond-Elkins on the number of oil drainback holes in the oil ring groove. Groove depth should be cut so that the face of the rings can be pushed .002-inch back into the groove—no more, no less. And don't mess around with any gas ports. I don't believe in plasma coatings, either;

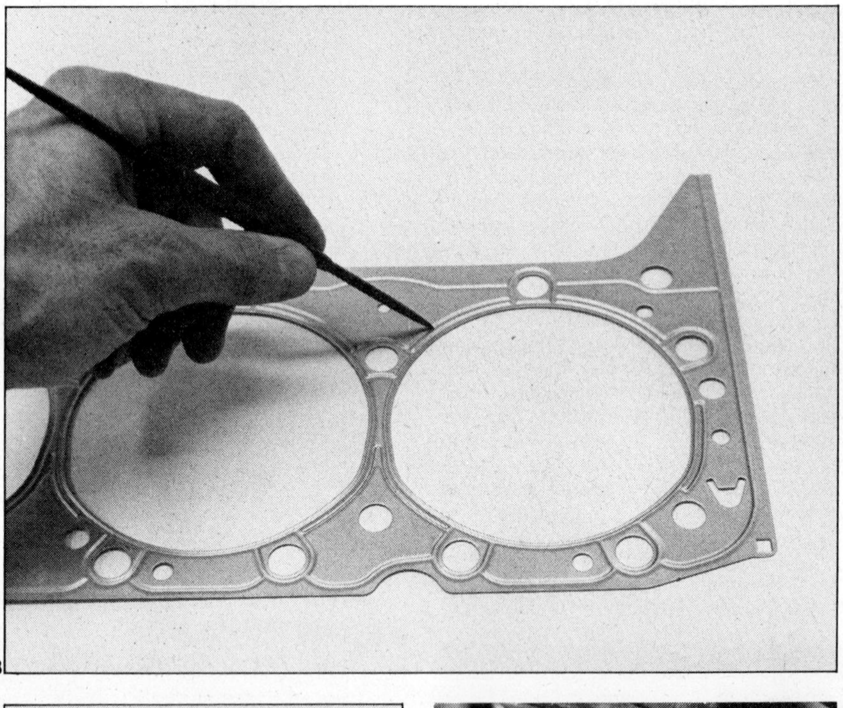

18

19

20

I've found the extra heat imposes durability problems.

The last piston consideration will be to get *at least* .080-inch piston-to-valve clearance before you blend and smooth the piston tops to eliminate any abrupt edges.

BORING & HONING PROCEDURE

For the actual honing operation, my opinion is that you can't beat the Sunnen CK-10 Honemaster machine for increasing bore size. This machine has got a lot of things going for it that I've never seen in any other boring or honing machine. With this setup, all the metal is actually removed by a honing process rather than a cutting or chipping process. The surface that results is much better.

For the final wall surface, 800-grit stones should be used. This will give a cylinder wall that is very smooth, and for my money you can't get the wall too smooth, regardless of whether you're planning on using moly or chrome rings. There's no compromise here! Just don't burnish the bores, and insist on a Sunnen CK-10 *and* somebody who knows how to run it. Go for a finished piston-to-wall clearance of .0065- to .007-inch, measured on the piston skirt at a point level with the wrist pin but perpendicular to it.

I feel so strongly about the quality of the job that the Sunnen machine does that I think a guy should load up his block and haul it a hundred miles if he has to to find a shop that has one of these Honemasters—that's if he's serious about what he's doing.

CONNECTING ROD LENGTH

With the cylinder bores properly sized and finished, it's time to worry about connecting rods and piston deck height. For the street, the 5.7-inch-long Z28 rod, if properly prepared, is okay. But if you want to make the most power, a 355-inch engine, such as we're building here, will like the longest rod you can put in it—at least 6 inches long—which obviously pretty much excludes the 5.7-inch stock rod. For my money, the best bet in rods is the Carrillo rod. Although these rods are expensive, with today's machining costs it'll cost you just as much to rework a set of stock rods, which still won't be as light or as reliable as the Carrillo rods, let alone the right length. Also specify the good S.P.S. rod bolts. Even using the reworked big-block rod forgings from Chevrolet isn't as practical as the Carrillo rod, because the big-block forging is prone to cam clearance problems on the 3.48-inch-stroke small-block. Whatever rod you use, stock or aftermarket, doweling the cap on a steel rod is a waste of money. Bearing crush retains the rod bearing just fine.

STOCK ROD PREPARATION

If you're committed to using the factory rods, they must be prepared properly. For the pre-'68 small-journal cranks, use connecting rod part No. 3864881, which has a pressed pin. For the later big-journal cranks, two Z28-style rods are available: The floating pin version (part No. 3946841) came stock on '69 Z28s and is available over-the-counter. The more common rod is part No. 3973386, which came stock on LT1s, '70-'74 Z28s and all Corvette L82s; it has a pressed pin, but up to 6000 rpm, you can use the pressed pin. But installing it is *not* a job for the amateur; it must be done in an expert machine shop with proper equipment. If you want to float the pin, don't bother bushing the rod, because a bushing weakens the small end of a stock rod. Just hone the steel rod, but you must provide a pin oiling hole through the top of the rod, and run .00135- to .00150-inch clearance in both the piston and the rod, with adequate provisions in both to prevent galling. Whether the pins are floated or pressed, it's a must to drill pin oiling holes from the piston oil ring land. The stock slots are inadequate.

To prepare any of these stock rods, a

18. Double steel shim gaskets are now used on the small-block when it is modified for high output. These gaskets have a rolled bead. Gaskets should always be installed with the concave side of the bead facing up, as if to hold water.

19. For wet sump applications, the oil pump pickup tube should be at least ⅝-in. inside diameter. This pump is the Z28 version.

20. The oil filter pad on the block should be machined off, as shown, and fitted for use with multiple remote oil filters.

21. Still one of the strongest street manifolds ever produced, the Edelbrock Smokey Ram features a massive internal volume which requires a large accelerator pump squirt for good response.

22. The original Z28 aluminum single 4-barrel manifold is also an excellent choice for street use, although it's quite expensive these days.

23. For most applications the heavy-duty AC mechanical fuel pump is adequate. Unlike the regular production pump, this unit can be disassembled for repair.

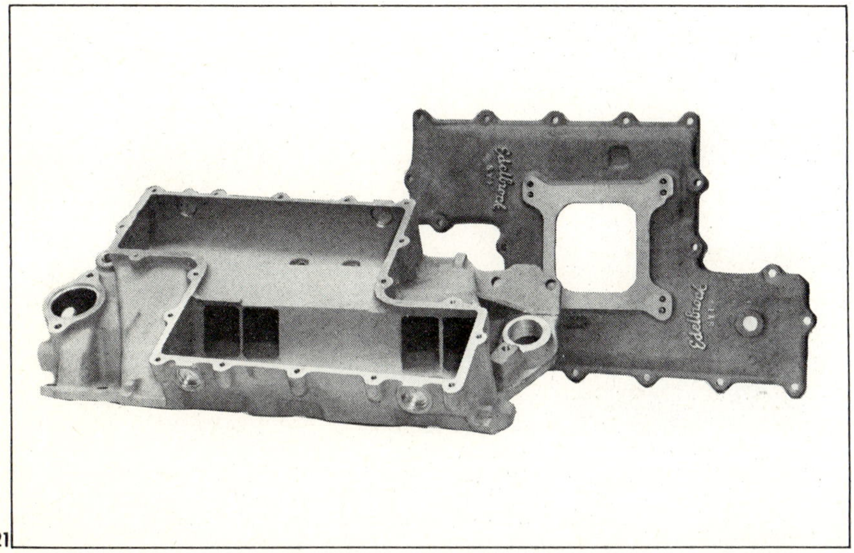

21

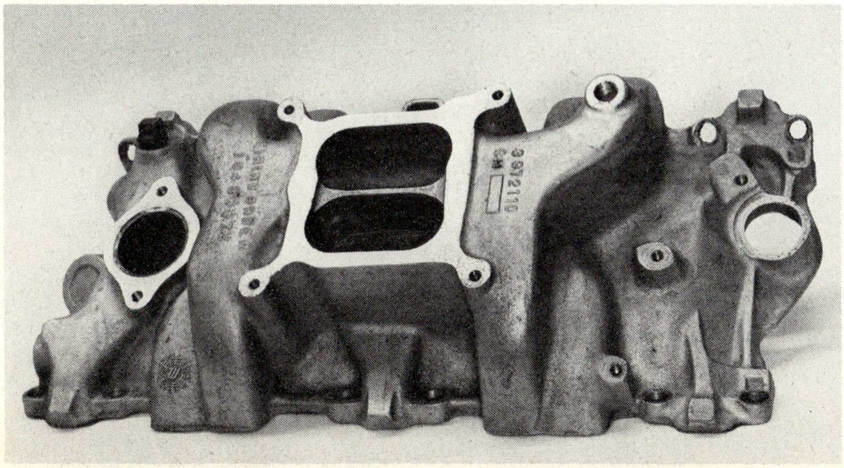

22

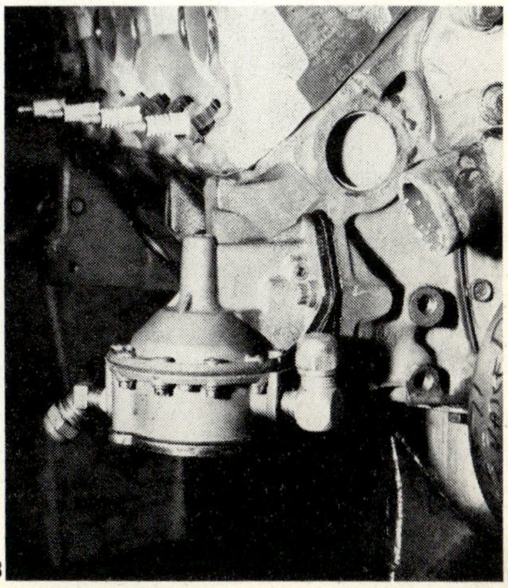

23

guy has to remove all the flashing from the beam of the rod. Grinding lengthwise, remove any sharp edges that might become stress risers and then polish out all tool and grinding marks. Next, the rods should be machined for a side clearance of .015- to .018-inch per pair. Then the rod and cap should be shot-peened and magnafluxed. There are quite a few shops that can supply these rods, ready to run, with all of the above modifications. Check with the guy who grinds your crank for the name of a reputable supplier.

PINS AND RETAINERS

With full-floating pins it's necessary to properly retain the pins in the piston. Under 6000 rpm, single Tru-Arcs or Spirolox retainers are okay. Over 6000 or in racing, use double .042-inch Tru-Arcs. If you can get them, use Tru-Arcs that are .001-inch oversize on the outside diameter for a tighter fit, and replace them with new ones each time the engine is disassembled for an overhaul.

Spirolox retainers are okay, except sometimes the heat treat makes the tang on the ends brittle. If the end breaks off, it can jam the oil pump. Double Tru-Arcs have the only system I've never seen fail. However, the double Tru-Arcs must be properly installed. First of all, the ends of the wrist pins must be machined to provide a .005-inch end clearance between the pin and the retainers. Any less will loosen the double Tru-Arcs and any more will allow the pin to hammer them out. Next, there's a right and a wrong way to install the double Tru-Arcs. When they make the Tru-Arcs they're stamped out by a die which leaves one side of the ring very sharp on the edges, and the other side is kind of rounded. Installed, the double Tru-Arcs should have the sharp edge of one facing the pin while the sharp edge of the other faces the cylinder wall—that is, the rounded edges will butt against each other to act as a type of bearing. Also, the end gap on the inside Tru-Arc should be 180 degrees opposite the outside one.

No matter what kind of retainers you're using, be sure to check the alignment of rod-to-piston. The rod must be exactly perpendicular to the crankshaft centerline and the pin must be perfectly parallel to the crank centerline. Sometimes the big end of the rod was drilled at a very slight angle; sometimes the piston pin holes don't line up just right.

As for the pins themselves, you should specify premium quality (not lightweight) wrist pins.

TRIAL FIT

At this point, we're ready for a trial fit of the crank, rods and pistons. Install the crank and the rod and piston assemblies and check the piston-to-deck height (the distance of the piston flat below or above the top of the block at top dead center). Right now I recommend a zero deck. In other words, the flat of the piston should rise no higher than the top of the block. This may require milling the piston flats to get all eight pistons identical within this deck specification. No part of the piston, except for its dome, should extend above this dimension. This is critical, since it is necessary to maintain a piston-to-head-clearance of .038-inch and piston-to-valve clearance of .080-inch. Don't go overboard, though. Excessive clearance here will destroy the quench effect of the combustion chambers. One other thing often overlooked is that the greater the piston-to-wall, or skirt, clearance, the more the piston rocks at the top of the bore. Therefore, for each additional .001-inch of skirt clearance, increase the deck clearance by .002-inch.

One other thing to check on this trial fit is adequate clearance between the bottom of the piston pin bosses and the crankshaft counterweights at bottom dead center. If there's any interference here you'll have to file the bottom of the bosses until you get .010- to .015-inch clearance.

OIL & WATER PASSAGE PLUGS

With everything disassembled once again, there's only one thing left to do before beginning our final cleaning prior to assembly of the short-block. The stock press-fit oil and water passage plugs sometimes blow out. Also on a used block, after the press-fit plugs are

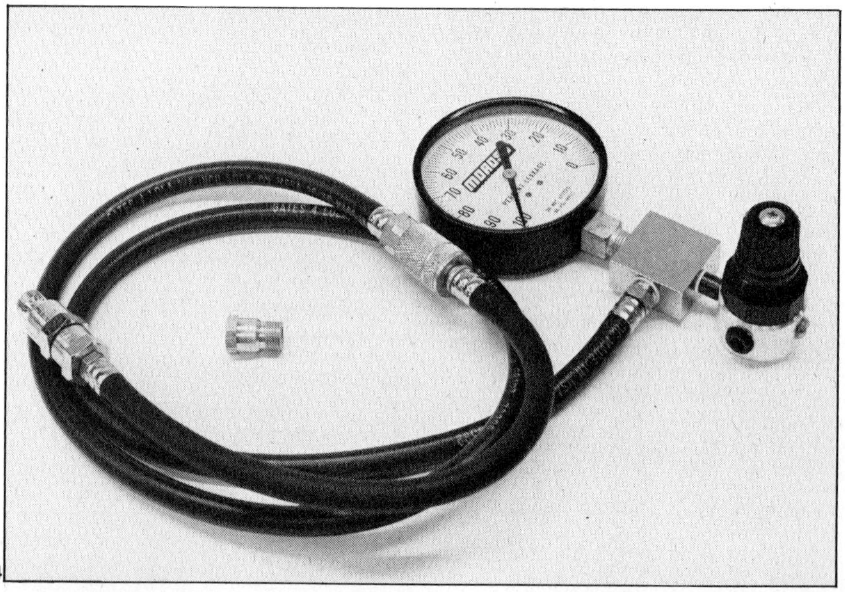
24

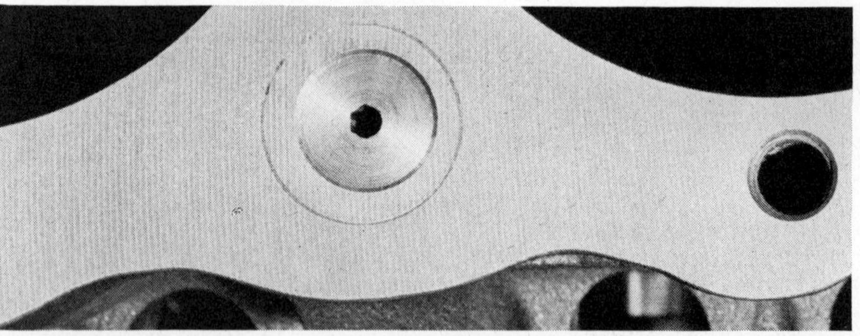
25

26

hammered in there a couple of times, the hole gets enlarged enough that the plug doesn't fit all that tight. From a real craftsman's standpoint, all of the oil and water passage pressed-in plugs should be replaced with screw-in plugs, but this means tapping all the holes. Actually, nowadays, only the oil galleries are tapped for screw-in plugs. The water jacket freeze plugs are simply installed with two-part epoxy. It works just as well.

FINAL BLOCK CLEANUP

Now that all the short-block machining is done, send out all the moving parts to the balancing shop and we'll do a final cleanup on the block. Go after all the internal water passages with bottle brushes, gun cleaning brushes, wires or whatever it takes to dislodge any core sand, dirt, rust, grit or flashings that might still be inside. Wash the cylinder bores with ketone. You can't do too good a job here, since we're after not only what's loose inside today, but also what might come loose tomorrow.

Of course, there'll be some internal passages that you just can't reach, and this is a problem. Not only is it common for chunks of core sand to remain lodged in a block, it's also common to find casting flashings that restrict or block off small passages. Such flashings are even more critical in the cylinder heads, particularly around the spark plugs. (Incidentally, everything we do here toward cleaning the block holds equally true for cleaning the heads and intake manifold later on.)

Since it is so important to remove these restrictions to ensure proper water circulation and cooling, it seems to me that the best way to accomplish this is to pour acid through the water passages. I guess you'd call this chemical milling. It's got to be the cheapest and fastest way to do it. Acid will cut loose anything that might get loose today or tomorrow. Understand, however, that this would just be a mild acid treatment, and it would have to be done by someone who's experienced with this type of process. I'm not saying you have to do this—in fact, this isn't even a widely accepted procedure at this time—but it is time that everybody thought about it for both blocks and heads. It might not seem like all of this is necessary for a street engine, but I think a guy should go by what he can afford to do. As far as saying that just brush-cleaning is good enough, who knows? Each block and head are different. Some are better than others.

PAINTING THE INSIDE OF THE BLOCK

Before we begin the actual assembly of the lower end, I'd like to say something about painting the inside of a block. While it's true that a layer of high quality, temperature-resistant paint will

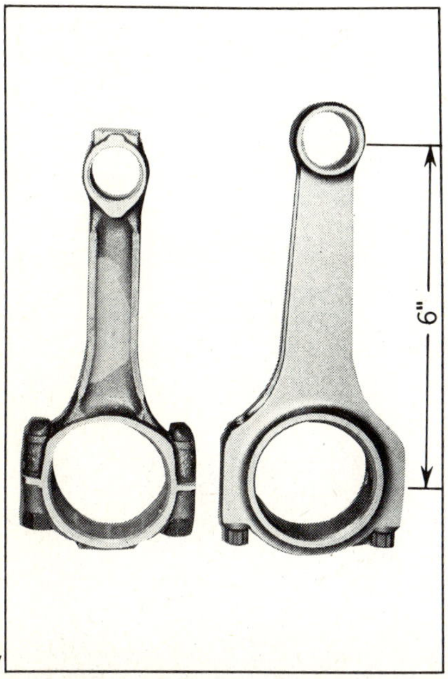

27

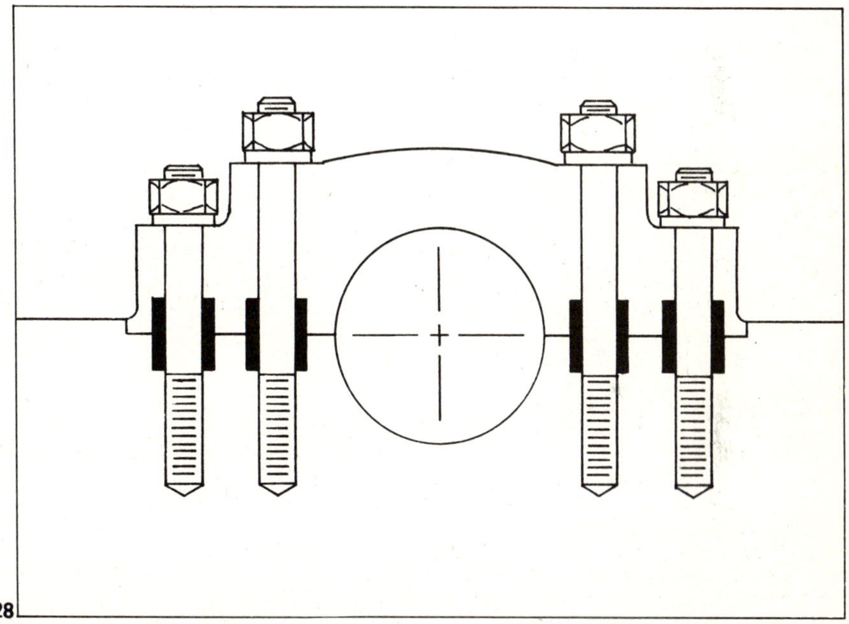
28

24. Cylinder leakdown checks are the best indication of engine sealing. A reading of 3-4% leakage is ideal, but at 7% take the heads off to find out why.

25. Plugs with ⅛-in. steam holes are installed in the decks for extra strength.

26. Additional water cooling holes are drilled beneath each bore in order to increase the flow of water.

27. Long rods (6.0 to 6.1 ins.) are used instead of the stock 5.7-in. rods. Smokey favors Carillo rods in terms of weight, strength and reliability.

28. By adding bushings around the main cap studs, the caps can be prevented from pinching in at the parting line under high load conditions for extra bearing life.

29. Care must be taken to ensure that the radii of the crank journals are correctly ground to prevent the edges of the bearing inserts from riding up on them, as shown (left).

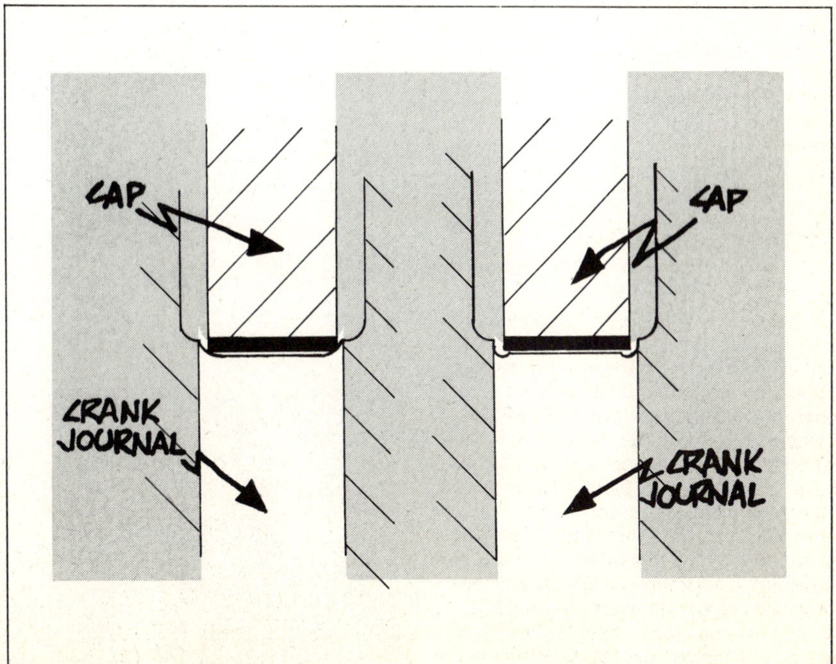

29

speed the return of oil to the pan and also trap any small metal grindings that might be in the porous surface of the cast iron, it's next to impossible to get a block clean enough to guarantee that the paint won't come off. Remember that the inside of the block is exposed to very hot oil that is being thrown off the crankshaft. At high speed this oil is being thrown against the block with a lot of force, so much force that it has a blasting effect on any paint in that area. If the paint isn't on there good, it'll be blasted right off. Then when it does come off, it doesn't come off in specks but rather in strips. It seems that the favorite landing spot of these strips is on the oil pump screen, or else they'll land on a rod end and be thrown up against the camshaft, where they're pressed between the cam nose and the lifter, destroying the cam. So if you do paint the inside of the block, and I don't think that it's important that you do, take extra care to clean the surface thoroughly prior to applying the paint.

BEARINGS & CRANK INSTALLATION

With everything all cleaned, we can bolt the short-block together now with a liberal coating of oil and high-pressure white grease. Start by installing new cam bushings, being careful to see that all of the oil holes line up precisely.

Next install the crank. If you find the main or rod cap bearing clearance is slightly off due to tolerance stackup, dress the cap or rod to increase crush and reduce the tolerance. Just get a flat piece of glass, lay some sand paper on top and run the cap back and forth several times. You can take out up to .0005-inch this way.

Wash the backs of the bearings with ketone or lacquer thinner to eliminate any oil film for maximum heat transfer. Assemble the bearings (and also the piston pins) with high pressure grease.

MAIN SEALS

Use the standard production rear seal. It seems to leak less than the Chevy high-performance seal. On the front, use the correct directional timing cover seal (it has a helix molded into the rubber) from Chevy.

HARMONIC BALANCER

Use the high-performance version with the malleable iron outer ring. Too many of the conventional ones have been flying apart lately, with some rather disastrous results.

RING COMBOS

Now you're ready for the piston and rod assemblies, but first you've got to get some rings. Most guys run chrome or moly rings, but you have to remember that one of the functions of a ring is to transfer heat from the piston to the cylinder wall. If they didn't, they wouldn't live long. If a ring is exposed to too much heat, you'll have an early ring failure. Now chrome is chrome, and nobody's chrome melts at any higher temperature than anybody else's, so that becomes a constant. Moly, on the other hand, is a mixture like an alloy that can be tailored to any desired temperature range. Hence, moly rings can be made that will surpass the temperature limitations of chrome rings. As far as I'm concerned, moly is better all the way around. Moly has some very desirable qualities: It's self-lubricating, better sealing and it doesn't encourage scuffing nearly as early, so you can run everything a little bit tighter knowing that if things are too tight the moly has the ability to wear away quite rapidly where chrome won't.

To get down to specifics, I like Sealed Power Speed-Pro rings (100 Terrace Pl., Muskegon, MI 49443). Use a 1/16-inch sprayed high-temp moly compression ring on top, a 1/16-inch reversion torsional-twist cast-iron scraper in the second groove and a 3/16-inch oil ring. For long life on the street, the second ring should still be moly, but to reduce friction for racing use a cast-iron second. Remember, the first biggest frictional loss in the engine is the rings. The second is the valvetrain.

End gap on the top and scraper rings should be .012-inch (minimum) to .014-inch (maximum), although the maximum tolerance isn't as critical on the scraper. Set up the gap about two inches down in the bore, below the bolt distortion area—then you don't have to have head plates on for this.

You don't need to run a low-tension oil ring. Today's standard oil ring is right for both racing and street. To combat oil, you must have a good scraper, and no pressure buildup in the crankcase.

PISTON & ROD ASSEMBLY

Before pistons are installed, grease is rubbed into the walls. Light oil (3-in-1) is used on rings and ring lands. This will keep everything well lubricated, especially if the engine sits awhile before fireup.

When it comes to torquing a rod bolt,

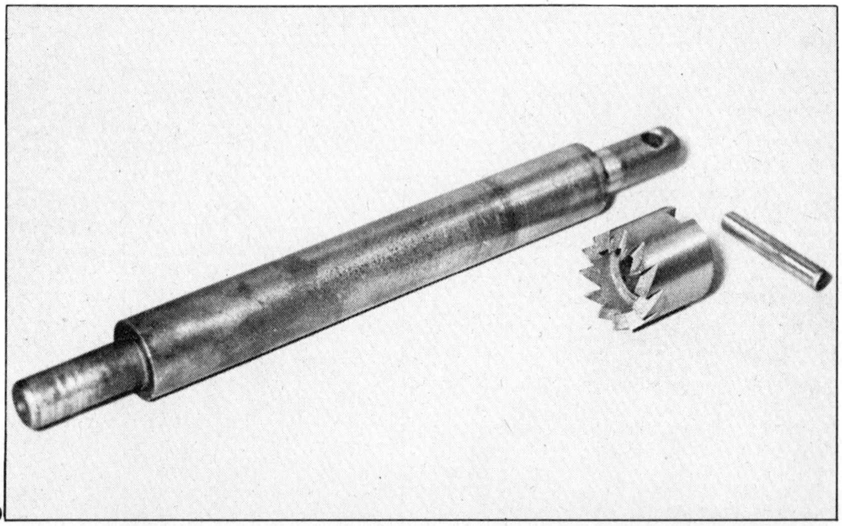

30

31

the best way to do it is to measure the stretch of the bolt. Rod bolt stretch tolerances are listed in the Chevrolet Shop Manual. You can check this stretch by using a micrometer before and after torquing the bolt, but there's more to it than that. Since the ends of the bolts aren't perfectly square, you need to machine a center in both ends and use a special point-type mike to check them. These mikes are readily available in tool supply stores. If you're measuring stretch, when you get to the proper stretch you won't be anywhere near the proper torque if that rod bolt is bad. All of them ought to stretch about the same amount. A bolt that varies more than five pounds either way should be discarded. Now this is a lot of trouble to go to, but unless you've got a lot more money than I have, it's the only right way to do it.

OILING SYSTEM

There's not really much that needs to be done to the oiling system. On the street, don't mess with it at all. For racing, the oil filter pad needs to be eliminated. There's two good reasons why you do this. First, you must get rid of that bypass valve, or you'll keep recirculating unfiltered dirty oil through the block. Second, filters are a big restriction and power drain, bigger than the filter manufacturers would have you believe. Depending on the filter, the pressure drop can be as much as 15-25 psi, so even dual filters won't really handle it. You need at least four remotely mounted oil filters, so tap the block for ¾-inch I.D. lines and run a *pair* of Fram dual remote-mount filter assemblies. Now, since you don't have an oil filter to attach to the pad, why not save some weight and grind it off? Follow these suggested recommendations and you'll get clean oil at a pressure loss of under 10 psi. It takes horsepower to push oil against a lot of restriction.

30. This tool, sold by Isky, is used to machine the bottoms of the lifter bores to clear the mushroom tappets. You can do this job at home with an electric drill motor.

31. Originally designed by Smokey, the Smokey Ram manifold provides excellent mid-range torque and top-end power, but careful attention must be given to fuel distribution.

32. This is the Webster gear drive for the camshaft. Special undersized crank gears are offered for align-honed engines. A reverse-rotation cam must be used.

Run as deep an oil pan as is practical and fit it with a full-length screen and scrapers in as close proximity to the counterweights and throws as possible, to help scavenge off excess oil from the rotating assembly. Some kind of pan evacuation system to create a negative pressure in the crankcase will also help immensely. Do not run a solid or sheetmetal-type windage tray.

You can run an aftermarket-type screen windage tray as close to the reciprocating assembly as possible. Use a heavy-duty oil pump with a minimum ⅝-inch I.D. pickup. A ¾-inch pickup is better. Lowering the pump in a deep sump application is better than a long pickup, because a pump is designed to push, not suck.

Oil pan research has also shown gains in specific rpm ranges by providing escape space and paths for oil in close proximity to the crank, especially when scrapers are used. The latest Moroso pan is said to be worth over 10 horsepower at 6500 rpm using these design features.

If you're building this engine to go racing, other than drag racing, you really should equip the engine with a dry sump oil system, if the rules and your budget permit it. In terms of reliability, there's far less chance of an engine failure with a dry sump system, especially if you have high lateral acceleration.

Other than this, don't mess with the oiling system at all. Leave the internal passages alone. Drilling the bearing holes to a larger diameter is definitely not necessary.

32

CYLINDER HEAD SELECTION

When it comes to heads, the late-model turbo heads with blocked heat riser (part No. 3965784) are the ones to use. Recently, this head has been further improved with tin alloy and extra cast-in ribs for improved strength, so try to get the newest head. (The part number has remained the same.)

Normally compression of at least 12:1 is best for power output with these heads. If a guy is using heads with the standard straight-plug design, anything above 11:1 will result in a loss of power. The maximum compression for these straight-plug heads, with 100+ octane gas, is 11.0:1. With the standard plug location, the plug is right at the bottom of the combustion chamber. So when the engine really starts to get with it to the extent that it's using a lot of fuel, the plug simply gets flooded. So if you go to more compression than 11.0:1, which necessitates using a piston with a higher dome, you effectively build a dam around the plug to flood it that much more. The angle-plug or slant-plug design raises up the spark plug so that it doesn't get flooded. Consequently, you can run more compression.

HEAD PORTING

You can't run these heads just the way you get them from Chevrolet. First you have to clean them up internally and then have them ported. The cleaning should be done just as thoroughly as the block, using the same process that I mentioned earlier. I can't overemphasize this business of freeing the water passages of any restrictions. It is quite common for passages to be blocked in a head, and when it does happen, the head forms a hot spot or possibly a steam pocket and the head will crack. So if you're going to spend the money to have the heads properly ported, protect yourself by cleaning those passages.

Porting is one of the most important things you can do to the small-block Chevy cylinder head. A set of well-ported heads is still one of the biggest bargains in performance today, and for our engine, porting is critical beyond belief. You should use somebody who really knows his business, like Airflow Research (11754 Roscoe Blvd. #6, Sun Valley, CA 91352) or Diamond Racing Engines, Inc. (3493 10 mile Rd., Warren, MI 48091). As with the crankshaft, there are probably other shops that can do really good work, but these guys I can recommend from past experience.

Regardless of who does your heads, here are some basic requirements: Right now, the biggest mistake is the overporting of the exhaust port. You have to know just what and how much to grind away in that exhaust port. You can totally destroy a cylinder head just by making the wrong move in there. Of course, once you grind it away, there's no way to put it back. The trick of knowing what you're doing in that exhaust port can easily make as much as a 40-hp difference between doing it right and just grinding it out. If you're a guy who likes to do his own work, and you want to be able to say that you ported your own heads, my advice would be for you to buy a set of good ported heads and then copy them. Now all this is mighty expensive, and a lot of guys might feel that they can't afford it. I can't see how they couldn't afford it if they're at all serious about building an above-average small-block Chevy—it's just that important. This is kind of a governing area. Without the heads the whole thing isn't going to be worth doing.

Although there'll be some argument and a lot of inflated claims about this, a really *good* set of heads will flow 225 cubic feet per minute at 28 inches of water with a .600-inch valve lift on the intakes and 210 cubic feet on the exhausts under similar conditions. Of course, these figures refer to simply flowing the head on a standard flow bench.

Another method that I prefer to achieve more meaningful flow rates is to install the head on a bare block, complete with intake manifold and carb, and then flow the intakes by attaching the flow bench to a modified oil pan. I similarly flow the exhausts by attaching the bench to the header collector. This method checks the total combination of pieces in a working configuration. Under such conditions, the same head as described above will flow 205 to 210 cubic feet per minute at the same depression and valve lift on *both* the intake and exhaust ports, even with the differences in the intake and exhaust valve sizes.

VALVES

And now that I've mentioned valves, you should use TRW or Del West Engineering (9440 Irondale Ave., Chatsworth, CA 91311) 2.050-inch titanium intake valves. The spring is in desperate trouble, so you must give it relief whenever possible. That's why titanium valves are very important. Lightening the tappets, pushrods and rockers is far less effective then lightening the valve. You can use the Chevy 2.050-inch intake valves for the street, but for racing the lighter and stronger titanium ones are a must.

The exhaust valve really doesn't need to be titanium unless you're running a super radical cam and very high rpm, like dragging. In most cases, a hollow sodium-filled stainless exhaust is better, to keep the hotter exhaust valve alive. Besides, most cams have milder acceleration profiles on the exhaust side. And since the exhaust valve is smaller than the intake, it's lighter to begin with. The exhaust valves usually come in the stock 1.600-inch diameter, but they should be cut down to 1.580. That way there'll be less obstruction in the port when the valve is open and better flow will result.

The real problem is the spring is just too small. It needs to be 1.625-inches in diameter. Extra length helps, too. That's why it helps to use .100-inch longer valves, which originally came about for higher lift cams, but should now be used even if the lift doesn't necessitate it to allow a slightly longer spring to be used. When using .100-inch longer valves, .100-inch shorter pushrods are required. Use the Chevy guideplates, and take the time to properly align them.

Stay with 11/32-inch valve stems. Have the stock guides *honed* with a Sunnen guide hone for .002- to .0025-inch clearance on the intakes and .00275- to .003-inch on the exhausts. Do not use those thin bronze guide inserts or ream the guides. The intake seats should have a three-angle cut—60, 45 and 20-30 degrees—while the exhaust seats should be a true radius.

Naturally you'll want a super-good valve job, so the best bet is probably to send the valves along with the heads when they go to the porter. This way he can do the valve job and cc the combustion chamber all at the same time.

VALVE SPRINGS

Exactly what spring to use depends on the cam type, profile and application. The hotter a spring gets, the more it is overworked. Efficiency drops, energy is lost and more power is required to overcome its sluggishness. In addition, springs heat the oil a lot. Now, you if make the spring longer, the wire doesn't have to twist so much; it doesn't have to work so hard. So friction is reduced and the heat is reduced. And, of course, the spring lasts longer. Ideally, the spring ought to be about two inches in diameter and four inches long! But that changes everything in the valvetrain geometry, not to mention the lack of space on the head.

I'll get to specific spring recommendations in a minute, but as for retainers, if you're going racing, titanium retainers are a must. For the street, aluminum retainers might be okay, but watch them. The stock Chevy keepers are adequate.

CAMSHAFT SELECTION

From this discussion, you should by now realize that the object is to reduce friction whenever possible, which translates into increased longevity and less power loss—in other words, greater operating efficiency.

So when it comes to a cam, run a roller whenever possible, along with roller rockers. Rollers just offer a better selection of profiles that are easy on the valvetrain in terms of reliability and durability without compromising performance. A roller cam also needs less spring pressure than a flat-tappet cam for equal lifter velocities.

In fact, I think you're going to see this combo in *production,* because obviously reduced friction means greater fuel economy. The slip and slide arrangement was wrong from day one, and was done simply for economy of production.

33

34

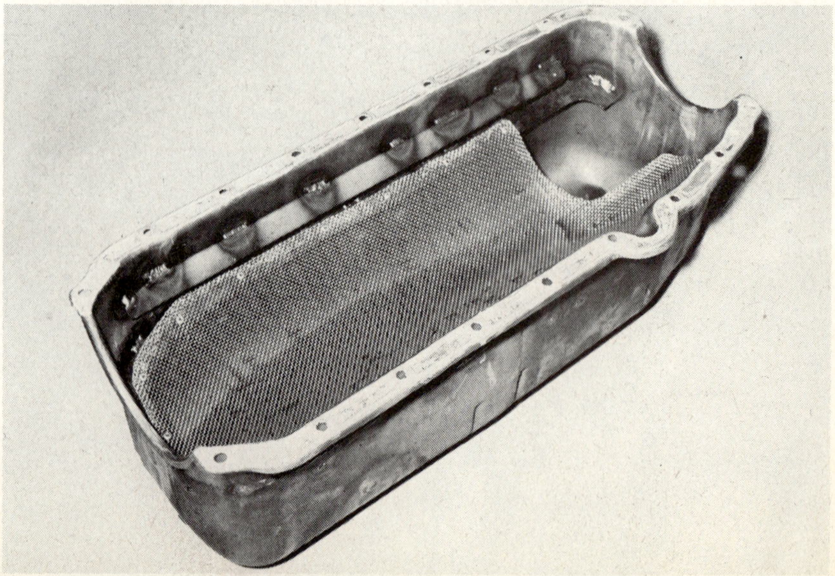

33. Mushroom tappets, such as these, provide optimum valve action with flat-tappet cams for racing where roller tappets are not allowed, such as NASCAR.

34. Full-length screens and scrapers, such as have been mounted in this pan, help control the amount of oil around the crankshaft for peak power output.

35. For peak performance, 2-in. diameter header tubes should be used on a 355-in. engine. To use such big tubes, a special flange plate must be attached to each head. These plates become an extension of the exhaust ports, so they must be in place when the heads are ported and flowed. Note the use of big-block rocker studs. The light-weight springs shown here are strictly for flow bench testing.

Remember, the valvetrain is one of the biggest power robbers in an engine, so for either street or track, use a roller.

Selecting a cam manufacturer isn't easy. But regardless of who you select, the profile has to fall within the working limits of the spring and valvetrain components. This generally requires computer technology.

For track racing, I'd like to see something in .650-lift for all-out super-speedway racing. Use a valve spring giving 165 pounds on the seat and 450 to 475 open. I have found the Crane R268/390-2S-6 to be a good cam for most track racing. This isn't to say this is the best—others might be better—but this one is good. Use Diamond No. 3306 springs or Crane No. 99893.

A street roller cam should have .550- to .585-inch lift on the intake and .530- to .540-inch on the exhaust, depending on the application. The valve spring should have 135 pounds on the seat and 350 to 375 pounds open.

If you're going racing where they won't let you run a roller cam, like in NASCAR, and you want optimum valve action, you'll have to use mushroom tappets with a .900- to .925-inch base diameter. To use these you'll have to machine the bottoms of the lifter bores (Isky makes a tool to do it) to clear the mushroom. The equivalent to the Crane R260 grind mentioned above in a mushroom profile is Crane's F-268/390-2S-6 grind.

No matter what the application, the exact duration and lobe center depends on the intended use, car type and compression ratio. Lobe centers are now being narrowed for high rpm running. High torque at low speed now calls for lobe centers of 114 to 118 degrees. The old 20-20 cam (intake opens 20 degrees BTDC, exhaust closes 20 degrees ATDC) will probably come back.

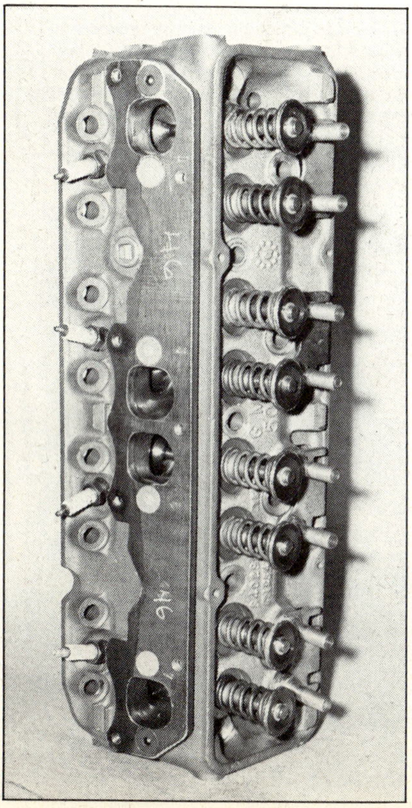

35

Other than these generalities, there's no blanket answer on which is the best cam, because of the wide range of vehicle weights, transmission type, rearend ratio, the kind of driving the guy is going to do, etc. A camshaft isn't flexible. You have a special cam for a special purpose. I believe that if you've got a cam that gives good high-rpm performance, there's no way that it's going to idle smoothly or have much snap at low speeds. I don't think that comes as a shock to anybody; everybody generally accepts it. I think it's impossible to have a good cam for horsepower that also gives good gas mileage.

VALVE STEM SEALS

Another critical area is valve stem seals. Use Perfect Circle Teflon seals on both the intakes and the exhausts, but be sure to machine the guides to provide at least .020-inch clearance between the top of the seal and the base of the retainer at maximum valve lift.

HEAD INSTALLATION

With the valves, springs, etc. installed in the heads, you can go ahead and bolt the heads on. The bolt pattern on a Chevy head is exceptionally good. Chevy uses 17 bolts to hold each head down, where Ford and Chrysler use only 10 bolts on their small-blocks. If you have a circle you want to seal, like a cylinder or four cylinders, and you use only 10 bolts, there's no way you can arrange those bolts to have more than four around each cylinder. But if you use 17 bolts, you can arrange them so that you have five studs or bolts around each cylinder. This gives a better seal than only four bolts. Replacing the bolts with studs will give a better seal yet, as previously related.

However, as good as the 17-bolt Chevy design is, there's still a problem. The manual says to torque 'em all the same (65 ft.-lbs.). Well, that's okay, but when you really get down to what's going on, 65 ft.-lbs. of torque on the shorter studs (the lower ones) has a lot more clamping force than the longer studs (the top ones) do with the same 65 ft.-lbs. of torque. This is because the longer studs stretch more. Besides that, the thickness of the head is greater where the long studs go through (that's why you have longer studs) and the column of metal in the head surrounding the stud is longer and it compresses much more than it does around the shorter studs. A longer column is just plain weaker than a short column. A longer column shrinks more when it cools and expands more when it's heated. Therefore, you have to tighten the long studs more than the short ones.

The short studs are also closer together, and they're in a harder part of the block. The top studs are all over the place in what is basically a shell area of the block that isn't too strong. Because of this you have to be careful not to overtighten the long studs. So if you torque the lower short studs to 65 ft.-lbs., you can only pull the top studs to about 70 ft.-lbs. With fuel, you can up the stud torque to 75 and 70. That's what will work. Actually, to get the same amount of clamping force on the longer studs that you get on the short ones at 65 ft.-lbs., you ought to pull the top ones to 80 ft.-lbs; but if you do, you'll simply pull the threads out of the block.

This is one area where the Chevy block has a problem. If you look at the center cylinder studs (or bolts), they have next to no support in the block, and if you weren't using studs, they'd literally twist hell out of the block when you really pull them down. We have to remember that these engines weren't designed for racing. At a stock compression ratio, 55 to 60 ft.-lbs. of torque did the job just fine. There was even a little bit of safety margin above that. But for racing or high-performance applications, you have to use all of the tricks that you can think of to get a proper seal that will last, especially when pushing 600 hp.

ROCKER ARMS

As I already mentioned, I now believe roller rockers are best for all applications if you want to reduce friction to a minimum. If cost is a consideration, you can stay with the stock Chevrolet rockers on the street (part No. 3974290), but you should weld a little sheetmetal cap over the top of the pushrod end of the rocker to stop the oil that's coming up the pushrod from going up against the valve cover.

On the small-block Chevy the stock 1.50:1 rocker arm ratio isn't the best. I feel that the best ratio is 1.70:1 or 1.75:1. That will give the most power while simultaneously allowing the gentlest cam profile, since the trick is to take some action out of the lobe and put it in the rocker, which makes for a stronger cam. Unfortunately, to use a rocker of this ratio it's necessary to relocate the rocker studs. The factory really ought to make the necessary changes to step up to 1.7:1 or 1.75:1 rockers, but that won't happen now, as small-block development is practically over with at the factory level. So in terms of practicality, use aluminum needle-bearing roller 1.6:1 rockers on the intakes and 1.5:1 rockers on the exhausts. It's not as much as I think you ideally ought to have, but it's better than nothing. Crane Cams (100 N.W. 9th Ter., Box 180, Hallandale, FL 33009) is one of the outfits that makes them.

Be a little careful on this ratio business, though. If you use more than the stock ratio, be sure to clay the pistons to check valve-to-piston clearance. Don't settle for any less than .080-inch. And just be sure the rocker roller isn't running off the edge of the valve stem. If it is, an adjustment in pushrod length will be required. It's things like this that make a trial assembly so important, so don't skip a trial assembly or two just to save a little time—it isn't worth it.

As rpm and cam lifts continue to climb, a return to shaftmounted rockers might be required to retain stability. It's something to consider.

ROCKER STUDS

Mount the rockers on big-block rocker studs. You'll have to cut .060-inch off the bottom of each one to prevent the threads from bottoming, but otherwise they screw right in, and they're a bunch stronger than the small-block studs.

STUD GIRDLES & GUIDEPLATES

As for stud girdles, I can't see the need for them on a cast-iron head. However, aluminum heads must have a girdle, because they flex more. Definitely use a stud girdle with adjusting nuts that feature a locking setscrew to eliminate thread sloppiness. Use the Chevy guideplates and take the time to properly align them.

TIMING GEARS

Whatever kind of cam you use, street or race, don't mess around with timing chains of any kind. Lately I've been seeing a lot of timing chain failures. Instead, go to a gear drive setup. The best one I know of is the Webster setup (Webster's Head & Engine Specialties, 4691 Cerritos Ave., Fremont, CA 94536). His unit is light and reliable, and he offers crank gears that are undersized to correct for align-honing and still maintain .003- to .004-inch backlash. The only problem with this setup is that you *must* use a reverse rotation camshaft. Unfortunately, there are almost no steel billets currently available for reverse rotation roller cams, so you'll probably have to run a three-gear setup to restore normal cam rotation, such as the Summers Bros. (530 S. Mountain Ave., Ontario, CA 91761) setup. I don't like the other designs as much as the Webster setup, but they're a lot better than a chain. Just make sure the three-gear setup, if you're using it, has a fixed idler; forget those drives with a floating idler. Regardless of the drive used, run the Isky cam button to control end play to .004-inch.

INTAKE MANIFOLDS

Intake and exhaust manifolds are next. Now of course in this department I'm going to be somewhat prejudiced, but I've got a lot of information to back this up.

If you're going racing, the Weiand Team G works good on the short track. Holley and Edelbrock have various track manifolds that work good for both short and long tracks, depending on which one you use. All require some massaging to obtain optimum performance. The manifold business is highly competitive, and what might have been the hot setup last month may be obsolete tomorrow. I'd check with my cylinder head porter for the latest trick setup when it comes time to buy the manifold.

For the street, I'll have to start off by saying that the cheapest and best buy is the Z28 intake manifold (part No. 3972114). If you've got the money, the manifold that I worked on, which Edelbrock now sells under the name of Smokey Ram, is across-the-board stronger with the best operating range and the best power. It's by far the best street setup, and works good on many race cars, too. It'll accelerate better than the Z28 and outcrank it.

This manifold takes some time and fiddling to get it right, but when it's right you've really got something. The problem with the Smokey Ram is that most guys don't think that they should have to mess with a new manifold. They think they can bolt it on and it'll go with no other changes. Now this is a free-flowing manifold; you don't have to struggle with it for flow. There's no restriction. It's got good distribution and good velocity, but it's a big box. It's much bigger in internal volume than the single-plane type intakes. Now let's suppose that an engine is running along at a constant rpm and that the manifold is full of air/fuel mixture at, say, 16:1. Now to accelerate, we have to go overrich—that's what the accelerator pump is on a carburetor for. So we go from 16:1 to 12:1. Now let's say that the carburetor is set up to work on a Chevy with a Tarantula manifold and that it takes a certain amount of fuel from the accelerator pump to change this ratio from 16:1 to 12:1. But the Smokey Ram is about three times bigger internally than the Tarantula, so we're going to need three times the fuel. Therefore, we have to triple the accelerator pump shot. That's the first problem.

HEADERS

The next problem is over on the exhaust side. With the Chevy or Tarantula street-type manifolds, which are not as free-flowing as the Smokey Ram or the specially-designed track manifolds, the exhaust headers have to help pull the air/fuel mixture through the intake manifold. But with the Smokey Ram that's not the case. So for this 355-inch Chevy, what's needed is a header with 2-inch primary tubes about 32 inches long. These dimensions aren't absolute, since the final header configuration depends upon the application, so I'd recommend you contact Jack Davis at Hooker Headers (Box 4090, 1032 W. Brooks St., Ontario, CA 91761) regarding your exact needs. At any rate, the 2-inch diameter tube will require adapter plates on the heads. These plates actually become part of the port, so have 'em on there when the heads are ported and flowed.

CARBURETION

Use the Holley 850 cfm double-pumper with primary and secondary idle circuits for improved idle and throttle response for racing, if the sanctioning body's rules permit. If using the Smokey Ram intake, be sure to triple that accelerator pump shot; one way to do it is by adding a 50cc accelerator pump to the primary side (there's already one on the secondary), and if that's not enough, using a bigger shooter and drilling out the restriction below the shooter. Now it may

36. These photos show how Smokey flows a set of heads for truly representative flow reading. Intake is pulled through an empty block, using a modified oil pan. An intake manifold and carb are also in place during the tests. On the exhaust side, a header is installed and the flow is pulled right through it, attaching the bench to the collector. That's doing it right!

37. Ignition crossfiring can be reduced by using Moroso insulating sleeves over the wires. A 45-degree cut on the head of the exhaust valve improves exhaust flow around the valve head.

36

seem that based on the theoretical volume of air that our 355-cubic-inch engine could use, an 850 would be too big. But with the gains in engine efficiency as we approach 600 hp, this carb works very well. The Smokey Ram definitely needs this size carb—no compromises; it likes an oversize carb, as long as it has symmetric bore spacing. A spread-bore just won't work.

Of course, if you're going NASCAR racing you'll have to run their designated cfm carb, reworked, of course, by an expert, such as Ralph Johnson, who works for me, for best flow and performance.

Otherwise, on the street or for any other kind of racing, the Holley 850-cfm double-pumper is still the way to go. The best advice I can give you on carburetor internal modifications is to leave 'em alone unless you really know your stuff. Most guys do more harm than good to a carburetor when they start messing with it. Another thing: Holley carbs, the performance models, are jetted about 5 percent rich the way they come out of the box. Therefore, if you find yourself having to increase that original jetting by more than five numbers, look elsewhere for the problem, such as a weak fuel pump or a line restriction.

Now I realize that's a pretty big carb, and gas isn't cheap these days, so some of you are probably figuring you should use a smaller 4-barrel or maybe even a 2-barrel. Well I'm the first to admit the 850 isn't an economy carb, but you shouldn't necessarily go to a real small carb either. Tests have shown that the closer to full throttle the engine is run at low rpm, the greater the potential fuel economy, because at small throttle openings the engine is pulling against itself. So if you're setting up for economy, what you want is a real low (numerically) rearend ratio, maybe 2.29:1 or even 1.95:1, if you can find someone who makes that kinda gear. With, say 8.25 or 8.5:1 compression, a Z28 intake and big carb, you're set up to where you're running 55 mph using ¾ throttle at about 1500 rpm. That's how you get economy. Of course, you use a real low 1st gear in the trans, maybe something in the three's, to get you going. More and more, that's what Detroit's doing now.

If you're stuck for some reason that you have to use a 2-barrel (like because of California's restrictions), you might use the Chevy cast-iron '78-'79 standard production 2-barrel intake. I understand that Edelbrock is workin' on an aluminum 2-barrel intake that should be out by the time you read this.

Whatever manifold you use, keeping it cool will help power. Warm it if you want mileage.

IGNITION

About all we've got left is ignition. I still think that the original GM transistor ignition system is good, but it's very expensive and hard to find these days. Stay away from the HEI—it just won't do the job above 5000 rpm.

The new Accel BEI III with Wiegand trigger is excellent, especially in terms of reliability and durability. The earlier BEI systems also work good . . . when they work. Above all, I wouldn't mess around with points.

Actually, none of these conventional ignition systems is really ideal. Unfortunately, it looks like we might be coming up against the existing limits of battery ignition in the high rpm ranges currently being run. At 6000 rpm, everything is happy, but now, at 7500 rpm, a longer duration spark seems to be needed. The proof is that when a magneto is substituted for an electronic ignition, the engine almost invariably picks up 15-30 hp at 7000+ rpm. Think about that for a while!

If everything is right, you can run a maximum advance for racing of about 36-37 degrees at 6000 rpm using Turbo heads. It's no longer necessary to run 42 or 43 degrees of lead. Optimum spark lead is a measure of combustion chamber efficiency. The poorer the chamber, the greater the amount of lead that's required. In any case, take a good look with that timing light to be sure you actually do have 36 degrees. If the timing is wavering or moving around, find out why and then fix it. With a low compression street engine, run about 36 degrees, if there's good enough gas available.

You can't use too good a plug wire. The big problem today is crossfiring induced by the high cylinder pressure and the very high firing voltages. Therefore, it is very important to use the best 8 mm silicone-jacketed solid core spark plug wire you can find, both to prevent crossfiring and to keep the wire from burning up against the headers. You must have good boots, too, for the same reasons. Be sure to keep the plug wires for cylinders that are adjacent in the firing order, especially on the same bank, separated. Actually I've yet to find any brand of plug wire that I consider to be completely adequate. However, Moroso Performance Products (Carter Dr., Guilford, CT 06437) has come out with an insulating sleeve that can be slipped over each individual plug wire to provide an additional 8000 volts protection against spark leaks. That means that between any *two* wires, there's an extra 16,000 volts worth of insulation, and I think you should do anything you can to get adequate protection. If this sounds like a trivial point, a cylinder that fires 90 degrees early while under power due to crossfiring will soon crack the cylinder wall or destroy the piston. Enough said?

SPARK PLUGS

As far as I'm concerned, the best spark plugs to use are the extended-core nose type. A racing-gap version will only hurt performance on this engine. The plug tip has to be up in there. A racing-gap plug will cost you about 15-20 horsepower. There's no one brand that's much better than another, as long as you have the right heat range for

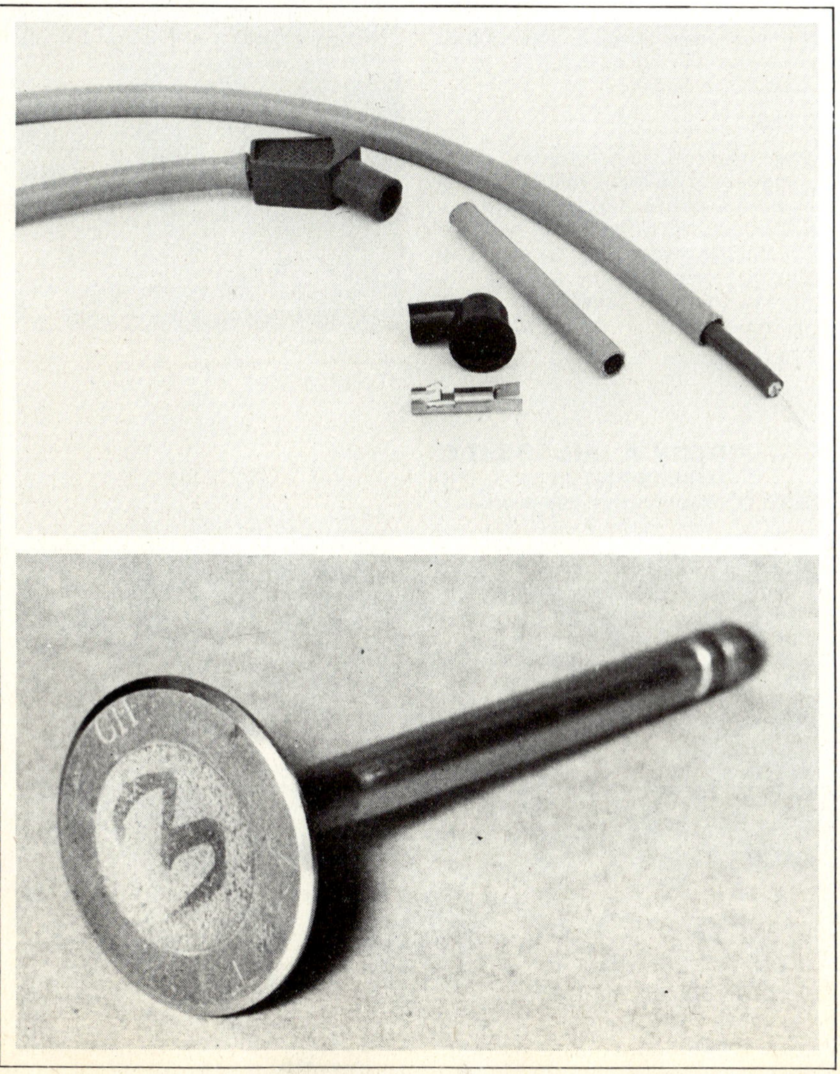

37

what you're doing. In Autolite numbers, something around 9 should work out okay. Champion BL-60Ys should also be in the ballpark. Use a slightly hotter plug with low compression street engines, with slightly colder plugs for all-out racing. Most guys run too cold a plug and that just causes fouling.

With the Accel Wiegand ignition, plug gap should be .050 to .060.

COOLING

The Chevy water pump only flows about 60 gallons per minute through the engine, and it takes about 10 hp to spin it at 6200 rpm. Water circulation is critical in the cylinder heads of these engines to prevent cracking. Consequently, I put a plastic top on a head so I could observe the flow patterns when the water pump was spun at speed. The flow turned out to be very turbulent and disrupted. By tapping the lower front of each head to form a new water outlet instead of taking the water out through the intake manifold, the turbulence decreased dramatically and flow increased to 90 gallons per minute. This means the water pump can be slowed down for less horsepower loss and still have better, and more even, cooling. Additionally, there is no heat transfer into the intake manifold, because the heads are now plugged to force the water out the new openings.

LUBRICATION

In the oil department, multi-viscosity oil is good, especially in multi-climate areas. If an engine isn't really abused, the multi-viscosity oil is generally the way to go. The only problem with multi-viscosity oil is that certain things are added to make the oil thicker as it gets hotter and thinner as it cools down. Now as the oil gets really hot it seems that there's some kind of process going on in multi-viscosity oil that is very much like the refining process of oil. This process seems to break loose the ends of the oil molecules. It's been my experience that when multi-viscosity oil reaches 300 degrees, this process takes place and the oil can't recover. I've pulled pans off engines where this has happened and the oil was so thick you could actually scoop it out by the handful. It had actually become a semi-solid. I guess that problem has been licked in the multi-viscosity oils now, but what it really did was point out that although multi-viscosity oils work well, they have to be watched when you approach what might be considered dangerous temperature limits. For instance, when you get to 300 degrees oil temperature, you're in bad trouble anyhow. I'd say that after you get past 280 degrees, you're headed for disaster. The bearings just burn up. A good range would be 200 to 240 degrees. This is hot enough to boil out any moisture that may have accumulated. You might even be able to run a little hotter than 240 degrees in the small-block Chevy, but each engine varies. As for weights, 20W-40 or 20W-50 should do just fine.

Right now, I have mixed feelings about synthetic oils. In terms of temperature tolerance and lubrication qualities under high stress situations, synthetics, especially those with a polyester base, seem to perform better than conventional mineral oils. But synthetics have been known to build up deposits over the long haul that could eventually cause problems. So right now, I'd stick with a good conventional oil like Valvoline Racing in most cases. If you do use synthetics, change the oil as often as you'd change conventional oil.

BREAK-IN

That completes the construction of our high-powered 355-cubic-incher. Take it easy on the engine during the first few hours of operation, and then go back and retorque everything to specs. If you've built *everything* the way I've outlined and used the best pieces, you'll have the satisfaction of knowing that you've got an engine just as trick as some of the big name racers—an engine that puts out in racing trim over 580 reliable—and *livable*—horses whenever you need them. That ought to be enough to handle just about any job you want to tackle, but in case it isn't, there's always the big-block.

—Smokey Yunick

CLEARANCES IN INCHES (or fractions thereof)

Piston-to-wall	.0065-.007, measure at centerline of pin hole and perpendicular to pin
Piston-to-pin clearance	.00135-.00150 in piston and rod for floating pins without bushing
Piston pin end clearance	.005
Rod bearing clearance	.0025-.003 (.0025 preferred)
Rod slide clearance	.015-.018 per pair
Main bearing clearance	.0025-.003 (.0025 preferred)
Crankshaft endplay	.004-.006
Piston deck height	Zero
Piston-to-combustion chamber	.035
Piston-to-valve	.080

Any other clearances and/or torque values remain as given in the Chevrolet Service manual unless otherwise stated in the text.

TRW PISTON AVAILABILITY

NASCAR PISTON

Piston is manufactured .005-oversize.

Part No.	Rod length	For cylinders	Compression ratio	Combustion chamber cc	Type of retainer
L2457F	5.7	1, 4, 5, 8,	12.5:1	68	Single .073 Spirolox
L2458F	5.7	2, 3, 6, 7	12.5:1	68	Single .073 Spirolox
L2459F	6.0	1, 4, 5, 8	12.5:1	68	Single .073 Spirolox
L2460F	6.0	2, 3, 6, 7	12.5:1	68	Single .073 Spirolox

LOW-COMPRESSION STREET-TYPE PISTONS

Part No.	Type	Compression ratio	Combustion chamber cc	Ringland oil drainback	Type of retainer
2417	Cupped turbo	8.9:1	64	Holes	Double .042 Tru-arc
2441	Cupped turbo	8.4:1	64	Holes	Double .042 Tru-arc
2417	Cupped turbo	8.6:1	68	Holes	Double .042 Tru-arc
2417	Cupped turbo	8.0:1	76	Holes	Double .042 Tru-arc
2441	Cupped turbo	8.1:1	68	Holes	Double .042 Tru-arc
2441	Cupped turbo	7.6:1	76	Holes	Double .042 Tru-arc
2256	Stock-type flattop	9.0:1	68	Slot return	Pressed
2304	.100 dome	9.5:1	76	Slotted on late motor	Pressed
2304	.100 dome	10.5:1	68	Slotted on late motor	Pressed
2304	.100 dome	11.0:1	64	Slotted on late motor	Pressed

Building a Blown 350

How to double the horsepower of the small-block with easy-to-install, bolt-on speed equipment.

This is the final chapter in our Engine Build-ups section, and just as in drag racing and football, we saved the best for last. This is the Top Fuel final and the Super Bowl all wrapped up in one: the world's best high-performance powerplant, the small-block Chevrolet. A lot of brand loyalty is prevalent with performance enthusiasts, but when you reduce the brand controversy to facts and common sense, the small-block Chevrolet is your best bolt-on high-performance bargain. A large part of the reason for the success of the little Chevy is the engineering behind the engine. It is a very well-designed engine that readily responds to basic performance tuning.

When people get together to bench race about their favorite engines, they like to throw around impressive figures. Big numbers make great conversation but they rarely do anything for everyday street driveability. Hot rodders who are honest with themselves know that it doesn't take 700 horsepower to make a very enjoyable street motor. Realistically, any motor that puts out over 300 horsepower and good bottom-end torque is a plenty potent street motor. And it is one that you can live with every day, the kind that is dependable and easy to operate and maintain.

When you talk about percentages, our 350 smogger Chevy made some very impressive horsepower gains. The improvement from stock to our final stage resulted in more than a 100-percent increase in horsepower and a 46-percent increase in torque. In terms of horsepower, our simple bolt-on modifications were worth over 200 ponies—not too shabby considering we started with a dead-stock 1978 350-cubic-inch smog motor, and we went no further into the engine than to change the camshaft. Everything else

1. Our series of dyno tests were conducted on the W.R. Grace Company's research and development dynamometer in Garden Grove, California. Here is the stock 1978 350-cubic-inch small-block Chevy as it was run in stock form. What appears to be an ultra-high-rise air cleaner is actually an air flow transducer, commonly used for dyno tests. Note that the transducer is rigged for test readings. The 350 was a smog engine even down to the smog-type, low-compression pistons.

2. For Step No. 7, a Holley spread-bore carburetor was installed on the Weiand two-plane manifold. The 650-cfm carb turned out to be better suited for the engine than the larger 850-cfm double-pumper that was installed in Step No. 8. A Mallory Unilite ignition was tried in place of the Chevy HEI system, and little, if any, difference was noticed.

1

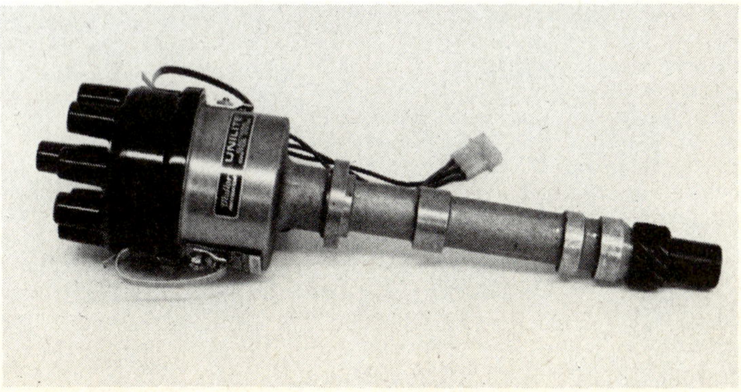

2

was bolted to the outside of the engine.

This chapter, besides proving the potential of the small-block Chevy, proved that a blower can simply be bolted on and run without modifying the engine. A properly set-up street blower like the Dyer Street Charger that we used is a bolt-on proposition, not some tenuous racing equipment loose on the street.

We conducted our tests at the impressive W.R. Grace Dynamometer Test Facility in Garden Grove, California. The test center serves as a research and development department for all Grace high-performance products like Mallory ignitions, Mr. Gasket, Brooks pistons, Blackjack headers, Team G manifolds, Hays clutches and Lakewood chassis products. The man in charge of our project was Art Chrisman, ably assisted by Gene Adams and Mike Chrisman.

The engine arrived in a crate from Chevrolet, the same engine you would find in any '78 Chevy equipped with the 350 option. There was absolutely nothing special about this engine. It was a standard late-model production engine. In fact, earlier versions of the 350 should be expected to perform even better, due to less-restrictive pre-smog parts on most of them.

STEP NO. 1:

STOCK ENGINE

Step No. 1 was to break the engine in and run it just as it had come from Chevrolet—dead stock. The engine was the same engine as any customer would receive from Chevrolet when ordering a factory-fresh motor. The engine was an 8.5:1 compression, unleaded fuel smog motor. The timing was set at the factory specs of 6° @ 650 rpm, which gives a total advance of 28° @ 4000 rpm.

The stock cast-iron manifolds were employed; but to make the dyno hookup easier, dual turbo-style mufflers were used with 2½-inch tailpipes. The stock air cleaner was used with an air flow transducer. The air flow transducer is a dyno device that raises the air cleaner about 18 inches to get an air supply that is more representative of normal driving conditions. Peak power achieved in step No. 1 was 186 horsepower at 4000 rpm. The peak torque was 288 ft.-lbs. at 3000 rpm. The peak horsepower produced by a similar engine with the catalytic converter and single factory exhaust would have been even less than the 186 horsepower achieved with the dual exhaust system.

STEP NO. 2:

HEADERS

For ease of modification and low cost, step No. 2 was a real winner. All that was done was to install a set of Blackjack headers and a Mr. Gasket low-restriction air cleaner. The Blackjack headers had 1¾-inch primary tubes that were 32 inches long with 3-inch-diameter collectors. These two simple changes really improved the breathing ability of the small-block, and peak power jumped to 224 horsepower at 4250 rpm. Peak torque was reached at 3000 rpm with a reading of 312 ft.-lbs. Giving the engine a little breathing room was worth almost 40 horsepower, or a 20-percent increase over step No. 1, proof positive that headers are one of the best performance buys.

3

STEP NO. 3:

MORE ADVANCE

Step No. 3 involved just the use of a timing light and a distributor wrench. The initial timing was advanced 8° to 14° for a total advance of 36° . Both power and torque went up slightly to 231 horsepower at 4250 rpm and 321 ft.-lbs. at 2500 rpm. No new parts were added, but the power was up 45 ponies with just the addition of headers, a dual exhaust system with turbo-style mufflers, and a low-restriction air cleaner, plus the simple act of advancing the timing. A nice gain for very little expense and effort.

STEP NO. 4:

NEW MANIFOLD

The fourth step involved the installation of an aftermarket aluminum manifold, in this case a Team G Weiand two-plane manifold. An adapter was used to mate the Team G manifold to the stock Quadrajet carburetor. Horsepower went up only 7 units, but the rpm range was extended as peak power occurred at 4500 rpm. Peak torque dropped a little to 314 ft.-lbs. at 3500 rpm.

STEP NO. 5:

CAM CHANGE

Changing cams was the most involved step in this series as far as getting into the engine goes. A cam change is still considered a bolt-on modification, because the change can be made with the engine in the car and without completely disassembling the engine. A cam change is definitely more involved than adding headers.

The cam used was a Chevy factory L-82 Corvette cam, part No. 3896962. All other factors were the same as before, including the timing and the AC part No. R45TS spark plugs. Fuel for all the tests was Daeco high-octane for consistent dyno results. Fuel pressure was 6½ pounds using the stock fuel pump backed with the dyno's electric fuel pump. Oil pressure was 35 pounds at idle and 48 pounds at 3000 rpm with the engine warmed up. Water temperature was held at 170°

The cam seemed to work quite well with the Team G manifold, as horsepower increased to a peak of 263 horsepower at 5000 rpm. Again the power range was extended. The peak torque leveled off at 312 ft.-lbs. at 3500 rpm.

STEP NO. 6:
NEW IGNITION

Step No. 6 proved that often the factory equipment is very good. The stock Chevrolet HEI (high energy ignition) system is a well-designed unit that provides excellent spark for any mild street engine. The HEI unit was replaced with a Mallory Unilite infrared ignition. Total timing advance was still 36°, with total spark obtained at 2800 rpm. The new ignition was good for exactly one more horsepower and one more ft.-lb. of torque than the stock Chevy ignition. The Mallory unit would undoubtedly be a good improvement over earlier engines that came with standard point-type ignitions, but in the case of newer engines with the HEI, it wasn't a real terrific horsepower-per-dollar value.

STEP NO. 7:
SPREAD-BORE CARBURETOR

A very typical carburetor change was made for step No. 7: the addition of a 650-cfm Holley spread-bore carburetor. This is a standard Quadrajet replacement carburetor commonly used with aftermarket manifolds. Horsepower increased to 278 at 5000 rpm, and peak torque increased to 322 ft.-lbs. at 3500 rpm.

At the end of step No. 7, the engine was at the stage very often matched by many bolt-on hop-up artists, and the power was up almost 100 horsepower over stock: a nice improvement that was easily obtained. An engine built to this stage won't win any Bracket 1 trophies, but it is a very responsive and enjoyable street engine, within the reach and budget of most enthusiasts.

3. The engine as it appeared in Step No. 7—your basic and very worthwhile hot-rodded engine. For the money invested, an engine in this state provides excellent performance. The Blackjack headers have a special aluminum coating which greatly improves the lifespan of the headers. Besides the stock cylinder heads on the left, a set of Chevy Turbo heads (right) were tried on the engine. The Turbo heads were modified by Cylinder Heads America.

4. The Dyer Street Charger uses a specially machined plate to adapt the blower to an aluminum dual-quad aftermarket manifold. The plate restricts the air flow in keeping with the street usage that the blower is set up for. It would be easy enough to make more horsepower, but Dyer knows through years of experience what works best on the street.

STEP NO. 8:
TOO MUCH CARBURETION

Steps Nos. 8, 9, 10 and 11 were done to illustrate what can happen when a motor is overbuilt. Very often, the biggest isn't always best, yet many enthusiasts think they will get gobs of power by simply installing a killer cam or a mega-cfm induction system on a relatively stock engine. These parts need to be used on an engine set up to accept their potential.

For step No. 8, a Holley 850-cfm double-pumper carburetor was substituted for the smaller Holley spread-bore carb used in step No. 7. Peak power stayed the same with 278 horsepower at 5000 rpm, and peak torque improved slightly to 335 ft.-lbs. at 3500 rpm. The whole power curve moved up slightly, but for all practical purposes the change didn't really improve the engine enough to warrant buying a bigger carburetor. In its present state, the engine wasn't able to use the increased capacity of the double-pumper carburetor.

STEP NO. 9:
TOO MUCH CAM

The installation of a big cam in an engine with stock heads and stock valvetrain can be a real waste, as we realized when one of the largest hydraulic cams available was installed. The cam was a Crane CCH-300-NC, which has a lift of 252 degrees and a duration of 300 degrees with a gross valve lift of .521-inch. Crane recommends that this cam be used in an engine with a large-cfm single carb or multiple carburetion, a single-plane manifold or a tunnel-ram, large-tube headers and modified cylinder heads. Crane knows what they are talking about because the cam wasn't made for our 350 in its present state. Power and torque both fell off through the low and medium ranges, and peak power wasn't any better than before the cam was installed. Peak power was 277 horsepower at 5250 rpm, and peak torque was 326 ft.-lbs. at 4000 rpm. Bottom-end horsepower was lower than that achieved with the stock engine.

STEP NO. 10:
BIG HEADERS

Crane said that big headers were needed for their CCH-300-NC cam, so the Blackjack street headers were removed and dyno race headers were bolted in place. The dyno headers have 1¾x32-inch primaries and 4-inch collectors. Very low-restriction dyno mufflers were also used. The power picked up to 290 horsepower at 5000 rpm, and peak torque jumped to 341 ft.-lbs. at 4000 rpm.

STEP NO. 11:
TRICK CYLINDER HEADS

Another recommendation for the Crane cam was modified cylinder heads, so a set of Cylinder Heads America NASCAR turbo heads were tried on the

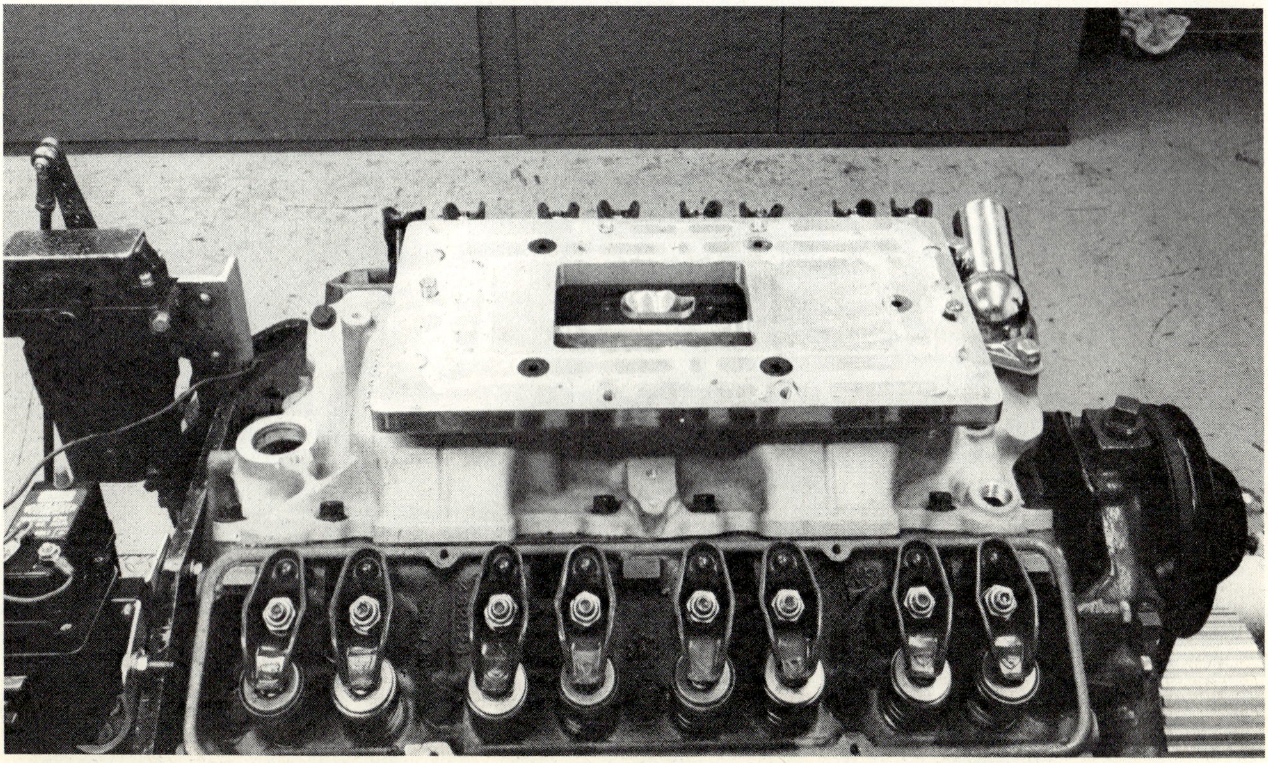

4

engine. The CHA heads had smaller combustion chambers, which raised the compression ratio to 9.5-to-1. The power curve was raised across the board, and peak power ended up at 324 horsepower at 5500 rpm. Peak torque was 364 ft.-lbs. at 4000 rpm.

STEP NO. 12:

BACKTRACKING

For step No. 12, the big cam was removed and the factory L-82 cam was reinstalled, as were the Blackjack street headers. The surprising finding was that power was about the same, with peak power actually 2 horsepower greater than in step No. 11. Peak torque was down to 342 ft.-lbs. at 3500 to 4500 rpm. Low- and mid-range torque were also down from that obtained with the big Crane cam.

Step No.12 was as far as we went with the normally aspirated engine. Power was up 140 horsepower, and the usable rpm range was extended from 4000 to 5500 rpm.

STEP NO. 13:

DYER STREET CHARGER

For the last part of the Chevy 350-cubic-inch small-block bolt-on dyno test, we decided to install the ultimate bolt-on performance part, a Dyer Street Charger Roots-style blower. The engine was returned to step No. 7 configuration and the blower was installed. The Dyer kit came complete with everything necessary to bolt on and go. The blower was underdriven 11 percent for street usage and came with two specially prepared 750-cfm Carter carburetors. There is a choke provision so the engine still started easily. The installation was straightforward. Dyer has taken all the guesswork out of installing a blower—it is really just a simple bolt-on operation.

Power and torque reading both took tremendous leaps with the installation of the Dyer Street Charger. The gains were substantial over the stock engine and even the modified engine as tested in step No. 12. Peak power was 382 horsepower at 5500 rpm, and peak torque was 416 ft.-lbs. at 4000 rpm. The spark advance was set at 35° total advance, and the spark plugs were changed to Champion BL57s. Water temperature remained at 170° The blower boost ranged from 4 pounds at 2200 rpm to 9.5 pounds at 5500 rpm.

STEP NO. 14:

LOOKING FOR 400 HP

Everyone was very pleased with the way the engine performed with the Dyer blower but felt that the engine was capable of putting out more power. In an attempt to gain entrance to the 400-horsepower level, the dyno headers were reinstalled. They boosted horsepower and peak torque slightly to 388 horsepower and 416 ft.-lbs., respectively. Remember that these figures were obtained with an engine that was totally untouched internally and had the undesirable smog pistons. It was felt that better pistons would easily have moved the peak power over the 400 mark.

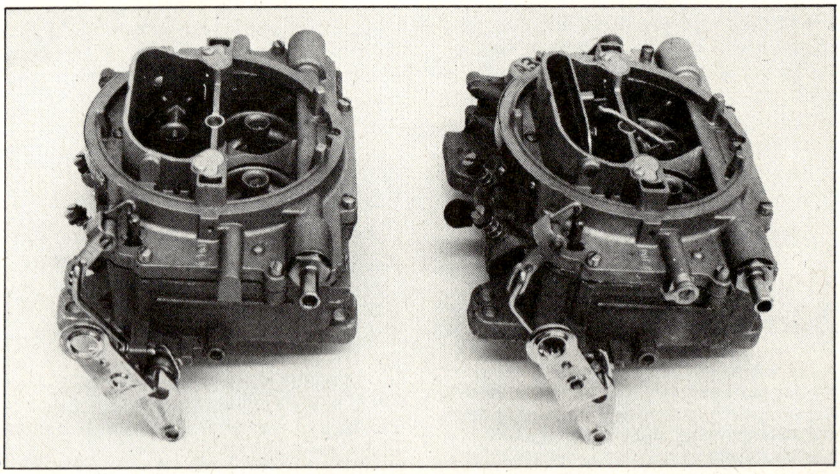

5

6

7

5. Dyer's Machine Service (7665 W. 63rd St., Argo, IL 60501) specially modifies a pair of Carter 750-cfm carburetors for use with the Dyer Street Charger blower system. The rear carburetor has a choke, so cold-starting is no problem. Dyer installs strip kits in the Competition Series Carter carbs.

6. The Dyer Street Charger is completely set up and ready to install as it comes from the crate. Dyer sets all tolerances with street driving in mind. Mike Chrisman and Gene Adams literally bolted the impressive blower to the top of the 350 Chevy and it was ready to run. The Dyer blower kit comes complete with everything needed for the installation.

7. You're looking at a virtually stock '78 350-cubic-inch Chevrolet smog motor that puts out almost 400 horsepower. The addition of a Dyer Street Charger blower and a few bolt-on speed pieces more than doubled the horsepower of the engine. The looks alone are worth the price of admission for the Dyer blower, but besides the looks and power, the engine is completely streetable. This blown small-block offers the best of all worlds.

Looking at the results realistically, the figure of 388 horsepower was very respectable and is as much power as most people can ever use on the street anyway. Gary Dyer admitted that it would be easy enough to push the blower much higher, but his many years of preparing blowers for street use lead him to this present combination. The idea with a street blower is to achieve a livable compromise between performance and driveability. The Dyer Street Charger is a well-thought-out unit that provides more than ample power when you want it, good dependable street operation (we know of many cases of Dyer-equipped cars making cross-country treks without incident), and more peer group status than any other single piece of performance equipment.

Looking back over the results obtained by Art Chrisman and Gene Adams on the W.R. Grace dynamometer test facility, three tests stand out: step No. 3, step No. 7 and step No. 13. Depending on your needs and finances, three very nice engines could be built following those three steps. Step No. 3 was a great starting point, easily accomplished for approximately $100 and a couple of hours. Step No. 7 was a well-balanced bolt-on aftermarket package within the abilities and budget of the average enthusiast. Step No. 13 was the ultimate bolt-on engine modification which showed how easy it is to bolt on over 200 horsepower.

This isn't a new revelation, but we will repeat it anyway: The small-block Chevrolet engine is an extremely well-engineered engine that responds readily to performance modifications. It is the best value in all of hot rodding, a distinction which the small-block Chevy has held for 25 years.

—Bruce Caldwell

HORSEPOWER

ENGINE SPEED	TEST PHASE 1	2	3	4	5	6	7	8	9	10	11	12	13	14
2000	100.4	107.6	113.9	112	103.2	101.4	103.7	108.9	82.1	82.4	102.2	100.9		
2200													144	151.2
2500	136	146.4	153	148	134.3	132	142.8	146.6	138.3	138.7	155.2	147.1	174.9	180.9
3000	164.7	178.8	182	177.7	176.9	175.8	180.9	188.5	184.8	193.2	203.6	192.3	226.9	231.1
3500	185.1	205	212.4	209.3	208.3	208.9	214.8	223.5	212	225.5	241.2	228.1	272	276.9
4000	186.5	223.7	230.5	233	237.9	236.8	243.3	253.4	248.5	259.8	277.7	260.6	316.9	322.7
4250	184.9	224.2	230.9											
4500	179	218.4	230.2	238.3	255.9	256.9	266.6	273.3	272.7	285.1	302.9	293.5	354.2	360.7
4750			225.3	231.4										
5000					263.3	264.5	277.7	277.5	276.7	290.4	313.2	315.6	375.6	380
5250					259				276.7	285.4		322.7	378.2	388
5500										285	324.4	326.5	382	
5750										274				
6000											316.3			

TORQUE

ENGINE SPEED	TEST PHASE 1	2	3	4	5	6	7	8	9	10	11	12	13	14
2000	263.6	282.4	299	294.1	270.8	266.1	272.4	285.8	215.7	216.3	251.4	264.9		
2200													345.7	360.7
2500	285.6	296.8	321.2	310.8	281.9	277.2	300.2	307.8	290.2	291	301.2	308.9	367.4	379.8
3000	288.3	312.9	318.4	311.1	309	297.1	316.9	329.8	323.4	338.1	356.2	336.5	391.6	404.4
3500	277.6	296.7	318.4	313.8	312.3	313.2	322.5	335.3	317.8	338.1	361.7	342.1	407.4	415.4
4000	244.8	293-5	301.8	305.7	312.3	310.8	319.6	332.5	326.1	340.9	364.5	342.1	416	423.5
4250	228.3	276.7	285.2											
4500	208.9	254.8	268.6	278	298.5	299.7	311.3	318.8	318.1	332.6	353.5	342.4	413.2	420.8
4750			249.2	255.7										
5000					276.4	277.7	291.9	291.3	290.5	305	328.9	331.4	394.3	399
5250					259.8				276.9	285		323	378.2	388
5500										272	309.7	312	365	
5750										250				
6000											276.8			

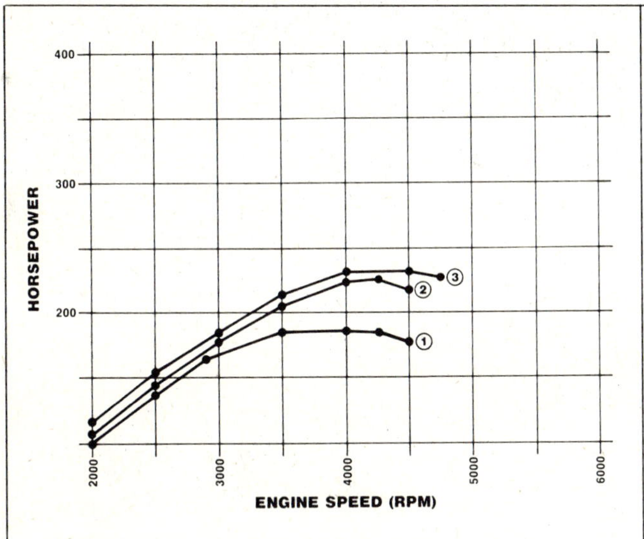

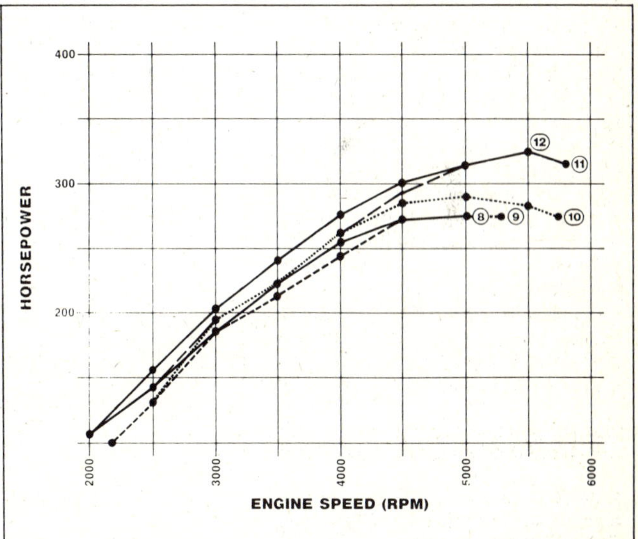

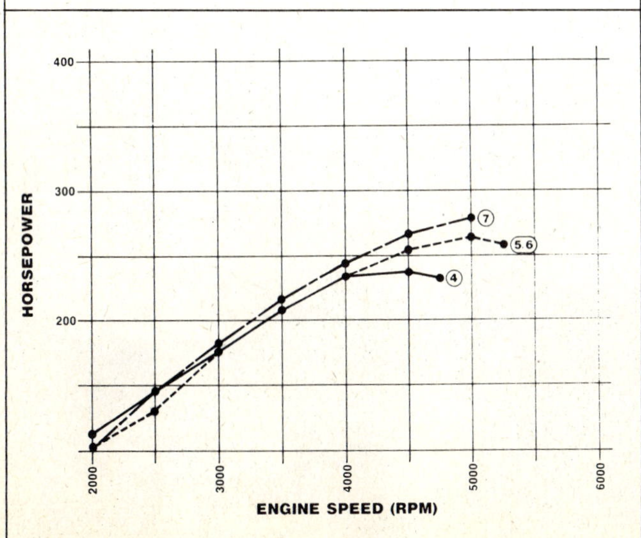

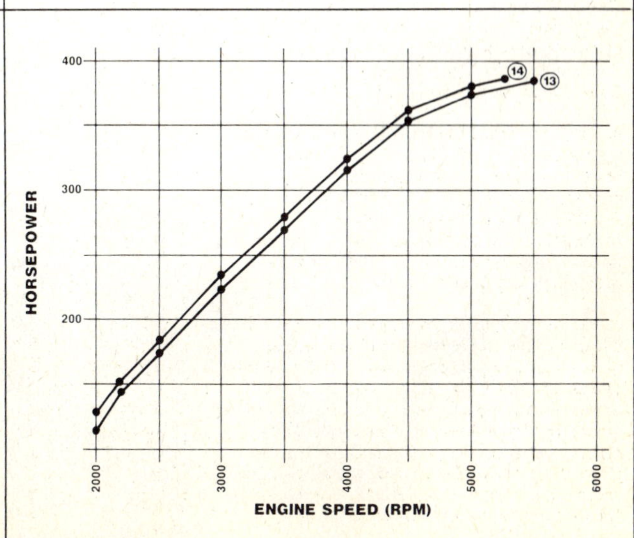

Engines You've Probably Never Heard About

And now, from the Glibness Book of Chevrolet Records, here is a quartet of "strange" unknowns.

300-MPH 3-LITER

The smallest and fastest of Chevy mouse motors belongs to two Bonneville racers, Don Cummins and Fred Larsen of La Mirada, California. The two have been racing together for over 20 years and have raced everything from highboy roadsters to their present streamliner. The "big" motor for the 'liner is a 183-incher which fits into International Class D, where the team is the record holder at over 275 mph. To arrive at this miniscule displacement, a '59 265 block was sleeved down ¼-inch to a bore diameter of 3.480 inches. A C/T Automotive ⅝-inch destroked crankshaft reduces the stroke to 2.350 inches. Early 283 cylinder heads, injector-fed alcohol and a 4-71 GMC blower powered the slippery 'liner to a phenomenal 291+ mph. That's nearly 300 mph with 183 cubic inches!

Now for an even smaller Chevy. The boys decided to run Class E also, so they built an externally identical engine to the 183. But once the heads are removed, you immediately notice something missing—four pistons! This time a 265 was sleeved 3/16-inch for 240 inches, and after a check with engine-balancing technician Gary Thompson for feasibility, every other piston in the firing order was removed; then suddenly Larsen and Cummins owned the tiniest mouse at a whopping displacement of 120 cubes! But don't let the lack of size fool you: This one has gone over 248 mph!

To summarize this effort: Two records were set—the D/Streamliner record at 289.508 mph and the E/Streamliner record at 227.178 mph—both accomplished with a "total" of 303 cubic inches.

500 HEMI

No discussion of special Chevrolet engines would be complete without a discussion of the Arias hemi. Conceived and engineered by piston manufacturer Nick Arias of Arias Pistons, it took countless hours over a period of two years to finalize its design.

The Arias hemi is based on the Chevrolet truck block which has .400-inch taller decks. The 4.250-inch bore is increased .060-inch to 4.310, and the 4-inch stroke is upped .500-inch to 4.500. The 6.510-inch connecting rod is .375-inch longer than stock, and all this yields a basic displacement of 500 cubic inches. The hemi-domed pistons are made only by Arias, to match his exclusive hemispherical chamber shape.

The hemi heads, interchangeable from side to side, are cast from 356 T6 aluminum alloy. They use 2.25-inch Chrysler hemi intake valves and Donovan stainless steel 2.070-inch exhaust valves. Arias designed the valvetrain to use 8-inch Chevrolet pushrods, 7/16-inch in diameter, with his own 1.6:1 ratio rocker arms. The intakes operate in normal fashion, but the exhaust side uses a 1:1 bell crank rocker arm. The exhaust rockers are also 1.6:1. The Arias hemi heads can be used with factory aluminum truck blocks with deck plates, or the new solid-aluminum Rodeck block. Both combinations have achieved success as drag boat powerplants. In recent dyno test sessions, a 500-incher pulled 760 horsepower on gasoline, and a 540-inch version made 867 horsepower at 6500 rpm. And how does 710 ft.-lbs. of torque sound? You can believe that when the throttle is opened on this unit, your neck had better be in the braced position.

KING RAT

The day of the really big, modern passenger car V8 engine displacement is all but over. The largest surviving engine for the 1977 model year, for example, was the 460-incher used in Ford luxury products. The 500-inch Cadillac was cut back to 425 cubic inches, and most car makers are working hard on small V8 and V6 powerplants. All this sounds fine for street performance and efficiency, but it'll never do in Southern California drag boat racing, where the big-block Chevrolet is the undisputed favorite for injected and carbureted boats. One builder, Ken Theis, of Ken's Automotive in Van Nuys, California, still believes that there is no substitute for cubic inches, and has built some of the biggest rat motors.

His most recent project incorporates an aluminum block, aluminum heads and Crower injectors, and measures a mind-boggling 622 cubic inches! The bore in the sleeved block is increased a full ½-inch over at 4.500 inches, and the stocker crank is ⅞-inch longer, at 4.875 inches. Ken has built dozens of big-inch rat motor variations, from 500 to 622 cubic inches, simply by varying the bore and stroke and using a selection of overlength rods and intake adapter plates. These low-revving, ungodly torquey motors are really popular on the Southern California waters, but not everyone admits to their size. And you can be sure it's a quick way to transport a whole family of skiers.

SOLID, JACK!

If the Chevrolet rat motor enjoys a resurgence of popularity in fuel racing, it will most likely be due to John Rodeck, a disenchanted Chevy racer who has re-engineered the Chevy 454. It's an all-aluminum, racing-only version he calls the Rodeck 481.

Rodeck's solid-aluminum rat (no water passages in the block and all oil galleys are drilled, not cast) is, in a word, beefy. It's solid A356 T6 high-purity alloy using massive 7075 T635 aluminum main caps and 110,000-psi ductile iron cylinder liners. Caps and heads are retained by 4340 steel ½-inch-diameter studs with hardened washers and special 220,000-psi 12-point head nuts with integral washers, and most of the studs are 8 inches long with 5 inches embedded in the block. The big beefy main caps register .300-inch deep into the block, as opposed to .120-inch stock-block register, and each cap is double-dowel-pinned on different centers. Across the lifter valley there are three ¾-inch-wide, 3-inch-deep stiffening ribs, and Rodeck adds an extra head bolt on cylinders No. 2, 3, 6 and 7 by going through the intake port, a la the old Can Am engines. The beauty of the Rodeck 481 (4.375 bore, 4.000 stroke) is that it's competitively priced with the aftermarket aluminum hemis but uses more stock components, such as crank, steel, closed-chamber Chevy heads, oil filter, oil pan, etc. And it's already run 5.90s at nearly 240 mph in early shakedown runs!

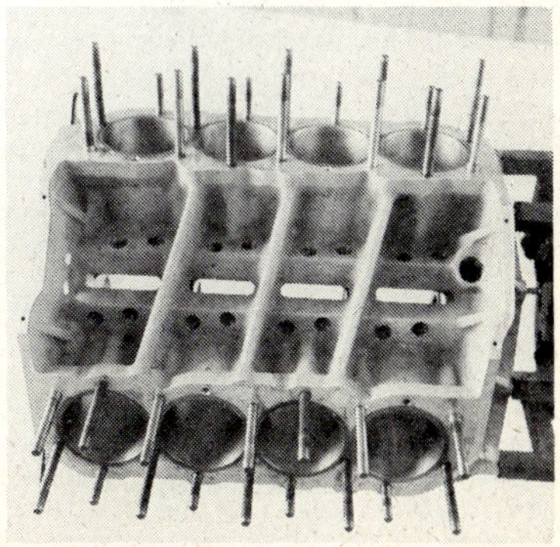

Proven Performance: The Big-Block Chevrolet

Detailing the exact combination of parts and modifications to make a reliable 650 hp from the 427.

The big-block Chevrolet is the second most popular performance engine in use by enthusiasts and racers today—exceeded only by the small-block Chevy V8. There are many reasons for this popularity, but the availability over the past 15 years of excellent high-performance factory parts, at comparatively low prices, for these engines has a lot to do with it. Additionally, making a big-block run and live is relatively easy. Even though its power output is not as efficient in terms of horsepower-per-cubic-inch as many other performance engines, its sheer size and breathing capabilities permit it to produce maximum torque and horsepower figures that are exceeded only by far more exotic or expensive engines—such as the Chrysler Hemi or the Ford Cammer. This makes the big Chevy one of the very best values in terms of horsepower-per-dollar.

The intent of this article is not to show the most exotic way to squeeze the last possible pony from the big-block, regardless of cost or reliability, but rather to show you an easily duplicated combination that has produced outstanding power with very low failure rates—in other words, proven performance. It is an engine typical of what a knowledgeable bracket racer might build, and it is legal for IHRA and AHRA competition in the Super Stock and Formula Stock classes.

The expense involved in duplicating the exact combination shown here is in line with what it costs to build most racing engines nowadays. Using all new components, such an engine would cost approximately $6000, depending upon the machine shop rates in your area and the amount of work you do yourself. Of course, if you're like most racers, judicious shopping for good used parts could permit you to lower that total considerably. The engine shown here also uses a number of lightweight components, such as aluminum cylinder heads, an aluminum water pump and an alumi-

1. The engine is topped off with an Edelbrock C-454 aluminum intake manifold that has been matched to the intake ports (not the intake manifold gaskets). Moroso aluminum valve covers cap the heads with homemade lightweight valve cover bolts. The carburetor is a Holley No. O-8082 1050-cfm unit. The plug wires are Accel 8mm solid-core versions modified with Belden spring terminals.

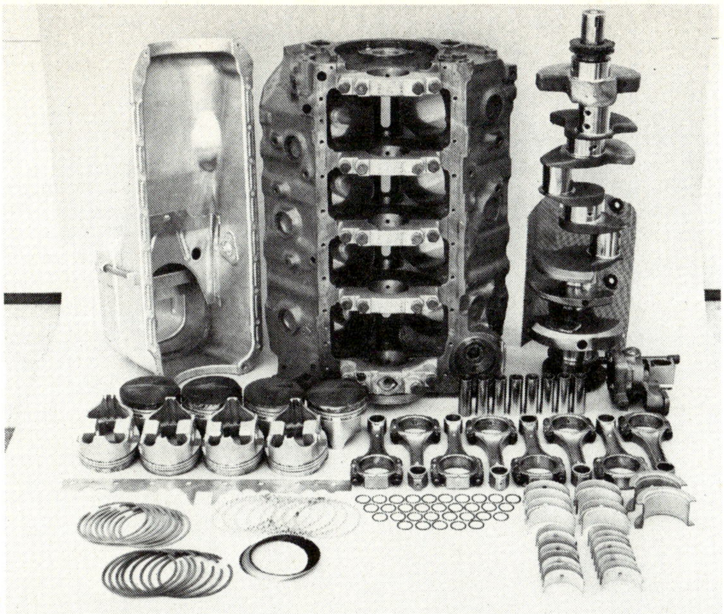

2

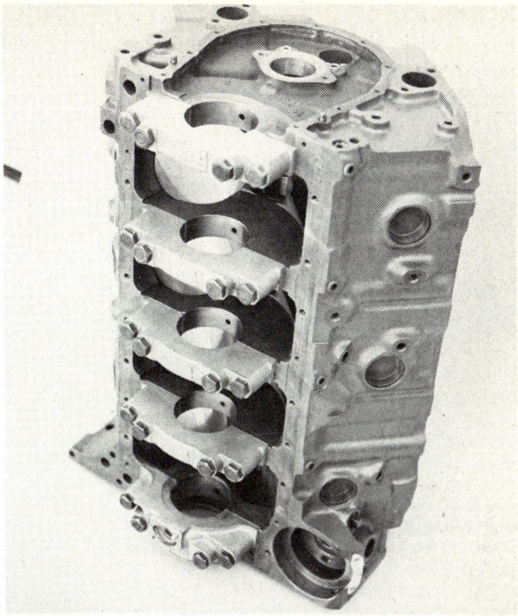

3

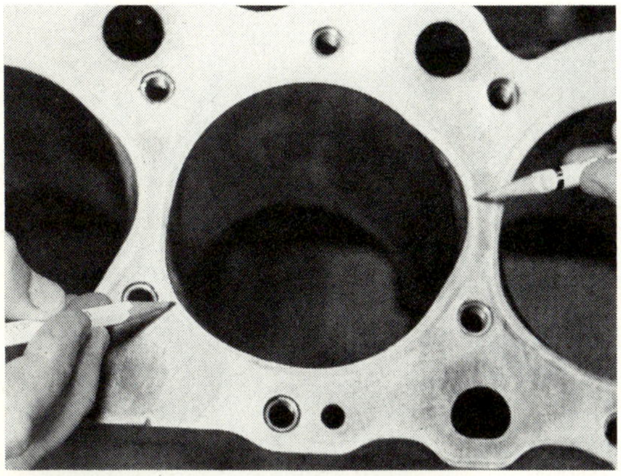

4

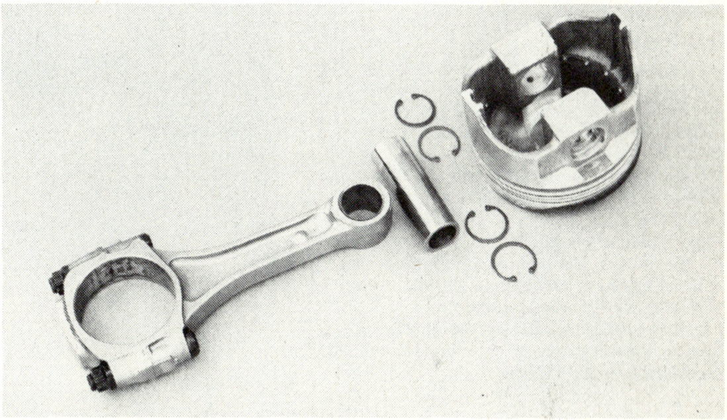

5

OPERATING AND TUNING DATA

Fuel:	Daeco 105-octane Super Premium
Fuel Pressure:	6½ to 7 psi
Oil:	Pennzoil 20W-40 Racing Oil
Oil Additive:	Molybdenum disulphide
Oil Pressure:	65 to 75 psi @ 7000 rpm
Thermostat:	165 degrees
Normal Operating Temperature:	180 to 210 degrees
Ignition Timing:	42 degrees (total)
Spark Plugs:	Champion N-64Y
Plug Gap:	1/16-inch (.062-inch)
Valve Lash:	Intake—.014-inch (hot), .008-inch (cold*) Exhaust—.016-inch (hot), .010-inch (cold*)
Carb Jetting:	No. 87 on both primaries and secondaries
Power Valves:	No. 35 on both primary and secondary sides
Power Valve Channel Restrictions:	.093-inch
Accelerator Pump Squirter Diameter:	.052-inch
Headers:	Hooker No. 2217 (2x35-inch primaries with 3½x10-inch collectors)

*with aluminum heads and titanium intake valves

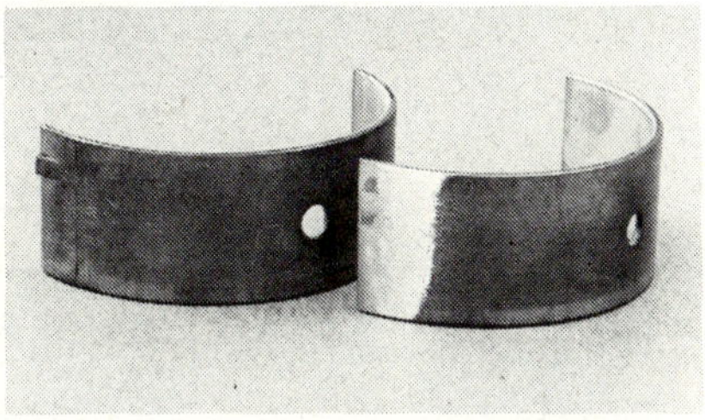

6

2. With all the machine work complete, trial assemblies and balancing complete, here are the short-block, oiling system and pan components, ready for assembly. All crankshaft, connecting rod and balance work was done at Hank the Crank's shop in North Hollywood, California.

3. The block used is the four-bolt high-performance version. It has been fully deburred, align-bored and align-honed, bored .060-inch oversize for a displacement of 440 cubic inches and honed with 600-grit stones to a very smooth finish and a .007-inch piston skirt clearance. The block has also been squared and decked so that the production TRW No. 2308AF-60 piston decks will protrude from the block .035-inch prior to piston machining. The modification to the oil filter pad is to simplify header installation in the Chevelle chassis. The pan rails have also been drilled and tapped to secure the windage scraper shown elsewhere. Prior to cam bearing installation, the valley of the block is drilled and tapped for the valley screen retaining bolts too.

4. Using the cylinder head combustion chambers (not the head gasket) as a template, relief notches are ground in the top edge of the cylinder walls to improve inlet and exhaust flow. Care must be taken to assure that this grinding does not extend down the cylinder walls into the area swept by the top piston ring.

5. Considerable attention is given to this assembly to enhance reliability. The rods are the 7/16-inch bolt versions. They have been polished and shot-peened, fitted for floating pins, and lower bearing retention pins have been installed in the caps. The rod bolts used are the special SPS versions, as are the nuts—this is a must. The pins are special premium alloy steel with .105-inch-wall thickness (marine version) from C&S Specialty in St. Charles, Missouri. The pins are retained with double .042-inch thick locks. Either Tru-Arcs or Spirolox are acceptable. The pistons are TRW No. 2308AF-60 forgings. Extra oil drain holes have been drilled in the oil ring groove and pin oiling holes have been drilled.

6. Because dowel pins have been installed in the connecting rod caps, the tang is filed off the lower bearing halves to assure correct indexing. The connecting rod bearings used are from TRW. A mix of CB955P and CB956P bearings is used to achieve a Plastigage indication of .002-inch bearing clearance on the connecting rods. A TRW MS2829P main set is used with a clearance indication of .0025-inch, also measured with Plastigage.

num oil pan, all of which could be replaced with far less expensive cast-iron or steel equivalents without affecting the power output one bit. However, the engine would weigh 85 pounds more.

What kind of performance could you realistically expect from an engine like this? Well, in my 3600-pound Chevelle fitted with a Powerglide transmission, an 8¾-inch torque converter, 5.57:1 gears and 13½-inch-wide slicks, this exact engine combination has produced elapsed times in the 10.50s at 130 mph under ideal conditions. This has been achieved with a maximum engine speed of 7800 rpm. The effective power range of this engine is 4500 to 7500 rpm. Extrapolating this same power output to different weight vehicles should produce the following performance, assuming optimum gearing, traction and weather conditions:

Weight	Projected Elapsed Time (¼-mile)
4000 pounds	11.10 seconds
3200 pounds	10.20 seconds
2800 pounds	9.80 seconds
2400 pounds	9.40 seconds
2000 pounds	8.90 seconds
1600 pounds	8.35 seconds

Periodic maintenance on an engine like this is relatively low, especially considering the power output, and that may

7. The crankshaft is straightened, Magnafluxed and micropolished prior to installation. The piston and rod assemblies are installed with a standard Speed Pro piston ring set (R-9745+.065). The 1/16-inch moly top ring is filed to .014-inch end gap and the 1/16-inch cast-iron second ring gets a .012-inch gap. The 3/16-inch oil ring is the standard tension version, as explained in the text. It is imperative that the rod bolts be tightened to a stretch of .009-inch, also as explained in the text. The stretch indicator shown is from King Engine Service (1755-F Wilwat Dr., Norcross, GA 30093).

8. When using factory rods, you're stuck with whatever the rod side clearance happens to be. Almost invariably it checks out in the neighborhood of .022-inch, which throws off a lot of oil. To help control the oil spinning around with the crankshaft and rod assembly, a Moroso No. 2583 crankshaft scraper is used. You have to cut and file this scraper to fit your own crank and rod assembly. To get it right (within .020-inch of the rotating assembly) will probably take you the better part of a day of tedious cutting and trimming—it did me. Locate the finished scraper with a couple of countersunk screws to assure it doesn't move.

9. Oil control is also enhanced by the use of a Moroso No. 2291 windage screen. The oil pump is a Moroso No. 2217 unit with bottom pickup. It is lowered into the sump with the Moroso No. 2330 1½-inch extension kit to position it properly in the Moroso No. 2148 aluminum oil pan. Note that some grinding was required, along with a little gentle hammer work on the pan, to eliminate oil pan interference.

10. The dipstick opening on the aluminum oil pan has been sealed and this oil level sight gauge installed, which allows monitoring of the oil level in the engine while it is running. This gauge is available from Yother Performance Products (618 Doolittle Dr., San Leandro, CA 94577).

11. Camshaft drive is accomplished with a Borg-Warner true roller timing chain and lower sprocket. The upper sprocket is an adjustable cam timer from Ed Hamburger's Hi-Performance Parts (1590 Church Rd., Toms River, NJ 08753). This is a great aid in degreeing a cam. The Racer Brown roller cam, 97RM-111RM-110, used in this engine is degreed 2½ degrees advanced without valve-spring preload. An Isky cam button bears against the inside of the timing cover to control cam walk.

12. This homemade device attaches to the back of the water pump to preload the engine front timing cover for precise cam walk control. The pump shown here is the Moroso No. 6352 aluminum version.

7

8

9

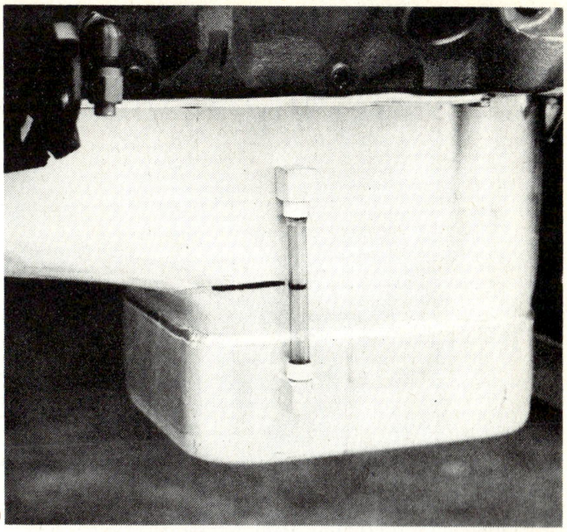

10

11

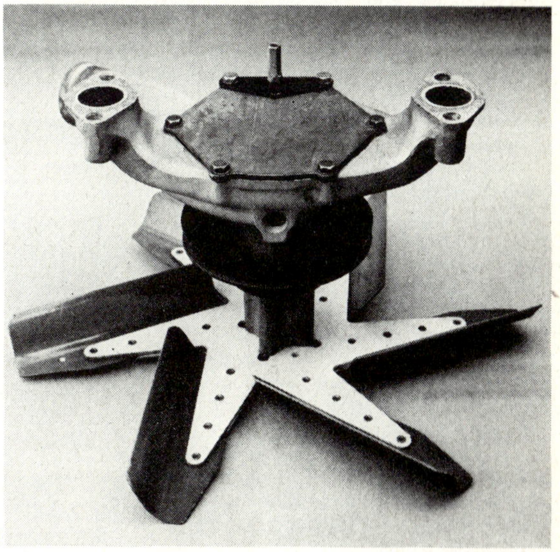

12

surprise a few people; but as I said earlier, this is a reliable engine well suited to bracket racing. Depending upon operating conditions and periodic cylinder leakdown results, 100 to 150 runs are normally logged between teardowns for a light cylinder honing, piston ring and bearing replacement, along with crankshaft and connecting rod Magnaflux inspection. In most cases, the bearings are replaced only as a precautionary measure, since they usually look perfect.

The valve job is freshened up at intervals of approximately 250 runs. Valve spring life is dependent upon the quality of the particular set of springs being used. Consequently, the springs must be checked religiously to maintain good power output *and reliability.* Some springs will lose pressure in as few as 10 runs, but an exceptional set of Vasco springs *might* last an entire season. Now, let's get down to the specifics.

The photo captions and data chart contain most of the information about this engine, but there are a few additional details that are essential to duplicating this combination. One critical area is assuring plenty of cylinder pressure and complete burning by maximizing the static compression ratio. This is done by positioning the piston dome high up into the combustion chamber and by making the combustion chamber as small as possible. The procedure for doing this is to deck the block until the TRW piston deck protrudes .035-inch. Then machine the piston decks .050-inch to produce a piston deck .015-inch *below* the block deck. The cylinder heads are then angle-milled to reduce the combustion chamber volume from a nominal 119-123cc to as little as 107cc. Next, a trial assembly is done to determine the piston valve relief machining that must be done to provide a .100-inch intake and .120-inch exhaust valve clearance and to make these reliefs parallel with the valve heads.

After the valve reliefs are cut, the piston domes must be carefully reshaped to provide a *minimum* of .015-inch clearance between every part of the piston top and the cylinder head when the head is positioned on the block *without* a head gasket. This done, radius all sharp edges on the piston top and lay back the edge of the intake valve relief to smooth flame propagation across the piston top. Also file a spark plug relief notch. Even with this notch, it will still be necessary to select fit spark plugs for each cylinder that will tighten with the side electrode opposite the piston dome to prevent closing the plug at high rpm. After completing all this piston work, have the piston tops coated with plasma-sprayed heat barrier material .015-inch thick. This will bring the piston deck height back to zero, as well as the gasketless piston-to-head clearance. Final assembly is done using the Fel Pro 8180PT-1 Teflon-coated head gasket, which has a compressed thickness of .045-inch, thus providing a .045-inch piston-to-head final clearance. After completing the final piston work, the entire engine rotating assembly must be balanced.

Standard-tension oil rings are used, because low-tension oil rings simply can't control the amount of oil normally thrown on the cylinder walls. When oil control is lost, oil finds its way into the combustion chambers where it fouls spark plugs, dilutes the fuel/air mixture and creates detonation which robs power. The slight power loss from the additional standard oil ring tension will be more than offset by the power gain achieved by keeping oil out of the combustion chambers.

I am also a firm believer in marine wall-thickness piston pins, SPS rod bolts and tightening the rod bolts by stretch rather than torque. The reason is simple: big-block connecting rods and those TRW pistons are *heavy!* The load on the rod bolts and piston pins is incredible as the engine approaches 8000 rpm, and good components add a lot of protection if you accidentally over-rev. They also add longevity to the package even if you don't over-rev. The reason for measuring bolt stretch is that bolt elongation is the best indication of clamping force and the preload needed to keep the cap tight, as well as the rod bolt nut from loosening. SPS bolts call for only 67 to 73 ft./lbs. of torque, but the stretch is .009-inch, which almost always requires a torque loading of 80 to 110 ft./lbs. to achieve! I can't overemphasize this. If you want that bottom end to stay together, spend the money for a stretch gauge and do it right—it's far cheaper than a new engine. I learned this lesson the hard way.

About the only thing left to dicuss is the valvetrain. This is a real problem on the big-block because the valves are big and heavy. The problem is further aggravated by a rocker arm ratio of 1.7:1 that multiplies the loadings on the valvetrain components. There is only one solution to this problem if you want to go fast and keep things together: You have to bite the bullet and step up to a roller cam, lightweight valves and the best valve springs you can beg, buy or steal.

The combination used in this engine is a Racer Brown 97RM-111RM-110 roller cam and kit. This cam has durations of 270 and 280 degrees at .050-inch tappet lift on the intakes and exhausts, respectively. It is ground on 110-degree lobe centers and has .723-inch net intake and .700-inch net exhaust valve lift. It is limited to a maximum engine speed of 8000 rpm, or 8500 rpm if a rev kit is used, if valve float is to be avoided, even with good springs. Whenever possible, the rev kit should be used for exactly the same reasons that I specified for using SPS rod bolts, rod bolt stretch and marine piston pins—over-rev protection.

The valves used in this engine are TRW No. V2713T titanium intakes and No. S2808H hollow-stem exhausts, which have diameters of 2.19 and 1.88 inches, respectively. These valves run in Winona bronze valve guides with Raymond valve seals on the ⅜-inch-diameter stems. The valve springs are installed at heights of 1.905 and 1.880 inches for the intakes and exhausts, respectively, for use with this cam.

That sums it up. If you don't cut corners, use all the same components and assemble everything as outlined, there's no reason why you can't achieve the same power levels and reliability that I have. See you at the races.

—C.J. Baker

13

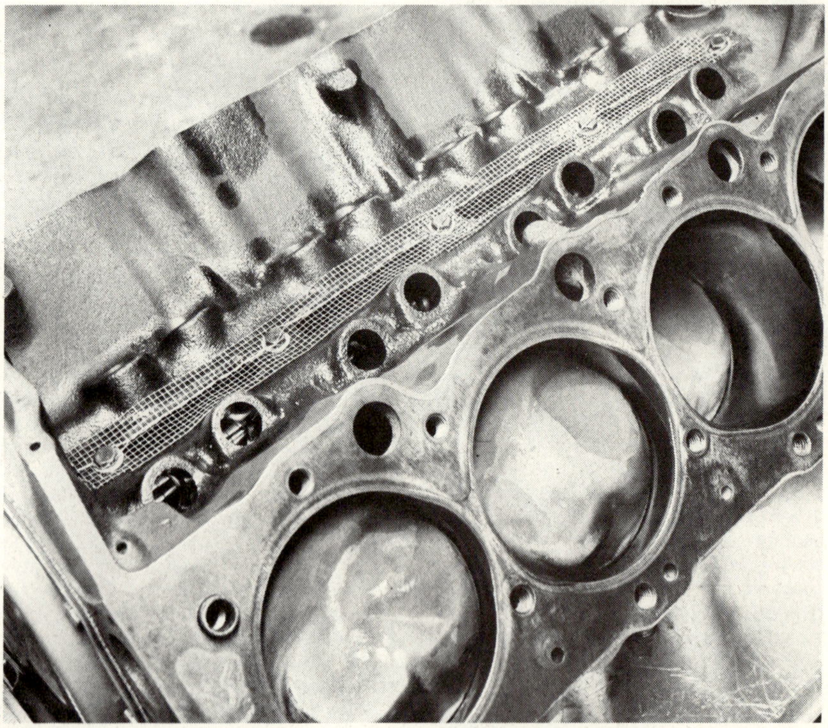

14

13. With the pump and No. 6361 pump spacer kit in place on the engine, the preload device is extended to force the timing cover against the cam button to achieve .005- to .010-inch cam end play. This is measured through the valley openings, or at the back end of the cam if the cam plug hasn't been inserted yet.

14. After the cam walk has been set, ⅛-inch hardware cloth is used to cover the valley openings to prevent any broken valvetrain parts from finding their way into the crankcase, where more serious damage might result. The retainer bolts have holes drilled in the heads and are safety-wired to prevent their loosening. Note the piston domes. All sharp edges on the domes have been smoothly rounded, a spark plug clearance notch has been filed and the intake pocket edge laid back. The piston tops have been coated with a zirconium oxide plasma heat barrier by Plasma Technology (1754 Crenshaw Blvd., Torrance, CA 90501), for maximum combustion heat utilization. Notice the even coloration of the piston tops, which indicates even burning across the entire piston top.

15. Shown here are the aluminum second-design open-chamber cylinder head (top) and the equivalent open-chamber cast-iron head (bottom). The cast-iron head is cheaper and will produce about 10 to 15 more horsepower, but the aluminum heads are 75 pounds lighter. Take your choice. The new aluminum heads that feature the fin in the bottom of the exhaust port may more than offset the power difference between the aluminum and cast-iron versions, but these latest heads have not been tested on this engine. See text for cylinder head details.

16. Study this photo closely, because it shows a lot of components that contribute greatly to the engine's reliability. Head studs are not used on this engine, but premium SPS head bolts are used with stock torque specifications. Stock 7/16-inch pushrods actuate Crane 1.7:1 needle-bearing aluminum rocker arms that reside on Moroso No. 6781 4340 chrome moly rocker studs. A Moroso No. 6727 rocker stud girdle is used with No. 6716 and No. 6717 hex-head adjusting nuts that feature positive-locking setscrews to prevent adjusting nut "walk" on the stud. Note that the crimp-type lock nuts on the girdle U-bolts have been replaced with Ny-Lok nuts to improve U-bolt life. Why is all this necessary? The valve springs used to assure absolute valvetrain control are Crower No.

15

16

17

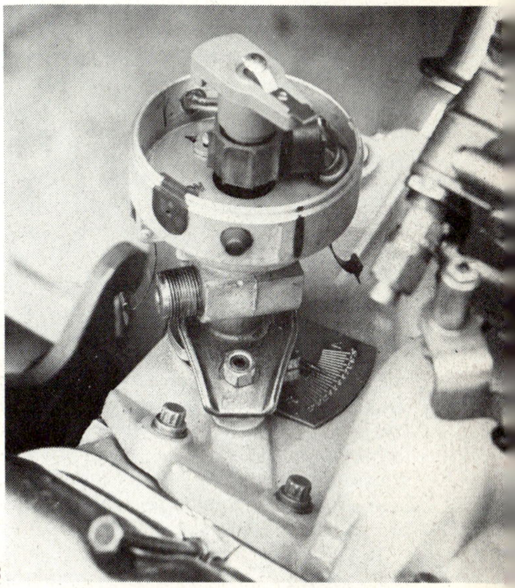

18

19

68555-1 Tungalloy Vasco "Red Stripe" springs. These 1.625-inch-diameter springs have 770 pounds pressure at a compressed height of 1.125 inches! The spring retainers are titanium versions from Crane.

17. A modified Chrysler oil separator attached to the valve cover is connected to the suction side of a mechanical fuel pump to provide some degree of crankcase evacuation at low engine speeds. A similar oil separator on the opposite valve cover is connected to the header collectors in a more conventional pan evacuation manner.

18. The distributor used in an Accel tach-drive model modified with a magnetic pickup which signals an Autotronics MSD-7A ignition system. A Moroso distributor degree plate and heavy-duty distributor hold-down simplify timing adjustments. The advance curve (24-degree crankshaft) is all in by 1500 rpm. Initial lead is 18 degrees, for a total of 42 degrees.

19. The MSD-7A ignition system puts out so much energy that vent holes are required in the back of the cap to prevent the air inside the cap from ionizing and conducting crossfire. Note the anti-slosh hoses on the carb bowl vents to prevent fuel slosh during acceleration and burnouts.

Smokey's 600-HP Road Rat

"The man" is back with clues on how to build a strong and reliable street engine. This time it's the big-block.

Earlier we persuaded Smokey Yunick to give you the lowdown on building a street-oriented small-block Chevy. Now the old master is back to give you the word on prepping a big-block for the street. Smokey's intent is to describe not an all-out competition engine, but rather one that will run hard when you want it to without being unbearable the rest of the time. If you read his discussion of the small-block, you already appreciate the logic of Smokey's commonsense approach to building strong, reliable engines. If you didn't read it, you missed something very worthwhile, even if you're not a Chevy fan, since a lot of Smokey's techniques can be applied to any engine.

Because Smokey went into great detail on such items as block and connecting rod preparation, cleaning techniques and such, we're simply going to suggest that you go back over the small-block feature rather than rehash it all again in this chapter. Instead, we've asked Smokey to concentrate on those big-block specifics that are most important.

By the time you finish reading this, you'll realize that Smokey's approach to power and reliability isn't cheap, but be forewarned: Cutting corners can be a lot more expensive. —Editor

When I discussed building up a little Chevy for the street, I mentioned that I'd pass along some suggestions on prepping a big-block in the future, so here goes. I'll start by saying that all of the machining steps and procedures I outlined for small-blocks hold true for big-blocks too. As far as that goes, they are the same for any engine, be it a Pontiac, Ford, MoPar, etc. Therefore I'm not going to spell it out again. I'll just say that if you didn't read the small-block piece, stop right here and read it. After you've finished, what I'm going to say about the big-block will make a helluva lot more sense.

Before I get down to specifics, let's get an idea of what we're going to have when we're done. We'll start with a basic 427-cubic-inch engine and overbore it .030-inch for a total of 433 cubic inches; you could also use a 454. It's a torquey, stump-pullin', gas guzzlin' killer and gives the most power per dollar, period, but you better have extra dollars for gas! Consequently, we'll stick to the 427—not that it's any economy champ, though. If you desire, unlike the small-block, most big-blocks will accept a .060-inch overbore for a displacement of 440 cubic inches, although such a large bore could cause heating problems on the street.

The engine will rely on a single 4-barrel carburetor. It will use a roller cam to reduce valvetrain frictional losses. If you can find (and afford) 100+ octane gas, power should be over 600 horsepower at 6800 rpm with around 525 ft./lbs. of torque at 5200 rpm.

From a reliability standpoint, the only block to use for a high-performance engine of this type is the 4.25-inch bore, four-bolt main bearing cap version (Chevrolet part No. 3963516). Just as in the small-block build-up, start with a good used block, check it for cracks, deburr it, remove any stress risers from

both the block and the main caps, and radius the juncture of the main webs to the block. Check the block for alignbore, have the decks milled parallel with the crank centerline and for a zero-to-.002-inch down piston deck height and, using a good head plate, bore and hone the cylinders to .030-oversize.

Of course, to do all this, you'll need the crank, rods, bearings and pistons. The best crank available for the 427 is the 3.76-inch stroke L88 version (part No. 3967811). If you're building a 454, use the LS7 crank (part No. 396354), and remember to use the appropriate externally balanced flywheel and harmonic balancer. Both cranks are forged 5140 steel, Tufftrided, cross-drilled and machined to accept the heavier 7/16-inch-bolt connecting rods that I'll cover later; so you can use them just the way they come from the factory. All you need to do is check them for straightness. Factory tolerances allow up to .007-inch of runout on the main journals. Naturally, I'd like to see it perfectly straight, but you can use a warped one, as long as it is within this limit, without any serious problems. Dollar for dollar, you can't buy a better crankshaft, providing you get a good one from the factory.

When it comes to connecting rods, there are several ways to go, and a guy can save a little money here, depending on what he wants to do. Ideally, the 427 likes about a .275-inch-longer-than-stock rod for ultimate power production. This means you have to go to a good aftermarket manufacturer, like Carrillo. However, the factory stock-length 7/16-inch bolt rod is more than adequate for any street use and (when properly prepared) okay from a durability standpoint for most types of competition. Chevy offers a big-block rod (part No. 3969804) that's fitted with good 7/16-inch boron bolts, shot-peened, magnafluxed, heat-treated and machined for floating pins. Some people refer to this as the L88 or ZL1 rod, and they sell for over $80 apiece. This is a good rod, but unfortunately the beam isn't deburred or polished, which lets a bad one slip through once in a while. Although the chances of getting a bad rod are fairly low, the consequences of a broken rod at 7000 rpm don't make this gamble worthwhile. I think a guy should grind and polish the rod beam, have it shot-peened and magnafluxed again to relieve any stress risers and doublecheck for cracks. There are a lot of shops that can do this for you, such as Hank the Crank (HTC) in North Hollywood, California. The problem is that you're paying to have the shot-peening and magnafluxing done twice, which really shouldn't be necessary.

A cheaper way to wind up with nearly the same thing is to buy the LS-6 rods (part No. 3963552), which sell for about half the price of the L88 rod. Although this is basically the same rod as the L88 rod, it isn't machined for floating pins, it hasn't been shot-peened or magnafluxed, the 7/16-inch rod bolts aren't the special boron versions, and it hasn't received quite as good a heat treatment. However, you can have HTC or somebody machine it for a floating pin, polish the beam, shot-peen and magnaflux it and install the good boron bolts. With the exception of the heat treatment, which really isn't important for a street engine, the final result will be a rod that's just as good as a reworked LS-7 rod at a much lower price.

If you want to add a little more dura-

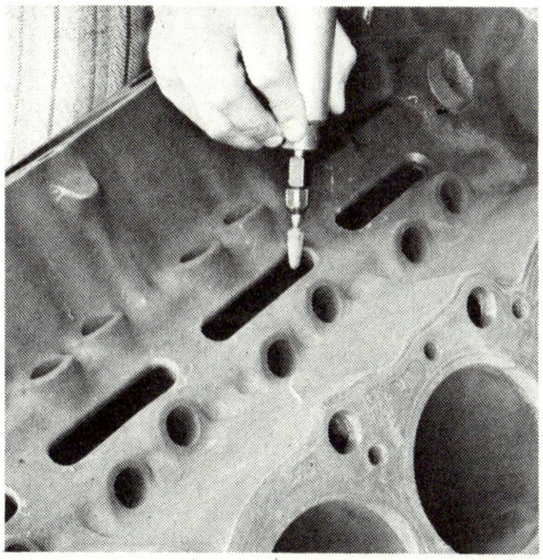

2

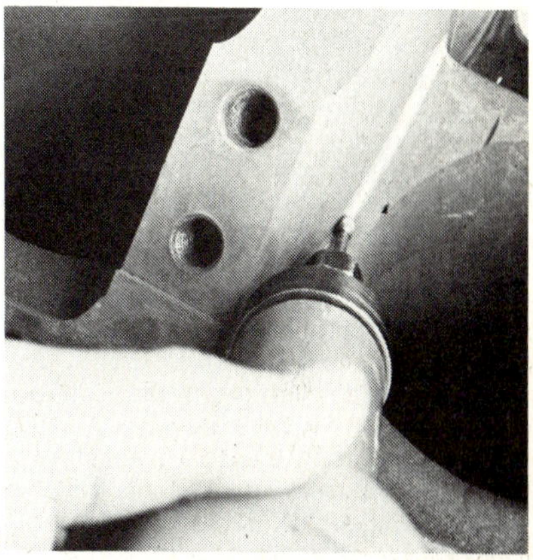

3

4

1. Smokey Yunick has practically no peers when it comes to getting both ultra high-performance and longevity out of Chevy's big-block, and he has long worked hand in hand with Chevrolet Engineering.

2. Grind away all casting flashings in the lifter valley. This eliminates any chance of their breaking loose at some future date and it also speeds oil return to the pan.

3. Carefully radius the juncture of the main bearing webs in the block. This helps reduce the possibility of cracks in this highly stressed area.

4. At left is the round-port aluminum head, and at right the rectangular-port cast-iron unit. The aluminum head will need the two additional studs, shown, to ensure proper head gasket sealing.

bility to either the LS6 or L88 rod, SPS bolts can be used (part Nos. 340289, 12-point aircraft nut; and 340285, SPS bolt), although the boron bolts are good enough for most uses.

As for bearings, I think the Clevite inserts are among the best. Shoot for .0020-.0025-inch clearance on both the main and rod journals. Side-gap should be .015-inch per pair and crankshaft endplay should be held to .005-.007 inches.

Install the bearings with a light coat of oil on their crank mating surfaces. Some guys are using dry lube coatings impregnated on bearings, piston skirts and timing chains. This is good for durability more than power.

The best *racing* pistons are the blank TRW forgings (part No. 2362 + .030 for the 427) for open-chamber heads. Have Diamond or somebody put the top ring .200-inch down from the deck, the second ring down about .160 from the top ring and the oil ring .125 down from the second ring. Shoot for a finished compression ratio of 12.5-12.7:1. The skirt contour and barrel grind problems you generally run into let you fit these TRW pistons a little tighter in the cylinder bores than anybody else's at .0060-.0065-inch, measured level with and perpendicular to the piston pin. Have the pistons machined for double Tru-Arc pin retainers as outlined in the small-block story and fit the floating pins for .00135-.00150-inch of clearance in the rod. In all other respects, leaving the piston alone is the best policy.

For the street, 9.0:1 compression is *more* than enough with today's gas. In the 454, this can be achieved with stock '71 LS6 forged pistons (part No. 6262979 for a .030-over motor) installed in open-chamber heads. There is no low compression factory forging for the 427. Compression ratios of around 10:1 can be achieved by using forged 11:1 closed chamber pistons with open chamber heads. Even with mixed gas, additive and water injection, this would be marginal, though, today. TRW plans to release a low-compression Turbo forging for both the 427 and 454, which should be available by the time you read this.

In selecting rings for these pistons, high-quality moly-sprayed rings, such as the SpeedPro rings, are the only way to go. Set end-gap on the compression rings at .015-.018-inch and .025-inch for the oil ring. For proper sealing with moly rings, it's important to finish the cylinder walls as smoothly as possible—the slicker the better.

Now that we've covered the short-block pieces, send them out to the balancing shop and we'll move upstairs and talk about cylinder heads and the valvetrain. These are two areas that'll make or break the big-block Chevy. We'll start with the heads.

I'll simplify things a lot by saying that the open-chamber heads are hands down the best in terms of flame propagation and power output. But for the street—and this is something a lot of guys might disagree with—those big rectangular intake ports on the Chevy open-chamber heads are just too big. They hurt low-end and mid-range response and torque, so maybe some thought should be given to using the small-port heads with oval-shaped intake ports on the street. It isn't until you go over 6000 rpm that the big-port heads are clearly superior. There are a number of different versions of the small-port heads. Some believe early '70s smog heads, which have a semi-open chamber design, may be the best of the lot. Check with your head porter.

In selecting the rectangular-intake big-port heads to use, there are three different versions worth considering. All are open-chambered—one is cast-iron, the other two are aluminum. All are much better heads than the old closed chamber versions, but even these open-chamber heads were never really finished. Originally, the open-chamber design was to have been completed in the late '60s with a spiral combustion chamber for a turbocharged application, but the second-design L88 open chamber setup was as far as the project got before development was halted.

All the big-port open chamber heads have 2.19-inch intake and 1.88-inch exhaust valves. All these heads (aside from differences in material) are almost identical except for exhaust port shape. The cast-iron head (part No. 6260482) has nearly square ports. The second-design aluminum L88/ZL1 head (part No. 3964072) has round exhaust ports. Recently a new aluminum head (part No. 14011076) with a fin or hump in the floor

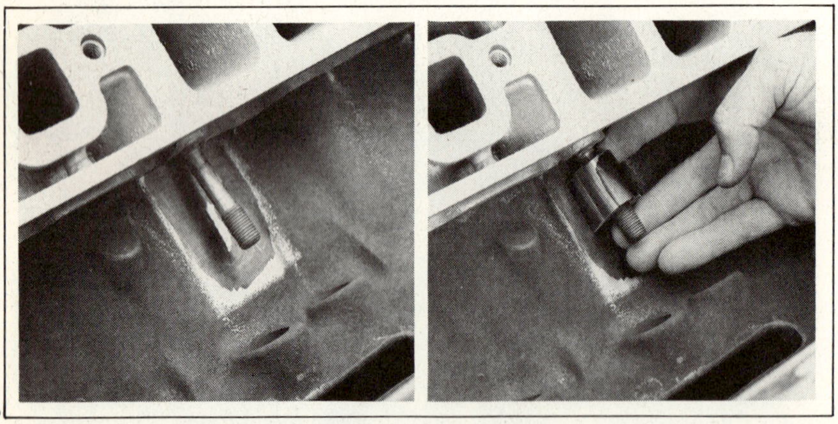

5

6

5. Mark the lifter valley of the block to determine correct placement of the special tabs needed for the aluminum heads. Machine the tabs from cast iron and have them welded in place as explained in the text.

6. Cam drive can be by either the stock Chevy pieces or by a top-quality roller chain such as this one from Cloyes. The Cloyes setup includes additional key ways in the crank sprocket to allow ± 4-degree cam timing variations. The shim goes between the cam sprocket and the block to prevent damage.

7. Valvetrain weight can be cut considerably by using lightweight valves such as the Manley 5/16-inch stem stainless steel valve (middle) or by the TRW titanium valve (right) in place of the stock one. Note the special keeper on the Manley valve; this product is designed to distribute rocker stresses over a bigger area for the small-stemmed valves. Rocker arm weight can be reduced by using needle-bearing rockers, such as the Crane version (right) instead of the stock steel setup. Lightweight retainers are also offered by Crane. The center one is hard anodized aluminum and the one on right is titanium. The retainer at left is the stock steel piece.

8. Deburr and radius the main bearing caps to eliminate stress risers. Note the chamfering and polishing of the oil passage compared to the stock cap.

9. The rod on left is an LS-6 rod that has been carefully reworked by Hank the Crank. Special boron bolts, such as the one shown, have also been installed. The rod at the right is the stock LS-6 rod. Which one would you rather have in your engine?

7

8

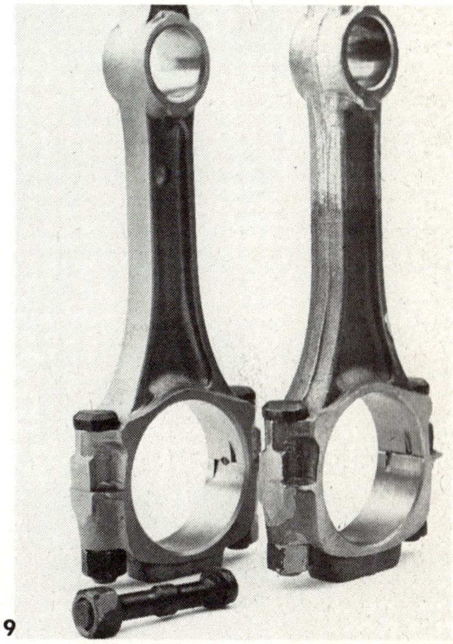

9

of the round exhaust port was released; its combustion chamber remains the same as the other two heads, although there is some very minor recontouring on the intake side. Normal stock or TRW pistons can be used. Neither of the aluminum heads have heat-riser passages.

To effectively use these big-port heads, engine speed is going to have to be fairly high to get good intake port velocity, and the exhaust ports have to receive the proper treatment. Just the way they come from the factory, the cast-iron exhaust port seems to flow a little bit better than the aluminum exhaust port, but after the porter reworks both heads they flow about the same.

The new finned-port head is superior to the other two heads, and with a good porting job, the margin of superiority increases even more. No matter which head is used—and this applies to small-port heads as well—the exhaust port is going to need some work by somebody who knows what he's doing, such as Air Flow Research or Diamond-Elkins.

So which head do you use? For the street, I'd go with a cast iron head with the small ports. If cost is not a consideration, go with the finned-port head for all-out competition. However, it's much more expensive than any other head (and the regular aluminum head is already three times the cost of the cast-iron version). On the other hand, both aluminum head designs are a bunch lighter (75 pounds) than the cast-iron heads. There's also a difference in power output on the round-port L88 head. Although this head and the cast-iron head flow about the same when ported, the aluminum heads dissipate heat from the combustion chambers much more quickly than the cast-iron versions, and that reduces power a little. However, the L88 head can be sealed to retain just as much heat as the cast-iron heads for equal power. The finned-port head, even with this heat loss, already makes more power than the cast-iron head, but if you seal them, you'll experience even greater power gains.

The cheapest way to seal the aluminum heads is to treat the water passages with sodium silicate, more commonly known as "water glass" and sold under the trade name of Irontite. It has many uses, such as sealing swimming pools. Eggs are even dipped in it to make the shells harder. It's available at most drug and hardware stores. To apply the stuff, bolt a plate to the combustion chamber side of the head to close off all the water passages. Then heat the head in an oven to operating temperature (250-300 degrees). When the head is thoroughly heated, pour about one quart of sodium silicone into the water passage opening on the intake port face of the head. Next, cap that water passage opening with a plate that has an air hose fitting welded into it. Pressurize the head to about 100 psi and rotate the head in all directions to slop the sodium silicone around to all the passages for about 5 or 10 minutes. This will let the sodium silicone fill all the pores under pressure. After you've done that, turn off the pressure, remove the plates, drain off the excess sodium silicone (about 1½ pints) and let the head slowly air-cool. This will produce a sufficient barrier to retain heat. Naturally, this should be done before the heads go to the porter. And in order for this to work properly, the water passages must be very clean. Ideally, the heads should be sonic-cleaned if you can find somebody with the equipment to do it.

There's still one more problem with the aluminum heads, and that is head gasket sealing. The big-block Chevy has six bolts surrounding two of the cylinders on each bank but only five bolts surrounding the other two. On a cast-iron head there isn't any problem, but the long span on those two cylinders where the sixth bolt is missing tends to leak when the less rigid aluminum head is bolted down. If you don't fix this problem, you've pretty much got a hopeless case. Fortunately, the aluminum head has provisions for two special studs to allow it to be tightened down. To use these studs, you have to weld special tabs into the lifter valley of the block. Since welding cast iron is pretty tricky and since the tabs have to lie at 6-8-degree angles, find a machinist and welder who really knows what he's doing. This again is one of those things that must be done before the block has any machine work done on it. These tabs, with the studs torqued to a mere 25-35 ft.-lb., are enough to cure the problem, if the conventional head bolts are replaced with head studs as well. (You might also use studs on the main bearing caps.) For the correct stud installation procedure, see my small-block chapter.

Earlier, I said that you should shoot for a 12.5-12.7:1 compression ratio for racing as long as you can obtain good gas. To get this kind of compression, the heads will have to be milled to get the combustion chamber volume down to around 114cc. Use the Victorcore composition head gasket (Chevrolet part No. 3969865), which has a compressed thickness of .040-inch. Since we're working with a zero deck height, don't try to use anything thinner.

Now we're ready to gather up the valvetrain components. Let me start out by saying that just like with the small-block, no one camshaft is going to be right for everybody, although I recommend a roller grind for all applications (for my reasons why, see the small-block story).

For a street combo with a stock rod

and 9.0:1 compression use a high torque cam. Figure on an rpm peak of 6000-6500—that's all you'll need. Keeping the rpm down will also keep you out of trouble with the stock heavy pistons. The grind required would have short duration and overlap to get fast enough action with low compression, but with fairly high lift, say .585/.550 intake/exhaust, or maybe even .600/.600. Naturally, cam requirements vary with car weight, transmission type and gearing, so maybe you'd be better off to consult good aftermarket cam grinder like Crane to get a cam and kit tailored to your exact needs.

A gear drive for the cam is best. I think a 2-gear setup is the way to go; they're available from the aftermarket and even from Chevy, but the required reverse rotation roller cam billets are hard to come by. The next best solution is a three-gear setup, but stay away from free-floating idlers. While I'm not too happy with chains, a good double-roller setup from Cloyes or TRW is probably okay for mild street use. Even the stock chain (part No. 3860036) and nylon cam sprocket (part No. 3856356), along with the steel crank sprocket (part No. 3860035) isn't too bad.

The remainder of the valvetrain is the biggest problem with the big-block engine. To begin with, both valves are much too heavy. If a guy doesn't turn the engine over 6400 rpm, he's got no problem, but after that he needs lighter valves. There are several different versions on the market. My guess would be that a titanium valve would be the answer for weight and durability, but they're pretty expensive. A sodium-filled exhaust valve would add life when the engine is constantly run hot, so it might be a good idea, especially in endurance racing or on the street.

Valve springs are another problem. Valve float tears more of these engines apart than anything else. A guy needs real good valve springs in this motor, such as Crane's or Diamond-Elkins's springs. Even for the street cam I recommended earlier, you need a minimum of 135 pounds on the seat—maybe more—with 375 pounds open.

The retainer is another place to save weight. Instead of using the stock steel retainers, I like the hard-anodized aluminum versions. If you're really bucks-up, you can get a little extra durability by using titanium retainers. The choice is up to you. Either way, the green Chevy keepers (part No. 3947880) will do fine.

The pushrods are no exception to the problem either. Those big 7/16-inch-diameter pushrods are strong, but they're heavy, especially when they're full of oil—which they are when the engine is running. I'd say a strong lightweight pushrod is the answer, and a guy might have to go somewhere other than Chevrolet to get it. You don't really need that big pushrod, especially if you get the valve, rocker arm and retainer weight down, just as you won't need as heavy a valve spring.

Getting the weight of the rockers down can be done by using lightweight needle-bearing roller rocker arms, or simply grinding about 1/16-to-⅛-inch from the top rim of the stock rockers (part No. 3959182). Then weld a small cap to each rocker as I described in the small-block story. Now that you've got all the pieces, slip things together for a trial fit and make sure you've got at least .080-inch piston-to-valve clearance.

For induction, when using the rectangular port heads on the street, I still believe the stock Chevy dual-plane aluminum high-rise intake (part No. 3940783) is one of the best manifolds you can buy, especially when the manifold ports are matched to the head intake ports. Use it with a Holley 850 double-pumper carb, preferably the new 0-8162 model with idle adjustment screws on both primary and secondary sides. The factory

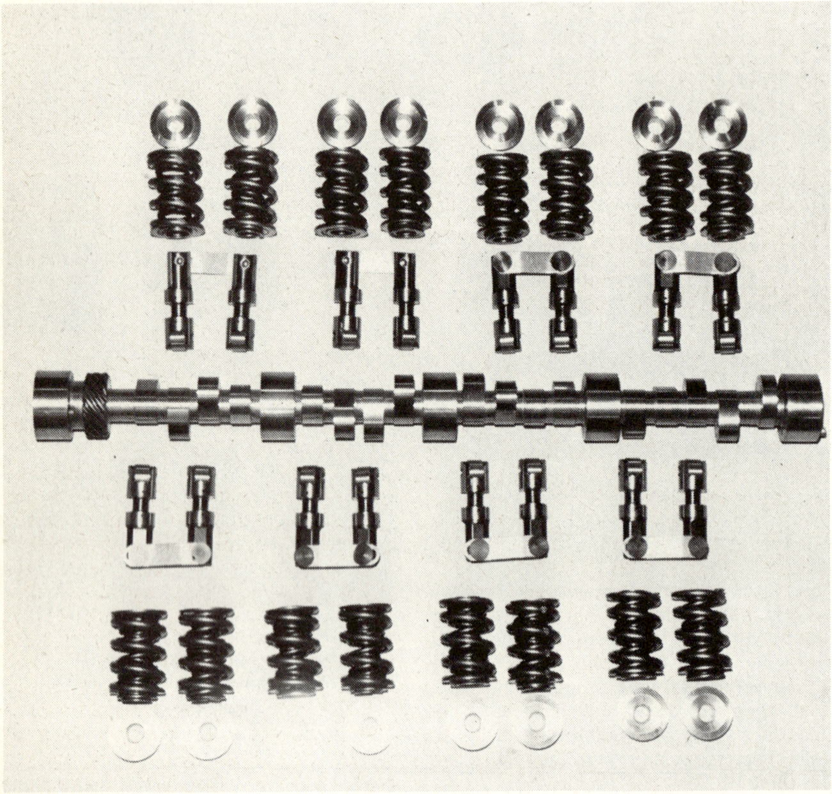

10

11

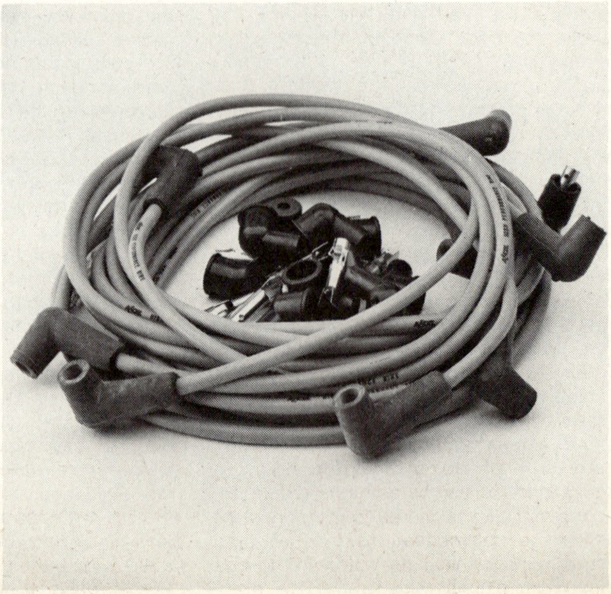

12

13

14

15

10. Regardless of the intended use of the engine (any Chevy engine, for that matter), Smokey recommends a roller cam as this setup by Crane.

11. The best single 4-barrel carburetor to use is the Holley 850 double pumper with mixture adjusting screws on both primary and secondary sides. You can convert an existing carb (as this one) or purchase one new from Holley.

12. A good silicone-jacketed spark plug wire is essential to prevent heat damage from the headers. Shown here is the Accel high-performance wire.

13. The Accusump is the poor man's dry sump. Containing about 1½ quarts of oil under pressure, it immediately releases the fluid as soon as it senses a momentary pressure loss.

14. This experimental swirl-chamber head is the best ever developed for the big-block. Designed by Smokey for Chevrolet in the late '60s to be used in turbocharged Can Am racing, the head would have required a special piston to work with its unique combustion chamber. Unfortunately, the project was dropped and the head was never released for production.

15. The Chevy dual-plane high-rise intake manifold may still be the best one available for hi-po street engines.

never made a manifold for the regular Holley to be used with small-port heads, so you'll definitely have to go aftermarket if you're using them. The Edelbrock C396 dual-plane is the best (I believe in dual-planes for the street), but it may be hard to come by. In that case, use an Edelbrock Torker or Holley Street Dominator. If you're going racing, consult the manifold maker, since manifold technology advances on a month-to-month basis, and any recommendations I could give you might very well be outdated by the time you read this.

Before installing the manifold, you should equip the engine with some sort of an oil shield, either the Chevy unit (part No. 3931093) or one of the aftermarket versions, to keep the hot oil off the underside of the manifold. Of course, if you use a rev-kit with your roller cam, there will be no room for this shield.

For ignition, I'm going to say the same thing I said about the small-block: Don't mess around with points. Get a good breakerless system, such as the Accel BEI III or Chevy magnetic impulse L88 setup (if you can find one). A magneto would be better yet. Keep the total spark advance under 40 degrees to prevent detonation. Run a high-quality 8mm silicone-jacketed sparkplug wire with Moroso insulator sleeves. Use spark plugs equivalent to the Autolite Number Nine heat range.

A free-flowing exhaust system is extremely important on the big-block, so get a set of good steel-tubing headers. Keep the primary tube diameter to a maximum of two inches for street applications. If you can find them, 1⅞-inch tubes seem to be ideal on the street. Save the big tubes for racing. Depending on application, primary tube length will vary from 28 inches to 36 inches.

When it comes to engine cooling, if you've cleaned all the water passages properly, the stock big-block water pump is more than up to the job. The stock oil system is also pretty good, but it's important to have plenty of oil capacity—at least six quarts—and proper baffling in a good pan to make sure the oil pickup never gets uncovered. You might help out the pan with an Accusump. It can really keep you out of trouble if, for any reason, the oil pickup did get uncovered momentarily. It can save the day, and save the cost of a super-trick pan. It's about the cheapest lube protection you can get.

But a dry sump is even better—it's better than baffling a wet sump pan any day of the week. A trick oil pan, tray, etc., might cost $500. A dry sump is $800. If you lose oil pressure for an instant with a wet sump, you've just bought the equivalent of four dry sumps; so which one is cheaper?

As far as windage trays go, screens work real well if there's a pressure drop below the screen, such as with a scavenge pickup on a dry sump. But it works on a wet sump, too; you just need to run it as close as possible to the reciprocating assembly.

On a wet sump, use the ZL1 oil pump with a ¾-inch pickup. Be sure the oil pickup tube reaches to within ⅜-inch of the bottom of the sump and tack-weld the pickup tube to the pump body.

Get rid of the bypass on the oil filter pad. Go to really high filter capacity: two double remote Fram filter mounts with four triple 10 micron filters. Don't mess with an adaptor. Install an A/N-12 inlet fitting, and, if a dry sump, a -12 outlet as well. It takes a lot of power to push oil through one filter, maybe 25 horsepower. Running through four filters, it may take only 12 horsepower. A pump pushes 11½ gallons/minute—that's a 55 gallon barrel in under five minutes! The proof of the inadequacy of the stock filter setup and bypass valve is that there's always grit in the bottom of the oil pan when you pull it.

For oil, use a good multi-viscosity oil on the street such as Valvoline Racing 20W-50. At this time, I'm not in favor of synthetics. Most additives are worthless, as well. Any additives must stay in suspension. The petrochemistry of current additive packages is such that it goes out the exhaust pipe.

That completes all the pieces. Get everything surgically clean and assemble the pieces with a coat of light grease. Take it easy for the first few hours of operation and then retorque *everything.* When this is done, you're all set. Remember to use nothing but the best gas, change oil and filter frequently, and you should have over 600 horsepower in racing trim whenever you need it.

—Smokey Yunick

Street Power

The simple facts about three straightforward bolt-on ways to boost street horsepower significantly.

Anyone who has an interest in cars knows that truly incredible amounts of power can be extracted from just about any engine by modifying it with exotic high-performance racing components, precision machining and careful assembly to withstand high-rpm operation. Racers have been doing it for years.

Now it's time for a different point of view: Such modifications require a great deal of effort, talent and lots of cubic money. But there are far cheaper ways to achieve similar power gains without resorting to irreversible engine modifications, without the talents of an experienced engine builder, and without any major disassembly of the engine. What's equally important is that such power gains can be achieved without hurting the basic streetability of the engine or killing its gas mileage during normal everyday driving. And the icing on the cake is that such methods of adding horsepower are far less expensive than building a fully modified, ported and polished, blueprinted racing engine, and the equipment can often be swapped from one car to another.

Sound too good to be true? Read on, because we're about to acquaint you with the basics of three simple bolt-on routes to street horsepower: nitrous oxide injection, supercharging and turbocharging. We'll explain how each system works, along with the advantages and disadvantages of each. One of these setups might be just what you've been dreaming of for your street machine.

NITROUS OXIDE

How much power an engine can make is dependent upon how much fuel it can burn efficiently in each cylinder at wide-open throttle (WOT). How much fuel can be burned is simply dependent upon how much oxygen is available in each cylinder to support combustion. All the tricks of bigger carburetors, special manifolding, radical cams, etc., seek to improve air flow into each cylinder, thereby increasing the amount of available oxygen, but nitrous oxide injection accomplishes the same thing by another method. Instead of trying to put more *air* into each cylinder at WOT, nitrous oxide (N_2O) is sprayed into the induction system, effectively displacing some of the incoming air. The payoff is that N_2O is extremely rich in oxygen, so the mixture of N_2O and air can support the combustion of far more fuel than just air alone. To provide the extra fuel for combustion, extra gasoline is simultaneously sprayed into the induction system with the N_2O.

The amount of power added by nitrous oxide injection will vary with the amount of N_2O and gasoline injected and the state of modification of the engine, but 50-percent increases in power output on street engines is not uncommon. In other words, if your stocker puts out 250 hp, adding N_2O injection can up it to as much as 375 hp!

The big advantages of N_2O injection are that it is fairly inexpensive (about $400); it can be installed in an afternoon without engine disassembly; only the carburetor need be removed to install a spacer plate, containing the N_2O and gasoline nozzles, between the carb and manifold; it can easily be swapped from one car to another; there usually aren't any hood clearance problems; it works only at WOT without upsetting normal engine operation; and it can be concealed, if so desired. Another feature of this modification is that it can be turned off completely to conserve N_2O (or while your girlfriend or mother borrows the car). The depressurization of the N_2O as it is dispersed into the induction system also provides a cooling effect to the intake mixture, thus reducing the likelihood of detonation.

The only disadvantages are: The N_2O tank must be refilled periodically. In some areas it is difficult to buy N_2O. The manufacturers of N_2O systems can generally supply the names of cooperative N_2O dealers in your area, but check this out before buying an N_2O system. Additionally, it is important to shut off the valve atop the N_2O tank whenever the car is parked, to prevent the possibility of N_2O seepage past the solenoid control valve. Such seepage could load the manifold with a dangerously lean mixture of N_2O and gasoline that could cause damage when the engine is started.

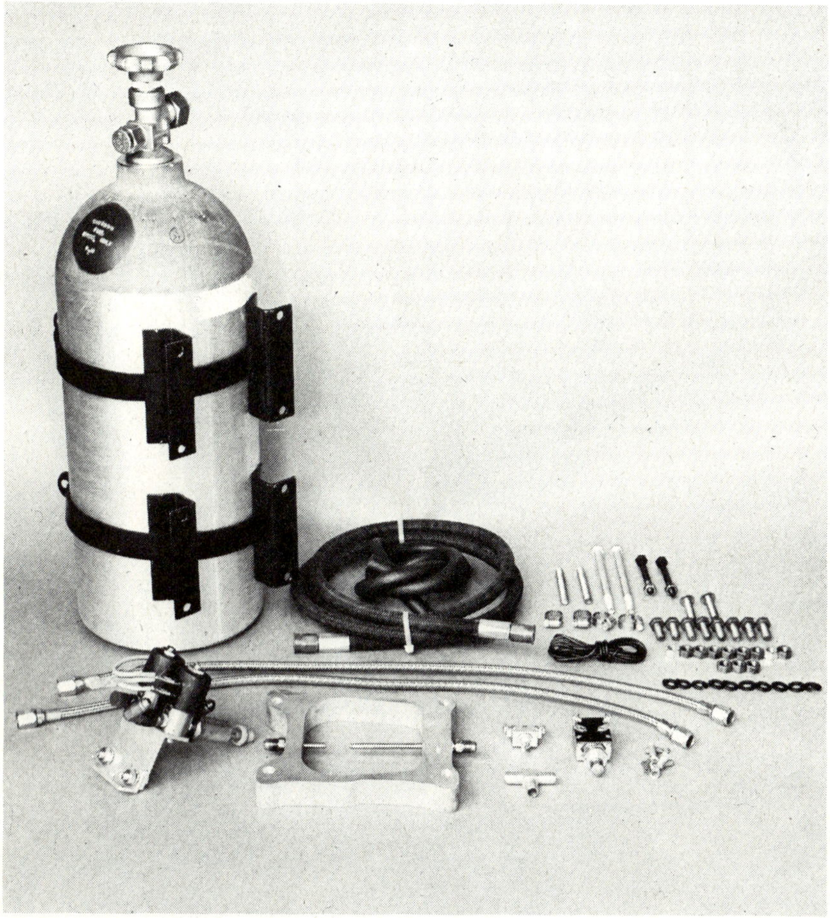

1

2

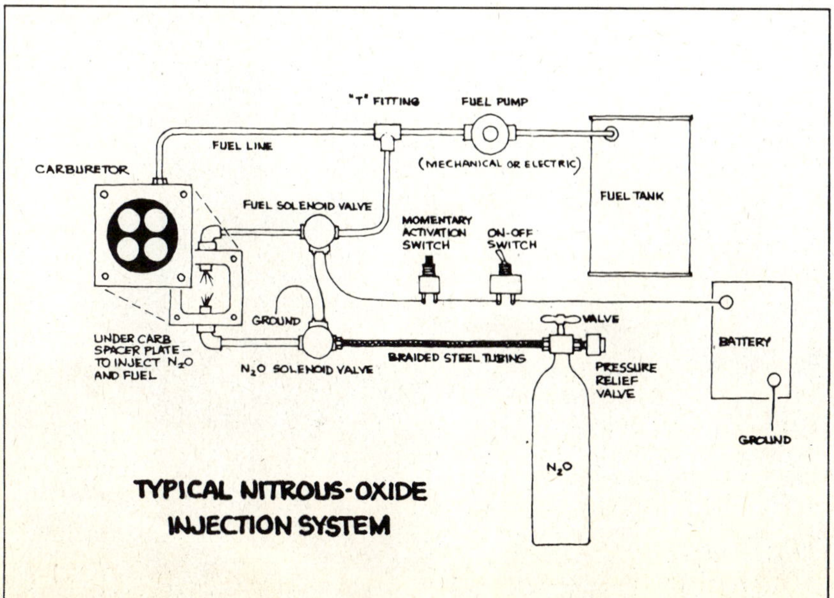

3

1. This is a complete nitrous oxide injection kit, as sold by Mary Miller Manufacturing, 7745 S. Greenleaf Ave., Whittier, CA 90602.

2. For those who wish to conceal the N_2O system, the underside of the intake manifold can be plumbed for direct port injection of both the N_2O and gasoline; instead of using the conventional under-the-carb spacer plate. Even the solenoids are concealed.

3. As this schematic of typical nitrous oxide injection shows, the system is very simple. The N_2O tank can be mounted in the trunk, the solenoids on the inner fenderwell and the activating switch on the carb linkage. The gasoline "T" fitting can be installed anywhere between the carb and the stock fuel pump.

SUPERCHARGING

The intent of supercharging is simply to force a greater quantity of fuel and air into each cylinder, effectively "fooling" the engine into acting as if it had a much larger displacement. A supercharger is really nothing more than a high-volume air compressor, installed between the carburetor and the intake manifold, that is driven directly off the crankshaft of the engine. Although there are axial and centrifugal designs, the most popular superchargers in use on automotive engines are of the "Roots" design, which incorporates two intermeshing rotors that positively move a given quantity of air with each revolution. In virtually all cases, a system of pulleys, idlers and belts is used to drive the supercharger directly off the snout of the engine crankshaft. Such an arrangement means that the supercharger is a *full-time* system, operating whenever the engine is running. Such supercharger kits, complete with drive assemblies, are available from a variety of sources, in various sizes, with corresponding intake manifolds.

The main advantages of a supercharger are these. It provides instant throttle response and good low-rpm torque. It is relatively easy to install, with everything located atop or on the front of the engine, although the intake manifold will have to be changed too. A supercharger is visually impressive, and depending upon the speed at which it is driven in relation to crankshaft speed, it can show power increases of 50 to 100 percent. Using again the example of a 250-hp stock engine, supercharging could increase output to between 375 and 500 hp. In some cases, a supercharger will even improve gas mileage during normal everyday street driving, thanks to the additional mixing action of the rotors on the air/fuel mixture.

The disadvantages of supercharging are: It requires an outlay of about $1000 for a complete setup; the hood on most vehicles will have to be modified to provide clearance; and the engine must have a sufficiently low static compression ratio to tolerate the boost pressure provided by the supercharger without encountering detonation. This last point is quite important, mandating a compression ratio of 9.0:1 or less and the use of premium grade fuel. Happily, most late-model engines already feature acceptably low compression ratios. One last disadvantage is that a supercharger draws all the power required to drive it directly from the engine's crankshaft. Consequently, it consumes much of the power it generates, but the net increases of 50 to 100 percent are still impressive. A complete supercharger setup is also quite heavy, adding about 75 pounds to the front end of a vehicle.

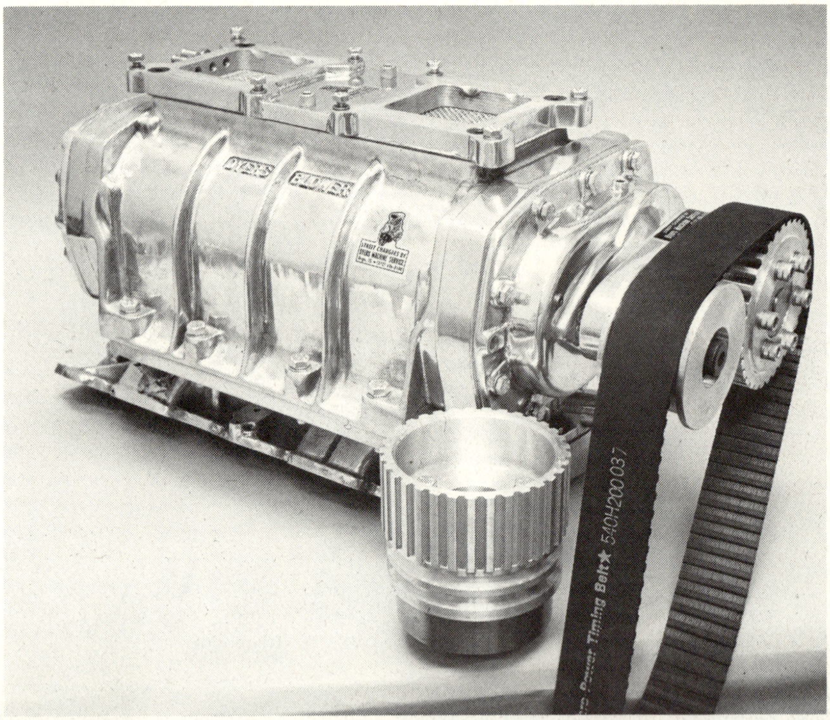

1

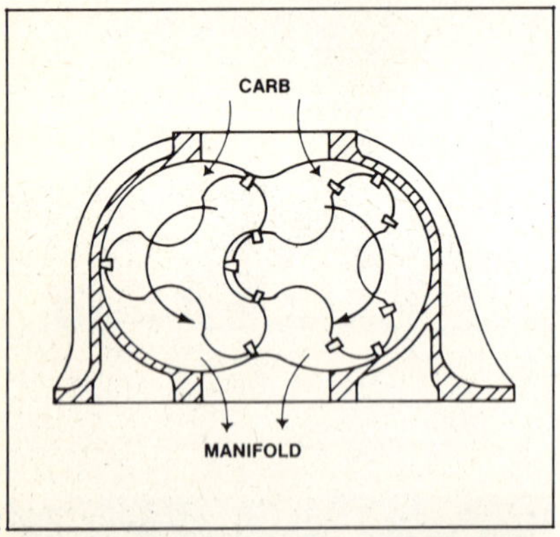

2

3

1. Polished performance. This supercharger kit, complete with drive pulleys, belt, manifold and carb adapter plate, represents a typical kit. This particular kit is from Dyers Machine Service, 7665 W. 63rd St., Argo, IL 60501.

2. Inside the supercharger case, the two tri-lobed rotors intermesh as they revolve, positively pumping air along the outer walls of the case and dumping it into the manifold. The carburetor(s) is mounted atop the case.

3. A typical installation of a supercharger on a small-block Chevy-powered street rod, this setup exemplifies both performance and cosmetic appeal.

TURBOCHARGING

Turbocharging is similar to supercharging inasmuch as it uses a high-volume air compressor to force additional air and fuel into the cylinders. Unlike a supercharger, which is mechanically driven off the engine crankshaft, a turbocharger uses exhaust gas heat (which would normally be lost out the tailpipe) to drive a turbine. The turbine, in turn, drives the compressor. Consequently, a turbocharger draws no power off the crankshaft, and so turbocharging actually makes an engine significantly more efficient thermally.

Turbocharging is just beginning to gain popularity, with the recent availability of kits to simplify installation, which requires modified exhaust and intake plumbing. Because of the complexity of such kits, the cost is quite high, beginning at about $700 and going up from there. However, turbocharging offers some very real advantages and the greatest potential power gain.

First of all, the amount of boost pressure generated by the compressor is relative to the heat energy of the exhaust. At idle or under light loads, the turbocharger is simply along for the ride, without altering normal engine operation in any way except possibly to improve fuel atomization as the mixture passes through the coasting compressor blades. As the engine load increases, exhaust heat increases and so does compressor output, raising boost pressure. In other words, a turbocharger works only when you need it—automatically. Properly matched to the engine, a turbocharger is more efficient than a supercharger, and because it draws no power from the crankshaft, the potential power increases are limited only by the turbo match and the detonation properties of the fuel. For the street, power gains of 50 to 100 percent are common, but highly refined systems with intercooling and water injection can more than triple horsepower output.

There are disadvantages to turbocharging too. As with supercharging, the static compression ratio of the engine must be held to a maximum of 9.0:1, with something in the range of 7.0:1 to 8.0:1 being even more desirable. Premium fuel is a must. To prevent hesitation in throttle response, the turbocharger must be carefully matched to the application. Installation requires considerable plumbing and ingenuity to keep everything below the stock hood on most cars. And the initial expense for a system and its installation is more than either nitrous oxide injection or supercharging; but it's still less than building a fully modified, normally aspirated engine, which couldn't begin to match the power output of a good turbocharged engine.

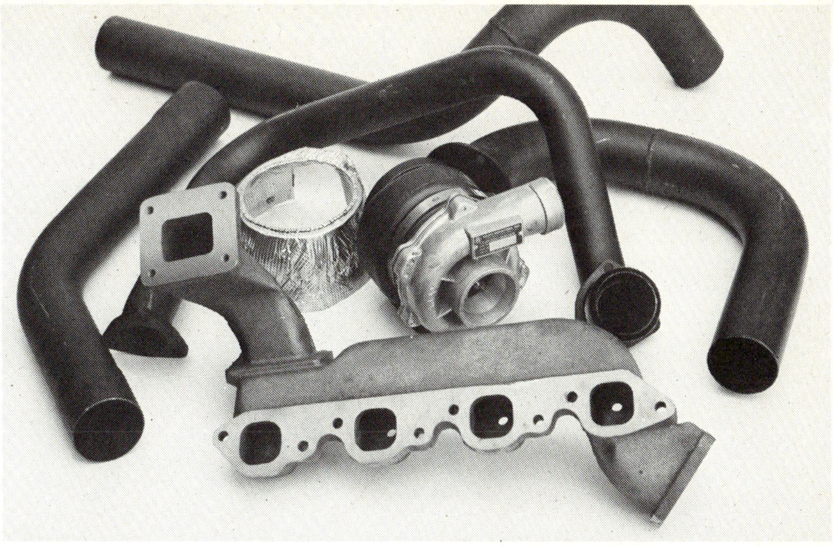

1

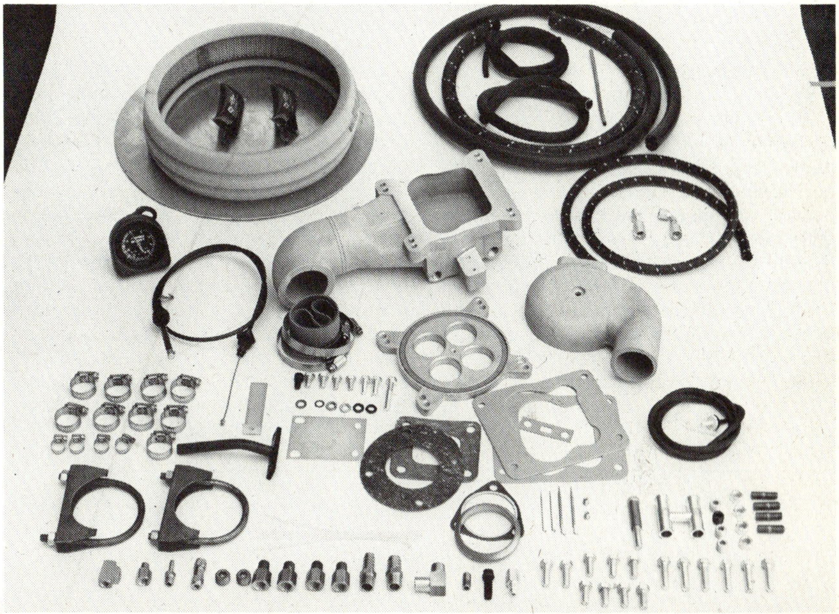

2

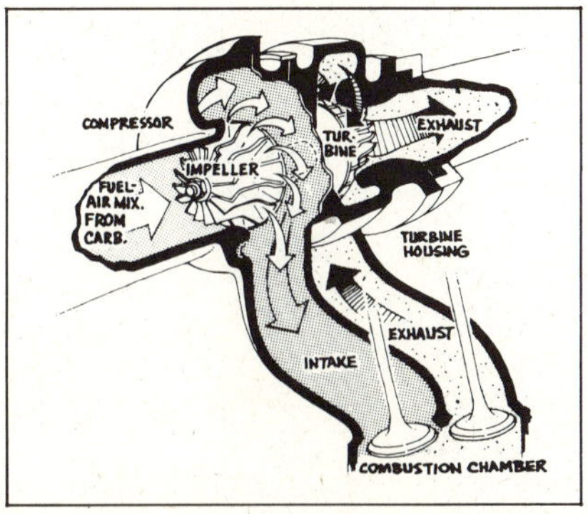

3

4

1. Connecting a turbocharger requires a certain amount of plumbing. Shown here is the exhaust plumbing for a 454-cubic-inch Chevy pickup. This is part of a kit from Spearco Performance Products, Inc., 10936 S. La Cienega Blvd., Inglewood, CA 90304.

2. Shown here is the induction half of the Spearco turbocharger kit for a big-block Chevrolet pickup. Such complete kits greatly simplify the installation of a turbocharger, and the turbo has already been matched to the installation.

3. The principle of turbocharging is extremely simple—exhaust heat spins the turbine, which is connected by a common shaft to the compressor, and the compressor produces boost pressure for the intake mixture. The hotter the exhaust gets, the more boost the compressor makes; and the more boost the compressor makes, the hotter the exhaust gets. Both sides feed off each other for incredible power gains.

4. Although slightly different from most turbocharger installations, this turbocharged 6-cylinder Mustang II was prepared by A.K. Miller, 9225 E. Slauson Blvd., Pico Rivera, CA 90660. This particular setup also features the Australian six-cylinder cross-flow head and a propane fuel conversion. Visible in the foreground to the left of the turbocharger unit is a waste gate, which diverts exhaust gases past the turbine when a predetermined boost level has been reached.

Turbocharging: Performance Key to the '80s

Building your own setup is becoming increasingly easy and affordable with the latest support equipment.

Turbocharging is probably one of the simplest but most misunderstood subjects in the automotive performance world. It is misunderstood partly because it looks complicated. Turbocharging also lacks the cosmetic appeal of an equally expensive but inherently less efficient (in terms of heat energy utilization) polished Roots supercharger. There are dozens of exaggerated stories about engine damage and lack of throttle response, told by self-appointed experts who have never even ridden in a high-performance turbocharged car, let alone worked on or owned one.

Critics also tend to lump all turbocharged cars together, citing any turbocharged car as indicative of all turbocharged cars, which is about as valid as comparing a DC-10 airliner to an F-15 fighter plane. After all, they're both powered by jet engines, right? Although offering considerable performance improvements over similar nonturbocharged vehicles, Detroit's turbo engines are extremely conservative packages combining emissions, mileage, smoothness and durability. They bear little resemblance to a twin-turbo small-block in a Camaro, and they are not representative of what can be achieved with a turbocharged installation that is aimed at performance.

Turbocharging is actually quite simple. The heat and flow of exhaust gases from an engine are channeled through a turbine which is connected by a common shaft to a compressor that forces air and fuel into the engine for increased cylinder filling and power output. The correct size and trim turbocharger(s) must be selected to match the engine size and intended application, the exhaust must be routed to the turbocharger, and care must be taken to prevent detonation due to the increased cylinder pressure. That's all there is to it, and an ever-

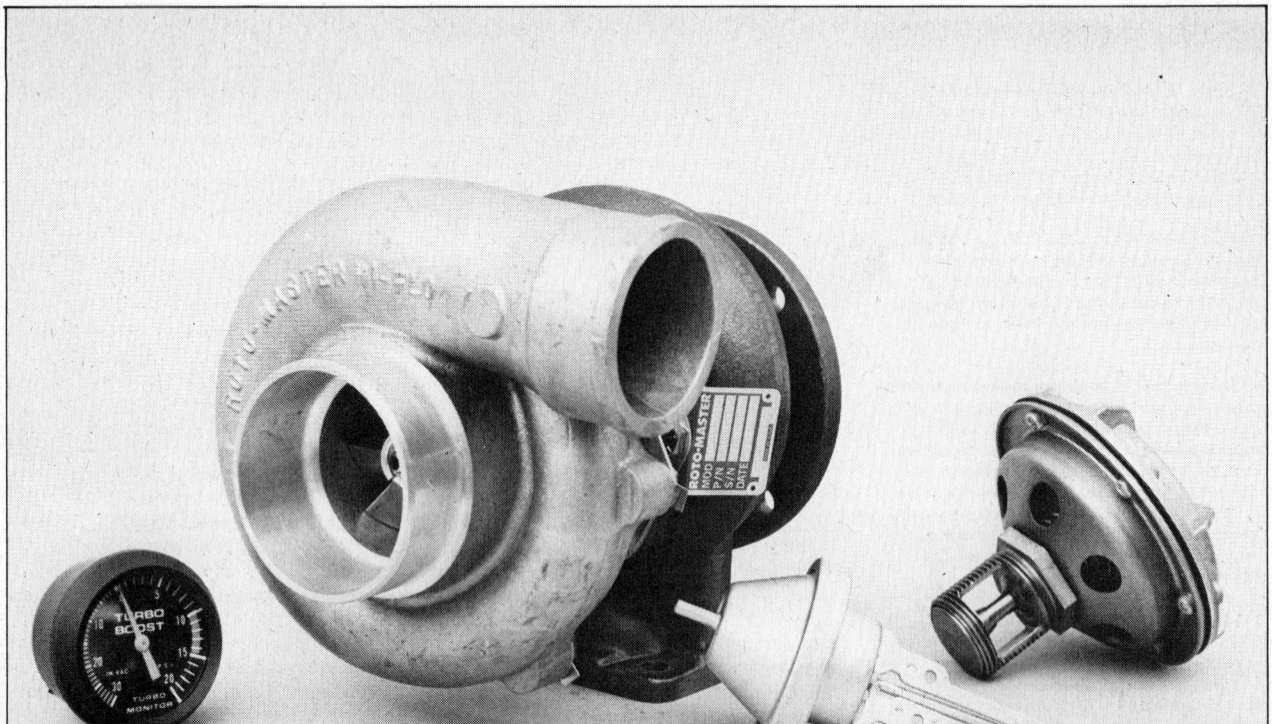

growing number of firms offer fully engineered kits that have these problems solved for you. Still other companies will handle matching and custom installation, again removing all the guesswork.

In a previous chapter we told you that turbocharging was the performance wave of the future. Enthusiasts are learning, and the acceptance of turbocharging is now like a snowball that has started to roll down a snow-covered slope—it's continuing to gain both size and momentum. Unfortunately, initial cost is one of the factors preventing even more rapid acceptance of turbocharging. Complete prefabricated kits generally begin at about $1000. Custom installation of turbo systems can run to two or three times that amount; and special state-of-the-art racing systems—complete with dual turbos, wastegates, variable boost controls and intercoolers—can exceed $5000. Consequently, many enlightened enthusiasts are beginning to buy individual components to fabricate their own systems, saving hundreds (if not thousands) of dollars. In fact, new companies that specialize in offering a selection of basic components, such as Turbo City (958 N. Enterprise, Orange, CA 92667) are beginning to appear to meet the demand.

To acquaint you with some of the equipment now available to simplify turbo installations, we've collected a sampling of components, along with descriptions of their functions, to help you determine if that's what you need. There are a lot of reasons why today's performance enthusiast should turbocharge; we hope the following information will strip away a few of the reasons why you haven't.

1

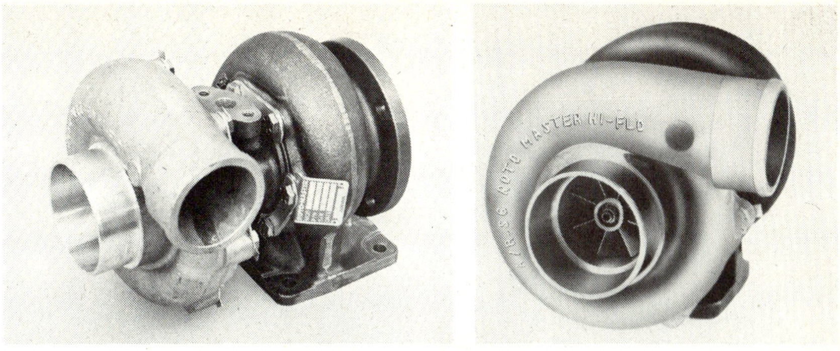

2

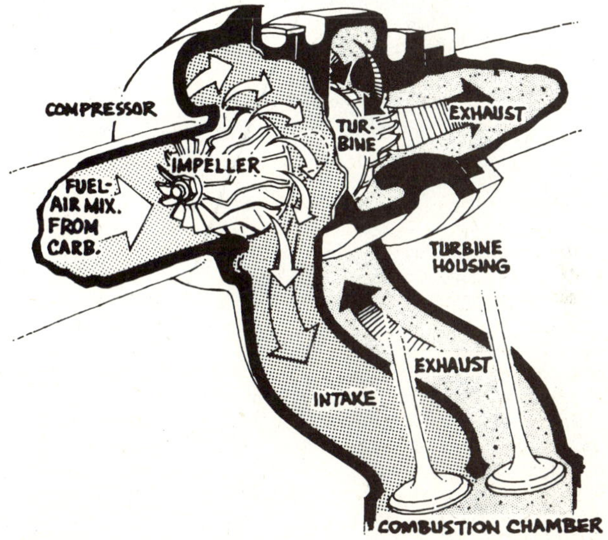

1. Increasing availability of turbochargers suited to specific automotive applications provides the enthusiast with a wider range from which to choose a good turbo match to his engine and application. These units are from the TurboSonics Division of Roto-Master, Inc. (7101 Fair Ave., North Hollywood, CA 91605). The left unit is a conventional design, whereas the right unit is their new Hi-Flo turbo designed for single turbo applications on large-displacement engines ranging from 375 to 500 cubic inches.

2. The principle of turbocharging is extremely simple—exhaust heat spins the turbine, which is connected by a common shaft to the compressor, and the compressor produces boost pressure for the intake mixture. The hotter the exhaust gets, the more boost the compressor makes; and the more boost the compressor makes, the hotter the exhaust gets. Both sides feed off each other for incredible power gains.

Tom Wyatt III is one of the driving forces behind Turbo Tom's (4090 Peachtree Rd. N.E., Atlanta, GA 30319). After doing countless turbo installations, both from scratch and using various aftermarket kits, Tom has compiled a brief list of common turbocharging mistakes that should help first-time installers avoid some of the pitfalls. The list is in no particular order, since any one or combination of these mistakes can detract from an otherwise good installation.

1—**Improper or inadequate carburetion.** The carb must be capable of supplying the engine under all boost conditions without flat spots or hesitation.

2—**Inadequate fuel supply.** Fuel line size and pump capacity must be capable of keeping up with the air flow demands. In most cases, ¼-inch or 5/16-inch fuel lines just won't cut it.

3—**Restrictive exhaust system.** Special turbo mufflers are a must. Most glasspacks simply won't handle the flow.

4—**No boost control.** Without boost control you have no reliability. It's as simple as that.

5—**Poor system design.** If engine and turbo system access aren't given prime consideration, service becomes an expensive, time-consuming headache.

6—**Inadequate materials.** Thinwall muffler tubing just won't handle high boost pressure downstream from the compressor with reliability. Pipe or castings should be used.

7—**No carburetor support.** Large 4-barrel carburetors are heavy. Support is needed to prevent engine vibration from eventually breaking the rest of the system if carb support isn't provided.

8—**Engine unsuitable for turbocharging.** An engine fitted with high-compression pistons, a long-duration or long-overlap cam, or an engine with poor oil control to the combustion chambers needs mechanical modification or repair before turbocharging can be expected to work properly.

9—**Improper ignition advance curve.** Excessive ignition advance will get a turbo engine in trouble every time. The ignition advance must retard under boost to compensate for the increased flame speed in the tightly crammed combustion chambers. The higher the boost, the more important this becomes.

10—**Inadequate ignition strength.** Firing the air/fuel mixture under boost becomes increasingly difficult as boost goes up. Most stock ignition systems aren't up to the task. A high-output ignition is absolutely essential.

Some of these mistakes are particularly significant on high-output turbo systems aimed at peak performance, which Turbo Tom's specializes in. They also do custom installations and offer an array of support components, such as special ignition systems, waste gates, boost controllers, water injection systems, and cast manifolds and plenums.

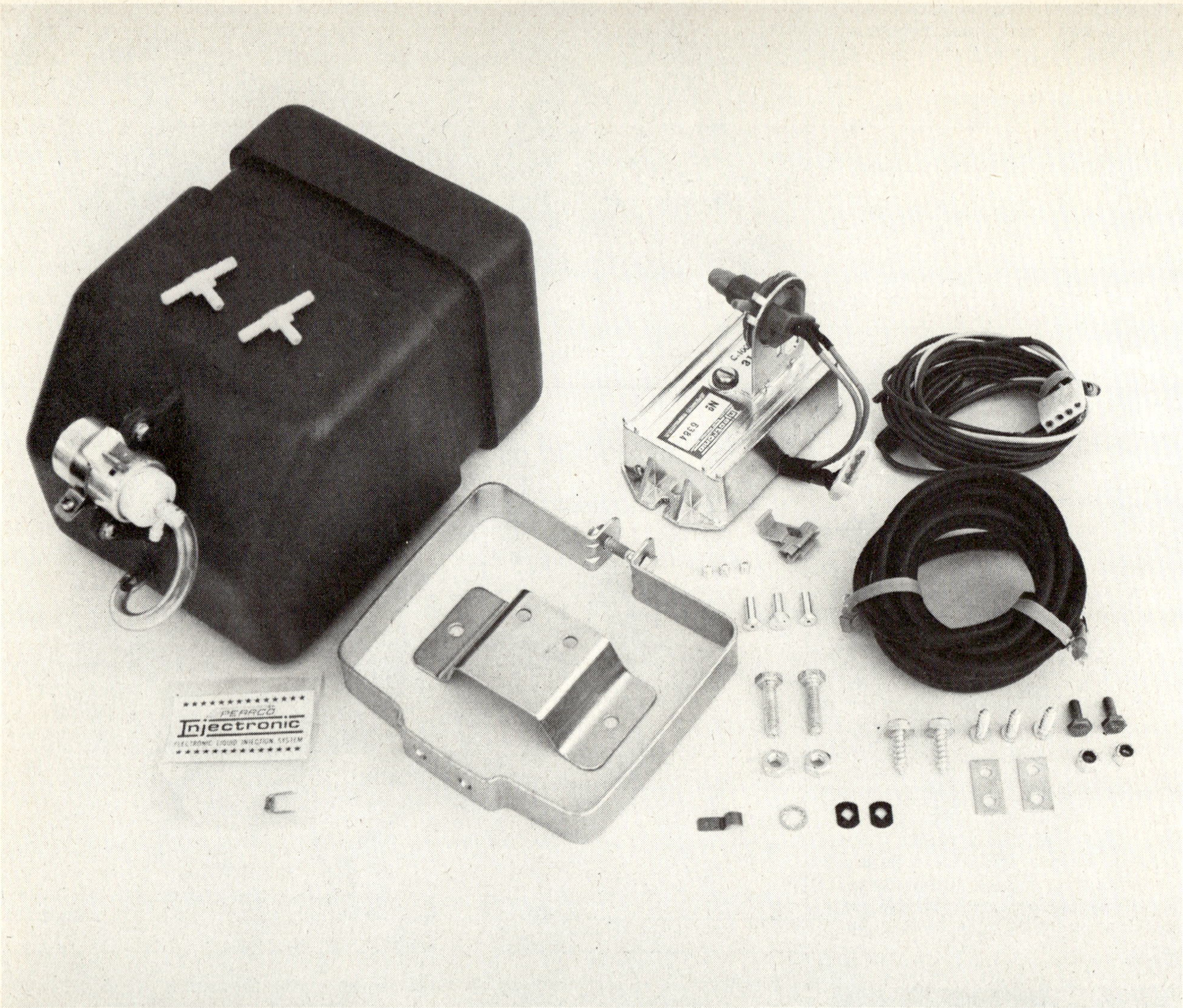

3

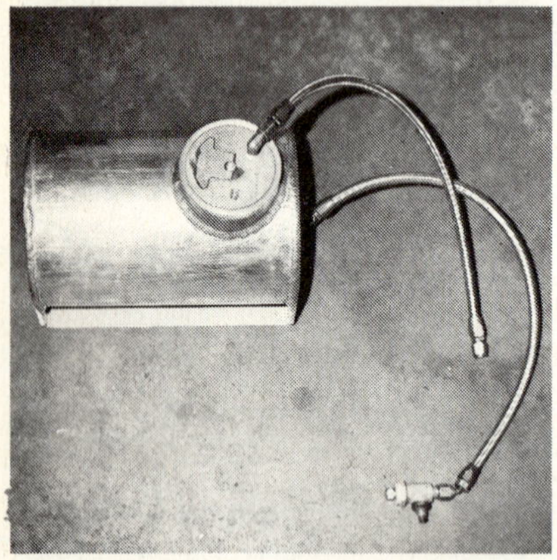

4

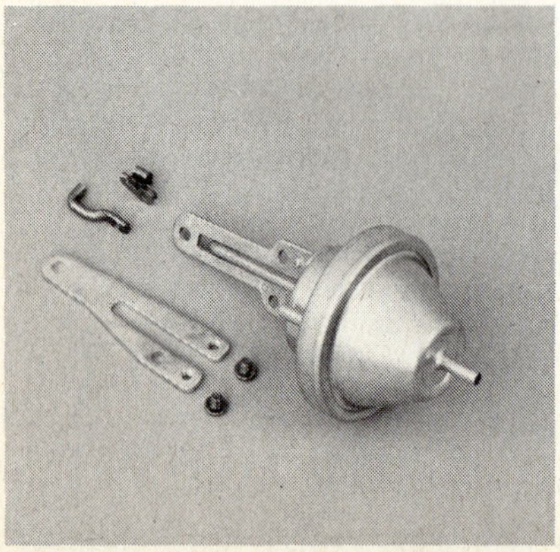

5

6

GAUGING THE PRESSURE

Before an accurate analysis of a turbo system's performance (or lack of same) can be made, it is essential to know the boost levels being achieved. In some cases, it is advisable to know the exhaust system temperatures and pressures, too. In other words, you must have good instrumentation. Most manufacturers of turbo kits or turbo equipment offer suitable boost gauges; however, John Anderson at AmTest Turbocharger Systems (2750 Jefferson, Napa, CA 94558) offers one of the nicest assortments of single and dual vacuum/boost gauges and pyrometers that we've seen, as shown here. One gauge even incorporates dual needles to indicate both boost and exhaust temperature on the same instrument face.

AmTest also offers custom installation services for both street and racing turbocharged applications.

DEFINITIONS:

Gauge Pressure—The pressure you read on a normal pressure gauge. This is the pressure above atmospheric pressure.

Absolute Pressure—The pressure you read on a normal pressure gauge plus atmospheric pressure (14.7 psi at sea level). This is the pressure above a perfect vacuum, such as above our atmosphere. Some gauges are constructed to read this directly.

One pound per square inch (psi) pressure equals 2.04 inches of mercury (the symbol is Hg); one inch of mercury equals .490 psi.

3. This electric water injection system is sold by Spearco Performance Products, Inc. (10936 S. La Cienega Blvd., Inglewood, CA 90304). It features electronic controls to vary water delivery with engine rpm.

4. Turbo Tom's (4090 Peachtree Rd. N.E., Atlanta, GA 30319) manufactures and sells this boost pressure water injection system which utilizes a 7-inch-diameter aluminum tank. Tanks are available in either vertical or horizontal configurations to fit varying space limitations.

5. This vacuum advance (pull)/boost retard (push) diaphragm can is sold by the TurboSonics Division of Roto-Master, Inc. (7101 Fair Ave., North Hollywood, CA 91605). Different versions are offered to fit most popular distributors, including GM's HEI distributor.

6. Precise measuring equipment was used by the engineers at the Rajay labs to measure the effectiveness of the Thermal Structures turbine heat shields.

DETONATION CONTROLS

Controlling detonation is essential to the durability, reliability and power output of a turbocharged engine. Since detonation is uncontrolled combustion created by excessive cylinder pressure or temperature that creates spontaneous combustion of the mixture, or simultaneous ignition of the mixture at more than one point in the combustion chamber, the primary means of controlling detonation are to reduce cylinder temperature prior to ignition, and also to control normal ignition so that it doesn't occur too early under boost conditions.

Two methods of reducing inlet air/fuel temperature are normally employed on turbocharger systems: intercooling and water injection. Intercooling, which amounts to passing the inlet mixture through a heat exchanger after it leaves the compressor to reduce the mixture temperature, is rarely used in automotive applications, due to the complexity of plumbing and equipment needed. Water injection, on the other hand, is commonly used, since it is both effective and cheap. Two basic types of water injection systems are commonly used. One system uses an electric pump to force the water (or a 50/50 mix of water and alcohol) into the induction system. The other method is to use boost pressure to force the water into the induction system. Both methods work equally well, although the boost pressure system avoids much of the complexity of regulating the flow to match the boost level. With such a system, the higher the boost, the more water it forces into the induction system. Shown here are examples of both systems. Many turbo kit manufacturers offer some sort of water injection system compatible with their kits.

Ignition control, or ignition retard, as it should properly be called, can also be accomplished electrically or through the use of boost pressure. The simplest method is to substitute a vacuum advance/boost retard can on the distributor. Such push/pull cans are sold by several manufacturers. Another method is to use a pressure-activated switch to trigger an electronic control box to retard the spark. Obviously, such a system is more complex than the push/pull diaphragm can.

7

9

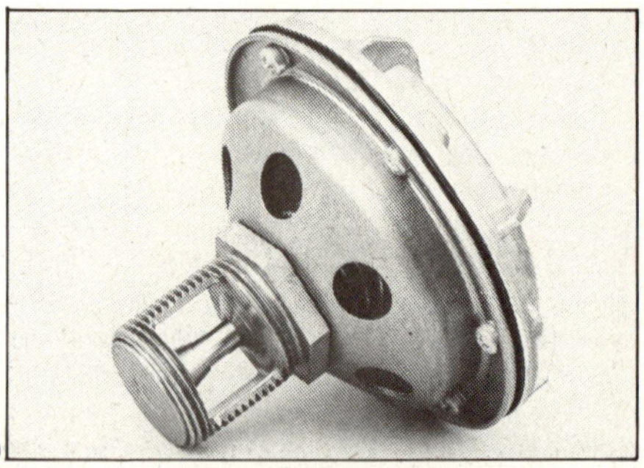

10

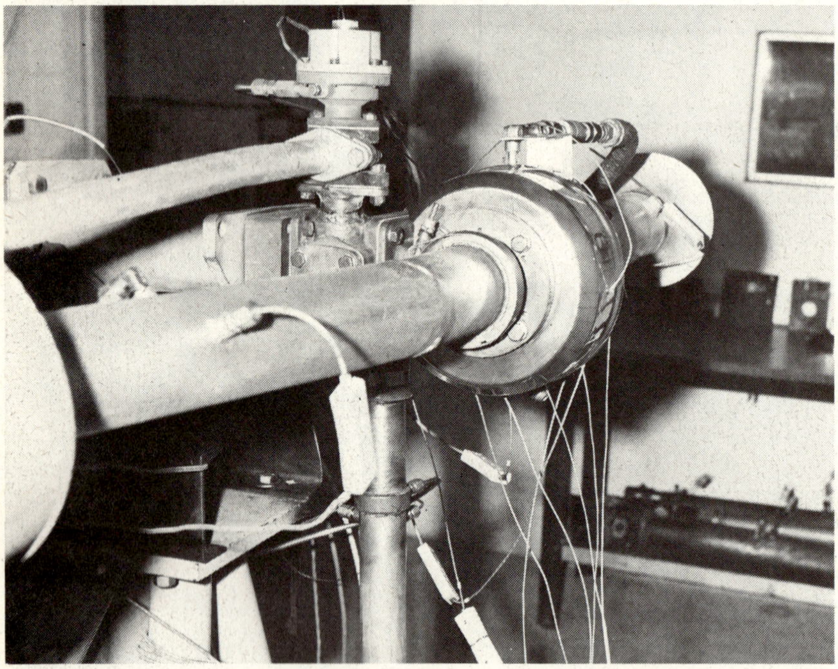

8

11

12

BEATING THE HEAT

Since turbocharging necessitates additional underhood exhaust plumbing, usually toward the top of the engine, underhood temperatures can become a real problem. Much of the extra heat can be contained or reflected by the use of turbine heat shields or special insulation. The materials and shields shown here are manufactured by Thermal Structures, Inc. (1141 W. Katella Ave., Orange, CA 92667).

Thermal Structures has used aerospace technology to develop extremely effective insulation. One of their AG600 insulated polished stainless steel air-gap turbo shields was tested on the hot gas fixture in the Rajay Industries test lab in Long Beach, California. Thermocouples were used to measure the temperature of the actual turbine housing and the surface temperature of the heat shield. With turbine exhaust gas inlet temperatures of 1650 degrees F., which is far hotter than would ever be experienced by a turbocharger in an automotive application, the turbine housing was measured at 1340 degrees, but the surface temperature of the shield was only 372 degrees, a drop of nearly 1000 degrees! Such a reduction in heat radiated from the turbine will go a long way toward reducing underhood temperatures.

For other items under the hood that are in close proximity to the turbine or the hot exhaust pipes, Thermal Structures offers ceramic fiber blanket-type insulation, which is covered with a neat corrugated aluminum foil. Fiber insulation sleeving is also offered for protecting wires and hoses. Two sizes of sleeving are offered, for diameters up to ½-inch and up to 1 inch.

Another recent addition to the Thermal Structures line is a bit of a departure from their insulation products. They have now begun offering cast-aluminum oil drains for Rajay turbochargers. Two styles are offered: one for direct ⅝-inch rubber hose attachment, and the other for -10 AN braided stainless steel hose. These oil drains seemed to be a natural addition to their turbine housing covers, since both improve the appearance of the turbo installation.

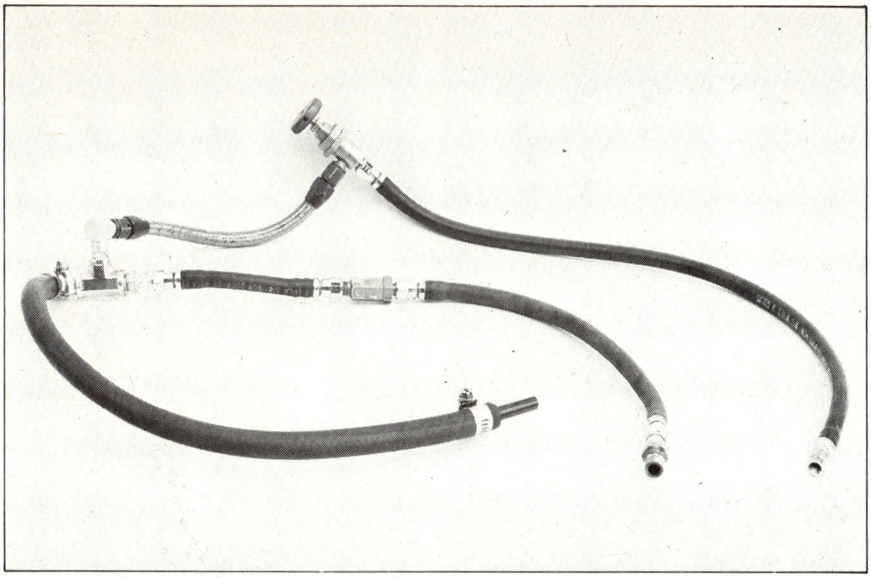

13

7. **Shown here are the Thermal Structures turbo heat shields, insulation blanketing and sleeving. Note that two types of turbine heat shields are offered, with and without the deep-drawn stainless steel cover. Wherever possible, the stainless steel-covered shield offers better heat insulation and appearance, but the non-covered shield is designed especially for limited clearance situations.**

8. **Numerous thermocouples were used to measure turbine housing and heat shield surface temps during the test, which used exhaust temperatures up to 1650 degrees F. Test stands such as this are used for testing various turbocharger designs for effectiveness and durability. It's far cheaper than doing it on an actual engine, especially if the turbo is tested to destruction!**

9. **These are the new cast-aluminum oil drains for Rajay turbochargers offered by Thermal Structures. The drain on the left is intended for use with -10 AN braided stainless steel hoses, and the other drain has a hose barb for ⅝-inch-i.d. hose.**

10. **This wastegate is from the TurboSonics Division of Roto-Master. Its compact size and relatively low cost make it a popular choice for many installations.**

11. **Shown here is the Moran Power Dynamics (1012 S. Branch Way, Forked River, NJ 08731) precision waste gate. This gate is adjustable. Two versions are offered, depending upon the intended flow rate.**

12. **Rajay Industries, Inc. (2700 East Wardlow Rd., Long Beach, CA 90807) has just introduced this new waste gate, which can be adjusted for boost pressures ranging from 3 to 25 psi. Rajay claims this new waste gate is compatible with engines ranging from 61 to 440 cubic inches.**

13. **The variable boost control is sold by Gale Banks Engineering (929 S. San Gabriel Blvd., San Gabriel, CA 91776). It is compatible with both fixed and adjustable waste gates, thus permitting the driver to select higher boost levels while the vehicle is in motion.**

14. **This waste gate and variable boost control assembly is sold by Turbo Tom's (4090 Peachtree Rd. N.E., Atlanta, GA 30319). It is compatible with virtually any turbo system that requires a waste gate.**

14

WASTE GATES—THE BEST OF BOTH WORLDS

One of the problems with many turbocharger installations is that in order for the system to be responsive, the turbine housings must be fairly small to react to low engine rpm demands. In such a situation, the small turbine sections become restrictions to free exhaust flow at high rpm, creating excessive exhaust back-pressure. The solution to this problem is the waste gate, which allows fast reaction of the turbine to develop a preset level of boost and then excess exhaust gas to bypass the turbine, maintaining the desired boost and simultaneously relieving some of the exhaust backpressure. The only drawback to waste gates is that they require extra exhaust plumbing and are usually expensive, due to the need for premium materials capable of handling the high exhaust temperatures without failure or malfunction.

Happily, a number of new waste gates are appearing on the market at greatly reduced prices (some of the first ones cost more than the turbocharger!). Shown here are just a few of those now available.

In-car variable boost controllers are another nicety now available for use with any turbocharger system that incorporates a waste gate. These relatively inexpensive controllers essentially bleed off boost pressure from the actuating diaphragm of the waste gate, allowing the driver to select high boost levels before the waste gate is actuated. Shown here is a universal controller adaptable to virtually any waste gate, and a waste gate with a controller specifically designed for use with that waste gate.

MANUFACTURERS: TURBO KITS & COMPONENTS

American Turbo-Pak
251 E. Stevens Ave.
Santa Ana, CA 92707

AmTest Turbocharger Systems
2750 Jefferson
Napa, CA 94558

Arkay, Inc.
14005C S. Crenshaw Blvd.
Hawthorne, CA 90250

BAE Division of Turdyne
3032 Kashiwa
Torrance, CA 90505

Gale Banks Engineering
929 S. San Gabriel Blvd.
San Gabriel, CA 91776

Barton Cars
Box 49175
Los Angeles, CA 90049

Crown Manufacturing Company
858 Production Pl.
Box 2860
Newport Beach, CA 92663

Custom Turbo Rebuilding, Inc.
11033 Shoemaker
Santa Fe Springs, CA 90670

Herbert Automotive & Marine
8792 Fruitridge Rd.
Sacramento, CA 95826

H-O Racing Specialties, Inc.
4708 W. Compton Blvd.
Lawndale, CA 90260

Kenne-Bell Enterprises
212 San Lorenzo
Pomona, CA 91766

Martin Turbosystems, Inc.
17421 E. Gale Ave.
Rowland Heights, CA 91748

Ak Miller Enterprises, Inc.
9236-38 Bermudez
Pico Rivera, CA 90660

Moon Equipment Company
10820 S. Norwalk Blvd.
Santa Fe Springs, CA 90670

Moran Power Dynamics
1012 S. Branch Way
Forked River, NJ 08731

Motion Performance Parts, Inc.
598 Sunrise Hwy.
Baldwin, NY 11510

Offenhauser Sales Corporation
5232 Alhambra Ave.
Los Angeles, CA 90032

Performance Systems, Inc.
Box 5226
Everett, WA 98206

Rajay Industries, Inc.
2700 E. Wardlow Rd.
Long Beach, CA 90807

R.V. Turbo
16810 Barker Springs Rd.
Ste. 200
Houston, TX 77084

Spearco Performance Products, Inc.
10936 S. La Cienega Blvd.
Inglewood, CA 90304

Thermal Structures, Inc.
1141 W. Katella Ave.
Orange, CA 92667

Turbocharger, Inc.
2544 W. McDowell Rd.
Phoenix, AZ 85009

Turbo City
958 N. Enterprise
Orange, CA 92667

TurboSonic Division
Roto-Master Inc.
7101 Fair Ave.
North Hollywood, CA 91605

Turbo Tom's
4090 Peachtree Rd. N.E.
Atlanta, GA 30319

IMMACULATE DECEPTION

Things aren't always what they seem . . . or, how to cheat with nitrous oxide.

Mention nitrous oxide and the inevitable question is, "How much faster will it make my car go?" It's not an easy question to answer, since it seems to be somewhat related to how efficiently the engine is operating in terms of its ultimate potential. For example, tests of nitrous oxide (N_2O) on a late-model showroom stocker with low compression and mild cam timing, etc., have shown fantastic performance increases, such as reductions in ¼-mile elapsed time of over 2 seconds and increases in trap speed of over 12 mph. On the other hand, a similar test on a well-flogged Pro Stocker will show improvements in the .25-second and 2-mph range. But that still leaves the question of just how much a sportsman class racer, such as in Super Stock, might gain if he were to install a concealed nitrous oxide system. In question too is the detectability of nitrous oxide when it is used, and what possible ill effects it might have on engine components.

To answer these questions, 303 Enterprises (319 N. Ontario, Burbank, CA 91505) modified an Offenhauser Port-O-Sonic singleplane manifold with a totally concealed port injection system for us to test. The test car was C.J. Baker's 396 Chevelle automatic. Since this car is a legal Formula Stocker, it seemed like a good candidate, and, of course, it was readily available.

The Offenhauser No. 6065 manifold was selected for its good performance on the rectangular port Rat and because it can be readily modified for concealed N_2O from the underside of the runners. Two-plane and air-gap (those that have an open air space between the bottom of the runners and the valley cover) manifolds present specific problems for concealed systems. Additionally, a new Holley 850-cfm double-pumper carburetor was used to verify that no specific carb tricks had to be employed to make the system work. The only modification to the carb was the removal of the choke mechanism and the choke plate for optimum air flow.

With the modified manifold and the No. 4781 double-pumper in place, we proceeded to Irwindale Raceway, where we dialed in the new combination without the use of N_2O to establish baseline performance figures. With that out of the way, we began making tests using the nitrous oxide during varying portions of each run. First we used the nitrous only in high gear. Then we mashed the button in both second and high gears. And finally we used it through all but the first 30 feet of the run. We had to wait until the car had initially launched before hitting the button to prevent instant wheelspin from the increased torque. The results were:

Baseline without N_2O	11.32/123.11
N_2O in high gear only	12.20/125.34
N_2O in second and high gear	11.14/126.05
N_2O from about 30 feet out	10.95/126.68

Using the N_2O through all three gears showed a gain of nearly .4-second and

3½ mph. Throughout the test session, the use of N_2O required no variation in carb jetting, ignition timing or spark plug heat range. However, it is advisable to use plug gaps in the neighborhood of .025- to .028-inch to prevent the possiblity of misfiring under the increased cylinder pressure. Additionally, the engine showed no indication of detonation or excessive heat. Apparently N_2O is quite easy on engine and drivetrain components in light of the power increases it produces.

As for its detection on the car, from the outside there was absolutely no indication that the system was installed on the engine. Of course, the unit would have been quickly discovered during a teardown, but there are ways to beat that, too (see sidebar). During actual operation, there was no discernible change in pitch from the engine when the button was pushed in either second or high gear, but hitting the button halfway through first gear did produce an immediate jump in engine rpm, just as if the secondary barrels on the carburetor had suddenly popped open.

It's a neat and relatively inexpensive way to go fast, especially for street performance and bracket racing where its use is legal. And for those of you who are considering the "less-than-legal" applications, remember that others besides you will be reading this article, and they're bound to be watching closely. And then there's always teardown!

— C.J. Baker

BELOW—From the top the Offenhauser Port-O-Sonic manifold looks completely stock in every way, but one look at the underside and you know better. Two lines go to each port, one for N_2O and one for gasoline. The solenoids to control the flow through these lines are also mounted here. All that has to be connected are the two hoses (N_2O and gasoline) and the single control wire, which can be concealed in many ways.

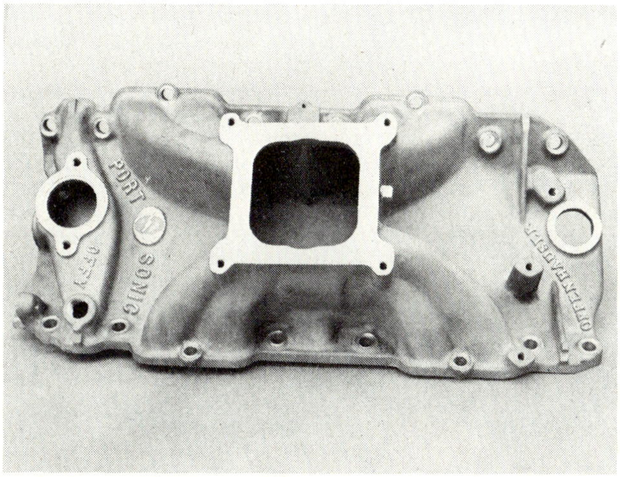

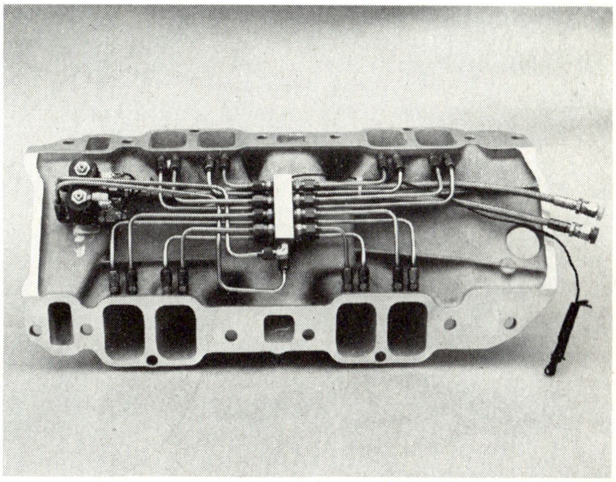

Possible places to hide a N_2O tank might include:

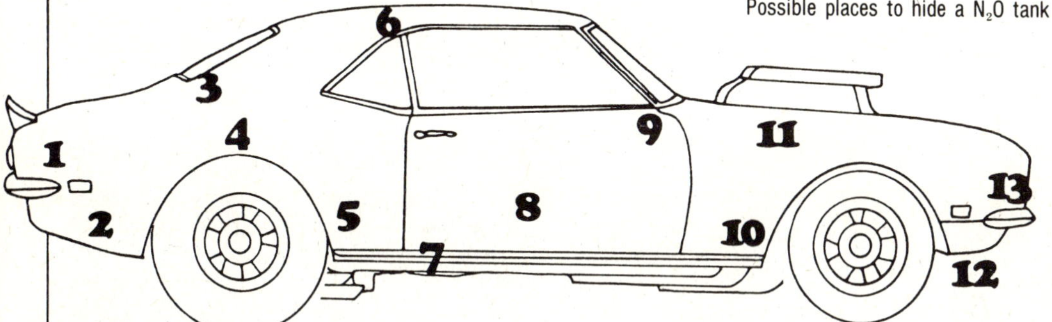

1—A bogus battery
2—The gas tank
3—Between rear shelf panels
4—Inside rear quarter panels
5—Bogus fire extinguisher
6—Rollbar or cage
7—Frame rails
8—Seat cushion
9—Behind the dash
10—Engine oil pan
11—Inside front fenders
12—Inside radiator tank
13—Inside front bumper

Nitrous oxide (N_2O)—We've talked about it before, its ability to produce significant performance gains and its practicality in terms of horsepower per dollar. But nitrous oxide is something else, too. In the hands of the unscrupulous racer, it is one of the most easily concealed methods of grabbing a quick edge on the competition.

The benefit of N_2O is its ability to support combustion of additional fuel in the combustion chamber, although by itself N_2O won't burn. With this in mind, only two things have to be achieved: The N_2O has to find its way into the combustion chambers, and additional fuel must be introduced into the mixture. Sounds simple enough, and it is. Obviously the most controllable way to do this is to inject the N_2O and extra gasoline into the intake manifold, either in the plenum or the runners, as we did on our test. However, since that method is easily found on teardown, other devious means must be considered, although they are bound to be less controllable and less effective in terms of power gained.

Half of the problem of concealment can be solved if the need for a separate gasoline supply can be eliminated. This can be quickly achieved by fattening up the carb mixture with bigger jets to run excessively rich. Then all that's left to do is to get the N_2O into the induction system. One way might be to feed it into the manifold plenum through a bogus vacuum line, but even that isn't actually required. All you really have to do is to turn the N_2O loose where it'll get picked up by the carburetor air intake. It doesn't take much to do the job. Just releasing the N_2O in the engine compartment would be sufficient, unless the car is equipped with a fresh air scoop. Then releasing the N_2O somewhere in the area of the grille or front bumper where it'll get carried up into the scoop by normal air flow over the car would be effective. It is by means similar to this that some NASCAR competitors may have achieved outstanding qualifying speeds and an extra surge on the last lap to take the checkered flag.

How do you turn on the nasty nitrous? Well, you could have a switch on the shifter linkage to do it automatically, or you can use the Line-Loc button, the horn button, a heater switch, etc.

Of course there's still the problem of hiding the nitrous oxide tank, but considering that a drag racer or a roundy-round racer needs only a short burst of power for a few critical seconds, the N_2O tank doesn't have to be very big—probably with a volume of less than one quart. When dealing with a tank that small, the number of places that it can be hidden are limited only by your imagination. But just for the fun of it, shown here are some of the more likely locations.

Blowin' in the Wind

While injection is nice, "wheezers" are "where it's at."

To the grocery-getting uninitiated—you know, the ones who think cars are nothing more than four-wheel holes in the asphalt that they put money into—the strange number 6-71 generates about as much enthusiasm as Einstein's $E = mc^2$. Maybe less. But to those who thrill to the audio-visceral aura which surrounds a highly polished case of a GMC "Roots-type" supercharger, the designation 6-71 continues to represent the epitome of performance.

Essentially, all Roots-type blowers are the same. They are nothing more than bisymmetrical cases containing two three-lobe, straight or epicycloidal (twisted) rotors. The instant these interlocking rotors are turned, built-in pockets on the rotors capture air at the inlet port, direct it around the inside of the case, and force it through the outlet side and into the intake manifold. The balance of the assembly is simply a pair of rotor-supporting end plates, a front cover and two helical-cut drive gears pressed onto the splined end of the rotors. Now "Jimmy" blowers—be they 3-71, 4-71, 6-71 or 8-71 (and these numbers simply mean they were designed to be used in conjunction with GM's diesels having 3, 4, 6, or 8 cylinders displacing 71 cubic inches per cylinder)—are nothing more than air pumps. Using a design pioneered by the Roots Connersville Corp. (hence the name Roots) for pumping foul air out of mines and subway tunnels during the late 1800s, GMC began mass-producing the familiar aluminum lung back in 1938.

Like the old man said, "Roots are where you find them." And that includes old and new Jimmy blowers. However, for the sake of brevity, we'll generally confine our coverage to the ever-popular 6-71 Jimmy. Now it isn't that there aren't plenty of 3-71s, 4-71s and the new Magnuson blowers around; they are common and provide a viable alternative to the big "six." But most blower experts will tell you the 6-71 is the most popular form of Roots supercharging and are quick to point out that most kits, parts and service are geared to the 6-71.

Where you find them is another story. Detroit Diesel will sell you a new one, but they cost an arm and a portion of your leg. You now have the alternative of buying a rebuilt blower or finding a core and having it rebuilt by a blower expert. Buying a rebuilt 6-71 from Larry Bowers, Mert Littlefield, Craig Railsback or Gary Dyer is a lot easier (if $650 is easy) than rootin' around for a core. But if you're into core-scrounging, we'll point you in various directions. You might start with racers who have worn but still street-rebuildable 6-71s for sale. There are always some swap-meet stuffers on the block, and almost every city has a diesel junkyard or a diesel repair shop that

1. To give you an indication of case size, we stacked a 3-71 on a 4-71, then placed the two over the blow-man-on-the-totem-pole, which is a 6-71. One thing about buying a disassembled blower is you can see how it looks inside—but not how it came apart.

2. This is what we mean about "how it came apart." There's a right and a wrong way to disassemble a blower, and this is the wrong way. Blower rebuilding shops, like Bowers', have special tools to keep from damaging end plates, rotors and cases. Little scratches like these won't hurt, but cracks will.

3. Badly galled or chewed-up rotors cannot be saved, so this blower will get a new set. Arrow points to where rotor has been pinned to keep it from turning on the shaft. Front, rear, rotor-to-rotor clearances are set loose to keep heat down.

makes for happy hunting grounds. According to Larry Bowers, a good core, where only the seals are gone, will cost around $250—maybe less. And Craig Railsback quotes a price of $355 to give your core a shave and a haircut. But unless you know what you're looking for—and that includes cracked and chewed-up cases, galled or worn rotors, elongated bearings, or warped end plates—you're probably better off buying a rebuilt 6-71 from a reliable blower repair shop.

A good rebuilt 6-71 Jimmy blower that has been set up by a Dyer or an Estrel—and that includes cleaning and sand-

1

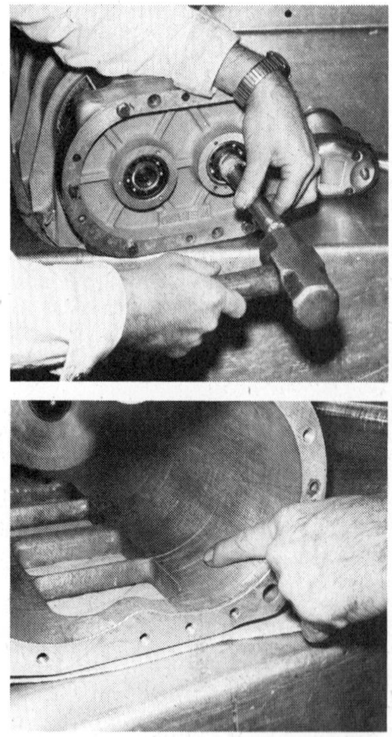

2

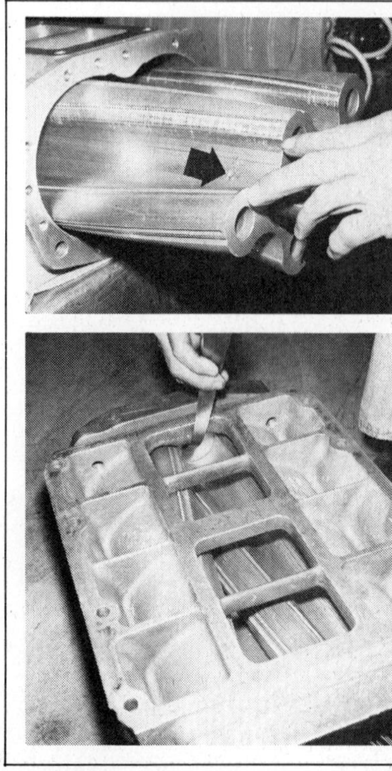

3

blasting (or polishing) the case, O-ringing the end plates, replacing all seals and bearings, pinning the rotors (to keep them from twisting on their shafts), setting rotor-to-case and rotor-to-rotor clearances and timing the rotors—will outlive the engine on which it's placed-. . . that is, if it's well maintained. If not, it will blow away in a hurry.

Other than using a good air filtration system to keep foreign objects from scoring the case or clearancing the rotors, the bearings and drive gears must be checked and lubricated periodically. Most of today's blowers use sealed rotor bearings in the rearend plates, so you don't have to worry about overlubricating these bearings. However, the two drive gears present a different maintenance problem. Some guys like to circulate engine oil through the front end plate to lubricate the drive gears, although that can lead to severe detonation if the front seal or bearing suddenly develops a leak. Consequently, most blower experts recommend the use of a good castor-based oil, which will not cause any detonation if it happens to be sucked into the engine.

Oil level is another thing you should know about. The oil level must be kept either one inch below the center of the front end plate or level with the bottom bolt hole of the triangular blockoff plate. And last, some experts recommend using a pressure relief valve, tapped into the center of the blockoff plate, to keep the internal oil pressure from building up and overriding the end plate seals and gaskets. Others claim that venting the pressure is bad, because the end plates and seals are designed to work under ambient oil pressure.

While you might call it a six drive (six as in 6-71), "don't mix parts," says Iskenderian's T. Willie Hutchins. "There is a good chance that your used blower drive will not fit the other guy's manifolds. So buy all your blower drive parts from the same kit supplier that sold you the intake manifold."

Today's tooth-drive belts will run 40,000 to 50,000 miles if the pulleys and idler are properly aligned. Use a steel ruler or plumb bob to make sure the blower drive is parallel and square, because if there is any misalignment or if the belt is cocked, it will want to walk back and forth, and eventually the ends will fray. The size and type of drive belts are up to you. In the past, V-belts were considered more than adequate. However, they have fallen into disfavor, because a matched set was impossible to find, they would slip if more than 8 pounds of boost were used, and they didn't look "racy" enough. The current crop of 2-inch (by ½-inch pitch) tooth-drive belts are more than ample, and if you really want to look "top fool," a 3-inch (both 8 and 14mm) Cragar setup is virtually bulletproof. "But," claims T. Willie, "the 3-inch belts do tax the bearings;

4. If you like the eyewash value of 3-inch belts and can afford them, these Cragar units are the ones for you. From left is the ½-inch pitch, 8mm and 14mm. Blower Drive Service offers both finned blockoff plate, pressure relief valve.

5. Two-inch drives will have to be used when you've got space limitation problems. Tony Nancy also relies on a 2-inch drive, but he uses a Dyer-modified manifold, based on an Edelbrock high-rise intake, so the stock water outlet can be used.

6. For blower intake manifolds which offer no thermostat housing provisions, Offenhauser has this remote aluminum housing. Unit can be attached to the block or head by a simple L-bracket. Cylinder heads are modified for water outlets in this manner: Hole for ⅜-inch pipe is drilled and tapped 1 inch above deck surface and 3 inches in from the outside of the head.

7. Blower kit suppliers are modifying aftermarket intake manifolds in two basic ways. BDS machines the carb pads, then welds a cast-aluminum adapter body onto the remachined surface. Adapter has a provision for a pop-off valve, and mounting studs are placed in the fingers. Dyer does it this way: Blower adapter plate is bolted to top of intake by means of tapered head screws. These "blowerized" manifolds fit most popular V8s.

8. Cragar and Weiand offer blower manifolds and drive kits for big- and small-block Chevys and early and late Chrysler hemis. The Cragar intake in the foreground is for a rat, and the unit behind it is Cragar's new blower manifold for small-block Chevys. Weiand has both street-style and competition blower manifolds. Arrow points to a 4-71 type, but the rest are for 6-71s.

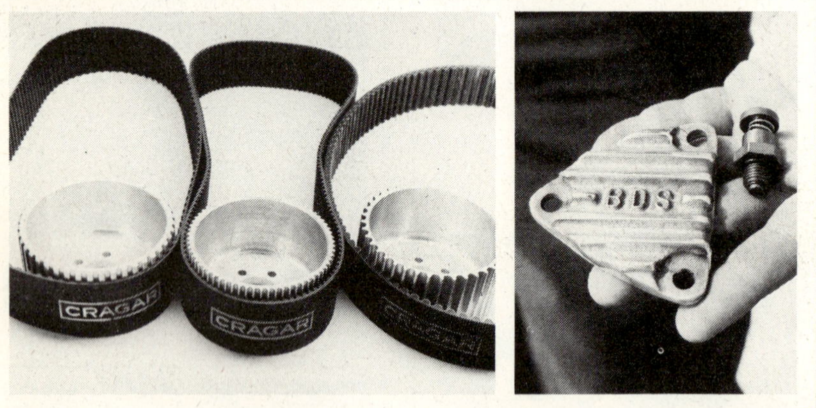

4

5

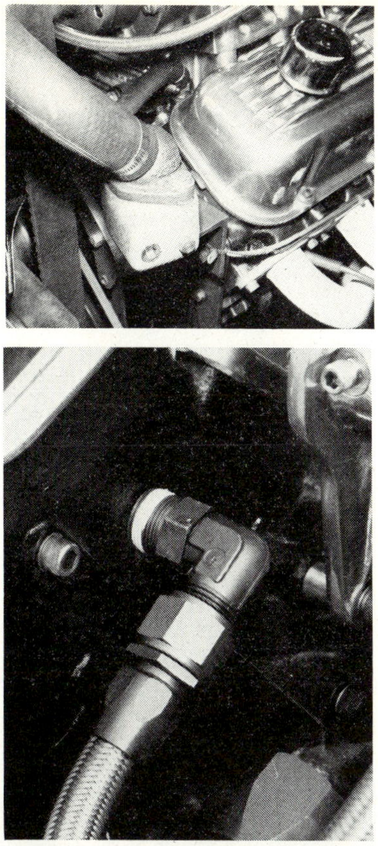

6

EDITOR'S NOTE: This partial list is offered as a guide to aid the reader in his quest for aftermarket sources of GMC superchargers and related components. However, we do not maintain that those listed in our guide will have what you want. We suggest you enclose a stamped, self-addressed envelope when you write to the companies listed below.

BLOWERS

Keith Black Racing Engines
11120 Scott Ave.
South Gate, CA 90280

Blower Drive Service Co.
12140 E. Washington Blvd.
Whittier, CA 90606

Larry Bowers
9763 Variel Ave.
Chatsworth, CA 91311

Gary Dyer
7665 W. 63rd St.
Argo, IL 60501

Ed Estrel
3402 N. Summit St.
Toledo, OH 43611

Magnuson Products
1020 N. Fuller St.
Santa Ana, CA 92701

Mert Littlefield
2773 E. 19th St.
Long Beach, CA 90804

Ed Pink Racing Engines
14612 Raymer St.
Van Nuys, CA 91405

Waterman Racing Engines
3410-C W. Fordham Ave.
Santa Ana, CA 92704

BLOWER DRIVES & KITS

Blower Drive Service
See Blowers listing

Larry Bowers
See Blowers listing

Cragar Racing Equipment
19007 S. Reyes St.
Compton, CA 90221

Gary Dyer
See Blowers listing

Iskenderian Racing Cams
16020 S. Broadway
Gardena, CA 90248

Weiand Automotive Industries
P. O. Box 65977
Los Angeles, CA 90065

INTAKE MANIFOLDS

Blower Drive Service
See Blowers listing

Cragar Racing Equipment
See Blower Drives and Kits listing

Gary Dyer
See Blowers listing

Edelbrock
411 Coral Cir.
El Segundo, CA 90245

Offenhauser Sales Corp.
5300 Alhambra Ave.
Los Angeles, CA 90032

Weiand Automotive Industries
See Blower Drives and Kits listing

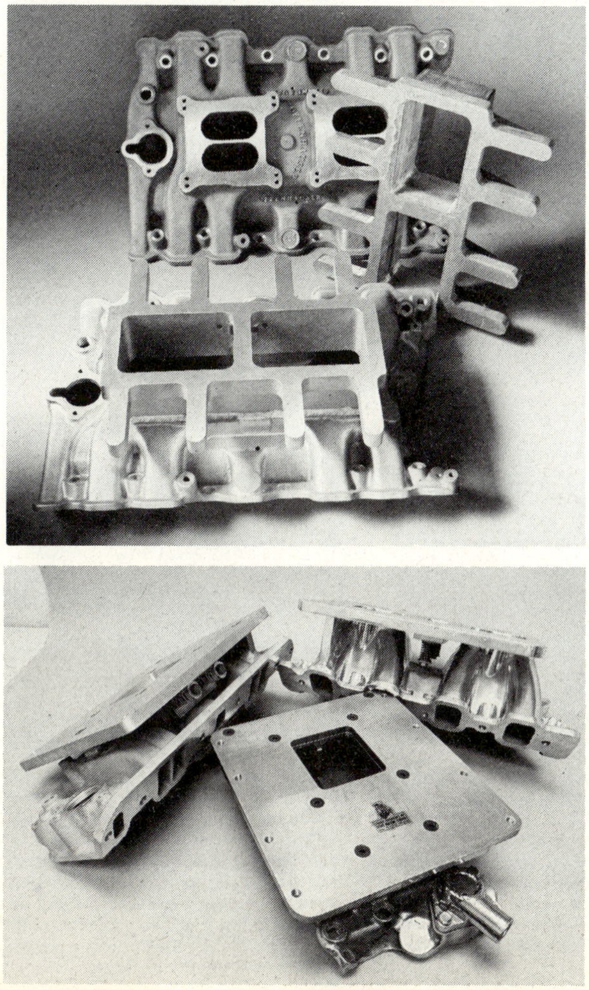

7

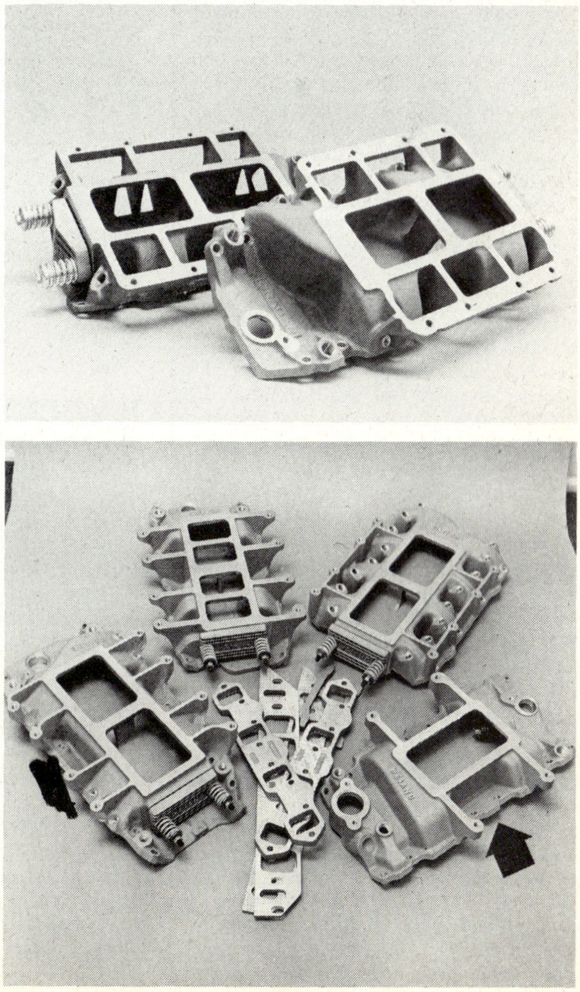

they don't run as free as the 2-inchers; and the $50-$150 cost differential isn't worth it."

Says Larry Bowers, "Use a 3-inch system if you don't have an underhood clearance problem. They look better and give you more reliability."

Whether to underdrive, overdrive or run a 6-71 one-to-one is up to you. But you should take these aspects into consideration. First is today's high-priced, low-quality gasoline. Then think about your engine's static compression ratio, displacement, cam timing and valve springs. And then there is a thing called "how tight is your blower." Bowers recommends driving a 6-71 10 percent under on an engine displacing less than 350 cubic inches, one-to-one on anything between 350 and 427 inches, and 10 percent over on anything larger than 427 inches. But if you want to play the pressure game, there is nothing to say you cannot underboost for economy on the highway and overboost (by changing pulleys) for stoplight cruising.

As T. Willie said, "You can bolt $2000 worth of 6-71 blower and its related components on top of any late-model V8 and it will run like stink—for a little while." What the loquacious T. Willie meant by that "little while" is quite simple. Most of today's crop of engines are not designed for Roots-style supercharging, the small-block Chevy 350 high-performance model a possible exception. Like it or not, crank-driven, positive-displacement blowers put a heck of a strain on an engine's reciprocating assembly. Even a paltry 7 or 8 pounds of boost increases cylinder pressure, and cylinder pressure produces heat. Consequently, any serious attempt to put together a Roots-blown engine must begin, as Bowers says, with "a high-performance short-block." Generally speaking, this includes a fully blueprinted and balanced package, starting with a four-bolt main block, augmented by a forged steel crank, good steel rods, 8.0:1 forged aluminum flat-top pistons (with a chrome or stainless steel top ring), high-quality main and rod bearings, a 270- to-310-degree cam (with .550-inch lift), a set of dual valve springs that will provide 120 to 150 pounds of seat pressure (good boost will cancel out 30 to 40 percent of the spring pressure) and a pair of O-ringed cylinder heads. If you are going to build a "reacher," then you have to think strength.

The second consideration is how to slough off the increased heat. A first-class oiling system—with a large-capacity oil pump, increased pan capacity and, perhaps, an oil cooler—is a must. And as for the water cooling system, Ed Estrel feels this way: "Articles on superchargers never mention the cooling system. I feel that it's one of the most important factors for both performance and reliability. Why else would diesel trucks run two thermostats at $40 each? They start to open at 173 degrees and are fully opened at 185. To compromise, I use a 180-degree thermostat in all my blown cars, and I never run an engine without a thermostat. A good radiator is a must, and the stock water pump is more than adequate. Use the largest possible fan

BOLTIN' ON THE WIND

. . . or what it takes to give your engine that breath of life.

Nothing is ever a bolt-on. Or as the violin player said, "You always have to fiddle around." However, Gary Dyer's complete kit, in addition to the expertise of the Zaino Brothers (2727 Stillwell Ave., Brooklyn, NY 11224, 212-449-8090) has kept the fiddling on Len Myers' '77 Can Am Hatchback Corvette to a minimum. Let's face it: Thanks to Jerry Kelley (Dyer's Machine Service, 7665 W. 63rd St., Argo, IL 60501, 312-496-8100), plus Glen and Lenny, who helped in the installation, this is one L-82 smogger that you don't fiddle with.

Start by removing all accessories, hardware and intake manifold, Make sure all gasket surfaces are clean and free of any residual gasket material, because a good seal is very important. Coat all mating surfaces with a good silicone sealer. Then cover the new intake manifold gasket with sealer and position and carefully bolt the new manifold into place, using Dyer-recommended torque specifications.

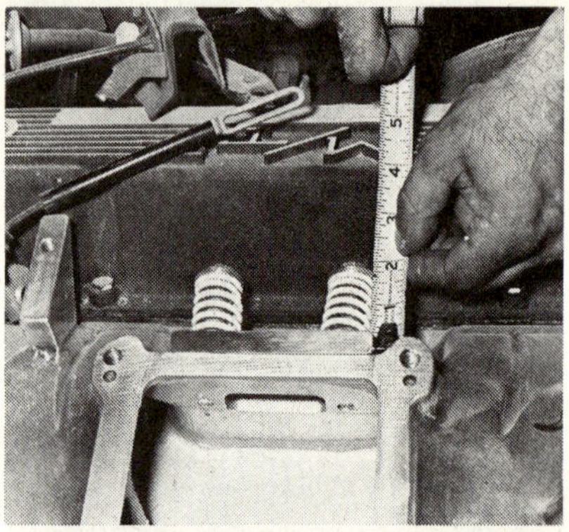

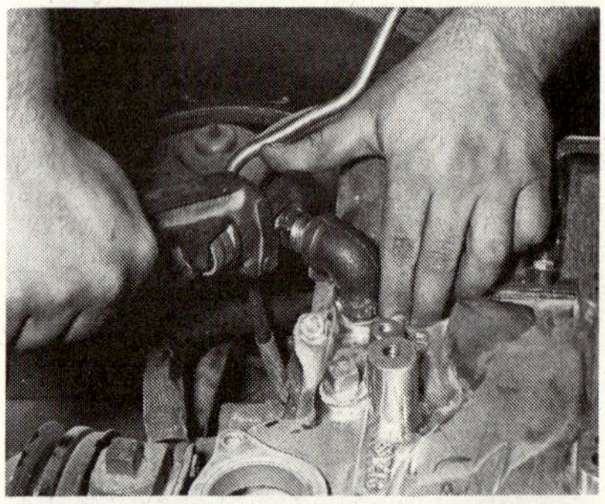

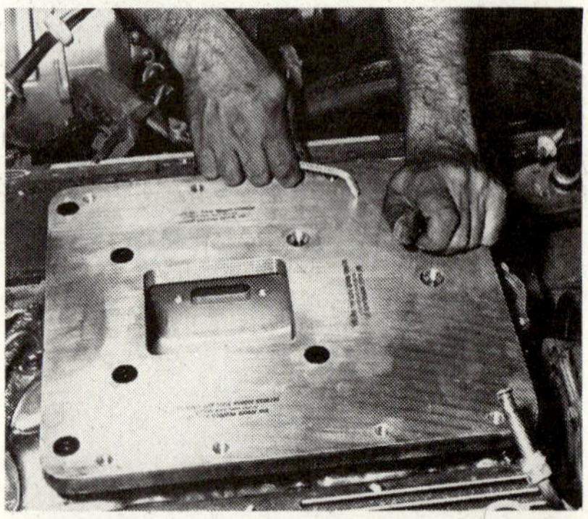

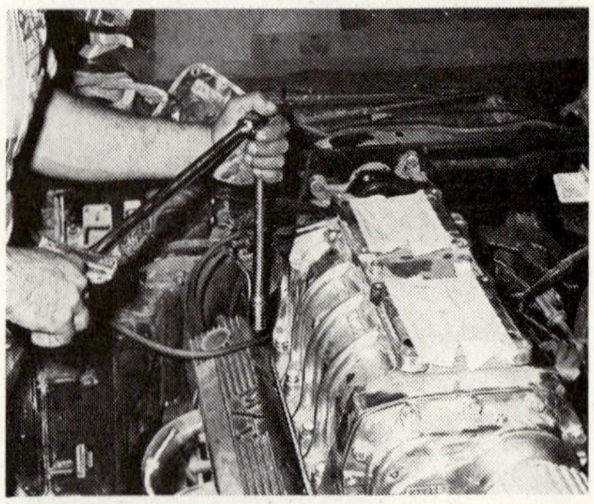

Next set pop-off spring height at 1½ inches; then attach both heater hose connection (a ½-inch elbow-and-nipple arrangement) and thermostat housing. Use plenty of silicone sealer, because housings have a tendency to leak. Use sealer, provided with blower kit, between the blower plate and the top of intake; then install the blower gasket and torque case to 12 lbs.-ft. Use a torque wrench here. Overtorquing will distort case.

Attach carburetors and tighten them evenly, because base plates break easily. Now place crank balancer and accessory drive on crank snout. Balancer is pulled on with four 4-inch and one 3-inch 7/16-20 bolts before crank nut is attached. Blower belt should be adjusted cold with about ½-inch clearance, as it is moved in and out, halfway up the drive side of belt. Adjust the idler arm forward and back to align the belt on drive pulleys.

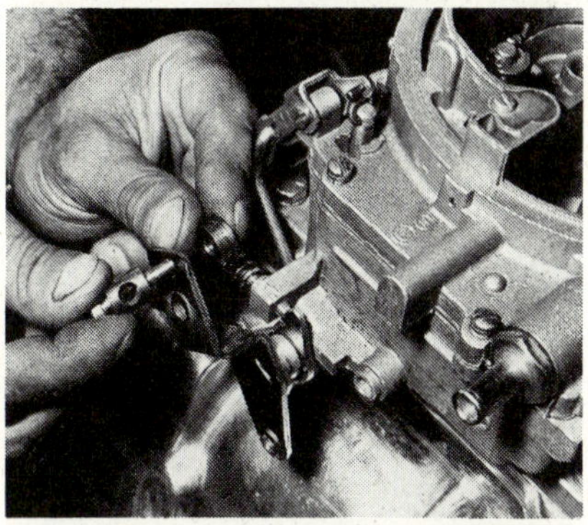

Next lube blower as per instructions. Modify original throttle bracket so it will bolt to the rear of the blower case. Now install the universal linkage kit and begin to hook up all the vacuum lines to the vacuum ports in the carbs. Make sure to keep all fuel lines free of hose rubber or carbs will flood. You may have to recurve the distributor (try a total of 34 to 36 degrees, with 12 degrees at 2000 rpm).

that will fit, and it should have six or seven blades."

While we're on the subject of adequate cooling, we must consider those blower intake manifolds that do not have provisions for water outlets and thermostat housings. There are two alternatives to this state of affairs. There is an increasing number of "blowerized" single- or dual-plane, single or dual 4-barrel aftermarket intake manifolds that have the water outlet and housing in the stock location. The other alternative is to tap into the water jacket on the front or rear of the cylinder heads and then pipe into a remote thermostat housing. Unless some sort of clearance problem makes it impossible to tap into the front of the cylinder heads, always draw the water from the front of the engine. If you don't, the front two cylinders are likely to run too hot.

Five years ago, finding a blower intake manifold for anything other than a Chevy (or a Chrysler) was almost impossible. Not so today. Cragar, Weiand and Edelbrock still produce "blower" manifolds. But guys like Gary Dyer and Craig Railsback (Blower Drive Service) are taking aftermarket intake manifolds and converting them to 6-71 use. These "blowerized" intake manifolds fit most popular cars. The conversion process is quite simple. The carburetor pad(s) and plenum chamber dividers are machined off. The manifold is then fitted with a bolted-on or welded-on blower adapter plate featuring some sort of pop-off or relief valve arrangement. And you can buy these adapter plates from Iskenderian if you want to do it yourself.

Evidently there is no hard set of rules for carburetor size or fuel delivery curves on a 6-71 blower. Cam timing, static compression, engine displacement, blower speed and amount of boost play an important part in blower carburetion and proper jetting. Craig Railsback doesn't even mess around with carburetors. He simply determines what his customer's induction combination is and then has a carburetor expert tailor the carburetion accordingly.

However, we can offer some basic guidelines so you won't "sneeze" your stuffer or "fry" your 8-cylinder lump. If in doubt, go rich! Under normal driving conditions, your engine is operating in a "vacuum" mode, so stock primary jetting is okay. On the other hand, when you're under boost, you've got to richen up the secondary and accelerator pump systems. How rich is up to what your spark plugs tell you. Guys like Estrel open up the secondaries .006-inch. Tony Nancy drilled out his secondaries .005-inch and reworked his accelerator pumps to deliver more fuel to eliminate a temporary off-idle throttle lag condition. T. Willie says, ".010-inch richer on the secondaries, and use a fuel flow regulator set at 4 or 5 pounds of pressure."

"When I tune a blown engine for street use, I just set the timing where the points are just beginning to open on the No. 1 cylinder (or about 10 degrees before top dead center). Then I advance the distributor until the motor is running nice and free before I lock it down." So sayeth Larry Bowers.

"Ignition timing is very critical on a

blown engine," writes Ed Estrel. "If you're backfiring through the pop-off valve, it's probably your timing or spark plugs breaking down. I run 37 degrees total advance with 10 degrees at idle. I run my spark plugs two steps colder than your engine calls for and gap them at .025-inch. A good ignition to use is a C.D., as it will fire your plugs even if they're fouled. Better yet is the multi-spark ignition systems which fire the plugs over a longer period of time."

"And," says T. Willie, "if you don't want to mess around with a battery ignition, why not try a Ronco Vertex magneto?"

Everyone that we've talked to agrees that you must use a full-flowing exhaust system, and that includes big-diameter headers or high performance cast-iron exhaust manifolds, Corvair turbo or Chrysler's low restriction mufflers and tailpipes.

Roots-type superchargers offer the rodder instant power, great flexibility and no real plumbing problems. While most hot rodders shop for looks, they don't understand that Jimmy blowers are dependable, driveable units. A standard blown 350-inch Chevy V8 with a cam and good valve springs will put out around 400 horsepower, no matter what the condition of the engine is, and it will idle like a stocker. But that extra horsepower doesn't exactly come cheap. No matter how you cut it, a first-class, turnkey "kompressor" kit can run as high as $2500. And if you're thinking of improved mileage, forget it. According to Larry Bowers, "To run a Jimmy, you've got to be on the ball. Don't put anything on your engine that you can't fix. You shouldn't be messing around with anything you cannot work on." *Caveat emptor*—or let the blower buyer beware—then take your money out and blow it, anyway. —*Gray Baskerville.*

9

9. Both Carter and Holley carburetors work well with Roots-type supercharging. These 750-cfm competition-series Carters on Chuck McConnell's 454-inch, 6-71 blown rat have stock primaries and No. 150 secondaries. Polished cast-aluminum carb base is built by Mike Kuhl and can be bought from Larry Bowers. Blower Drive Service offers both single- and dual-carb adapter plates for Holley carburetors.

10

10. Iskenderian's complete kit includes everything but T. Willie's T-shirt, and if you send cash, they'll include that, too. Small-block Chevy kit is based on an Edelbrock intake, Littlefield lung and Isky 2-inch drive.

Fast and Flashy

The Larson Engineering Corvette turbo package looks as good as it goes.

The sad truth about Corvettes is that they aren't as quick as they used to be. Over the years, the Corvette has evolved into a luxury GT, with more emphasis on comfort and styling than on performance. This move away from blinding straight-line performance hasn't hurt sales one iota, but there are still die-hard Corvette enthusiasts, like Robert Larson of Detroit, Michigan, who believe that late-model Corvettes can still burn rubber with the best of them. Larson runs Larson Engineering (26121 Van Born Rd., Taylor, MI 48180), a company that specializes in custom fabrication, particularly of turbochargers and exhaust systems. He does development work for the major auto manufacturers, as well as for racers, such as John Greenwood. With this involvement, it's only natural that Bob chose turbocharging as the best way to provide his '76 Vette with that old kick-in-the-pants feeling.

The Larson Engineering Corvette turbocharger kit is known as a flathood kit because it fits easily under any stock hood. The kit bolts onto any L-82 or L-48 small-block 350-cubic-inch Chevrolet engine without any welding required and is worth an additional 100 horsepower. If 100 horsepower isn't enough boost, a second turbocharger can be added to the system at a later date, although the dual turbo setup is primarily a drag strip combination that is designed to work at its optimum at between 4000 and 8000 rpm.

Larson uses an AiResearch turbo

1. The entire Larson turbo kit fits under any stock Corvette hood. The massive air cleaner is an industrial unit.

1

(model No. T04), which works fine with a stock engine when run at 12 pounds of boost. A new manifold casting for the right side, which mates to the turbo, is included in the kit. The carburetor is a 650-cfm Holley. Larson converts the carburetor to a pressurized model for use with the turbocharger, and solid carburetor floats are used so there is no danger of the floats collapsing under pressure.

A special low-profile bonnet is used on top of the carburetor to clear the stock hood, and an O-ring is used on the base of the bonnet to form an airtight seal with the carburetor. The bonnet also has a pop-off valve, which is set at 12 psi. The pop-off valve acts as a safety device in case the engine backfires.

An additional fuel pump (in this case a Holley electric unit) cuts in when the turbocharger hits two pounds of boost. A Holley regulator, used in conjunction with a Hobbs pressure switch (part No. M4006-4), senses the boost and starts the auxiliary fuel pump. Larson prefers the pressurized carburetor because it is easier to start in cold weather, and since he drives his Corvette all year around (even during the cold Michigan winters), cold-starting ability is very important to him. Larson also feels that the pressurized, or blow-through, system is best for low-profile installations, as opposed to suck-through systems.

The air filter mounts on the front of the AiResearch turbo. The unit Larson uses is a huge (6½-inch diameter and 8-inch length) unit originally designed for tractors and manufactured by AC under part No. Type A67C. The back end of the turbocharger is connected to a stainless steel exhaust system, and 2½-inch-diameter stainless tubing is used all the way to the rear bumper. Stainless steel is used for the exhaust system because it won't rust (remember, Larson drives his Vette during the winter when lots of salt is used to keep the snow off the roads). The exhaust system is used with either two Corvair Turbo-style mufflers or Larson Engineering's special extruded-aluminum side pipes.

The turbo kit comes with a water injection kit to prevent detonation. Larson is also working on a warning-light system that will tell the driver when to add more water. The way water injection systems are now, they need to be checked quite often. Another item Bob is working on is an adjustable boost system that is operated from inside the car. Currently, the adjustment is made at the turbo with an Allen wrench. There is a boost gauge for the instrument panel.

Other items that come with the Larson Engineering turbo kit include a distributor retard kit which retards timing 8 degrees, special brackets for the alternator and air conditioning and an alternator bracket for '76-and-older Vettes. The air-conditioning bracket moves the air-conditioning pump inboard on '77-and-newer cars and moves the unit to the opposite side for '76-and-older Vettes. Proper oiling is important in a turbocharged vehicle, so Larson includes a custom-built oil pan featuring an external return provision. Location of the return tube is critical, so the pans are made in a special jig. Although not included with the kit, a heavy-duty oil pump and Chevrolet heavy-duty hydraulic lifters are recommended for Corvette owners who do a lot of *hard driving*.

Larson's personal test car has recorded 0-80-mph times of 7 seconds with a Turbo 400 transmission and 3.08 rear gears. The response is very good, yet the car is completely tractable in traffic.

For customers who want a little quicker response than the standard unit, Larson offers a waste gate which plugs into

2

3

2. The single turbocharger system can be had with or without a waste gate (arrow). The waste gate is a modified motorcycle unit, which improves turbo response. A small glasspack muffler quiets the waste gate.

3. For the power-hungry, Larson offers a dual-turbo kit. Twin waste gates are featured on this installation. Also shown is the center-mount air conditioning for '77 and later Corvettes.

4. The exhaust system is all stainless steel. The arrow points to the external oil return on the custom-made pan.

5. The low-profile carburetor bonnet features a pop-off valve and an O-ring to seal the pressurized carburetor.

6. The fuel system consists of a pressurized Holley 650-cfm carb and an auxiliary fuel pump (and regulator), activated by a pressure switch at 2 psi of boost.

the supplied exhaust manifold. The waste gate is a modified Blake unit originally designed for big-bore Kawasaki motorcycles. The waste gate has an isolated diaphragm, which means less leakage, less chance of popping off and smoother operation. It uses a small glasspack muffler to keep the noise level down; for even more performance a second turbocharger can be added. The twin-turbo application is great for high-rpm power, but the engine suffers under 4000 rpm.

Turbochargers are the coming thing in high performance, and the Larson Engineering kit is an ideal setup for any late-model Corvette. Corvette owners can have strong performance, along with good looks and comfort, thanks to turbocharging.

Larson Engineering isn't concerned solely with acceleration. Bob Larson likes his Corvettes to look as good as they perform, so the approach he has taken for improving his Corvette's looks is golden; that is, all of the trim parts and accessories are done in gold. His personal '76 Corvette is a rolling example of his golden touch.

Probably the most noticeable are the gold-anodized side pipes. Side pipes have always been big with Corvette owners, and these are some of the most handsome ever produced. The side pipes are made out of extruded aluminum that has been anodized gold. The pipes are also available in chrome or clear protective anodizing. Mufflers are used inside the side pipes, but there is an air space between the mufflers and the side pipes. This keeps them cool to prevent burned legs.

Another noticeable feature is the gold-tinted T-top panels. The light gold T-top panels are see-through and surrounded with a 2½-inch border of gold paint. The paint gives the top a finished look. The rear window louvers are also gold-colored. The Western wheels are gold-anodized, which not only improves their looks but eases maintenance.

Every trim piece imaginable is available in gold anodizing, including side mirrors, door handles, windshield trim, antenna, door locks and all the Corvette and Stingray insignia. A very popular trim item is the gold version of the 25th Anniversary front emblem.

Bob Larson went one step further with the gold treatment on his personal car! He used gold vinyl spray to get a gold padded steering wheel, dash pad and center console. The gold-trim treatment looks very handsome on Larson's pearl white car. Bob has added gold paint to the sides, back and front of the car, along with matching pinstriping that makes his creation as good as gold.

—Bruce Caldwell

4

5

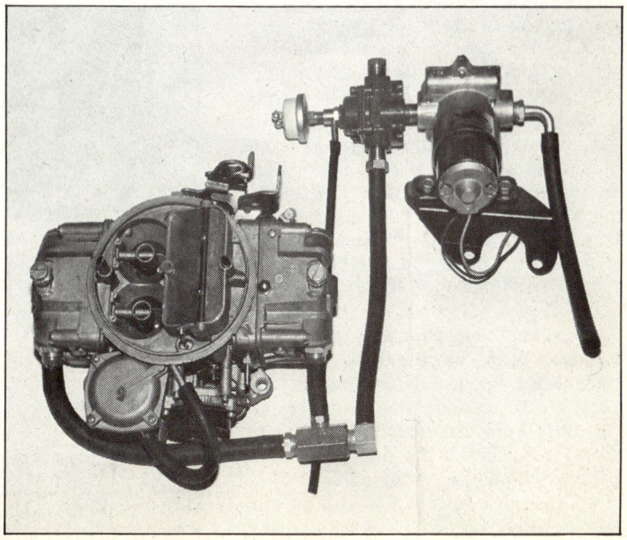

6

Greater Than an 8

Down on cylinders, up on performance; a turbo'd Chevy 6 comes alive with a vengeance.

Five or six years ago, finding a performance vehicle was a simple matter—you just went down to the dealer and ordered that car or truck you wanted with the biggest V8 offered. The turban-and-harem boys have seen to an end of that, as we all know now; Detroit is hardly even offering *low*-performance V8s, let along the "fat-blocks." Just as it was before the Detroit musclecar era, the maxim today is "high-performance vehicles are built, not bought."

We've all seen countless articles on turbocharging V8s as one highly practical route to this kind of performance, but with the cost of fuel going higher every day, there's more to be considered than just brute power. A turbocharged 6-cylinder Stovebolt, such as we have here, may just be the answer for you; in it, you get economy and driveability combined with full power when you need it. Our Chevy C-10 project delivered a respectable 17.5 mpg with its smog-laden 250-cubic-inch six, but performance left more than just something to be desired, particularly in a workhorse vehicle.

The turbo "answer man" we sought was John Anderson "Napa John" of Amtest Turbocharging (2750 Jefferson St., Napa, CA 94558), the czar of boost for the Northern California wine country. Having once campaigned a dragster powered by a big-block Chevy with no less than *four* turbos, John has since calmed down only slightly, doing custom installations on everything from Maseratis to Datsuns.

John was happy to do the installation on our truck, since he is usually asked to fit a turbocharger and attendant plumbing into something as cramped as an air-conditioned Corvette compartment. The acres of room around our six-in-a-row made it a comparative dream to work on. As on any other installation, there were six major connections to be made, the *in* and *out* plumbing of oil, intake and exhaust. Crafty John always has all the necessary tools at his disposal, and after a short skull session with the bare Rajay turbocharger from Spearco Performance Products (2054 Broadway, Santa Monica, CA 90404) and a look at the stock manifolds on the six, John began cutting, welding, machining and fitting our installation.

The exhaust was plumbed right out of the bottom of the stock exhaust manifold to a four-bolt flange for the turbo to bolt to, and a new exhaust pipe was fabricated to duct from the rear of the exhaust housing down under the truck. A Spearco aluminum adapter mounted the Holley 650 spread-bore carb to the inlet side of the compressor housing, and some quick work with a talented pair of tinsnips and welding rod made a neat tubing loop between the compressor's outlet and the hole on the intake manifold left by the removal of the stock one-barrel.

Oil supply for the turbo's single center bearing came from a line John tapped off near the oil pressure sender on the block, and reinforced ⅝-in. I.D. hose was used for the oil drain. This can be critical for proper operation of a turbo, because of the high volume of oil and the aeration that takes place in that 100,000 rpm maelstrom. To avoid having to drill, tap or weld on the stock oil pan, John used a neat hole punch and epoxied a short length of copper pipe into the pan, just above the oil level and tilted at a slight downward angle to avoid any possible restrictions to the oil flow.

Two problems that crop up in a turbo installation are destructive detonation and boost control. The former situation results from using today's lower-octane gasoline; most new vehicles are on the ragged edge of ping-and-knock even in stock form—add a few pounds of boost and it sounds like the crank is ready to drop out on the pavement! In our case, just a little over 3 pounds boost had the

1

2

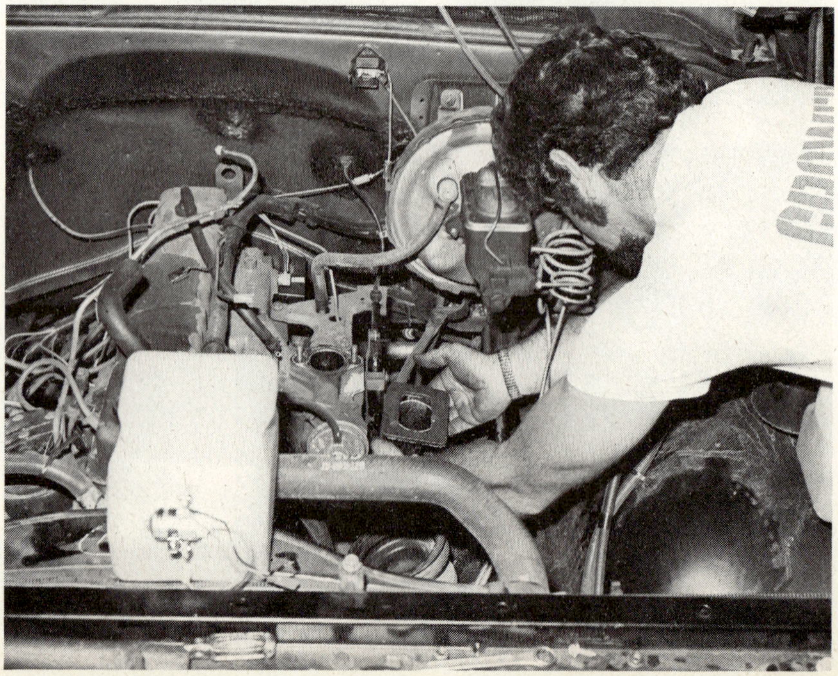

3

1. With the stock carburetor and exhaust system removed from the two manifolds, John Anderson surveys our compartment with the Rajay turbo in hand. Six-cylinder vehicles offer a lot of working room for turbocharging.

2. Using the steel flange from the old exhaust pipe, John welded up a few short lengths of tubing to duct the exhaust from the cast manifold to the turbocharger's turbine housing.

3. The round 2½-in. tubing had to be hammered to fit the rectangular hole in the flange plate that will mount the turbocharger. John tacked them together on the engine, then removed the piece for final welding.

4. One of Amtest's exhaust waste gates was considered a practical accessory for our C-10, giving us a variable boost control at the twist of a bolt.

5. The waste gate dumps the excess exhaust gases when it senses your preset level of boost. Thus, it must be mounted somewhere in the exhaust supply, ahead of the turbo. John added a short length of tubing and a special flange (with a machined valve seat) for the waste gate.

6. With the pipe and "squirrel-cage" mounted on the engine, things begin to take shape. Here John is checking the Spearco carb adapter for level. It mounts to the front of the turbo's compressor intake with hose and clamps.

7. To steady the adapter for the weight of the Holley spread-bore, a support bracket was made to bolt to the intake manifold. The lower half of the adapter is heated by engine water to prevent carb icing.

8. From a different vantage point, you can see three portions of our turbo plumbing. The "A" indicates the exhaust, which bolts to the stock manifold, "B" is the exhaust pipe *out* of the turbine housing (which will connect to the rest of the exhaust system) and "C" is the oil line that drains from the turbo to pan.

5

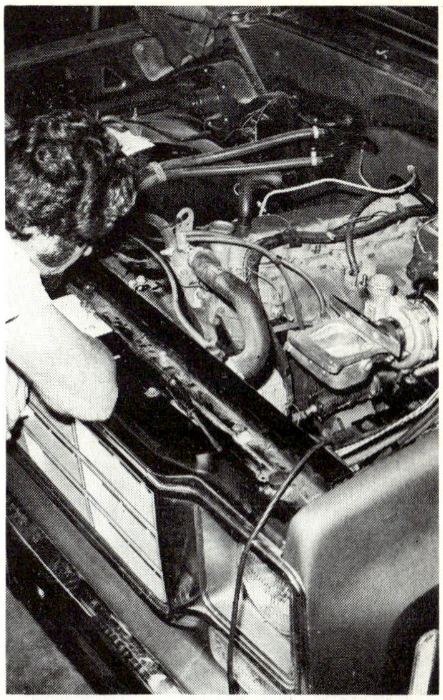

6

7

4

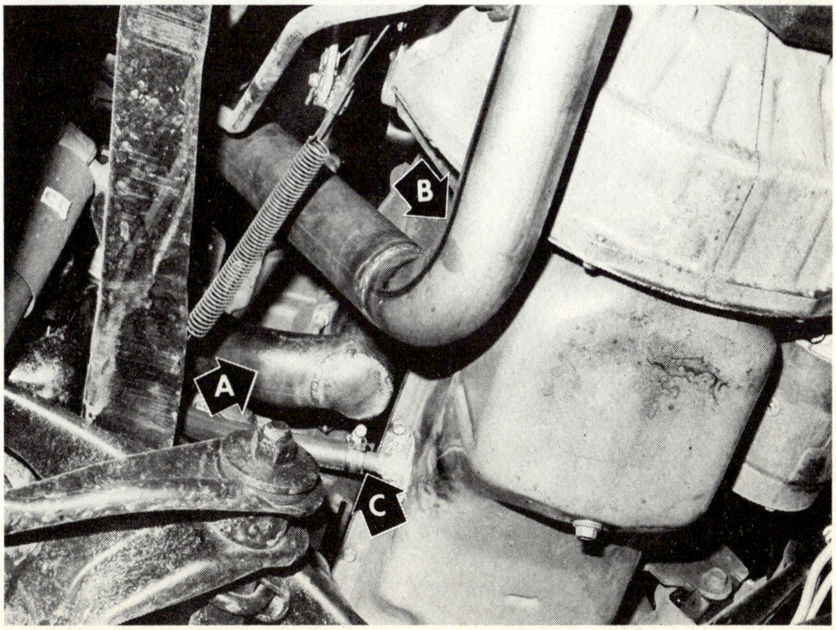

8

stock six rattling loudly, even without a lot of ignition advance. The remedy supplied by Spearco Performance Products was in the form of their Injectronic water injector. Unlike most water injectors, this one has a mini-computer that senses manifold vacuum/pressure and engine rpm and determines just when engine load requires a squirt of water to the air cleaner. This unit worked flawlessy, allowing "big-boost" John to dial us in at as much as 15 psi boost without knock. Boost control was handled by adding an Amtest waste gate, which is the unit John used to dial in our pressure.

A flange and short length of tubing welded to the exhaust pipe just before it enters the exhaust turbine served to mount the waste gate, which contains a spring-loaded valve that opens when a preset level of boost is achieved, bypassing the excess exhaust to slow down the turbo. A single adjusting bolt, the "horsepower screw" just like the one Indy cars use, changes the boost level at which the waste gate opens, making light work of our test program, in which we experimented with different boost levels and carburetor adjustments.

It might seem at first glance like the 650 Holley 4-barrel is overkill for a 250-cubic-inch motor; however, the small primaries kept fuel economy reasonable and the big secondaries provided that extra mix required when the boost is flowing. Although the Holley worked well right out of the box, the secondary jets had to be enlarged to cure a mid-range ping, and the secondary vacuum spring was exchanged for a lighter one. The Chevrolet HEI ignition provided adequate spark, but the distributor's vacuum advance can was swapped for one from a '67 Pontiac, a dual-diaphragm type. This allowed normal vacuum advance; under boost it would retard about one degree (to a total retard of 8 degrees) for each psi of boost, just what we needed to prevent detonation.

We had a ball testing the C-10, using a private byroad in the Napa Valley.

9

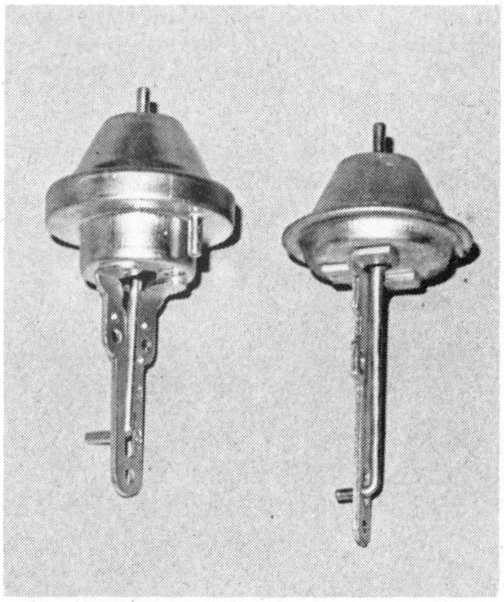

11

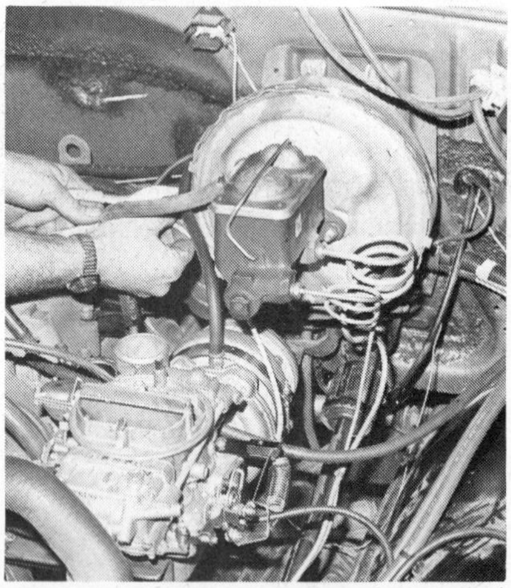

12

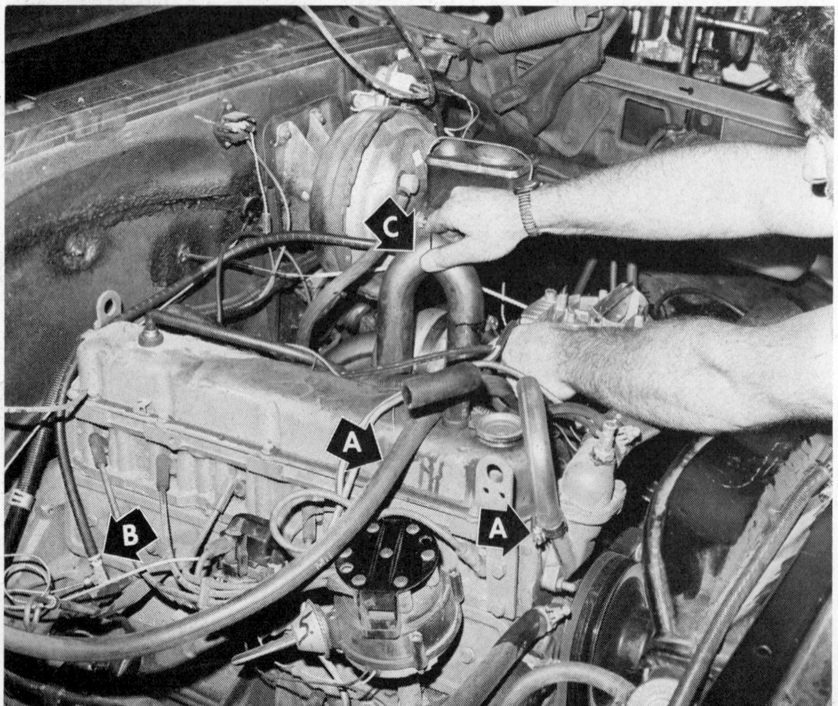

10

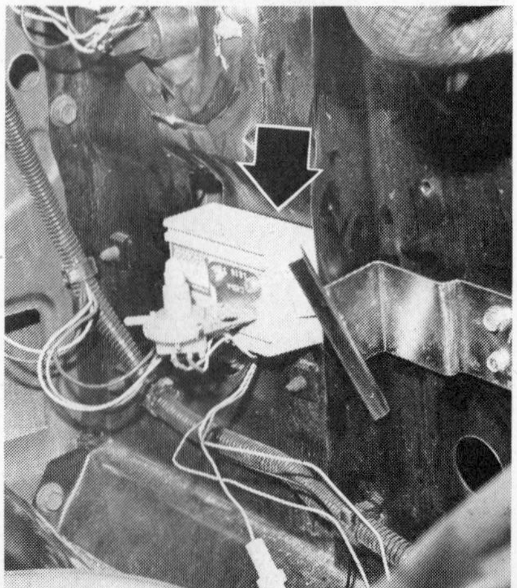

13

Driveability was smooth, we didn't have the extra front-end weight of a big engine and the truck could leap tall grades with a single effortless bound. The chassis dyno confirmed our gut reactions to the new-found power; where the best stock power level had been 78 hp at 2500 rpm, the turbo'ed engine climbed to 115 @ 2500 rpm and 160 hp @ 3000 rpm, all with a load of only 5 psi boost. Napa Muffler constructed a new free-flow exhaust system of 2½-inch pipe with a single Corvair turbo muffler, and the free-flowing system was quieter than stock! The California Highway Patrol book specified a 91-decibel level for this truck (our stock system gave a 100-decibel reading) and the "after" version only put out 89 decibel.

Though Detroit is finally on the turbo bandwagon and is installing them on a number of small cars to regain lost performance, it isn't legal in some states to install them yourself. If it *is* okay in your area, *do it.* You might make enough in trading away your thirsty V8 truck or car for a six-banger that you could pay for the turbo installation, and then we wouldn't be the only ones driving around with a supressed smirk for having the best of both economy and power.

—Jay Storer

9. Napa Muffler finished our exhaust system with 2½-in. pipe and a Corvair turbo muffler by Walker, which was made even more efficient by capping the stock offset outlet (arrow) and welding in a new one at the center.

10. The heater hose at "A" has been rerouted to the Spearco carb adapter to aid vaporization. You can see at "B" where we picked up an oil supply for the turbo. Here John is fitting the last two pieces of pipe ("C"), that will plumb the compressed air/fuel mix to the stock intake manifold.

11. A '67 Pontiac dual-diaphragm vacuum advance can was modified to fit our 77's HEI distributor. The arm had to be extended to the same length as the stocker (at right).

12. Hot rodding is often a series of cut-and-tries. After a clamp bracket was made to hold a cable throttle and mounted off the rear of the carb, John spliced a new Teflon-lined cable to the stock pedal inside the cab.

13. Spearco's Injectronic found a good home behind the left headlight bucket. The arrow indicates the mini-computer that senses engine load; the plastic tank mounts at right.

14. A larger air cleaner should be used, but this is what we started with. The tubing going into the air cleaner top is from the water injector and a precise orifice controls feed.

15. Like a tach on a Sixties supercar, our formerly mundane pickup sported an impressive gauge on the steering column. The TurboTest gauge has both pressure/vacuum readings and exhaust gas temperature (pyrometer) in one unit, just like the 18-wheelers have.

16. The final step before taking it out was to add water (or water mixed with methyl alcohol) to the water injector, which we found indispensable to our installation. Napa John likes to refer to this as adding "turbo wine."

14

15

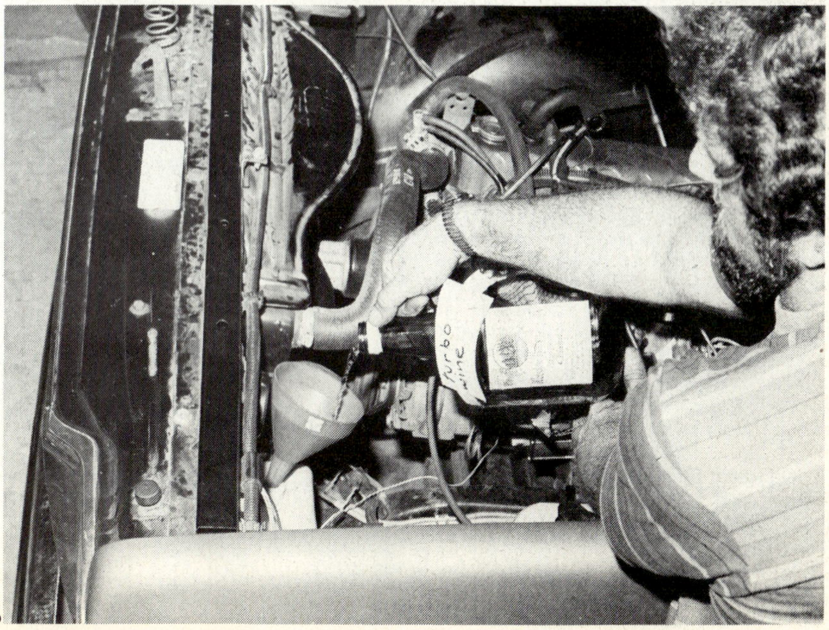

16

500-HP, Blown Small-Block Build-Up

You don't have to be a math major to figure out what happens when you mate a Jimmy with a small-block—the sum is better than 11.5 secs. at over 117 mph, in a 3000-lb. vehicle.

There are two ways to build an engine: One is the complete-blueprint, high-compression, naturally aspirated route, which necessitates a megabuck investment and usually results in an engine that is just about impossible to live with on the street. The second is to combine a GMC 6-71 supercharger and a late-model, low-compression, Chevy, factory high-performance small-block.

Blower setups can be found for around $800. Outfits such as Dyer's Blower Service (7665 W. 63rd St., Argo, IL 60501) have blower kits (supercharger, manifold and drive) for nearly every engine, costing between $800 and $1000. And you know how many small-blocks are lying around.

But, you say, blowers are temperamental, and there's no way a stock bottom end can stand up under all that power. Nonsense! This is just one of the many myths that will hopefully be dispelled by this article. When the engine is correctly set up, you can have a reliable 500 horsepower on call, and at a usable rpm range as well, eliminating the need to buzz the motor. On top of this, there is the awesome torque put out by a blown small-block.

What about turbocharging? Turbocharging, which works off the engine's exhaust, has the advantage of more top-end horsepower, since none is "wasted" turning the rotors of a supercharger. But a turbo takes a while to get up to full boost, whereas with a Roots-type positive displacement blower the power is there *right now,* as soon as you shove your foot to the firewall. Also, if a turbocharged engine is running too lean, chances are the first warning you'll have is a few melted pistons. With a supercharger, if something is not kosher in the fuel system you will get plenty of warning from backfiring and roughness.

Now if what we've just said makes sense and you feel it's your turn to go out and build a blown engine, we're going to tell you how to do it and make it live—and not break your wallet.

Start with a good low-compression (1971 and later) LT-1 or L-82 engine. These engines feature four-bolt mains on all bearings; forged crank, forged rods and 9:1 pistons, and big valve/big combustion chamber 2.02 intake heads. The main difference between these two engines is in the cam and valvetrain. Low-compression LT-1s have the good Chevy street mechanical cam, while L-82s have a milder hydraulic cam. LT-1s are found in 1971-72 Corvettes and Z28s. L-82s were installed in 1973-77 Corvettes and 1973-74 Z28; 1977 Z28s do not have L-82s. These engines should be fairly plentiful in the junkyards. If they are not, you can still get a low-compression short-block LT-1 (which is preferred over the L-82 because of better camshaft) from Chevrolet under part No. 3997472, but the factory is charging big bucks for their engines.

If the engine is in good shape, practically all you have to do is plunk the blower on top of it. There is no need to blueprint at all. However, some modifications will have to be made to the oiling system. A larger capacity sump, along with a windage tray, should be installed in order to increase capacity to at least seven quarts (including filter). Use the LT-1 pump and pickup, but after the pan is deepened, be sure to extend the pickup to compensate for the deeper pan. It's also good insurance to weld or braze the pickup to the pump. An alternative pan/windage tray setup is shown in the photos. Fill the pan with 20W-50 heavy-duty oil, and do not use racing oil unless you're willing to change it every 1000

1

2

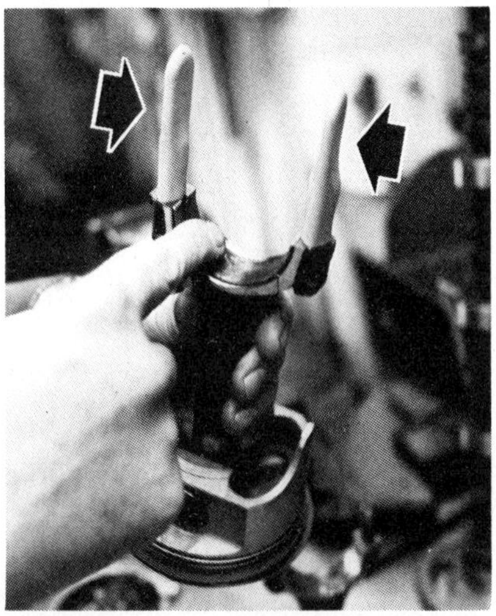

3

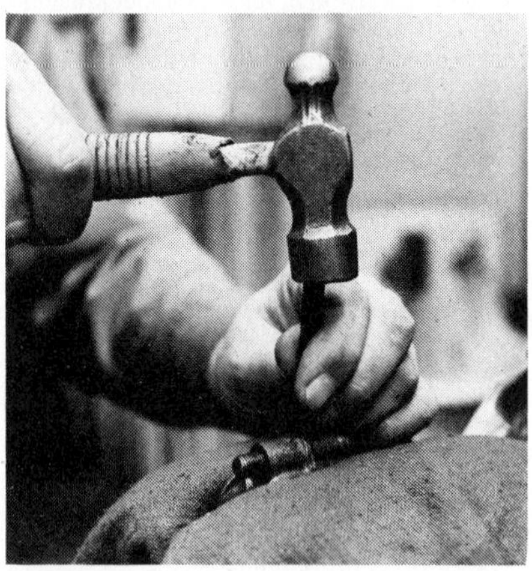

4

1. Isky blower drive designed for a Gilmer belt, was used. Gilmer belts are stronger than conventional V-belts and don't have their tendency to jump pulleys. Friction-driven water pump pulley "slips" at high rpm, saving horsepower; avoid it, if cooling is marginal.

2. A steel or copper head gasket .025-.030-inch thick should be used. The gasket Lee chose for his engine was a Mr. Gasket copper piece. He used 9:1 TRW "eyebrow" forged pistons.

3. TRW Clevite 77 bearings were used, but stock GM HD bearings are okay. Rod and main bearings should be prelubed with graphite or moly. Place protectors or rubber hose over rod bolts to avoid scratching crank journals.

4. Rods and rod caps should be numbered; use metal dies to avoid mix-up. Isky blower manifold has no provision for water outlet because blower takes up too much room, so both heads must be drilled to enable proper coolant circulation.

miles.

It is not necessary to run oil lines to the blower, although some people do. We prefer a sealed end plate and self-contained lubrication system to avoid the possibility of engine oil contamination. The rear bearings need only be lubricated once before the end plate is sealed. The front bearings should run in degummed castor oil, and the front case plate should be vented. This eliminates the problem of pressure build-up, which often causes grease to blow out the front snout.

No tricks need be performed on the bottom end. If it's in good shape when you get the engine, leave it alone. If you're rebuilding, set it up as it came from the factory, using stock clearances. The stock factory "eyebrow" pistons are good forged 9:1 pieces made for Chevy by TRW. They come stock with pressed-in pins; there is no need to convert to a full-floating configuration. We used Perfect Circle rings on our engine, which was assembled at Ron Waldrop's Machine Shop, Taft, California, but any other high-quality rings may be used when rebuilding. The rings are of standard configuration and width. The top ring is moly, the second cast iron and the third a chrome oil ring. They were installed using stock ring gap specifications contained in the factory overhaul manual.

If the engine has many miles on it, the timing chain should be replaced, since the worn unit's slackness causes the timing to retard at high rpm. The TRW-Cloyes double roller chain is preferred.

The stock LT-1 mechanical camshaft is a good grind, but the L-82 (smog) hydraulic bumpstick is not the best setup for overall performance. If the camshaft is worn and needs to be replaced, performance can be upgraded and periodic valve lash adjustments eliminated by installing an Isky 300-degree Hi-Lift hydraulic cam which specs out at 300-degrees duration and .485-inch lift, a little more radical than the LT-1 cam. You can run a wild cam with a blower, since it tends to "tame" the camshaft. With the Isky cam, use anti-pump-up Isky lifters. You can keep the stock LT-1 springs, pushrods and rocker arms. If you opt to use the Isky spring that comes with the cam kit, the correct installation height should be 1.780-inches intake, 1.812-inches exhaust. The intakes are installed with more pressure because the blower tries to push the intake valve off its seat due to its high pressure. Install a cam button to prevent the camshaft from "walking" forward and causing unwanted timing changes due to the pressure from the rear-mounted distributor drive gears.

The best heads are the 1971 and later LT-1/L-82s (part No. 6272068). They have a larger combustion chamber than the 1970 LT-1 heads, service package angle-plug heads or the new "Turbo" heads. The larger chamber is desired to keep the compression below 9:1. Even though these heads and flat-top pistons are being used, always check piston-to-valve clearance whenever new pistons and/or camshafts are installed.

Although this particular engine was O-ringed, it's not necessary to O-ring as long as no more than 7 to 8 pounds of boost is run (which is all you should run anyway, if you want the engine to live). Many machine shops have trouble with this operation, and if the O-rings in the heads and block aren't cut in *exactly* the right place, you'll never get a decent seal. If you insist on O-ringing, buy the head gasket first and then have the O-rings cut, to lessen the chances of a mismatch.

Speaking of gaskets, a composition or steel shim gasket should not be used, since the high cylinder presures will cause them to blow. Instead, use a .025- or .030-inch steel gasket and install it with a good grade gasket sealer.

Valve seal is important. For longevity, a standard valve job is preferred. Even if a valve job is not needed, the guides should be machined for Perfect Circle-type seals.

A blower manifold is available from Ed Iskenderian Racing Cams (16020 S. Broadway, Gardena, CA 90248.) Port-matching the manifold to the heads is not required, but the manifold-to-blower surface should be O-ringed.

We mentioned that boost should be limited to between six and eight pounds to ensure reliability and longevity. To determine the amount of boost, you'll need to install a boost gauge, since boost varies from blower to blower and depends on many variables, such as how tight the unit was set up. Underdriving the blower 17 percent should be just about right.

For an exhaust system, use whatever headers will fit. One-and-⅞-inch-diameter primaries work fine. Run them back through 2¼- to 2½-inch pipe to Corvair Turbo mufflers (available from Chevy under part No. 3858442, or from aftermarket sources). If you run really hard, watch them, since the tremendous exhaust pressures generated by the blower have a tendency to blow off mufflers.

A magneto is considered the best ignition for a supercharged engine, since it delivers plenty of spark to fire the highly compressed mixture. The Ronco Vertex mag is a good choice because it comes with an available stop/start key-activated relay. As for the advance curve, it should start at 350 mag (700 engine) rpm and should total out at 13 degrees mag (26-degrees crank) at 2500 engine rpm. Initial crankshaft advance should be four degrees for a total of 30 degrees crankshaft. This will enable the car to run well on 95.5 octane gasoline, the highest available on the West Coast. In some states better gas is still available, enabling more advance to be run. At 36 degrees total advance, you can expect 40 to 50 more horsepower, but you can also expect to melt the motor faster.

We recommend that carburetion be limited to one 4-barrel, such as an 850 double-pumper Holley (list No. 4781). This helps to limit boost, eliminates synchronization hassles and generally presents half the problems. If an 1150-cfm Dominator is used, the boost is automatically increased because the engine can breathe better. So one way to control boost is to control carb size. A fine-mesh wire screen (available from most diesel repair shops), should be placed between the blower and carb adapter to prevent foreign objects from entering the blower and damaging the case and rotors.

Both an electric (mounted as close to the tank as possible) and a mechanical fuel pump should be used. Fuel-line diameter should be no less than ⅜-inch. A pressure regulator should be installed and adjusted to provide no less than four pounds and no more than seven pounds of pressure at wide-open throttle.

As far as jetting goes, it's better to begin richer than leaner (and perhaps have to replace a piston or two). A ballpark figure is primary, 98, and secondary, 125. Since the biggest jets available are 100s, you'll have to drill them out to the correct size. Final jet combination will depend on many factors, including altitude, atmospheric pressure, intended usage and individual engine peculiarities. While you can come down on the primaries, you'll want to keep the secondary side fairly rich, because the blower really needs that fuel under full-throttle conditions.

If there were no cooling problems before, there shouldn't be any now. But if cooling was previously marginal, be sure to choose a type of blower drive setup that prevents the water pump from freewheeling. Water outlet holes (see photos) will have to be drilled in the heads, since the blower manifold has no provision for the stock water outlet neck.

Now that the package is together and ready to go, a word of caution: There is a right and a wrong way to start a car equipped with a blower. The right way is to floor the throttle to give the engine a good shot of gas before starting, then close the throttle and hit the starter. *Never* crank the engine over with the throttle open; this will create a lean condition in the plenum, possibly resulting in—you guessed it—a blower explosion. Also, after dialing-in the combination, be sure to recheck total advance! All it takes is a slight over-advance and a bad load of gas. . . .

The recommended redline for a motor of this type is about 7000-7200 rpm, but with both the LT-1 and the recommended Isky cam, power starts to fall off beyond 6500.

This combination costs about one-third the price of a fully blueprinted engine (less if you already have the engine). With a 6-71 blower and 350 engine, set up according to the above conservative specs, you'll get an honest 500 horsepower, and with no other changes besides good cylinder head porting, over 525 reliable horsepower is possible. Even without the headwork, the combo will easily move a 3000-pound car with slicks and 5.12:1 gears through the quarter in under 11.5 seconds at better than 117 mph. And it will last as long as any normally aspirated motor, if you drive it reasonably.

—Marlan Davis

5. Because this particular engine was destined for a marine application, a big-block pump was used for extra insurance. For most applications, the LT-1 pump will work just fine. Note the trick oil pump pickup tube, which can never fall out of the pump, because it bolts to the main cap studs. It (as well as the trick windage tray) are available from Dooley Enterprises (1320 N. Miller, Anaheim, CA 92086.) They offer oiling systems for almost any need—street, strip or marine.

6. A Dooley 7-quart pan was chosen because of excellent baffling. Its front sump precludes use on many cars, however. Fitting for oil temp gauge can be installed in place of one of pan's drain plugs. Temperature should stay below 250.

7. Graphite or moly-base lube should be used during cam installation, as was done with this Isky cam. TRW-Cloyes double roller timing chain is mandatory.

8. End valve cover sealing problems with these Corvair clamps. Lightweight Competition Engineering valve covers (5714 Chapel Rd., North Madison, OH 44057) were used because they can easily be modified to clear blower manifold.

9. Electric and mechanical fuel pump combo necessitates use of Holley fuel pressure regulator. To prevent disaster if Gilmer belt breaks and cuts ordinary rubber fuel hose, all lines are aircraft-quality braided steel.

5

8

6

7

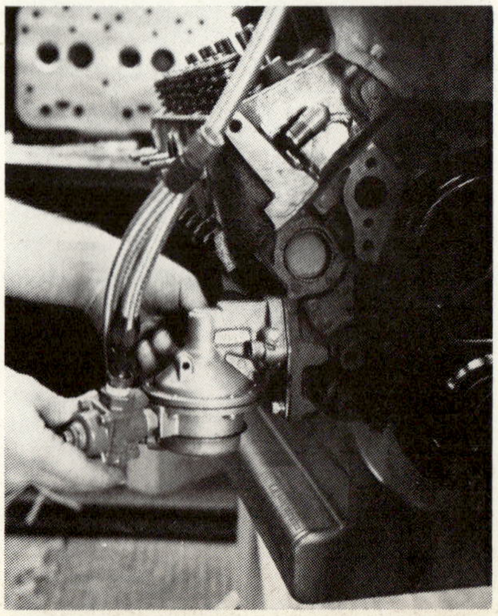
9

Double Trouble

Serious racers take note: Gale Banks' intercooled twin-turbo package for all-out off-road racing efforts is now ready.

In the previous chapter we showed you one of Gale Banks' prototype turbocharger installations on a 454 Chevrolet Suburban. In that chapter, it was mentioned that Gale Banks Engineering (GBE) has developed some "maximum-effort" turbocharger packages precisely engineered and calibrated to reflect the current state of the art.

At this point, it must be restated that this package is *not* intended for any kind of street use. Instead, the system is designed for racing only, with up to 35 pounds (!) of boost on the very best racing gasoline. It's not something for beginners, nor is it cheap. The pricing has yet to be finalized, but the bad news *will* be something over $5000.

With that kind of price tag, it's only reasonable to ask what is included. First of all, the package contains virtually everything needed, with the exceptions of the intercooler water hoses and fittings, the exhaust and waste-gate pipes, the ignition system and the matched camshaft. However, GBE will offer an optional ignition system and can supply camshafts matched to GBE's system and the application.

The next thing that is unique about this twin-turbo system is the special modification of many of the components. For example, most production turbochargers are designed for maximum exhaust temperatures of 1600 degrees F., but for all-out racing applications with high boost pressures, exhaust temps can reach 1900 degrees F. Turbine shaft speed is also higher in racing applications. Therefore, the TO-4B AiResearch units used in the GBE system have modified housings, special alloy impellers and oversized shafts.

Other items of interest in the GBE setup are boost switches for an auxiliary electric fuel pump and for an electric choke which is added to the 850-cfm Holley R-4781 double-pumper. The carburetor is further modified for the correct accelerator pump shot, and the power valve is referenced to true intake manifold vacuum/boost conditions. The internal bronze fuel filters are discarded in favor of high-flow in-line filters, and the float levels are calibrated. In some applications, a 50cc primary accelerator pump is also added.

What makes the GBE system truly unique is the use of intercooling. There is no disputing that intercooling is essential to any current-state-of-the-art turbocharger system designed for ultra-high output. It hasn't been used previously on most automotive applications only because of the extra expense and plumbing required in situations where hood clearance is frequently a problem.

Intercooling cools the fuel/air mixture after it leaves the turbocharger compressor, thus increasing the charge density. It is charge density that dictates how much air and fuel is packed into the combustion chambers to produce power; boost alone isn't the key. Another benefit

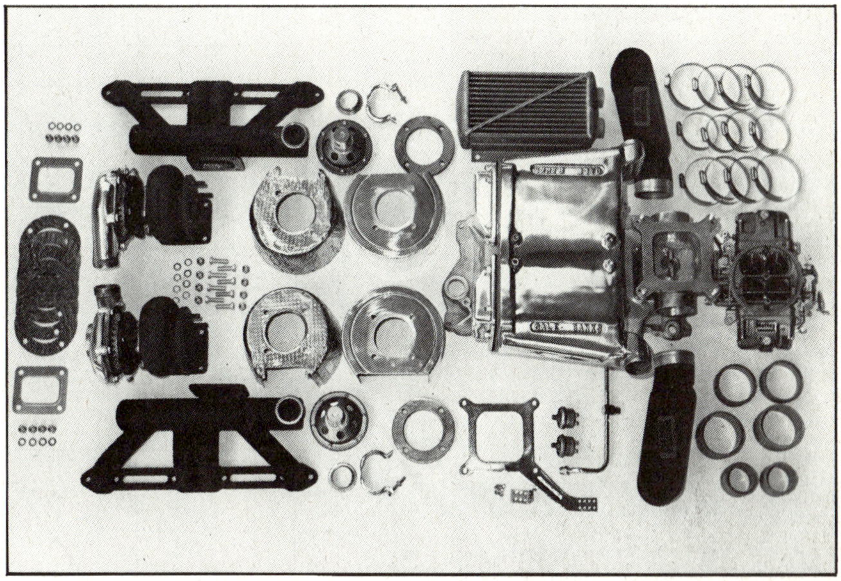

1

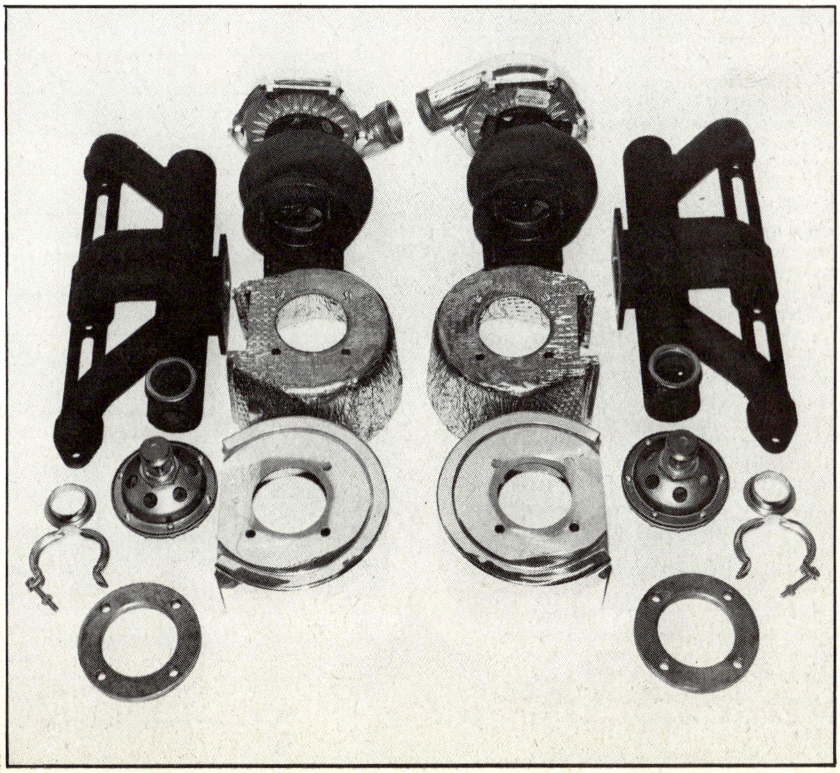

2

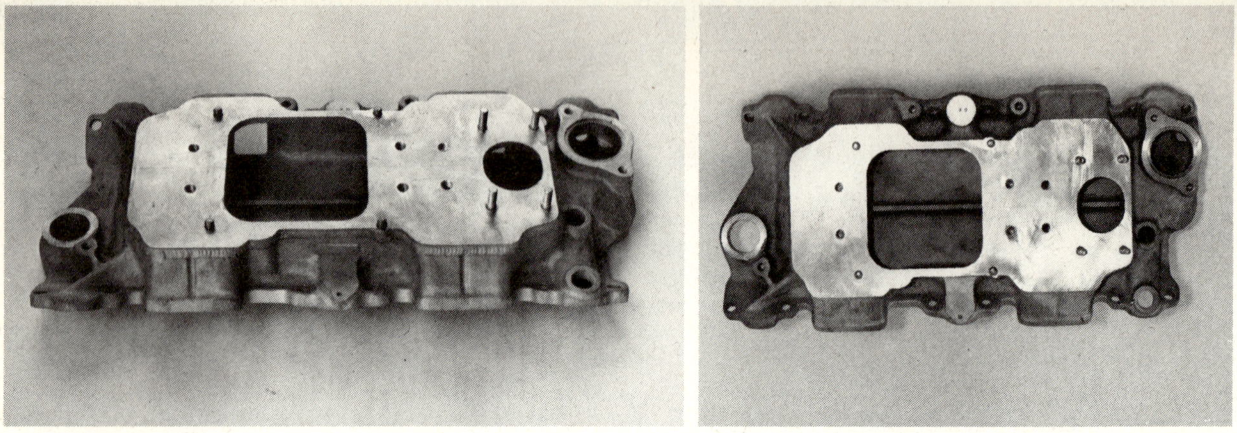

3

of intercooling is that it resists detonation, thereby permitting even more boost and a greater spark lead.

The GBE system uses a common housing for twin intercoolers. The coolant for the intercoolers is circulated at a very high rate, separately from engine coolant, by a belt-driven auxiliary pump (not shown). The coolant flows through either one or two Harrison NASCAR heat exchangers, depending upon the application.

Just how effective is intercooling? Well, to give you an idea of what can be accomplished when the intake mixture is cooled, the GBE system installed on a 302-cubic-inch, 7:1-compression, small-block Chevrolet V8 road racing engine produced over 900 hp and 750 ft.-lbs. of torque at 30 pounds of boost with 32 degrees of ignition advance!

—C.J. Baker

1. The complete kit includes virtually all of the necessary hardware for installation. Fabrication of the exhaust pipes and waste gate by-passes are left to the car builder. The apparent simplicity of the kit belies the extensive engineering and testing that went into it. The only critical components not supplied with the kit are the ignition system and the camshaft. Both of these items must be matched to the application, but they are custom-tailored options from GBE.

2. The exhaust side of the system includes stainless steel manifolds with built-in waste gate provisions. The waste-gate by-pass attaches directly to the side of each manifold with the V-band clamp provided. Exhaust pipe flanges are also provided. The header flanges are .400-inch thick to prevent distortion under high heat. The dual waste gates are specially modified Rotomaster units with double the stock flow. Boost can be regulated by the driver from 10 to 35 pounds maximum. The twin TO-4B AiResearch turbochargers are modified by GBE, as outlined in the text. Heat insulators and decorative chrome covers are included for both appearance and increased turbine efficiency.

3. The aluminum intake manifold is modified to incorporate an open plenum. Four dowel pins locate the intercooler, and the front four studs secure the carb plenum. The front opening is for idle fuel and vacuum sourcing to the carburetor. Shown here is the first-design manifold. Subsequent designs permit near perfect distribution.

GALE BANKS' TWIN-TURBO DYNO TEST

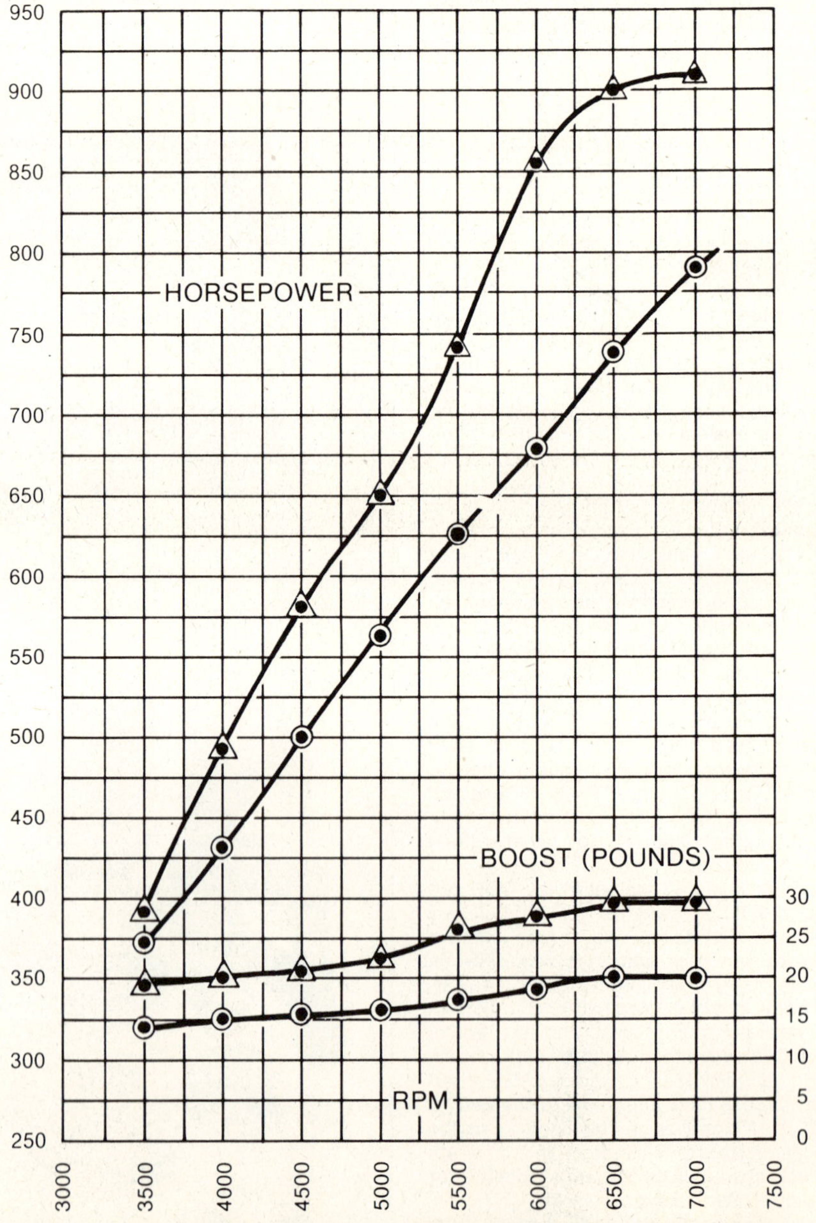

4

4. Multiple views of the Gale Banks twin-turbo, dual-intercooled, variable-boost competition package reveals the fine craftsmanship and planning that went into the component layout for maximum efficiency, accessibilty and minimum overall dimensions. Note the ease of spark plug access, the low intercooler profile for cowl induction ducting and the carburetor and distributor access. Note, too, how compact all plumbing is for instant throttle response and minimum heat loss.

5. The carburetor plenum actually serves as a divider between the inlet and outlet sides of the compressor, since there is no direct path for the air/fuel mixture through the plenum into the manifold. Instead, mixture is drawn through the carburetor, where it flows out both sides to the turbocharger compressors. The compressed mixture then enters directly into the intercooler housing. There is a small orifice in the base of the plenum for idle fuel passage and vacuum signaling to the carburetor. The plenum is also water-heated to prevent carburetor icing. A carb linkage attachment spacer is included, thus simplifying installation.

6. Unique to this automotive system is the use of dual intercoolers in a common housing. The actual heat exchangers are specially made for GBE's marine system from Cupro-Nickel to resist saltwater corrosion. Consequently, nobody will ever be plagued with rust or scale build-up in automotive use. The four long bolts pass completely through the housing to secure it to the manifold.

6

5

Light Touch

Milodon's aluminum "mouse-block" is just what the weight-watchers ordered.

From mighty fuel-burning seven-liters grow five-liter offspring, or in the words of Milodon's "Spider" Razon, "After we produced our 426 hemi for the last five years, the racers asked us to do the same for the small-block Chevrolets." What the racers didn't figure on was that Don "Dooner" Alderson was way ahead of them. The president of Milodon had just sweet-talked Mickey Thompson out of his old Chevrolet-engineered, Indy-type small-block patterns and completely updated them for today's market.

As you might have guessed, the old/new Milodon five-liter cylinder block is a near duplicate of a cast-iron Ed Cole. The only deviation from stock is that it has no provisions for the block-mounted fuel pump or oil filter. Naturally, this means that the owners of these all-alloy replicas are going to have to use an electric fuel pump and a remote filter. Other than these two minor problems, Dooner's dandy is a bolt-on delight. Your stock cranks will fit, the oil galleys are the same, and the bolt patterns—including cylinder heads, bellhousings and motor mounts—are all to original equipment specifications.

In essence, the block is cast out of the same aircraft alloy (A-356) as Milodon's top fuel "seven-liter." The raw castings are fully machined on a numerically controlled tape mill to ensure that each block will be as accurate as a totally blueprinted stock-block. As is common with today's generation of aluminum blocks, the stepped dry sleeves (dry because they do not come in direct contact with the water in the block) are spun out of centrifugally cast, and then heat-treated, nodular iron. The sleeves can be ordered in an infinite number of bore sizes, including a hogged-out 4-inch, plus .030" over. As for stroke, Razon says that the block can accept a ½-inch-over 350 crank. The five-liter will also include heat-treated four-bolt main caps, 1¾-inch stud bosses, and ground and rolled thread studs heat-treated to 180,000 psi.

While the Milodon five-liter is approximately 70 pounds lighter than its cast-iron counterpart (160 pounds for the stocker, 90-95 for the Dooner), it will not suffer from a corresponding reduction in strength. For instance, the ductile iron sleeves are said to be three times as strong as the stock cast-iron cylinder walls. With those 1¾-inch "blind" bosses, an engine builder can take full advantage of the increased thread surface. Like the sleeves, the deck surface is considerably thicker, and if there is a corresponding reduction in water jacket capacity, it would only amount to a pint or so. For all these advantages, there is but one disadvantage—its price. The Milodon five-liter cylinder block—complete with fully machined casting, eight ductile sleeves, four-bolt main caps, heat-treated studs and preinstalled cam bearings—will run around $2500. But as Joe Anahory points out, "The cylinder walls stay rounder, it weighs half as much as an iron block, you don't need to blueprint our block, and everything built for a small-block Chevrolet fits like a glove and makes it a winner." Not only that, but it'll look like gangbusters when it's given the full-polish treatment.

—Gray Baskerville

1

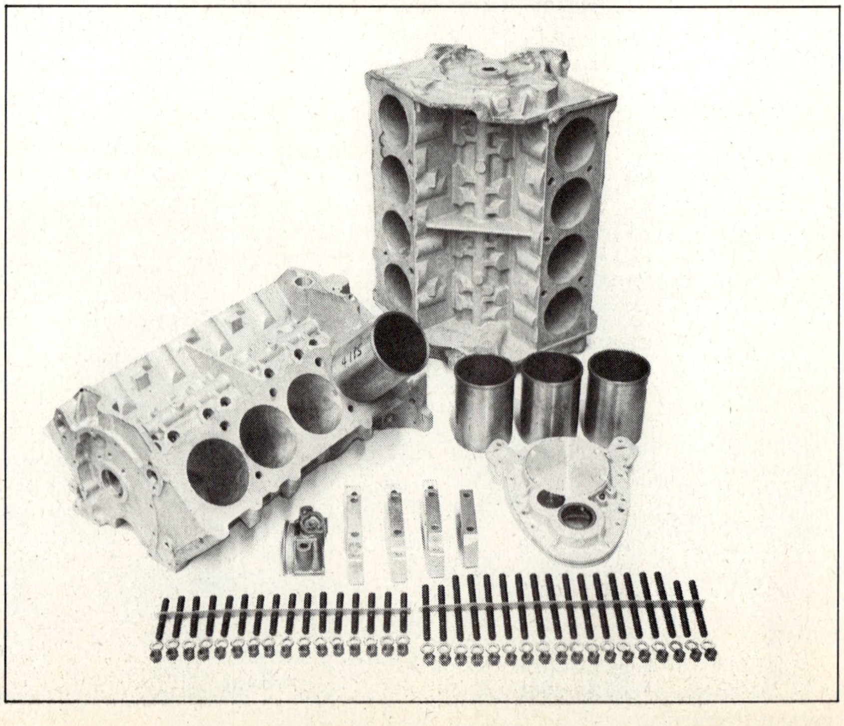

2

3

4

5

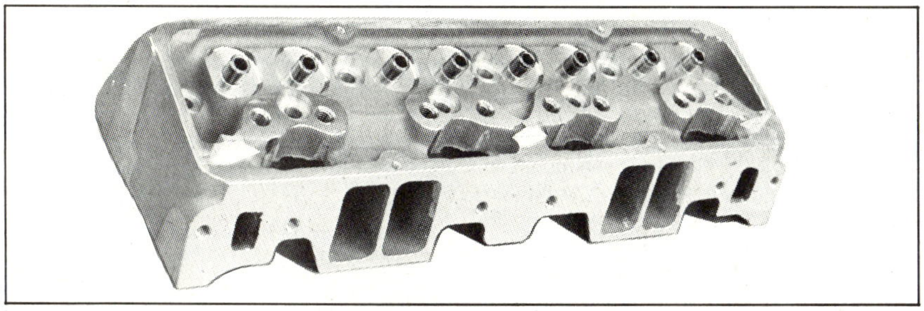

6

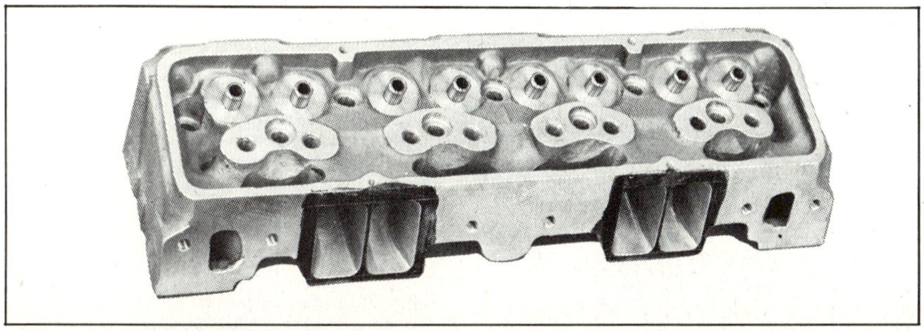

1. Don Alderson is as big and heavy as his all-alloy five-liter mouse is small and light.

2. "Muldoon's" basic kit will be composed of a blueprinted cylinder block (line-bored, decked with all the mating surfaces milled and ground), fitted with ductile iron liners, cam bearings, four-bolt main caps and all the necessary studs and fasteners. The front cover is extra, as will be a new pan and deep-drawn rocker covers. Total assembly will tip the scales at between 90 and 95 pounds and can be "sleeved" and stroked to 400 inches.

3. Heat-treated ductile iron main caps feature angle-in bolt holes. This allows the forged fasteners to be placed in shear and to screw into blind bosses to take advantage of full thread contact.

4. What makes the Milodon "fly-weight" so interesting is that you can vary its displacement by simply changing liners and pistons. Liners (or sleeves) are made from heat-treated, centrifugal cast iron which is said to provide three times the inherent wall strength of a stock-block.

5. This is the late version of the very well-designed Brownfield aluminum head. These heads are essentially ready-to-run just the way they come out of the box. Even the valve job is already done.

6. Shown here is the Brodix aluminum head as it is normally prepared for circle-track racing. This one has been fully matched and ported by Reed Cams in Decatur, Georgia.

LIGHTHEADED PERFORMANCE

The all-aluminum small-block is now within reach, because not only has the aluminum block become a reality, aftermarket aluminum heads are also readily available. Shown here are the two leading contenders in the world of alloy alternatives, the Brodix aluminum head and the Brownfield aluminum head. Both versions feature a reduction in weight, improved flow and an increase in strength in critical areas. The accompanying specifications will give you a good idea of what you get when you lay down something in the area of $800 for a pair.

SPECIFICATIONS:

	BRODIX	BROWNFIELD
Weight:	21 pounds	19 pounds
Material:	Premium alloy, heat-treated to Brodix specifications	356-T6 alloy, metallurgically enhanced to achieve maximum strength and durability. Heat-treated
Ports:	Modified ultra-wide intakes for use with offset rocker arms and redesigned exhausts	Intake and exhaust ports redesigned. Unported, they equal or exceed the flow of the best welded and ported iron heads.
Guides:	Hardened iron or aluminum bronze guides available for either 11/32- or 5/16-inch stem diameters	Aluminum bronze guides, 11/32- or 5/16-inch stem diameters offered. Tops machined for P.C. seals
Valve Sizes:	2.02-inch intake, 1.60-inch exhaust, hardened steel seats	2.02-inch intake, 1.60-inch exhaust, hardened steel seats
Chamber Size:	67cc. Special milling available	67cc. Special milling available
Deck:	.450-inch minimum thickness with reinforcing ribs in critical areas	.563-inch minimum thickness, reinforced
Spring Pads:	Reinforced, machined to customers specified spring height and o.d.	Reinforced, machined for 1.80-inch installed height with .030-inch shims. Installed heights up to 1.90-inch possible.
Rocker Stud Pads:	Reinforced and Heli-Coiled	Reinforced, interconnected and Heli-Coiled
Spark Plug Placement:	Angle plug, 14mm, ¾-inch reach	Angle plug, 14mm, ¾-inch reach
Comments:	Each head pressure tested at the factory to eliminate leakers. Large exhaust flange can be drilled to accept special bolt patterns. Special machining available	Redesigned water passages for controlled heat transfer. Porting seldon required. Enlarged exhaust flange will accept big tube headers without an adapter plate.
Source:	Brodix Aluminum Products P.O. Box D Mena, AR 71953 (501) -94-1075	Brownfield Company 3062 Lima St. Burbank, CA 91504 (213) 846-2227

Panning for Gold

A scoop in oil pan modifications is worth a fortune to the serious racer.

Someone has uncovered more usable horsepower in a competition engine. While everybody else was frantically trying to find another 10 or 15 horsepower in trick cylinder heads, oddball manifolding and slick piston ring combinations, Bill Jenkins was playing with oil pans and related components. While what he's discovered may not be news to some people, like the guys at Chrysler who have been sorting oil pan combinations for years, the potential power gain will help many small- and big-block Chevy racers. And the best news is that these components are available over the counter.

Admittedly, many racers have found power gains through such devices as pan evacuation systems, knife-edging crankshaft counterweights, streamlining the lower ends of the connecting rods, louvered or screened windage trays and even dry-sump oil systems. But why have these devices really increased power? At first thought, it seems obvious that there's a lot of oil being thrown around in the crankcase as it drains back from the upper portions of the engine and is slung off the crank and rods. It's difficult to determine what path all this oil takes inside the crankcase of an engine turning 10,000 rpm, so theories can only be proposed and modifications made based on those theories.

If oil is flung off the rotating parts as we first imagined, does it really bounce off the sides of the crankcase and surface of the oil in the pan with sufficient force and volume to create a significant drag on the rotation of the moving parts? Perhaps, but let's consider a second theory. Let's assume that a large portion of the oil coming off the rod and main journals, along with that draining back from the top of the engine, is retained and spun in a "taffylike" mass and that the contents of this mass could equal nearly two quarts, before being released. This is the most recent concept, and the puzzling pieces seem to fit when applied to this theory. For example, it explains why large-capacity oil pans are required on high-rpm engines, including those where oil flow to the top of the engine has been greatly restricted. It also explains why knife-edging works, since the rods are actually slicing through a restrictive mass of oil rather than just oil spray. It also explains why dry-sump systems and other such devices that help pull oil away from the crank show considerable power gains.

If we accept the theory that some oil gets carried around by the crank and

1. Moroso supplies a universal small-block scraper that can be trimmed to within a fraction of an inch of the rotating parts.

2. Eight windage tray studs are used to support the screen. Screen is shaped to catch oil flying off the rotating assembly before it strikes either oil in the sump or sides of the pan.

3. With both the screen and scraper in place, the relieved areas in the pan below and at the side of the screen can be appreciated.

4. The Jenkins/Moroso pan is angled to the right and to the rear to direct oil to the scavenge pickups. Note the dual drain plugs (necessary because of an interior baffle), the relief just under the pan rail and the tubes for the bolts.

5. Here are the two pickup tubes before they're welded into the pan. One end of the tube will be plugged. Note the shape of the opening (slotted) and the screens which slip into the outer end.

6. Interior shot shows the relief under the rail, positioning of the pickup tubes and the anti-surge baffle halfway back. Pickup is about ⅜-inch off the floor.

1

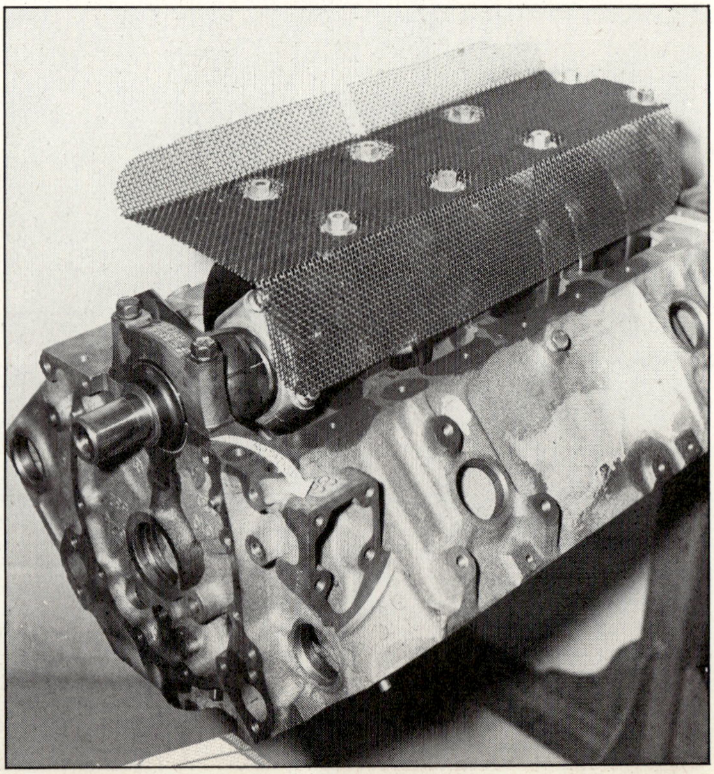

2

horsepower is lost due to this frictional or fluid drag, then the first thing to do is to remove the conventional solid windage tray, since that helps contain the oil and air around the crank even more. Of course windage trays have been shown to be beneficial, but by today's standards, only on engines that do not operate at ultra-high rpm. This is the area where we must modify the theory to say that oil splash is significant and that a windage tray does help until crankshaft and connecting rod speed gets sufficiently high to trap the oil around it. The rpm at which this happens is determined by a number of things: how much oil is being thrown off the crank journals, the amount of oil draining back from the top of the engine, the overall diameter of the crank and whether or not the crank and rod edges have been streamlined.

For theoretical purposes, let's assume an optimum engine speed of slightly over 6000 rpm. What would be needed here is a system that prevents splash both below and above 6000 rpm without trapping the mass of air and oil around the crank above 6000 rpm. The current answer is a screen, positioned as close as possible to the crankshaft and rods without touching them. Such a screen, running the full length of the crankshaft, has been tested, and it works. Score point one for theory two.

Effectively, a screen allows the oil being thrown off the crank at high velocity to pass through, and yet because of a pressure differential, the oil bouncing off the sides of the pan and the oil surface below the screen does not manage to find its way back through the screen to the crank. The crank, spinning just a fraction of an inch above the screen, creates a high-pressure area above the screen, while the oil pump pickup (below the screen) creates a low-pressure area, and thus the oil tends to move off the crank and into the sump. In any event, sufficient space below such a screen must be provided in order to collect the escaping oil.

To improve upon the scavenging action of the screen, a rigid scraper, installed along the oil pan rail, more effectively cleans oil off the rotating assembly (rods and crank). Such a scraper is easily positioned along the side of the pan and trimmed to extend within a few thousandths of an inch of the crank and connecting rods. Once this was tried, performance again improved, but not as markedly as anticipated. The basic problem was that so much oil was being scraped off the crank that the scraper was being overwhelmed. By altering the side of the oil pan to provide additional escape for the oil being caught by the scraper, the expected power gains were achieved. In fact, further modifications to the pan to relieve it in all areas where quantities of oil exit the crank showed still more improvement.

The power increases we are talking about amount to possibly 50 horsepower on a 331-cubic-inch small-block Chevy, but this is greatly dependent on the overall execution of the entire system. Of course development work continues. In fact, Jenkins has worked with Moroso Performance Sales to construct the prototypes and revisions. Moroso has dry-sump small-block Chevrolet pans incorporating all these features for distribution, with wet-sump versions incorporating similar designs to fit other classes that disallow dry sumps.

Before we end this discussion, it might be worth noting that there is a lot of work to support this approach to bottom end oil control. Several years ago, Chrysler conducted extensive tests that proved the importance of relieving the pan to reduce oil splash and, where possible, to lower the oil level in the sump from the proximity of the crank. More recently, working independently, Smokey Yunick has arrived at similar conclusions for use in high-rpm small-block engines used for circle-track racing. In fact, if you could take a peek inside the oil pans of some of the top-running NASCAR Grand National cars, we suspect you'd find a wide variety of screens and scrapers. Maybe it's time you hung a screen door and scraper of your own.

—C. J. Baker

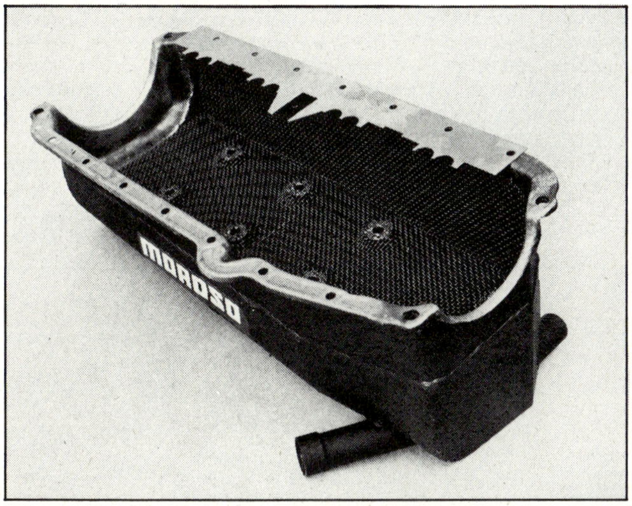

3

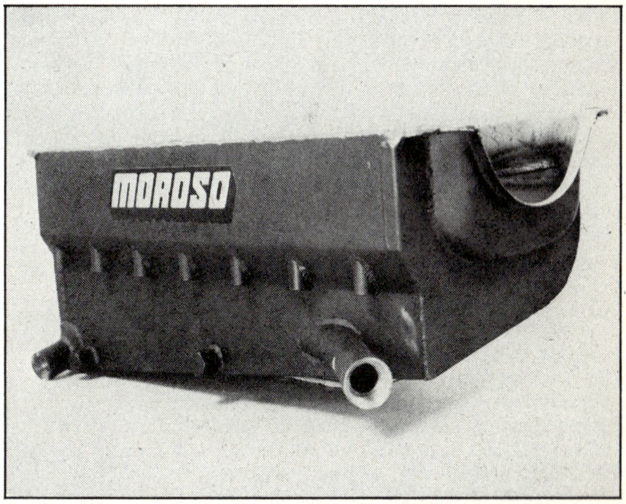

4

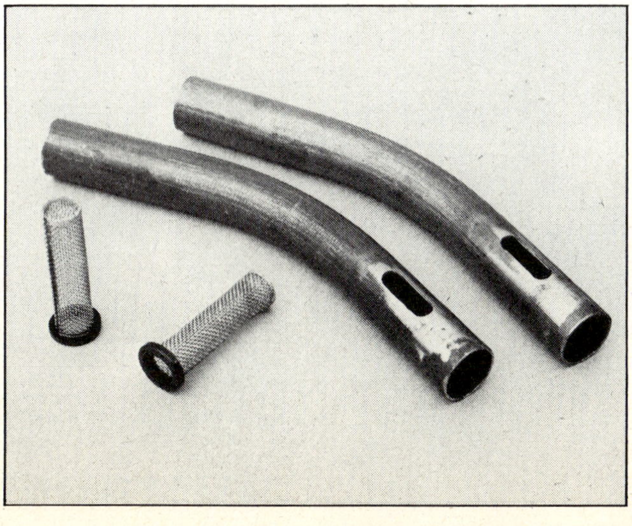
5

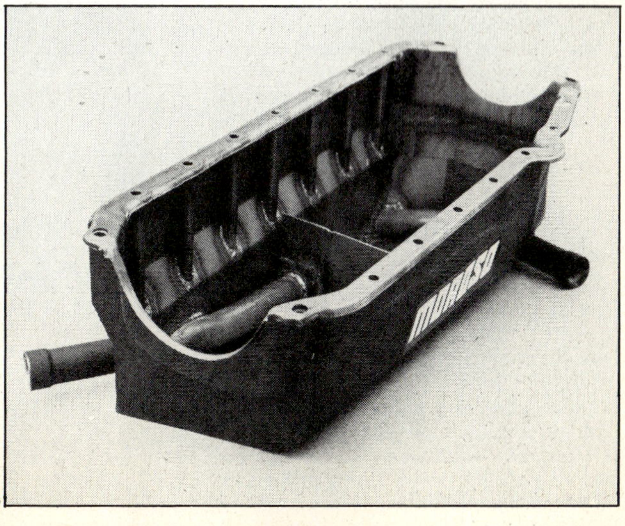

6

The K Rod

Here's a fully-finished, polished and shot-peened rod from Chevrolet.

The switch to small-block engines in NASCAR-style roundy-round racing has stressed many areas of weakness in many different engines. On Chevrolet engines, it became apparent that the durability of the connecting rods, even the expensively reworked heavy-duty versions, left something to be desired.

In an effort to offer a durable connecting rod to meet the full range of off-road engine builders' needs at a reasonable price, Chevrolet introduced their semi-finished rod (part No. 343710). This rod is actually based on the massive big-block forging, but with a big-end diameter of only 2.1 inches to fit the '68 and later small-block crankshafts and with the small end left blank to permit the builder to finish the rod to any length between 5.7 and 6.1 inches. Additionally, a larger balance pad was included and adequate stock was left on the small end for the installation of pin bushings. And because of the massive dimensions of the forging, it was sufficiently over-designed to negate the need for exotic steel alloys, like 4340 steel with its attenuating forging problems, such as inclusions, and its more expensive machining costs. Instead of using this, SAE 1038 steel is used.

On paper, the semi-finished, no-pin-hole rod looked like a good idea, but it seems that Chevrolet overestimated the abilities of some machine shops around the country and the engine builders' understanding of proper connecting rod preparation. A number of builders committed a wide range of errors in finishing the rods, ranging from excessive material removal at the pin end, failure to reshot-peen or properly shot-peen the rod after machining to simple failure to check rotating clearances during assembly. Consequently, there were numerous reports of rod failures and destroyed engines, which gave the rod a bad reputation. And if that wasn't enough, by the time a racer had paid a shop to finish the rods, they became almost as expensive as even the best aftermarket rods.

So once again Chevrolet has come up with a solution aimed at both durability and economy. The solution is the new *fully* finished off-road connecting rod (part No. 366230). Essentially, this is the same forging as the 343710 rod, only now it has been fully finished by Chevrolet. The small end is drilled for a standard pin and a center-to-center length of 5.7 inches. All excess material is contoured from the small end and the cap, although there is still adequate stock for pin bushings. The beams are polished to remove forging flash, roughness and nicks. The completed rod is Magnafluxed and shot-peened to eliminate surface imperfections and stresses. And of course, it is fitted with the premium-quality 185,000-psi tensile strength 7/16-inch-boron bolts.

Because such premium rods represent a very small portion of Chevrolet's total connecting rod production, quality control and low unit costs would have been difficult to achieve within their own facilities, so Chevrolet has contracted Kiekhaefer Aeromarine in Fond Du Lac, Wisconsin, to handle the finish machining and processing. Consequently, we've dubbed this new rod the "K-rod" to readily identify it.

Because the K-rod is considerably more massive than conventional small-block rods, there are a few things that must be considered when installing them. The K-rod is nearly 200 grams heavier than a stock small-block rod, so the rotating assembly will have to be rebalanced accordingly when using them. And since the big end of the rod is physically larger, care must be taken to verify that the rod bolt nuts clear the lower end of the block and that the rod bolt heads clear the camshaft lobes. Normally, this problem arises only on those engines with a 3.48-inch or longer stroke. It is not a problem on 3.00- or 3.25-inch-stroke engines. If a block clearance problem exists, it can usually be corrected by using the 12-point aircraft (SPS) nuts (part No. 340289) and/or by grinding clearance notches in the block. If there is a cam clearance problem, it can usually be corrected by grinding the bolt head at the same angle as is now incorporated in the head shape.

List price on the K-rod is projected at $56.95 each, which means it'll go over the counter to most racers at just under $50 each—ready to run. At first that sounds like an expensive rod, but when compared to the costs of fully machining the semi-finished no-pin-hole rod or special aftermarket rods, it's a bargain. And Chevrolet later brought out the K-rod in a 6.0-inch length. For the street, for the short-stroke small-blocks, for the racer with a limited budget or for those with no ready access to a shop qualified to correctly finish connecting rods, the K-rod has got to be one of the better things to come across the local dealer's parts counter in a long, long time.

—C.J. Baker

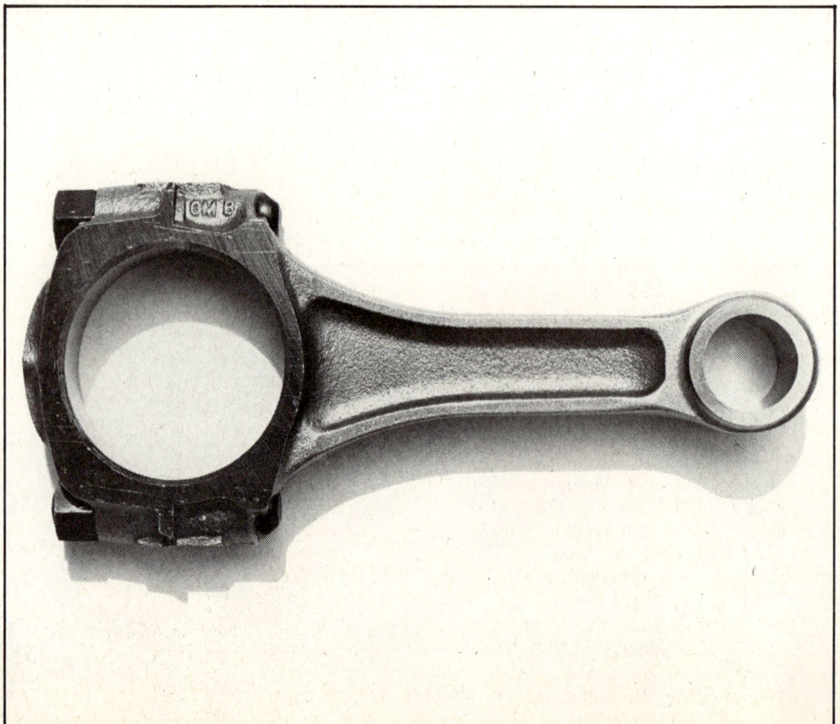

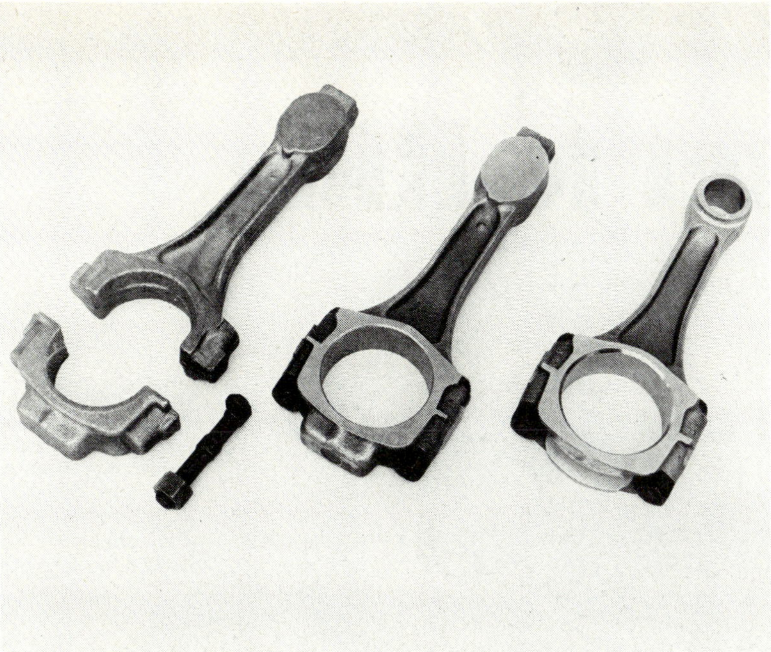

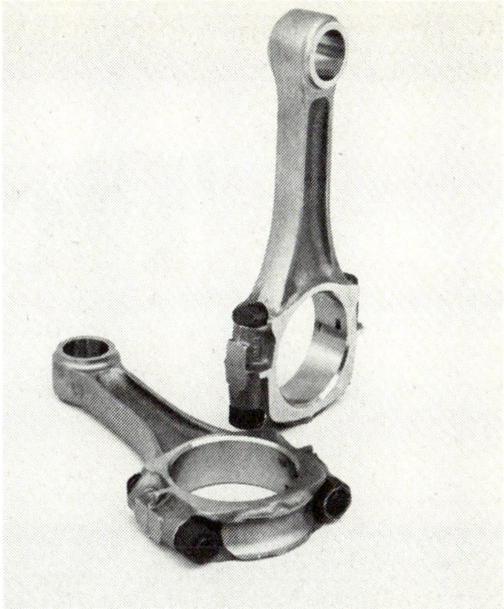

Both the No. 343710 semi-finished no-pin-hole rod (middle) and the new No. 366230 finished K-rod (right) start life with the massive 1038 steel forging (left) and the premium-quality 7/16-inch boron rod bolts. The K-rod is contour-machined on both the small end and cap to remove weight and possible stresses.

Testing connecting rods is very expensive. It requires incredibly expensive hydraulic apparatus and fixtures that are beyond the financial capabilities of most independent rod fabricators. Consequently, most aftermarket connecting rod manufacturers rely upon their customers to test their rods for them—in their racing engines!

The large automobile companies, such as Chevrolet, do have the resources for such testing equipment. In fact, due to warranty considerations, the OEM manufacturers must test their production components to ascertain their fatigue life before allowing them to fall into the hands of their customers. And of course, these same test fixtures can be used to evaluate heavy-duty high-performance components as well, and that's a definite plus for the racer.

The test fixture, control panel and hydraulic power supply unit subjects the rod to alternating compression and tensile loads. The rod is installed with bearing inserts in place, rod bolts torqued to specification and pressurized lubrication to the rod bearings. Additionally, the rod is bathed with oil, just as it would be in an engine, to dissipate heat. The combination of one compression load and one tensile load is called a cycle. The amount of these loads, which is dialed into the control panel, is determined by the component weights, anticipated rpm and cylinder pressure. Therefore, the compression loads, which include the anticipated cylinder firing pressure, are higher than the tensile loads. Examples of the test loads would be 9500 psi tensile and 14,000 psi compression for a stock small-block connecting rod. With stock piston and pin weights, that would be roughly equivalent to 7600 rpm in a 350 engine. In fully reworked condition, which includes polishing and shot-peening, the heavy-duty small-block rod (part No. 3946841) will normally survive from 1 to 2 million cycles before failing at these loads.

The fact that 3½ hours' running at 7600 rpm, such as in a NASCAR 500-mile race, puts nearly 1½ million cycles on each connecting rod points up the need for a stronger rod, such as the new K-rod. Chevy tests the K-rod at higher loadings, 11,400 tensile and 15,000 compression, due to its heavier weight and in anticipation of higher engine speeds. This is roughly equivalent to 8100 rpm in a 350 engine and over 9000 rpm in a 302 engine. Under these loads, the K-rod was still in perfect condition after 10 million cycles, at which time the test was terminated as completely successful.

Of course, this machine is designed to test fatigue life of the rods rather than ultimate strength, so it is quite feasible that the rods will withstand higher loads, such as generated by even higher engine speeds, for short durations of time, as in drag racing. As a final note, Chevrolet has tested various aluminum rods with eye-opening results—some failed in as little as 150,000 cycles!

Ticket to Ride

A round-trip tour to sticking a big Chevy valve train into a small-block engine.

1

The advances produced by the requirements of Pro Stock have been a bellwether for all other users of modified racing engines; that is, all other users who wish to remain competitive. Insatiable camshaft lift and 10,000-rpm engine speed tend to produce a valvetrain with more untoward movement than the Keystone Kops.

The small-block Chevrolet seems to be champ at this, but there are ways to allay the problem. Though he claims no credit for the idea, Dan Jesel (and Competition Machine Service, Freehold, New Jersey) have realized a workable solution, using valvetrain parts designed for the big-block Chevrolet.

Danny looks at it this way: When a camshaft with an actual lift of over .600-inch is used, rocker arm geometry begins to bend out of shape rapidly. It's like visualizing the old seesaw: The farther the rocker arm tip is from the fulcrum of the rocker, the lesser its angle is at any given camshaft lift. And the lesser this angle is, the lesser the "sweep" (full crossing of the valve stem by the rocker tip from start to end of lift) is across the valve. When you can do this, your rocker arms will be able to retain their ratio for a longer period of time.

The CMS aluminum rocker arm pictured here uses a 1.55:1 (exhaust) and 1.65 (intake) ratio, with a .120-inch-offset pushrod cup on the latter. The very size of the big-block rocker allows use of the large 1⅝-inch-diameter valve spring without notching the arm.

Dan prefers to use shortened ⅜-inch ZL-1 pushrods while running the roller lifter with as much as .200-inch offset. This allows more pushrod clearance in the cylinder head too, as the pushrods are not as close to the combustion chamber as with conventionally offset rockers.

So what we're talking about here is valvetrain rigidity: a meaty rocker arm; shorter, stouter pushrod; large journal billet camshaft; fat valve springs and a cam-walk stopper. But none of this will jibe until the LT-1 or Turbo cylinder head is fitted with the offset big-block rocker studs. And you can find out how that's done right here.

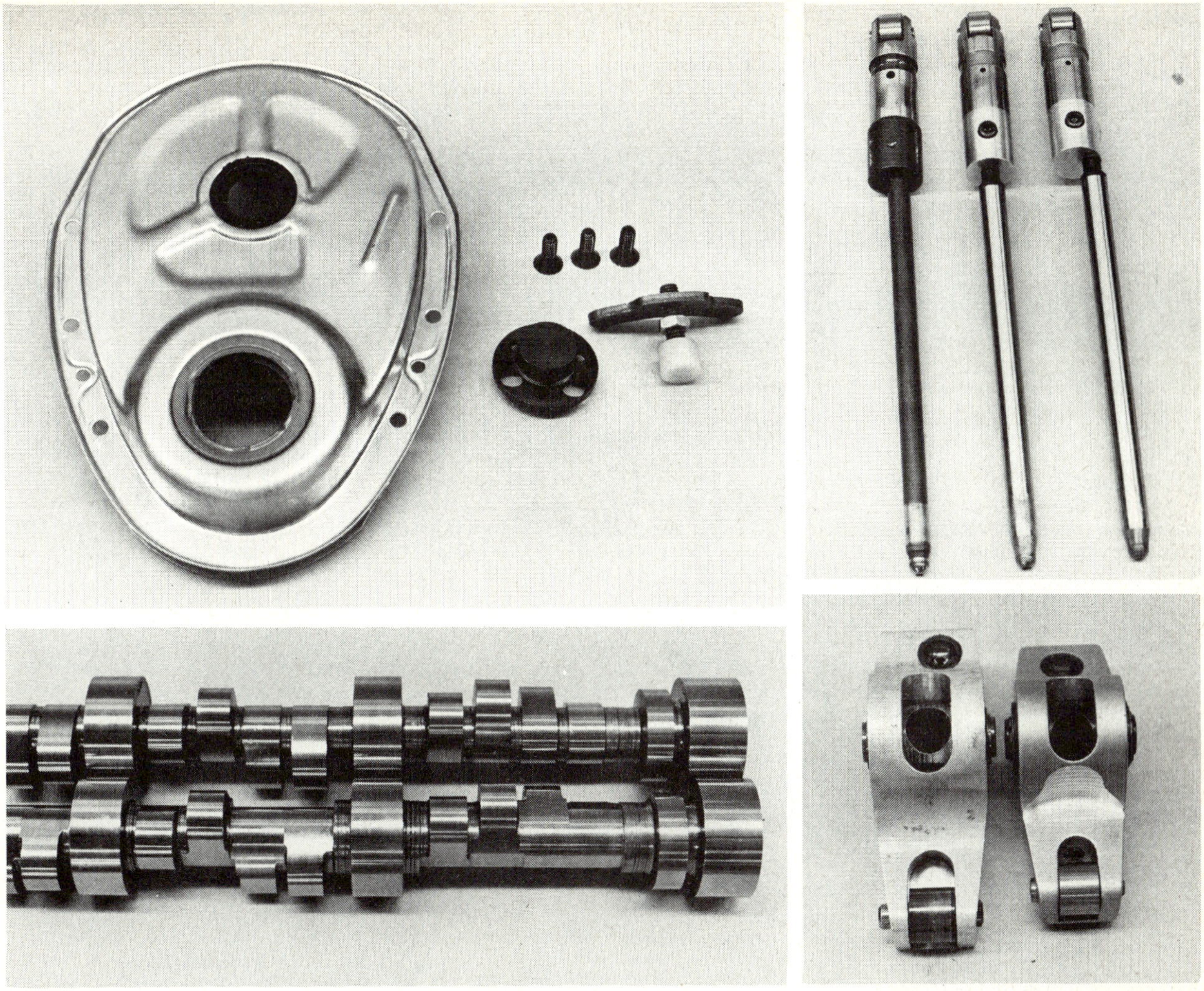

2

3

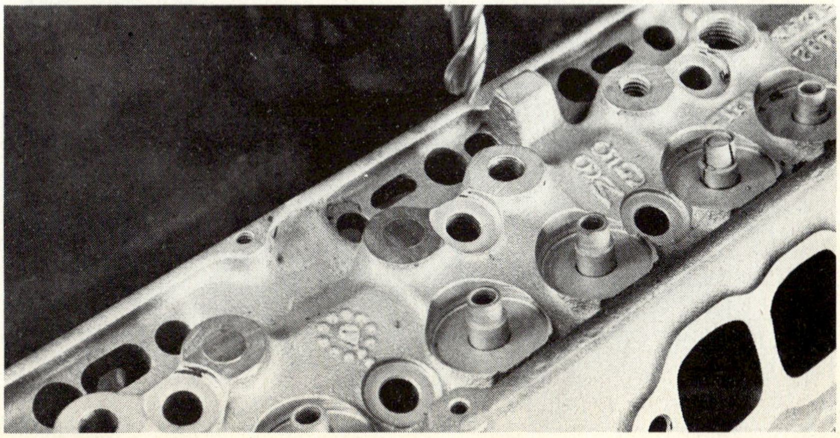

4

1. In order that Jesel's stud boss drilling jig (or one of your own design) can be used, drill the threads out of the two inner stud holes. Since the big-block studs are larger than the small-block ones, the newly drilled hole can be further cleaned by replacing the drill bit with a reamer.

2. Competition Machine Service's cam stop prohibits roller camshafts from walking. Dark button goes on the cam, protrudes through the timing chain cover. The Teflon piece attaches to the back of the water pump, and the two meet and become basically immovable. Both camshafts have the same specs, but the lower cam is much beefier billet stock and helps keep deflection to a minimum. The offset Isky lifter at right (soon to be fitted with a steel pushrod cup) and shortened ZL-1 pushrods hate deflection, too. General Kinetics big-block rockers offer full-width offset roller pin, longer fulcrum and offset pushrod tip.

3. The old tap and die routine revisited: The reamed stud boss is now tapped with a fine thread to accept plugs, which will be Irontite with a .490x1-inch dimension. In effect, the plugs fill the enlarged holes so that a completely new, offset hole can be drilled.

4. Screw the appropriate end mill into the Bridgeport chuck and cut the Irontite plug flush to the stud post.

5 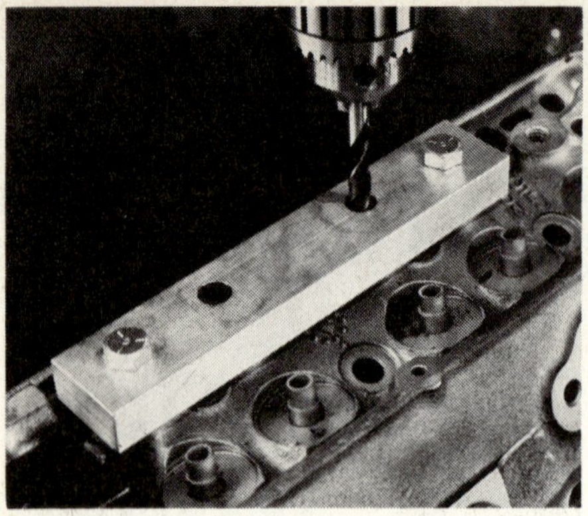

6 

5. The "inner hole" jig is bolted to the end stud holes (still untouched at this point) and a new .368-inch-diameter hole is drilled. Do the same trick on the exhaust side with a jig made for that purpose.

6. For final prep, new thread is tapped for the big studs. Stud bosses may need additional milling to make room for bottom arc of larger rocker arms.

7. Jesel's own pushrod guide plates are enlarged and set at appropriate angle with the new studs installed.

LARGE-BLOCK ROCKER ON SMALL-BLOCK CHEVY

The product of increased hp effected by small changes in parts or part configuration is an item long sought after by engine builders. The use of big-block rockers on the small-block Chevy is a typical example of effective changes. The four sketches attempt to shed light on some of the benefits of this modification.

Figure 1 describes the sweep or surface travel of the rocker over the face of the valve stem as it makes its arc while pushing the valve open. This loading tends to produce a number of undersirable conditions, including the following:

CAUSE	RESULT
A. Surface drag	**Friction, hp loss**
B. Side loading on valve stem	**Friction, hp loss and worn valve guides**

Using a big-block rocker reduces this travel, due to the longer arc, thereby causing less actual sweep. When you consider that a short haul down a quarter-mile strip can produce as much as 2000 inches of actual travel (or drag) across the valve stems, you begin to see the buildup.

Now look at **figure 2.** This illustration shows the different angles that the rocker must approximate to produce the same valve opening. As you can see, the small rocker needs to move more in order to produce the same amount of valve opening. Less movement in almost any case means less wear. Also, because of the shallow angle of movement, there is less of a decreasing ratio of movement, as illustrated in **figure 3.** Since the big rocker has a longer (moment) arm or larger arc to cover, it produces less of a ratio decrease (when compared to the small rocker), while opening the valve an identical distance.

Another rocker-related rework worth mentioning is the inward relocation of the rocker mounting stud to allow pushrod clearance of the enlarged intake ports. This practice generally necessitates off-center location of the rocker pushrod cup (a), as illustrated in **figure 4.** It is also important to relocate the bearing (b) on the opposite end to offset the imbalance and maintain an even radial load on the center bearing.

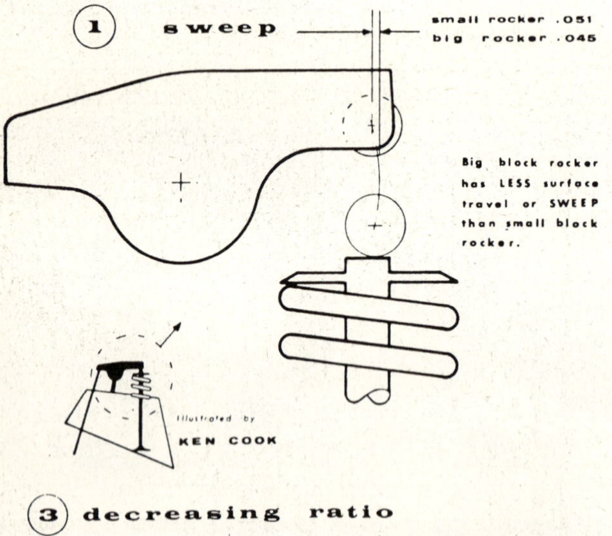

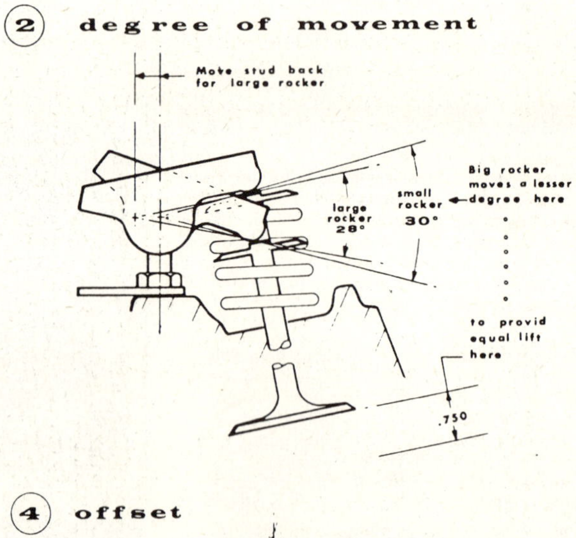

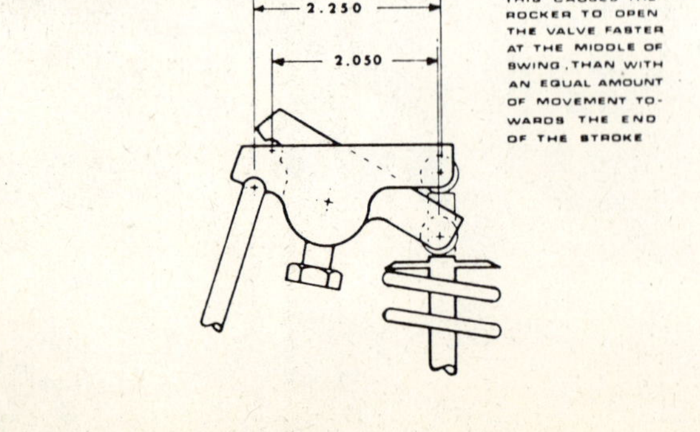

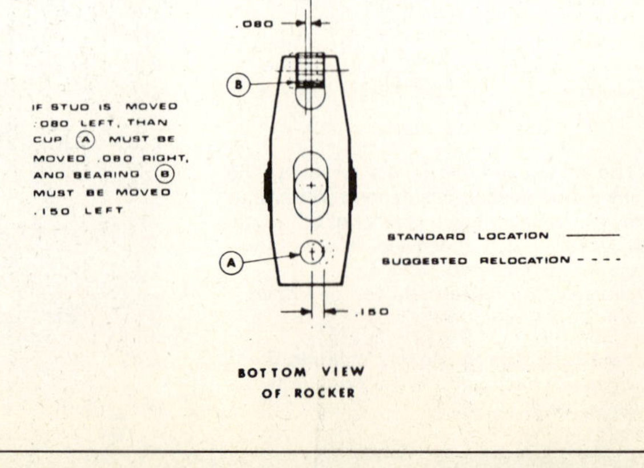

7

The Littlest Porcupine

You're looking at factory Chevy heads, but they're not for the big-block and they never made it to the parts counter.

Take a close look. These are rare Chevrolet aluminum cylinder heads and an intake manifold that few people have ever seen or knew existed. These look-alike big-block heads are actually for the small-block engine. A close inspection reveals that they are of a canted-valve (porcupine), semi-hemi combustion chamber design. You will also notice that the placement of the intake ports is unlike any other small-block Chevrolet head.

All other factory heads have the intake ports siamesed into two groups of two, but these heads have the intake ports evenly spaced. The idea was to provide identical cylinder entry angles for each intake port, thus assuring relatively equal flow and mixture swirl characteristics. But why such a departure from the traditional design of the small-block engine?

These heads, and the special manifold, were originally cast by Chevrolet in 1968 as a bolt-on replacement for production heads in an effort to get the heads approved for Trans-Am racing. They never were. You may also remember with some amusement that this was a point in Chevrolet's history when they were supposedly uninvolved with racing.

Actually there were only 25 sets of these heads ever produced. Because of the unusual intake port placement, special intake manifolds, again of a twin 4-barrel cross-ram configuration intended for Trans-Am use, were also produced in similar limited numbers. Other associated pieces included special camshafts, with the lobes in the correct order to match the intake and exhaust valve locations in these heads, and special cast-aluminum seven-bolt valve covers. But the story doesn't stop there.

You may wonder whatever became of this design. Surely there were other performance or racing applications where these heads could have been used? The truth is that the heads suffered from some serious design problems. Although the desired intake flow improvements had been achieved while retaining 2.02-inch-diameter intake valves, the Trans-Am requirement of retaining the stock small-block cylinder head bolt pattern necessitated the use of a rather poor flowing L-shaped exhaust port, which was the design's biggest shortcoming. Many attempts were made to rework the exhaust ports so that they would flow at least as well as those in the conventional production heads, but those efforts were to no avail. Additionally, the canted placement of the valves required longer stems, thus adding to valve weight and also restricting revving capabilities.

Chevrolet never released these heads, and only a few of the topnotch engine builders back in '68 ever saw them. They are rare, but at least a few of those original 25 sets still exist. So if you ever happen across a pair of them, as we did in Smokey Yunick's back room, rest assured that you have found a true collector's item—but it's doubtful that they'd make your car go any faster.

—C. J. Baker

1

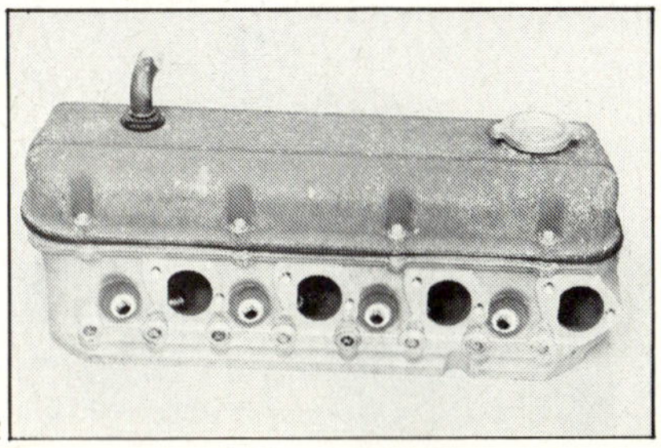

2

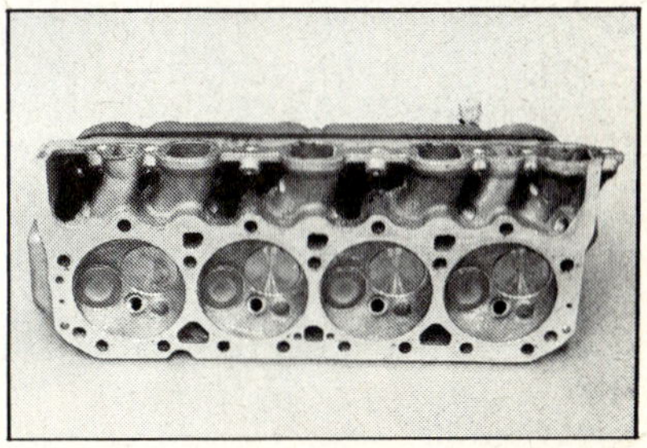

3

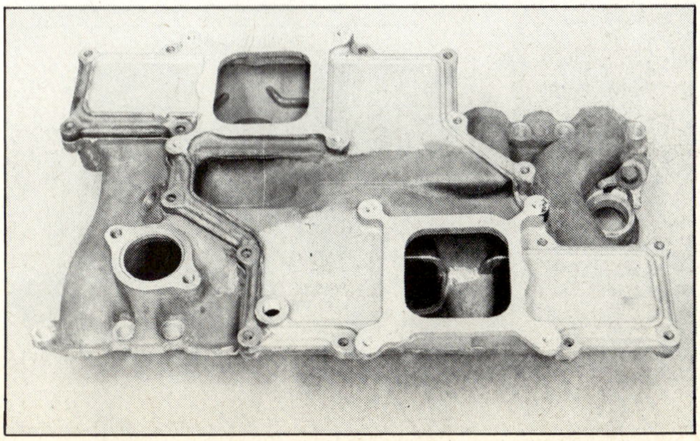

4

1. No, the camera isn't deceiving you; the valves are canted. Note also the placement of the intake ports and the spacers under the exhaust rocker studs.

2. Here is the downfall of these heads, the exhaust ports, which required special headers. Note the cast-aluminum valve covers and count the number of bolts.

3. Not exactly your standard small-block Chevy combustion chamber. Note the semi-hemi design and the entry angle of each intake port.

4. This is the special twin 4-barrel cross-ram intake manifold specifically designed to fit these heads. A special camshaft was also required because of the revised valve placement in these heads. These pieces are collector's items.

The Offset Stud

A little more rock to make your small-block roll.

It's no big revelation that high-lift rocker arms (those with a greater multiplication ratio than stock rocker arms) generally increase the performance of an engine without necessitating a camshaft change. This seems to be particularly true on the small-block Chevrolet V8, and numerous approaches to increasing the rocker arm ratio have been developed. Unfortunately, most of these approaches are quite expensive or difficult to accomplish, requiring relocation of the rocker studs and/or expensive roller-tip rocker arms.

But now the folks at H-O Racing Specialties, Inc. (4708 W. Compton Blvd., Lawndale, CA 90260) have come up with an economical kit that uses offset rocker arm studs and special stamped steel rocker arms, balls and nuts to accomplish the same result without expensive machine work. In fact, the heads don't even have to be removed from the engine for installation on late-model engines with screw-in studs.

The rocker arm studs in the H-O system incorporate a slight offset which decreases the distance from the pushrod to the center of the pivot ball and increases the distance from the pivot ball center to the contact point on the valve stem for an effective ratio of 1.66:1. That's nearly as much ratio as can be achieved by relocating the studs to accept the 1.70:1 big-block rockers, without all the hassle.

The kit includes the special 7/16-inch offset studs, with lock nuts to hold them in proper alignment, self-locking adjusting nuts, new stamped steel rocker arms to fit the new dimensions created by the stud offset, and new heat-treated rocker arm balls. The new rocker arm balls are required, since the H-O rocker arms use a slightly larger radius ball than the stock Chevy rockers. Optional heat-treated grooved rocker balls, poly locks and double lock adjusting nuts are also offered for heavy-duty applications. Racer net on the base kit, which includes everything needed to convert all 16 valves to the 1.66:1 rocker ratio, is just under $75. For those who only want to put the high-lift rockers on the intake valves, half kits are also offered, and individual service pieces are listed.

Is it worth it? We think so. For example, if the lobe lift on the stock cam is .290-inch, gross valve lift with the stock 1.5:1 rockers would be .435-inch, but with the 1.66:1 rockers, lift jumps to .481-inch, and that's generally enough to produce power you can feel in the seat of your pants. Of course, this setup must be accepted for what it is—an economical design for street and bracket performance, with valve lifts of less than .650-inch and open spring pressures of less than 400 pounds. It isn't intended for wild roller cams and King Kong valve springs. Instead, it's a quick and neat way to increase performance, and that's what hot rodding is all about.—*C.J. Baker*

1 

2 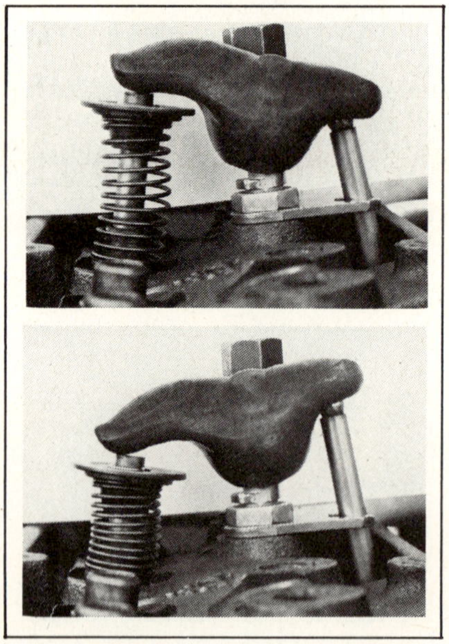

3 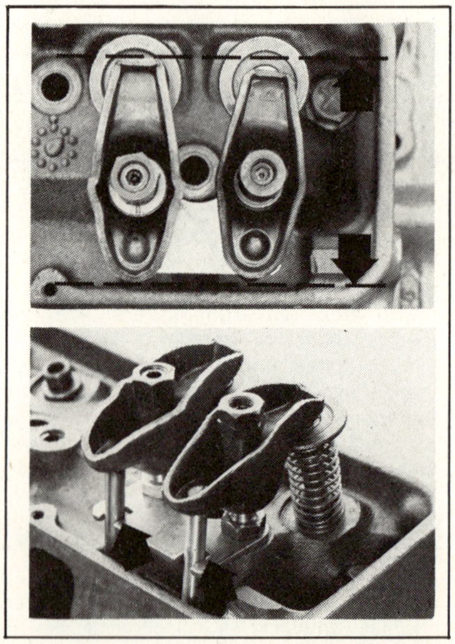

4 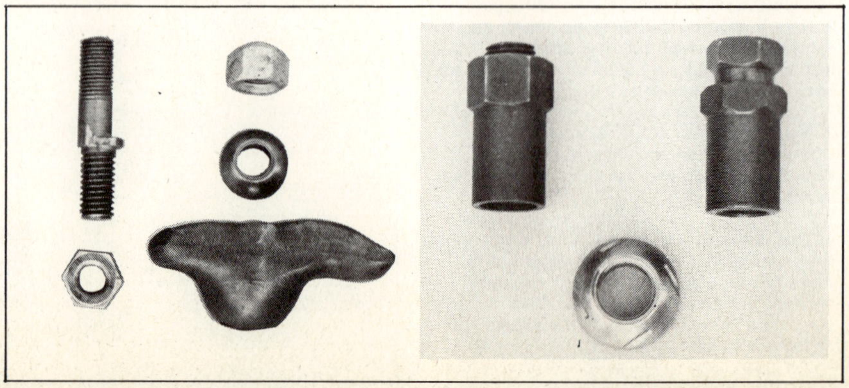

1. The stock 1.5:1 rocker arm (left) has stem contact very near the edge, but the 1.66:1 rocker (right) is nearly centered.

2. Close-ups of the 1.66:1 rocker contact at both full lift and zero lift confirm proper geometry.

3. This comparison of the 1.66:1 rocker arm (left) to the stock 1.5:1 rocker shows its greater length and the larger radius rocker balls. Although the pushrod no longer rides at the bottom of the guide plate slot, it still passes through the head openings without interference.

4. The kit contains the stud, nuts, ball and rocker. Special heavy-duty lock nuts and balls are optional.

Brownfield Heads

Warren Brownfield's small-block head began as an optimum port design, with aluminum poured around it.

This aluminum head is the result of extensive development work by Warren Brownfield, the driving force behind the well-known Air Flow Research. Warren is adamant in pointing out that this head was a completely new design based on optimum port configurations, and not simply an aluminum recasting of existing Chevrolet designs. For example, flow bench tests have shown that a stream of air wants to flatten out when negotiating a right angle turn, such as it must do just above the valve seats. Consequently, the Brownfield ports are shaped to accommodate this broadening of the flow in these critical areas—something that cannot be effectively done on the stock iron castings. Cross-sectional area control of the ports has also been maintained to eliminate the need for small stem valves and to compensate for frictional losses as the air moves through the port. In fact, substituting 5/16-inch stem valves in place of the stock 11/32-inch stem valves shows absolutely no improvement in flow in Brownfield's heads. Valve centerline spacing has also been increased to permit use of 2.02-inch intake valves without disruption of the flow adjacent to the chamber walls, while simultaneously increasing the strength between the inlet and exhaust valve seats, thus increasing valve seat life. Oversized 2.05-inch intake valves that have been used on modified stock heads aren't required here and, in fact, they don't show any flow improvement either.

Other features include spring pads that are ⅛-inch thicker than stock for additional support and to permit machining for oversized (1.550-inch) big-block springs. The spring pad is also 1/16-inch deeper than stock to accept increased spring heights. Combustion chamber configuration is identical to the stock design so that the new heads will be compatible with existing aftermarket piston dome configurations.

The intake and exhaust valve pockets in the combustion chambers are *bored* to give absolute control of chamber wall proximity to the valve and for more consistent volume control.

When dealing with aluminum heads, the question of rapid heat dissipation and a subsequent loss of power always arises. Brownfield has solved this problem through the use of a unique water jacket design that actually removes 40 percent of the water surrounding the ports, exhaust valve seats and combustion chambers that is normally present in the cast-iron head. This is possible with an aluminum head, whereas cast-iron would probably crack. Such a design reduces heat transfer and simultaneously strengthens the combustion chamber. In fact, tests have shown that the water temperature of an engine equipped with the Brownfield heads actually runs cooler than when cast-iron heads are used.

The basic intent of this design was to offer an extremely good head that picks up where the best ported heads leave off, and do it for less money. The result is an exceptionally versatile head that works, as is, on any small-block of 302 cubic inches or more. Since this is a very versatile design, you may wonder how it stacks up against the super-exotic welded cast-iron heads for maximum output engines. The Brownfield head can be further modified by selective porting to out-perform the most exotic heads currently available. *—C.J. Baker*

SPECIFICATIONS

	Stock (292 Head)	Brownfield Head
Material	Grey Cast Iron	356 T6 Aluminum Alloy
Weight	44 pounds	22 pounds
Intake Port Size	2.05 x 1.2 inches	2.13 x 1.3 inches
Minimum Intake Restriction	1.0 inches	1.17 inches
Exhaust Port Size	1.38 x 1.38 inches	1.2 x 1.55 inches
Minimum Exhaust Restriction	1.38 8inches	1.63 inches
Valve seats	cast iron	steel step-seat
Valve Sizes	I-2.02 E-1.6	I-2.02 E-1.6
Valve Stem Size	11/32-inch	11/32-inch
Combustion Chamber	68cc	68cc
Intake Flow (stock)	285 cfm	360 cfm
Intake Flow (ported)	385 cfm	400 cfm
Exhaust Flow (stock)	140 cfm	270 cfm
Exhaust Flow (ported)	285 cfm	330 cfm
Spring Pad Diameter	1.435 inches	1.450 inches
Minimum Deck Thickness	.188 inches	.500 inches*
Spark Plug Location	stock/angle design	raised .060-inch/angle design
Spark Plug Reach	⅜-inch tapered seat	½-inch reach flat seat
Heat Riser Passages	Yes	No

*Big-block Chevy stock aluminum head has .375-inch thick deck.

STOCK-TO-STOCK DYNO HORSEPOWER COMPARISONS CONDUCTED AT JENKINS COMPETITION, MALVERN, PENNSYLVANIA

Dyno Operator: Bill Jenkins

SERIES # 1

Single Holley 830 cfm four-barrel carb on an Air Flow Research modified Edelbrock Scorpion intake manifold.

RPM	292 Chevrolet Head	Brownfield Head	Horsepower Increase
4500	321	331	+10
5000	392	401*	+ 9
5500	439	453*	+14
6000	473	494*	+21
6500	498	530*	+32
7000	510	546*	+36
7500	504	542*	+38

SERIES # 2

Dual Holley 4500 carbs on an Air Flow Research modified Weiand tunnel-ram intake manifold.

RPM	292 Chevrolet Head	Brownfield Head	Horsepower Increase
5000	389	402	+13
5500	441	461	+20
6000	490	524	+34
6500	523	562	+39
7000	536	582	+46
7500	528	588	+60
8000	512	579	+67

*Note that the single four-barrel readings with the aluminum heads are higher than the dual four barrel readings with the stock cast iron heads at comparable rpm. All tests were run on the same day under identical conditions. The engine was a 350-cubic-inch Chevy using TRW 12.1 two-ring pistons, stock rod length, Acecl BEI ignition, 2″ x 31″ Hooker headers and a General Kinetics cam with 108-degree centerlines, 321 and 329 degrees duration, intake and exhaust and lifts of .620 and .632 inches. (This cam favored the breathing of the stock cast iron heads rather than the aluminum heads. Even so, the aluminum heads produced the best power.) Both sets of heads used the same 1.55-inch G.K. big-block valve springs and G.K. rocker arms. Combustion chamber volume was 65cc for both sets.

High C

Factory heads for the big-block provide more than 100 percent of the engine's flow requirement.

After many years of patient waiting by enthusiasts, Chevrolet has finally introduced a new aluminum open-chamber, big-block head (part No. 14011076). With a recontoured exhaust port shaped like a "C," and minor reworking on the intake side, the new head is reputed to offer significantly increased flow potential compared to the 3965784 open-chamber, round-exhaust-port head it's designed to replace.

To verify these claims, HOT ROD Magazine had Pete Incaudo of Integrated Engineering, Inc. (13505 Shoup Ave., Hawthorne, CA 90250) run some tests on the two heads' exhaust ports, where it was felt the greatest potential for flow differences existed. The results, summarized below, reveal that the new head's flow advantage lies principally in the high-lift range above .500-inch. Below .500, there is little significant difference in the total flow of the heads. However, the smaller cross-sectional area of the new exhaust port should show greater low-end torque production in the "real world," due to the higher flow velocities achieved, although it's impossible to ascertain on the flow bench. This should make the new heads well received on the street, as well as by torque-hungry professional road racers.

As far as drag racing goes, strip tests on one competitor's Super Modified car, where no welding or radical head mods are allowed, have shown as much as .2-second e.t. reduction solely from swapping heads. And since radical welding mods that reduce head durability and increase the chance of porosity and water leaks are no longer necessary, the new heads ought to last a lot longer in competition. As is typical of Chevrolet's philosophy of parts interchangeability, intake manifolds and headers designed for other high-performance rat heads will bolt right up to the new design.

So much for the good news, but there is always, unfortunately, the flip side of the coin. The heads are coming through with Stellite valve seats, a material particularly hard on valve grinding tools. Our particular casting also lacked enough meat around some of the valve spring pockets. Such a flaw would preclude the installation of ultra-heavy racing valve springs, but Pete informed us this is not a common occurrence—he's seen six of the new castings so far, and ours was the only bad one. Nevertheless, you should inspect the head carefully before accepting it.

TEST 1 & 2

Out-of-the-box round and C-port heads (photo A) were used with stock steel 1.88-inch exhaust valves in tests 1 and 2, respectively. Although results varied slightly, depending on the overall diameter (o.d.) of the headers used (more on that later), the C-port flowed slightly better at extreme low valve lift and much better once .500 lift was passed. The old design was marginally superior in midrange flow capacity; apparently, the "hump" in the port causes some flow restriction in this range. Nevertheless, the higher velocity achieved by the new casting and consequent increased torque production should more than make up for any flow deficiency.

TESTS 3 & 4

The C-port head was modified in two stages.

Test 3 included a 3-angle valve job (30°-45°-60°) and the unshrouding of the exhaust valve (photo C, compared to stock chamber in B). Slight flow increases over the stock port were observed (see graph on opposite page).

Test 4 was what really separated the men from the boys. The new casting includes additional material added near the valve guide and the short side of the exhaust port to enable the valve guide to be properly trimmed and the short side correctly radiused. With the old port, additional material had to be welded on, then ground away, to achieve the proper shape (photos D through G), but the new C-port already looks like a welded round port (photo F), so all you have to do is get out the ol' hand grinder. (We'd show all this on a new casting, except they're too scarce to section for the camera!)

A two-angle valve job (45°seat, 30°top angle) was performed on the seat, with the valve head ground correspondingly. This gives a straighter shot for the escaping gases, when combined with the radiused port and trimmed guide.

As you can see in the graph, significant flow improvements over the unmodified heads were obtained. In fact, as modified for test 4, the head's ability to flow air showed almost no tapering off

A

B

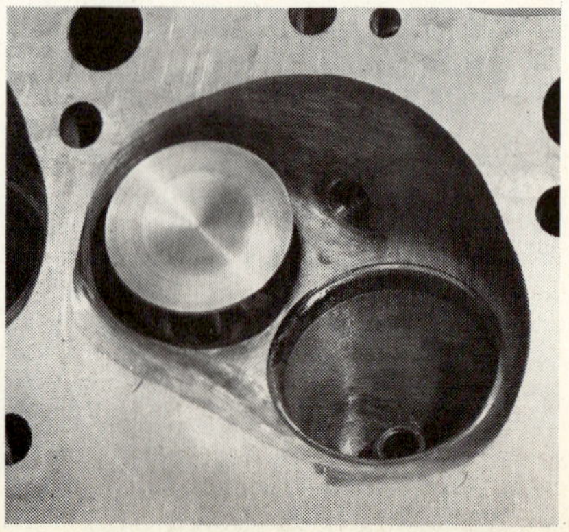

C

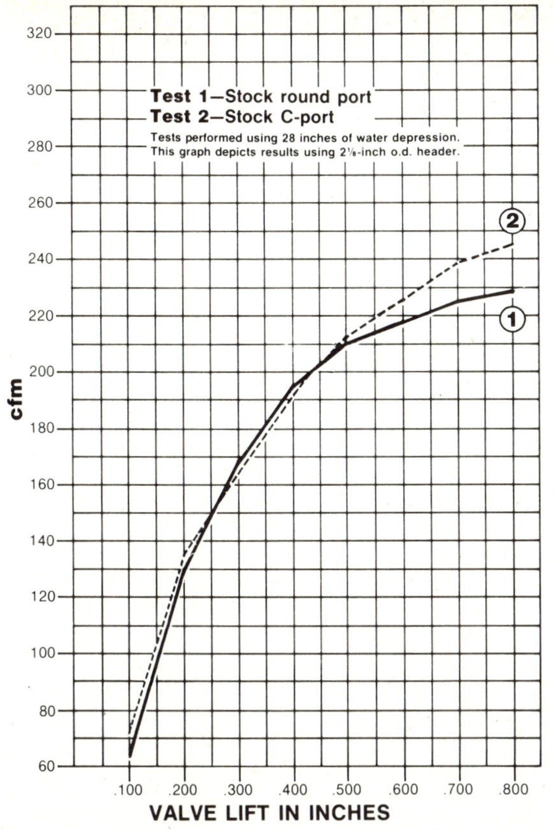

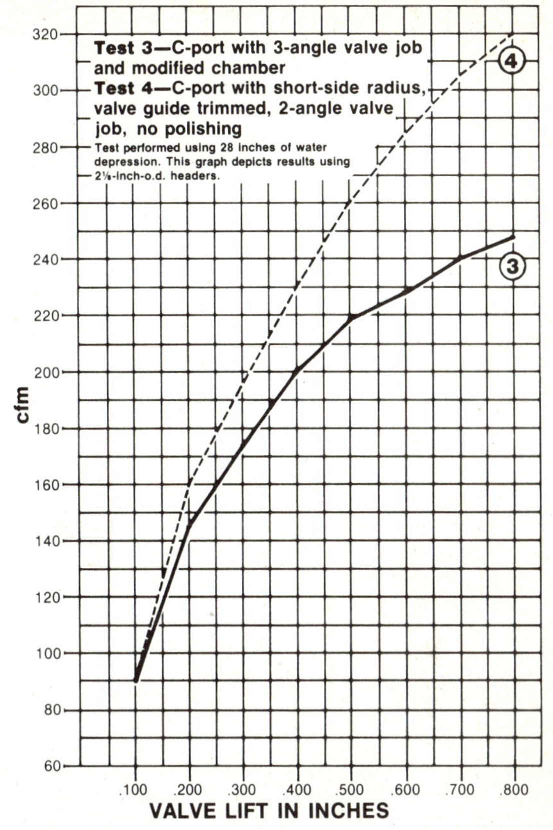

as .800 lift was passed. Since with state-of-the-art technology, valve lifts much beyond .850 are impractical in terms of component breakage, it appears that the valvetrain, and not the head itself, can now be considered the limiting factor as far as ultimate flow capability goes. But, since much higher lifts aren't likely until better valve springs are developed, the task of the porter will be to play around with the head and see if he can bring the huge potential high-lift cfm flow rate down into the usable lift range where it can be effectively utilized. Of course, this may involve a sacrifice of some potential ultra-high lift flow potential, but it can't be used anyway.

Although the best ultimate flows were achieved with 2⅛-inch-o.d. header tubes in all tests, both heads and all modifications were tested with 1⅞- and 2-inch tubes as well, supplied by Jack Davis at Hooker Headers. The best all-around combination in most cases appears to be a 2-inch pipe—it flowed about as well as the 2⅛ size down low but with much higher velocity. Although it began to fall off rapidly at about .400 lift, the high-velocity low-end flow means greater low-end torque, so automatic trans-equipped drag cars, or road racers on courses with tight corners, might well consider using the smaller pipe. This seems to be borne out in the "real world," since HOT ROD Magazine Executive Editor C.J. Baker's 440-inch rat-motored, Powerglide-equipped Chevelle runs its best times with 2-inch-o.d. tubes. On the street, 1⅞-inch tubes would appear to be the tip for even greater low-end power response.

It should also be pointed out that flow velocity is also related to displacement and piston speed; translated into English, this means a 427 might require a 1⅞-inch header while a 454 would run best with a 2-inch tube, all other things being equal.

Since the new heads will supersede the round-port configuration entirely when present stocks are depleted, an interesting question is raised: Will NHRA accept the new casting as a "replacement part" in Stock and Super Stock classes?

—*Marlan Davis*

D

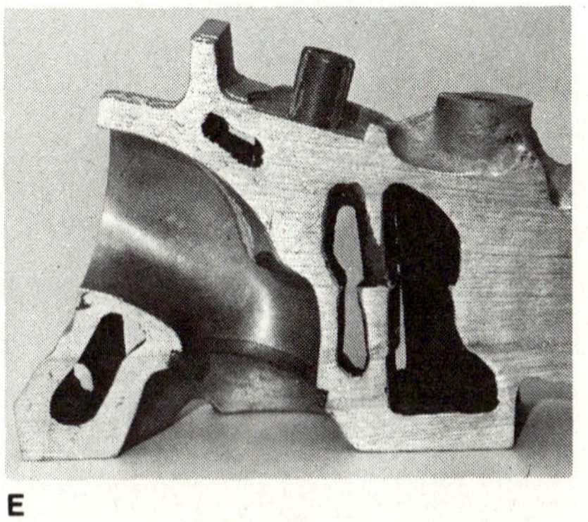
E

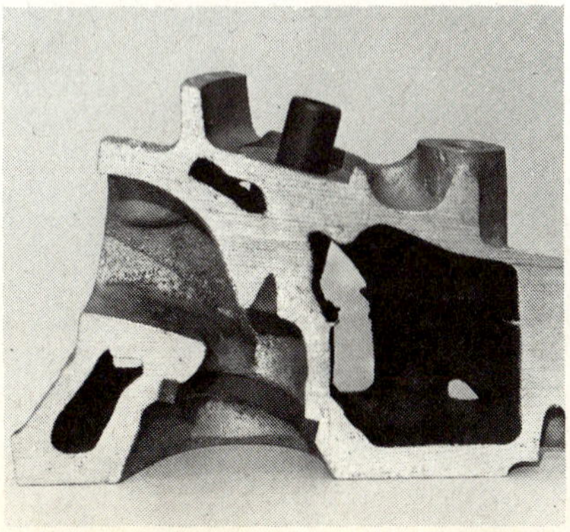
F

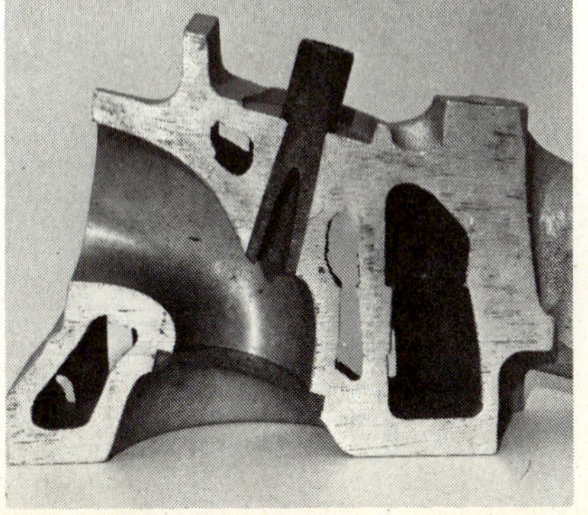
G

Clamp Down

Sealing up big-block Chevy head leaks.

A good feature of the Chevrolet big-block engine is that it was designed with six bolts surrounding each cylinder for positive cylinder sealing and control of combustion pressures. One of the worst features of the Chevrolet big-block engine is that when all of the production problems had been worked out on the original design, intake port configuration forced the elimination of one of the six head bolts on two cylinders on each bank.

Anyone who races a big-block engine with fairly high compression ratios or high cylinder pressures is well aware of this shortcoming, since during disassembly there is inevitably evidence of cylinder leakage at the very locations where the two head bolts have been deleted. The problem is particularly acute where aluminum cylinder heads are used, since the reduced rigidity of the aluminum permits more distortion of the head across the gap where the bolt is missing, thus increasing leakage. Chevrolet is equally aware of the problem, and on the special aluminum blocks, two special bosses were added to each side of the lifter valley to accommodate two studs that screw into the bottom of the offending intake ports, thus providing the additional clamping force required to seal the head to the block.

Over the years, racers have tried several approaches to sealing the cylinder leakage problem on cylinders 3 & 7, and 2 & 6. On engines that use aluminum heads, some engine builders welded the additional bosses into the valley area of the block to accept the extra head studs, but this is an expensive and risky operation, since welding to cast iron is difficult

1

2

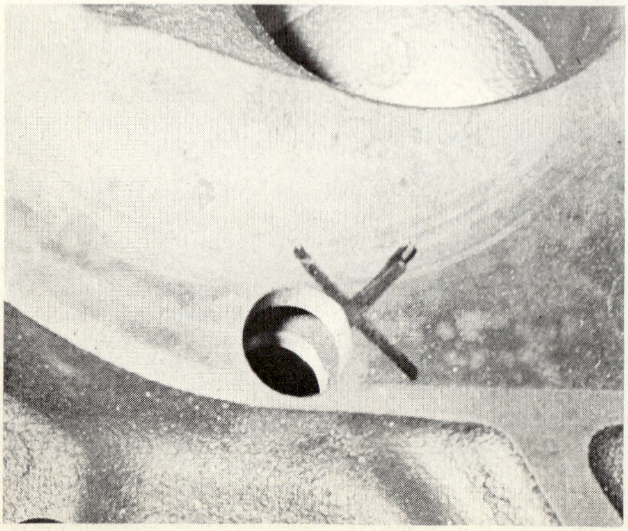
3

4

and requires special procedures to prevent cracking. And such welding always causes some block distortion, so remachining of the block after welding is necessary. Another approach was to pantograph either the heads or the block to accept iron or copper O-rings. This procedure is also expensive. The problem with O-ringing the engine is that it limits steel-core head gaskets that can be used to either copper shim or composition sandwich gaskets. If a conventional composition gasket is used, which is the most popular style of gasket used in racing big-blocks, the O-rings will cut the gasket, making the overall seal less effective than if no O-rings were used.

Now there's a better approach that is inexpensive and does not limit gasket choice. It has been developed by Evans Speed Equipment (2550 N. Seaman, South El Monte, CA 91733). Simply stated, the procedure consists of modifying the heads so that two countersunk bolts can be inserted into the floor of each of the two offending intake runners, screwing directly into the block deck surface, which of course must be drilled and tapped to accept the extra bolts. Cost of the complete modification is $54 for both heads, which includes the special 5/16-inch countersunk bolts. And if it weren't for the fact that there is a water passage running beneath the intake ports that must be sealed off to prevent water leaks, the procedure is so simple that it could be done at home. But the water jacket problem must be handled correctly, and that requires hauling or shipping the heads to Evans Speed Equipment. Take a look at the accompanying photos and art and you'll see that, once and for all, the case of the missing bolts has been solved.

1. Indicated here are the locations where the two missing head bolts have been removed due to the presence of the intake ports directly above.

2. Evans has fabricated a special fixture to position the head for accurately locating the two new bolt holes. They're positioned to miss any O-ring grooves that may have been previously pantographed into the head.

3. After the hole has been drilled, this close-up reveals the water passage (note double wall) between the intake port floor and the head surface.

4. When the holes have been drilled and tapped, a solid cast-iron plug is coated with a ceramic sealer and screwed into place so that it just barely protrudes into the intake port.

5. After the plug is installed, a hole is drilled through it to accept the 5/16-inch head bolt. The drill also extends to drill through the roof of the port to locate the position for the access plug.

6. The roof of the port is then drilled and tapped for a plug to permit access to countersunk bolt on the port floor. The floor of the port is also countersunk at this time so the bolt will lie flush with the floor.

7. The final operation is to mill off the cast-iron plug flush with the head surface. All that remains is for the customer to drill and tap his block and head gaskets to accept the new bolts, which need be torqued to only 20 ft.-lbs. to seal the head securely to the block.

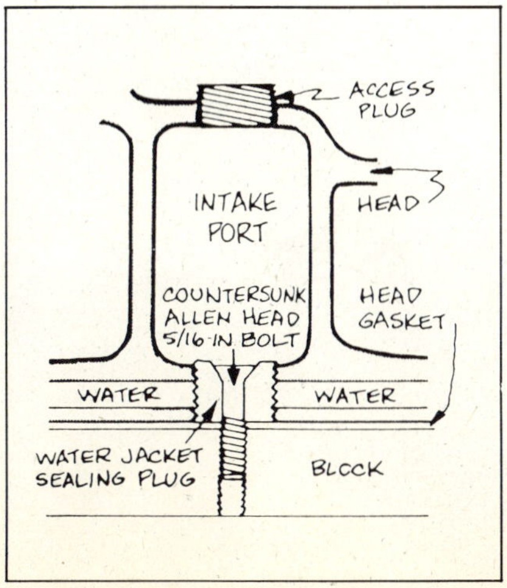

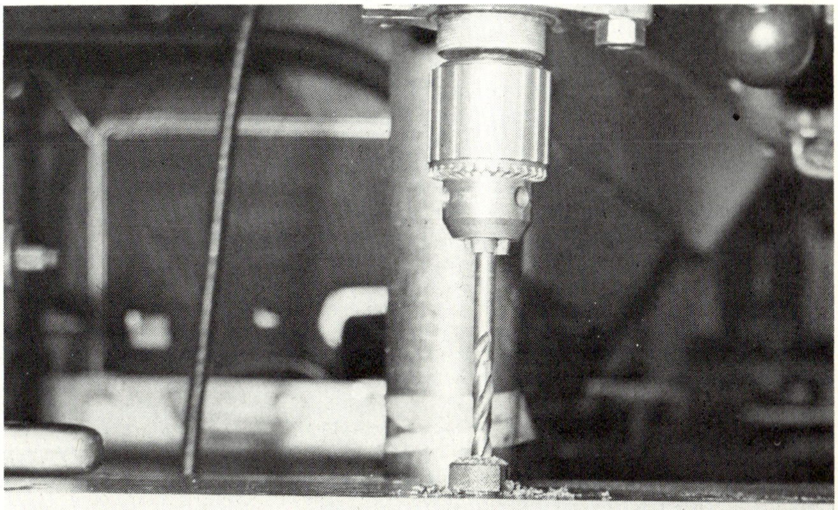

5

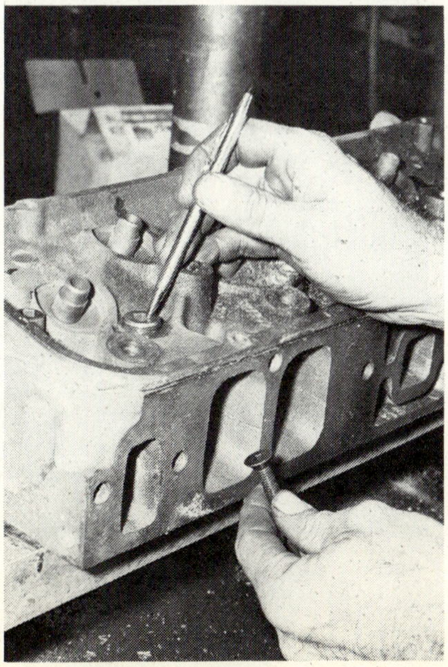

6

7

Fine Tuning

Getting a camshaft in key with your engine.

There's a lot more to wringing optimum performance from a camshaft, any camshaft, than simply degreeing it in and setting valve lash to manufacturer's specs. If this comes as a surprise to you, sit back and read this piece with care, because it is *not* a discussion on how to degree-in a cam—you can find that in almost any cam manufacturer's catalog. Instead, we'll be talking about how to *dial-in,* or tune, a camshaft, which is a special subject in itself.

We'll assume you have a cam, stock or aftermarket, that is matched to your needs. This doesn't mean that you must have *the* very best grind available, only that it is at least close. Second, we'll assume that you have properly installed the cam and completed the initial run-in as specified by the manufacturer. And last, we'll assume that your vehicle is in suitable running condition to produce consistent operation that can be used to evaluate the changes made. In other words, if the car has ignition, carburetion, transmission or traction problems, those will have to be solved before we can do any meaningful testing.

Essentially, what we'll be doing is changing the camshaft timing and valve lash to produce optimum output and to match the engine's power curve to your specific needs. On a drag car, this will be in terms of traction, elapsed time and trap speed. On a roundy-round car or marine application, the goal may be quicker lap times, power coming off the turns, straightaway rpm, etc.; and on a street car, this same procedure can even be used to optimize gas mileage and throttle response.

Before we get into the actual tuning specifics, a couple of preparations are in order to simplify the procedure and to avoid any likelihood of engine damage. First, since we'll be varying the camshaft timing, much of the work and time required to change cam timing can be saved by replacing the camshaft sprocket with a Quick Change Cam Timer, manufactured by Ed Hamburger, at Trik Race Car Products in Bricktown, New Jersey. This unit permits accurate external adjustment of cam timing in a matter of minutes. Of course, if you're running a gear drive setup, or there isn't a cam timer offered for your particular engine, you can still use offset dowel pin bushings or offset crankshaft keys to make timing alterations, even though they require considerable effort to change.

The second item that *must* be checked when altering cam timing is the piston-to-valve clearance, since it changes when the camshaft is either advanced or retarded. Experienced engine builders check piston-to-valve clearance with the cam both advanced and retarded up to 8 crankshaft degrees, in addition to the straight-up position, when they install a new cam. If you haven't, do it before going any further. Most cam manufacturers recommend an absolute *minimum* piston-to-valve clearance of .080-inch for intake valves and .100-inch for exhaust valves on steel rod engines and .100-inch and .125-inch, respectively, for aluminum rod engines. In checking this, you may find that you'll be limited to something less than 8 degrees of advance or retard before reaching these minimums. If so, write down your observed limits and be careful not to exceed them during your testing.

Everything should now be in readiness. Begin with the cam degreed-in as specified by the manufacturer and with the recommended valve lash. Make several runs until you have a consistent baseline performance figure. Then start tuning by varying the valve lash, unless of course, you have a hydraulic lifter cam. Changing the lash has a definite effect on the duration of the cam. A rough rule of thumb is that each .001-inch change in valve lash changes the duration by 3 degrees (crankshaft). Start with the intake valves only, loosening and tightening them .004-inch at a time until the best results are obtained.

With the intake lash determined, repeat the procedure on the exhaust valves. With the lash roughly set in this manner, go back and try settings .002-inch looser and tighter, again starting with the intakes, to fine-tune the lash setting. Tightening the lash increases the duration and usually builds more top-end power while sacrificing some low-rpm torque. Loosening the lash has the opposite effect. Generally speaking, lash shouldn't be varied more than .008- to .010-inch from the manufacturer's recommendations. If you need to go more than that, consult with the manufacturer.

With the lash tests completed, it's time to vary the cam timing. Changes should be made in 4-degree increments, not to exceed a total of 8 degrees or the limits imposed by your previous piston-to-valve clearance measurements. Advancing the cam picks up the low-rpm response and torque at the cost of some top-end horsepower. Retarding the cam scrubs off some bottom-end power and may boost high-rpm output. As with the lash, fine-tune the cam timing in two-degree increments after you've found the best rough setting. If you're a real perfectionist, you might want

1

1. Degreeing-in a cam can be simplified by using a Trik Race Car Products cam timer in place of the cam sprocket for quick adjustment.

2. A small access plate added to the top of the timing cover permits subsequent cam timing changes in minutes without disassembling the engine. Window in the cover is optional.

3. Offset cam bushing, offset keys or multiple keyway crank gears can be used to make the same changes, although a lot more work and time are required.

4. A rocker arm stud girdle and a cam button improve valvetrain stability by keeping the parts rigid and in proper alignment. Note special nuts.

5. Rev kits, such as this one, are usually used with high-revving roller cams, but use with flat-tappet cams also helps keep the tappets on the cam lobes and extends the rpm range of the engine before valve float occurs.

to go back and try varying the lash at the optimum timing location, but gains here will be minimal.

In conclusion, keep careful records of the exact changes you make and the results. Such information can be a big help to your cam grinder when recommending another cam even better suited to your needs. Don't forget to reset the ignition timing each time you vary the cam timing (an excellent indicator of exactly how much you've changed the cam), and keep all other factors, such as track conditions, air density and temperature, vehicle weight, jetting, shift points; etc., as constant as possible during the testing to assure meaningful results.

—C. J. Baker

ON A MORE SERIOUS NOTE

The best cam grind and all the fine tuning in the world is a waste of time if the cam can't accurately control the valvetrain. It is essential that the valvetrain be as stable as possible to minimize deflection and harmonic oscillations. A lot of effort goes into designing cam profiles that are as smooth and stable as possible, but the engine builder must do his part too. For example, if the end play on the camshaft is excessive, the cam can move fore and aft, causing rapid variations in ignition timing, accelerated timing chain wear and even damage to the block and tappets. This is especially crucial with roller tappet cams which have no provision for dynamically loading the cam toward the rear of the block. Consequently, it is a good idea to install a cam thrust button or bearing. They're available from most cam manufacturers. Hold cam walk (endplay) to less than .010-inch.

Rocker stud flex (on those engines that do not use shaft-mounted rockers) can also cause valve action to vary. Rocker stud girdles to reduce this flex are available from either Jomar Engineering or Moroso. Such girdles are almost a necessity on ultra-high-lift cams with radical valve spring pressures used in today's modified drag racing engines. Just be sure to specify the latest adjusting screws which feature a locking setscrew similar to a polylock to assure a secure grip on the rocker studs.

And although they're inconvenient to install, use a rev kit whenever possible. Not only will a rev kit increase valvetrain control, but it also provides an extra margin of safety against valve float during an engine overspeed condition.

All of these things are valvetrain stabilizers, and all too often they're taken for granted. But it's that attention to detail that really matters.

2

3

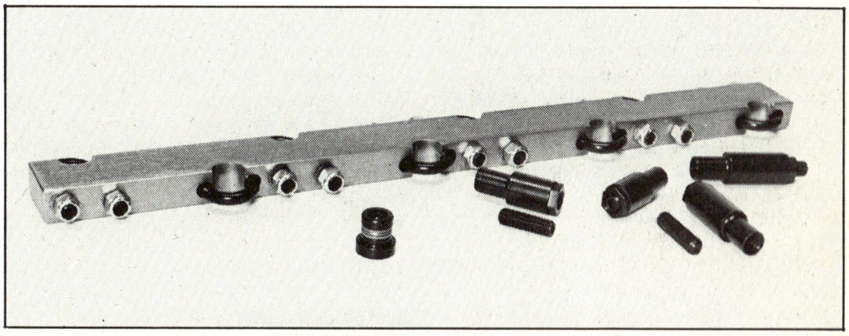
4

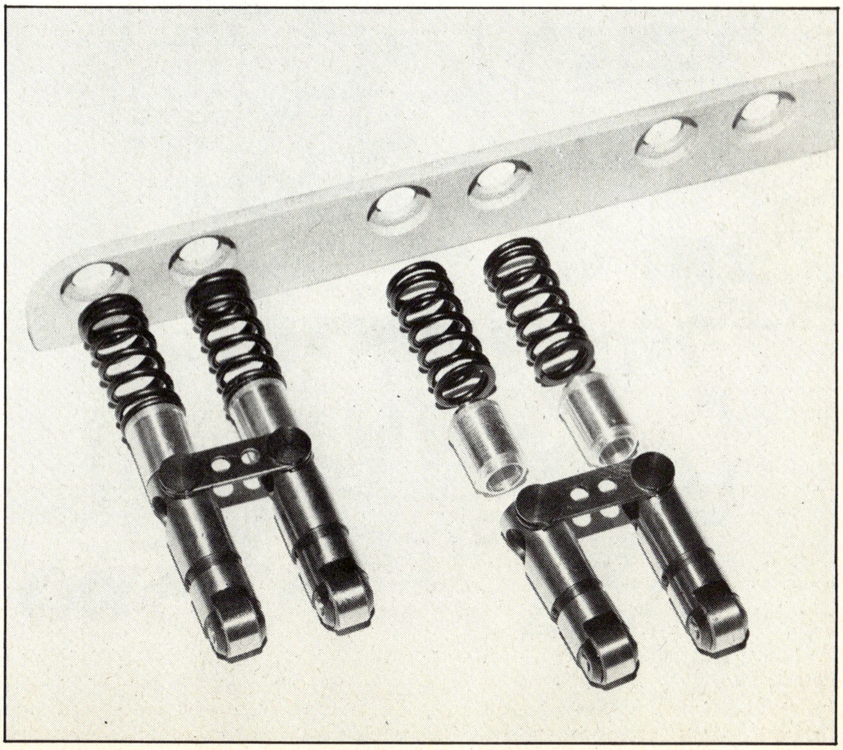
5

Holley Carb Mods

A lot of so-called experts are charging a lot of money to rework Holley carbs for competition use, but do you really get what you pay for?

It isn't any secret that even the competition-designed versions of Holley carburetors can be reworked, cleaned up and modified from their "out-of-the-box" condition to achieve increases in performance. The racers know this, and so does Holley. The engineers at Holley continually evaluate those things that are being done to their carburetors to determine if those mods are truly effective. If they are advantageous, and if such improvements are feasible from a production standpoint, they are often incorporated into revisions of existing models. And even if those modifications are prohibitive on a production basis, Holley still passes the information along in Tech Clinics, periodic service bulletins and their Holley Carburetors & Manifolds manual (part No. 36-73) which describes system operation, disassembly, reassembly, specifications and modification of various models.

But the story doesn't stop there. Lately Holley has been seeing an increasing number of their carburetors that have been altered by various people who have made a very lucrative business out of such modification. Unfortunately, sometimes the modifications being performed, while costly, are of questionable value in terms of performance gained and quality of workmanship. Some of these mods, however, do work quite well. When there is a problem, it stems from the fact that those modifications are un-engineered and each carburetor is rarely tested to assure consistent results. Many of the aftermarket modifiers are not presently equipped to wet-flow carburetors to simulate the actual flow of both air *and* fuel. Instead, they measure only air flow and pressure drops, if they check anything at all. But fuel flow is just as critical. At least one firm, Braswell Carburetion, has constructed a wet-flow bench.

Because obtaining the ultimate in performance from a carburetor is so important to many racers, Holley has taken a firm stand on current modifications—what's good, what's bad and what's only eyewash, regardless of cost. They have begun pointing out the real value of what is being done, and more important, they are offering recommendations on how to do those modifications yourself, or to obtain equivalent results by different or cheaper methods. Holley has made this information available to us, and we have condensed and summarized it for you here. Remember, modifying carburetors requires careful planning and a certain level of technical proficiency. Don't attempt to perform a modification until you understand its precise purpose and whether it is correct for your application. Above all, proceed slowly and make changes in small increments,

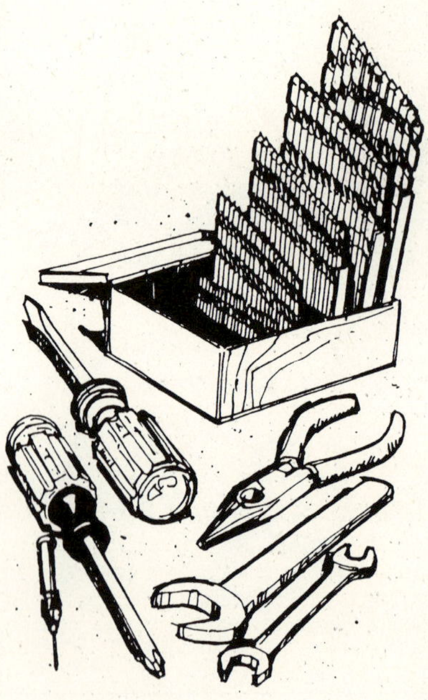

TOOL SCHOOL

Working on and modifying carburetors requires specific tools such as those shown here for all but the most complex alterations. At a minimum, you should have a complete set of wire drills, a pin vise, needle-nose pliers, screwdrivers, 1-inch, 3/8-inch and 7/16-inch open-end wrenches. Also useful for cleanup work is a set of cheap jeweler's files for removing casting flashing.

checking your progress as you go to see if what you're doing really helps.

THE 4150 SERIES

The Model 4150 Holley carb was introduced in 1957. This series includes all of their standard-flange, double-pumper performance carburetors in use today. These carburetors are currently being modified by numerous people, with the tab running between $185 and $525, plus the carburetor! Any way you look at it, that's a lot of money. Various tricks are employed by each modifier. In most cases the tricks are of *some* benefit, but in many cases Holley feels that equivalent gains can be achieved by cheaper and easier methods, or you can make the same modifications yourself. Here's what is being done.

The idle circuit usually gets considerable attention, in an attempt to provide finer tuning capability, prevent loading and sharpen throttle response. One way of doing this is to install a primary metering block on the secondary side, along with machining the required idle passages and extending the secondary transfer slots. This modification does permit greater flexibility in fine-tuning the carb, but in many cases, it's not really necessary. Another approach is to drill holes approximately 3/32-inch in diameter in each of the throttle plates to permit extra air flow under low-idle vacuum conditions with high-overlap camshafts. This modification works well, and you can do it easily with a drill motor. It is also being incorporated on some production Holley carbs. In still other cases, the idle air bleeds are removed and the passages tapped for replaceable bleeds in various sizes. Others simply drill or resize the existing air bleeds, but Holley feels this is a poor way to alter calibration, since it also affects the metering signal. Idle mixture is better controlled by sizing the idle *fuel* restriction, if required, to bring the idle mixture adjustment screws into an operational range.

Idle quality is also affected on many competition engines by the power valves (if the carburetor is so equipped), due to the oscillations in the manifold when a high-overlap cam is used. Some modifiers block the vacuum passage to the power valve(s), effectively opening the power valve(s) at all times. This is a definite mistake, frequently causing loading during staging and immediately after burnouts. A power valve supplements fuel flow from the main jets during very low vacuum (high engine demand) conditions and prevents overrich mixtures during high vacuum (low demand) conditions. Consequently, power valves should be used whenever possible, although they must be correctly matched to the normal idle vacuum of the engine. The power valve should open 1- to 1½-inch Hg. below the observed vacuum level of the engine at idle or while staged, whichever is lower.

Most modified 4150-series carburetors have the choke shroud (air horn) milled off. This is generally a desirable modification, if the machined edges are radiused. This aids air entry to the carburetor and increases air flow slightly, especially on larger venturi carbs. Still others polish the venturi. Any polishing beyond the simple removal of casting flashing is generally a waste of time. Other approaches to improving air flow include slabbing the throttle shafts and screws and/or the installation of larger bore

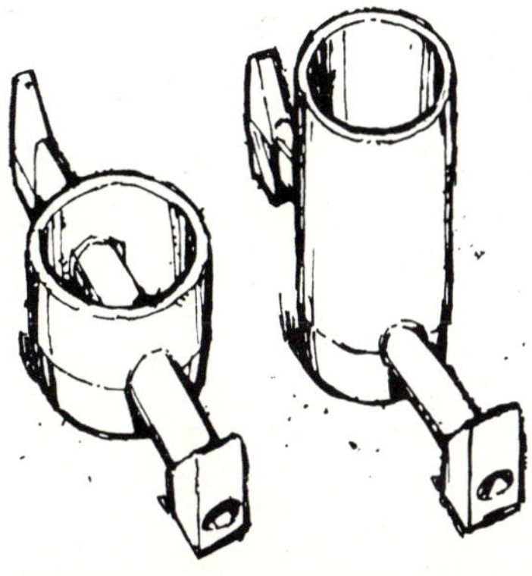

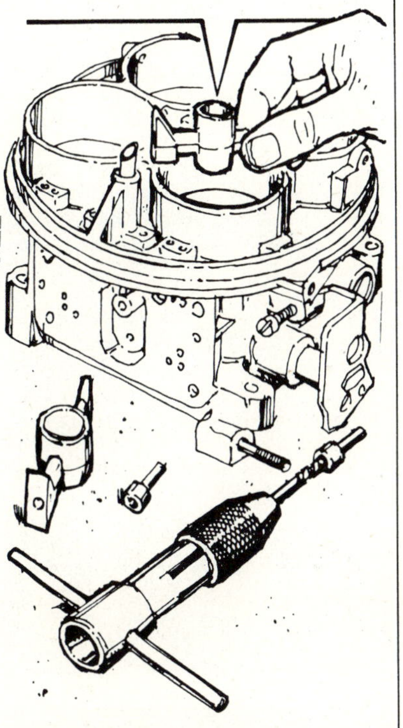

TIME FOR A CHANGE

Changing the booster venturis, as recommended on certain models and applications of 4500-series carburetors, isn't particularly difficult, but it does require a stud extractor and replacement fuel passage brass bushings (part No. 14R-826). The booster venturi is held in place by the brass bushing. Using the extractor, simultaneously twist and pull the bushing from the main body. The booster can then be removed. A few gentle taps with a wooden dowel from below may be required to free the booster from its alignment slots. To reinstall a booster, simply tap into the alignment slots until the fuel passage hole in the booster lines up with the hole in the main body, and then tap the new bushing into place with a plastic hammer.

1

1. Idle quality can be improved on radically cammed engines by drilling a 3/32-inch hole in each throttle plate, as shown. The hole should be on the same side of the throttle shaft as the idle discharge port, meaning the upstream side.

2. Replaceable idle-and high-speed air bleeds can be used to vary calibration of these circuits, but Holley recommends varying the fuel restrictions for these circuits rather than the air bleeds.

3. Changing the size of the idle fuel restriction bushing can bring the idle adjustment screws into an operational range. Enlarging the bushing richens the idle mixture. To lean the idle mixture, small pieces of wire may have to be inserted in the orifice, as shown in another illustration.

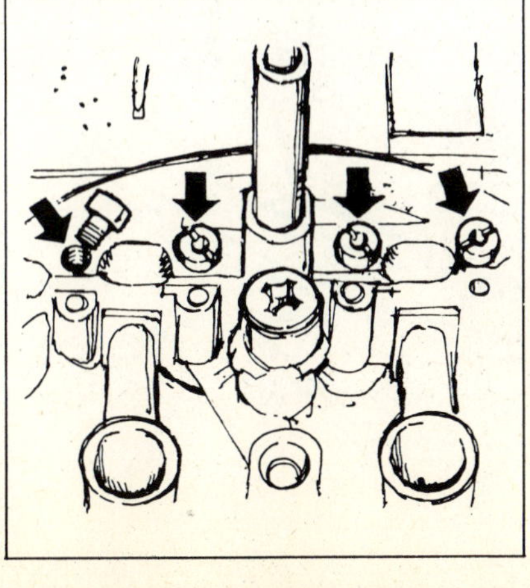

2

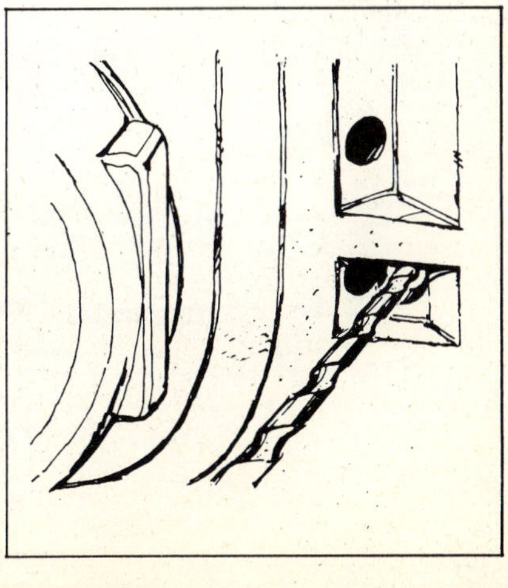

3

throttle plates with the venturi blended in accordingly. Tests indicate that such procedures, when properly executed, do increase air flow appreciably. Some modified 850-cfm carbs showed flow rates approaching 1000 cfm, but it must be remembered that increased flow to produce top-end power usually reduces the metering signal at lower air flows, often creating lean spots or "holes" in the fuel flow curve, which can result in an overall performance loss. Size alone does not guarantee better performance.

Because leaning and flat spots are a problem, one modifier installs the metering blocks from either the 6214 or 6464 Dominator carburetors incorporating not only the idle circuits on both primary and secondary sides, as previously discussed, but also intermediate circuits to smooth the transition into the main metering circuit. Corresponding passages and air bleeds are also added. Such changes require wet-flow testing of each and every unit to assure consistent and accurate results, which does not currently seem to be the situation. A better solution here is a revision of the accelerator pump circuit to achieve the same results. For example, on the 850-cfm double-pumper, Holley recommends the installation of a 50cc pump on the primary side. The secondary side already has a 50cc pump. The accelerator pump shooters are then opened to .042-inch on the primary and .047-inch on the secondary sides. This approach is far easier and cheaper than adding an intermediate system, and the results are comparable in most cases.

In concluding the modification of 4150-series carbs, several other things are being done which are either of no benefit or detrimental to carburetor performance. These mods include: chamfering of all main body passages, installation of 1:1 gear-type linkage, adding strengthening struts to brass floats, cutting down Nitrophyl composition floats, and even opening of the power valve vacuum chamber to add extra float bowl volume on those carburetros where power valves aren't used. Holley has not seen any benefit from these modifications.

THE 2300 SERIES

The 2300 series includes the 500- and 650-cfm performance 2-barrel carburetors used in IHRA and AHRA class racing. Two of the companies modifying these carbs are C&S Specialty and Braswell Carburetion, at a cost of about $200 to $265, plus the cost of the carburetor. Considering the extent of the modification, including installation of larger annular discharge venturis and larger throttle plates along with a complete reworking of the fuel circuitry, and the improvement in flow and performance that is achieved, these carburetors are among the best values in modified carburetion currently available—if you run in such a class. Both units seem to work very well as delivered, although Holley field technicians report a reduced metering signal at lower air flows on some applications with the C&S unit, which can present difficult calibration problems. In such cases, the problem can generally be remedied by using .045-inch accelerator pump shooters, a 4.5-inch-Hg. power valve, and drilling the main jets to .144-inch, as No. 100 jets (.128-inch diame-

SIZING UP THE FLOW

Although it isn't completely accurate in terms of wall friction and boundary flow, a close approximation of fuel flow through a passage can be achieved using cross-sectional areas. The area of a circle is calculated from the formula $A=\pi r^2$, where A is the area, π is 3.1416 and r is the radius (half the diameter) of the circle. For example, a No. 100 jet has a diameter of .128-inch. Therefore, its radius is half of that, or .064-inch. Putting that into our formula (thank heaven for pocket calculators), $A = 3.1416 \times .064^2$. This comes out to .012868 square inches.

Now let's suppose we want to calculate how big the fuel passages must be in the metering block and main body of a carburetor to adequately handle all of the fuel passed by a No. 100 jet and a power valve restriction of .113-inch. We already know from our previous example that the area of the jet is .012868 square inches. Performing similar calculations on a .113-inch orifice yields an area of .010029 square inches. Adding these two areas together gives a total area of .022897 square inches. Therefore, every part of the fuel passage must have a cross-sectional area at least that big to pass the fuel. Working backward in our formula, we can find the radius of a circle with that much area from

$$r=\sqrt{\frac{A}{\pi}}$$

$$r=\sqrt{\frac{.022897}{3.1416}}$$

Again plugging this into our calculator, It all comes out to r=.08537-inch. That means the diameter of the fuel passage must be .171-inch.

This same procedure can be used for determining just how much power valve restrictions should be enlarged or for determining accelerator pump passage restrictions, etc. It all seems a bit complicated at first glance, but it's a snap with any pocket calculator that has a square root function on it.

4

4. Milling off the air horn generally increases the total air flow of the carb and aids air entry to the venturis. Do not mill off the boss around the vent tube. Smooth and radius the edges after machining.

5. Slabbing the throttle shafts usually shows a slight improvement in air flow, but unless you really need that extra 5 cfm or so, it usually isn't worth the trouble. There's more to gain by installing larger throttle plates in most cases, often as much as 40 cfm.

6. Mechanical secondary kits that employ gear drives are definitely not recommended, as they may cause premature wear, bending or binding of the throttle shafts.

7. Strengthening struts on brass floats is totally unnecessary. There's no way the fuel forces within a carb can bend one, so all the strut does is add undesirable weight to the float, making it less responsive to fuel movement.

ter) seem to be too lean.
Editor's Note: *Personal experience with this carburetor on large-displacement engines with radical cam timing produced the above conditions. An alternate fix that provides greater tuning flexibility is to enlarge the power valve restrictions to .113-inch. Then conventional Holley numbered jets in the No. 95 to No. 100 range can be used to achieve the optimum air/fuel ratio and tune for varying altitude and atmospheric conditions. Such changes have already been incorporated on the latest versions of the C&S carbs for big-inch applications.*

5
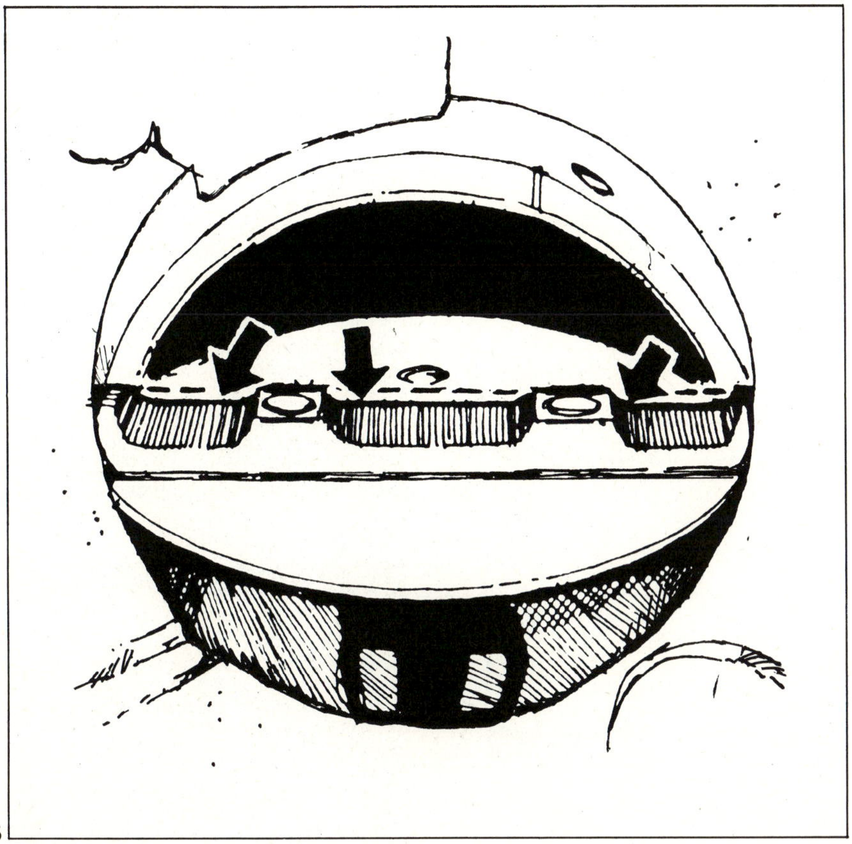

6
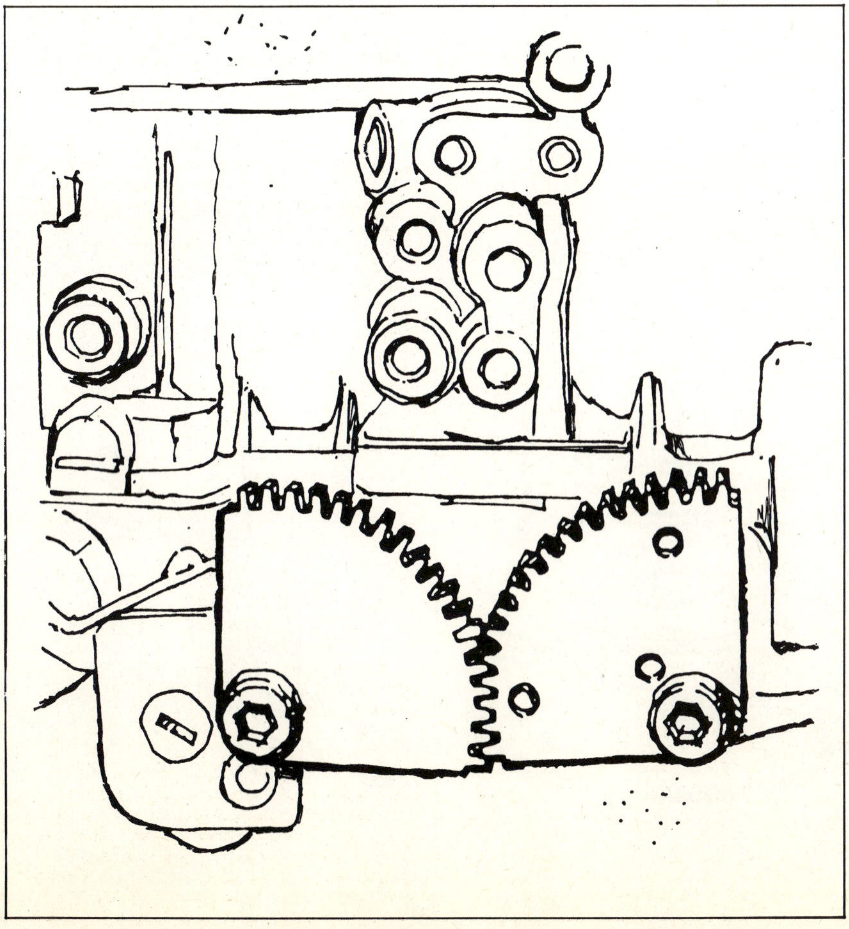

THE 4500 SERIES

The 4500 series includes all of the wide-spaced-flange, large-bore-venturi Dominator carburetors. Modification prices for these units range from $185 to over $700, plus the cost of the carburetor. The most expensive mods include enlargement of the throttle bores from 2 to 2⅛ inches for a flow capacity in excess of 1300 cfm. Such modification is extensive, requiring remachining of the venturis, movement of fuel passages, changes of the booster venturis and a general cleanup of the unit. As with similar modification to the 4150-series and 2300-series carbs, increased total air flow decreases low flow metering signals, which are already marginal on the 4500-series carbs. And since it is doubtful that the manifolds and cylinder heads on most Econorails, for which this modification is intended, can use the extra top-end flow, the decrease in low-end performance may actually hurt overall performance.

Lesser modifications of Dominator carburetors generally include little more than removal of the air horns, replacement of the booster venturis, enlargement of certain fuel passages and a general cleanup of the unit, since most 4500-series carbs already have idle circuitry on both primary and secondary sides. Additionally, the Dominator is offered in a variety of configurations, with and without progressive linkage and with and without intermediate circuits. In other words, optimum performance from a 4500 carb amounts to little more than selecting the right model and performing simple do-it-yourself mods on it. Consequently, Holley has the following recommendations for specific applications.

For Econorails or single Dominator applications, use either the R-7320 (1150-cfm) or R-8082 (1050-cfm) version with 2.5-inch-Hg. power valves on both primary and secondary sides without modification to the power valve channel restrictions. Drill the accelerator pump

7

shooters to .042-inch and use the 85BP-5229 hollow shooter screws. Begin with No. 88 jets all the way around and fine-tune jetting from there. The idle system should also be leaned by inserting a .010-inch wire, ½-inch long, in each idle fuel restriction. Adjust the idle so that all four adjustment screws are backed out the same amount, usually starting with them out 1½ turns. Both the primary and secondary throttles should also be opened the same amount at the desired idle speed. And don't bother milling off the air horns in this application.

On all applications where two 4500s are used, the air horns should be milled off and the entry to each venturi smoothed and radiused.

On Weiand-equipped Chevrolet 396 and 402 big-block engines with dual 4500s, use the R-6464 carbs. Only the air horns need be milled off here, since stock jetting, accelerator pumps, shooters and the idle circuitry are all acceptable as is.

Big-block Chevrolet engines of 427 cubic inches or more with the Weiand manifold should use the R-6214 carbs with machined air horns, the 21R-718 booster venturis, 50cc pumps, .035-inch shooters and jetting somewhere between No. 88 and No. 95. Begin with No. 90 or No. 92 jets and tune from there. Additionally, the intermediate discharge passages need to be restricted to .046-inch with brass plugs, and the idle air bleeds should be enlarged two wire drill sizes.

Beyond the specific modifications listed here, Holley carbs need little more than a careful cleanup of casting flashing in the venturis and booster venturis, along with a close general inspection to see that all passages were correctly drilled at the factory. Now that you have the facts, those of you who are budget-minded might do it all yourself.

—C.J. Baker

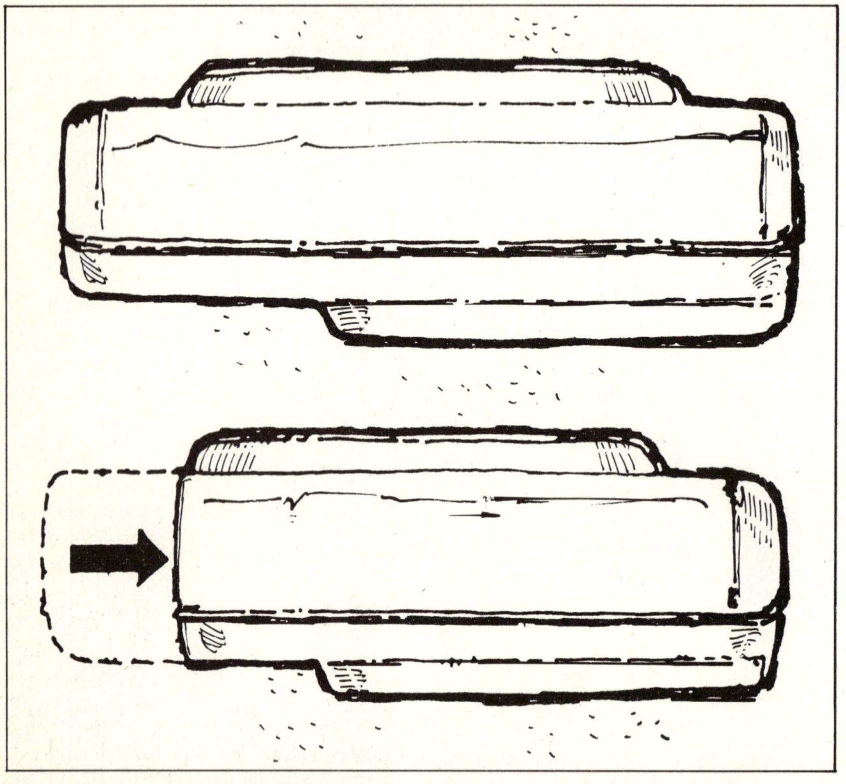

8

9

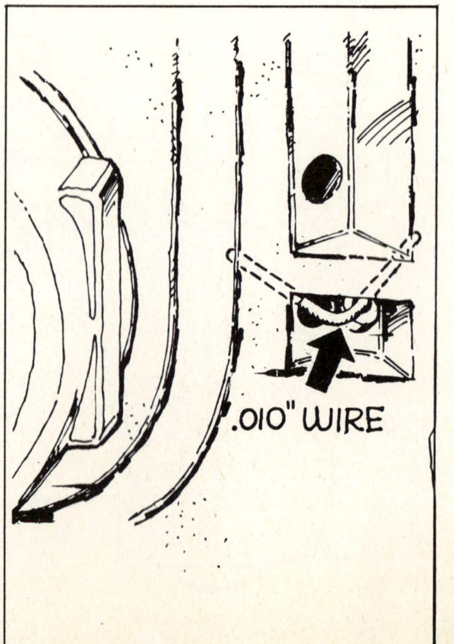

10

8. Another detrimental modification is cutting down the Nitrophyl composition floats. This trick can work in special oval-track and some drag applications where specific fuel slosh conditions may exist, but usually it does nothing more than reduce the float.

9. If additional fuel flow is needed beyond the capacity of No. 100 jets, the power valve channel restrictions can be enlarged. Make enlargements in small increments until the desired jetting is achieved. Also make sure the fuel passages can handle the total flow of the main jet and power valve (see sidebar).

10. To change the venturi boosters in the 4500, remove the brass bushing from the main body with a stud extractor. Where recommended, the shorter 21R-718 boosters should be installed. These shorter boosters are more efficient than the long booster used on some 4500 models.

Kitchen Table Carb Rebuild

Any four-barrel requires a periodic rebuild, and a qualified shop is as near as your kitchen—just sit down and dig in.

The kitchen table, usually nothing fancier than chrome tubing and formica top, has become a significant part of the American home. It has always been much more than just a place to eat corn flakes. Traditionally, it's a place for playing poker, building models, doing homework and filling out those dreaded tax forms. For the person without a driveway or a garage, it's also a great place to rebuild a sick carb, so move over, cereal; make room for the Holley.

Before beginning, there are a few precautions that must be taken. Remember to drain all fuel from the float bowls before taking the carburetor into the

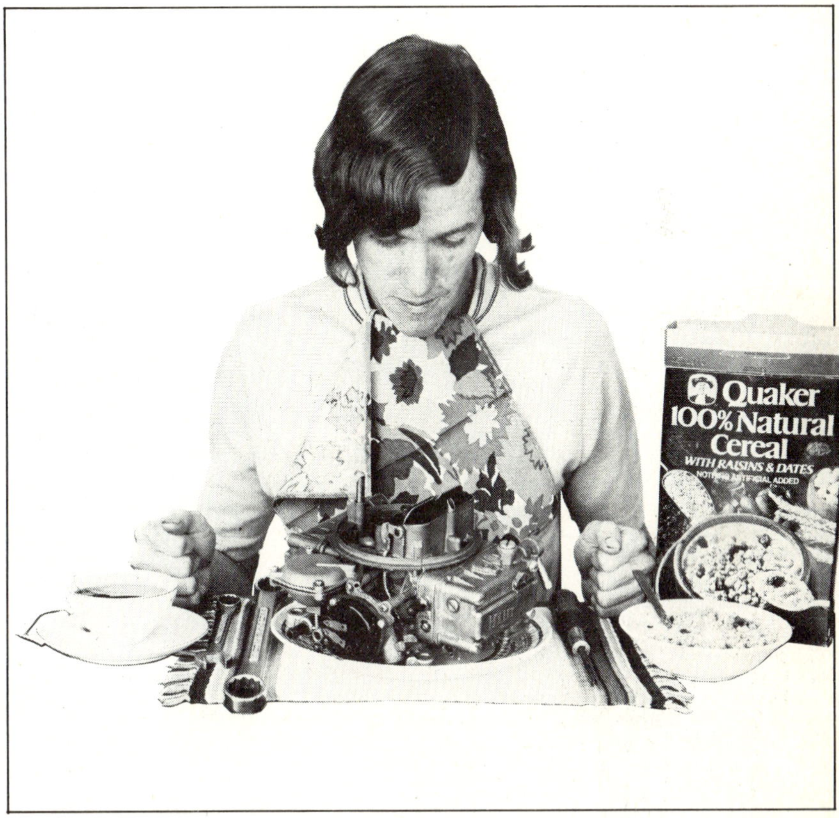

1. The exploded view of the Holley provided with the repair kit, combined with these external shots, will help should you misplace a part or two. However, as the following photos will illustrate, any mistakes can be avoided by confining your repairs to only one section at a time. With only minor differences, these external photos apply to the majority of 4150 and 4160 Holley carbs.

2. Remove the primary float bowl from the front of the carb and, from it, remove the components shown in the photo. Inspect float for pinholes and replace if any are found. Soak bowl in cleaner, rinse in thinner and return float and spring to bowl.

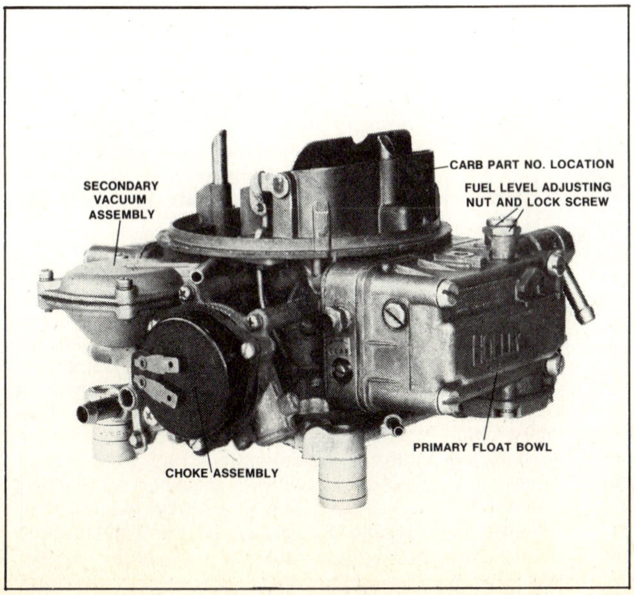

1

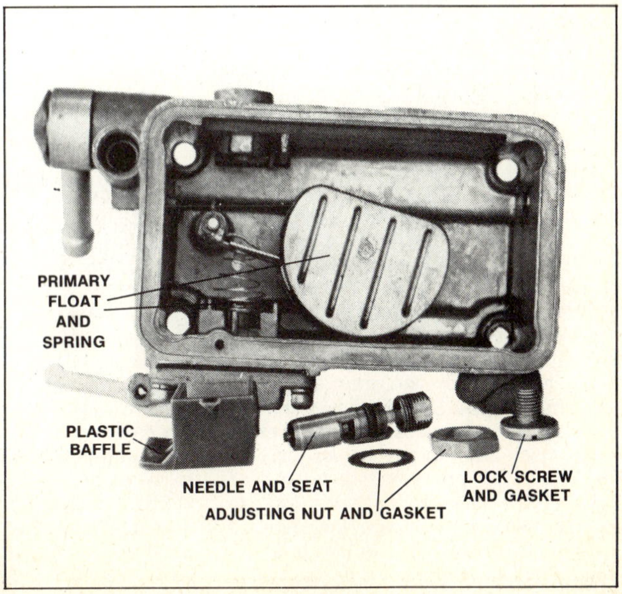

2

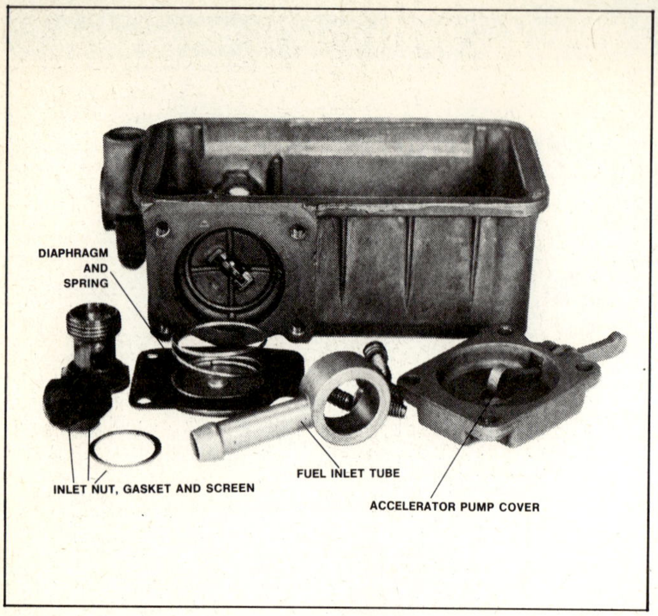

3

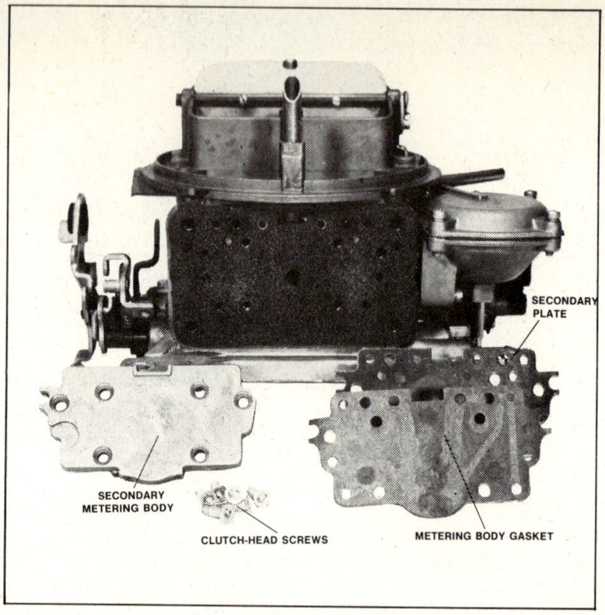

4

house. Gasoline on the rugs will not promote family peace. The job is going to require a carburetor cleaner containing chlorinated hydrocarbons, and prolonged exposure to its fumes is not recommended. Food, plants, pets and the human body are all affected, and for that reason, all carb soaking should be done outside. The carburetor should be disassembled one section at a time, soaked in the cleaning solution, and then rinsed in paint thinner or water before any parts are replaced. The cleaning agents are highly flammable, and to use them near a stove or other source of heat would be asking for trouble. It's also suggested that gloves be worn to protect your hands against the harsh chemicals. Carburetor cleaner is nasty stuff, but it sure does its intended job.

For this project, we've chosen one of the more popular examples of Holley's breed of smaller 4-barrels—the 6619-1. For years, the majority of street machines breathed through large double-pump, dual-feed Holleys. Most of these cars were overcarbureted, and it took nothing less than a gas crisis to make this a reality. As fuel prices rose, the big 4-barrels were ultimately shelved for smaller, more practical alternatives, and with a '74 introduction, the 6619-1 filled the bill nicely. At 600 cfm, the single-feed, single-accelerator pump carburetor is the perfect size for most V8s. Throttle response is smooth and crisp, economy is good and emissions are easily brought into line.

The No. 6619-1 is Holley's identification (or list number, as Holley calls it) for this specific 4-barrel carburetor, just one of many small-cfm carbs included in the 4150 and 4160 model series. This series has been around for a number of years, and all the carbs within this particular family are very similar; most parts are interchangeable. Consequently, the rebuilding procedure described applies to many more part numbers than just the 6619-1. The parts number list of 4150/4160 Holleys is much too long to include here, so compare the outward appearance of your carb to the photo in **figure 1.** If they appear to be the same, with only minor differences (for example, some carbs in this series have a secondary metering block like the primary, complete with power valve and jets), then the following how-to also applies to you.

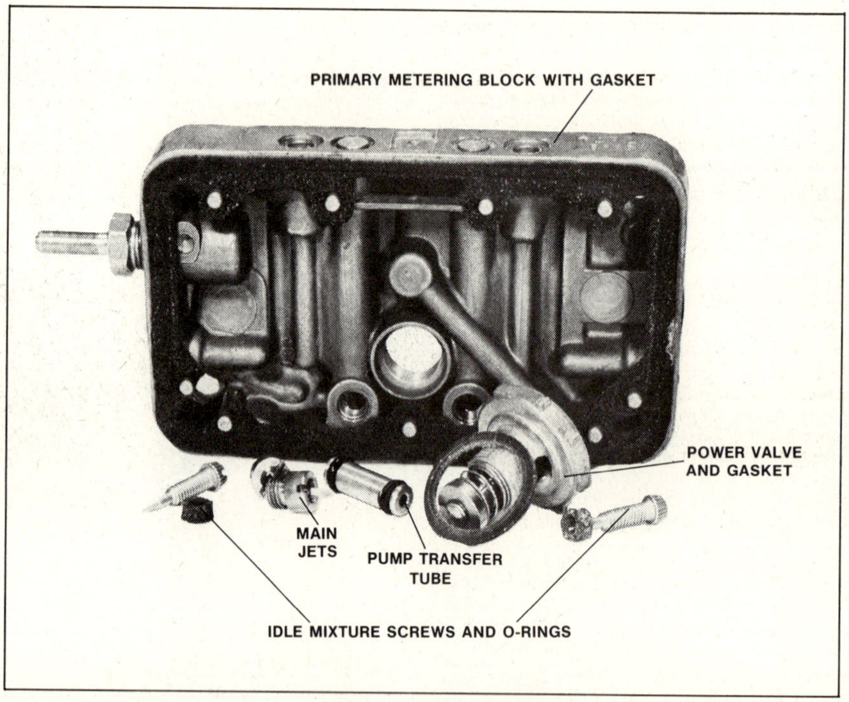

5

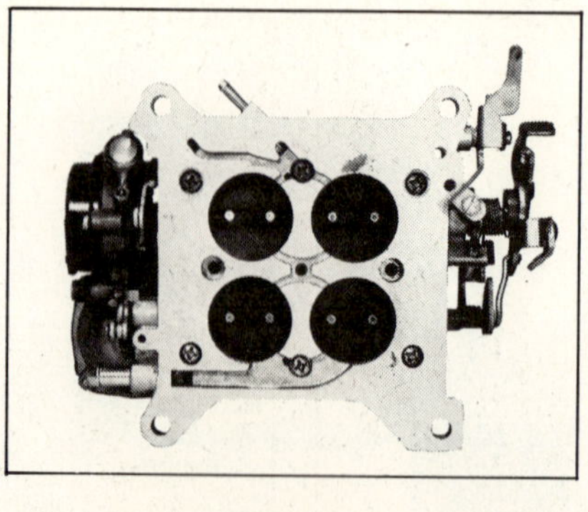
6

Every carburetor requires some periodic repairs, and after two or so years of service, the 6619-1 may exhibit symptoms of wear for some of its owners. If the engine has recently been tuned but continues to idle roughly, hesitate during acceleration or use excessive amounts of fuel, then the carburetor may be the problem. Fuel leakage around the carb, flooding, eye-burning fumes of unburned gas at the tailpipe and insensitivity to fuel/air adjustments are further symptoms of a tired carburetor.

To many, the carburetor is a complex, mysterious instrument containing many little components, and for those reasons, many people are afraid to tamper with it. Despite its complexities, the Holley 4-barrel is actually quite simple to rebuild (probably the simplest of all 4-barrels, it has the great advantage of external float bowl adjustment), if the project is handled systematically. That is, don't be overeager with its disassembly. Taking something apart is always easier than putting it together, so don't get carried away. A professional carburetor rebuilder completely disassembles a carburetor, throws all the pieces into a basket, and within minutes, the unit is back together and operating smoothly. Of course, that kind of expertise comes with lots of experience, though. For the beginner, such an optimistic approach would yield nothing more than a basket full of unfamiliar parts. The safest way to rebuild a carburetor is to work on one section at a time.

As you will see, the first step is to remove the float bowl. The bowl is cleaned and then each part is removed—one piece at a time—cleaned or replaced and then returned to its place in the bowl assembly. When the bowl has been serviced, the next section (the metering block) is removed, cleaned and serviced. Although doing the job in this manner may take longer, it ensures that there will be no mistakes. Cleaning should always precede disassembly, because it's so easy to forget what went where while parts are soaking.

The entire project requires a clutch-head screwdriver (about $3) to remove the secondary metering body (not necessary for carbs with secondary metering blocks) and just a few basic hand tools: a wide- and a thin-blade screwdriver, a small- and a medium-point Phillips screwdriver, a 5/16-inch open-end wrench, 7/16-inch, 5/8-inch, 11/16-inch and 1/2-inch (for the carb-to-manifold bolts) boxend wrenches. If you own these tools, or if you can borrow them, then your monetary investment is relatively small. You will need to buy a carburetor repair kit for a Holley 6619-1, a 1-gallon can of paint thinner and a 1-gallon can of carburetor cleaner. The repair kit includes a complete gasket set, needles and seats, a new power valve, an accelerator pump diaphragm and all the required O-rings. Instructions are also included, but they usually aren't too helpful. The same kit is used for many 4150/4160 models; this means you'll end up with some extra gaskets, so don't panic. For a Holley carb, the best kit is made by Holley and sold by most speed shops. While there are several other kits on the market, try to stick with the popular name brands—they're usually best.

Carburetor cleaner is available at all auto supply stores. The 1-gallon can is a good investment, because it's usually large enough to accommodate the entire carburetor for thorough soaking, and it often includes a wire-mesh parts basket. Since cleanliness is critically important to carburetor operation, you should choose a good cleaner.

The beginner should allow a full Saturday for the job and a small investment, for a $30-$40 saving and that feeling of satisfaction from doing the job himself . . . right on his kitchen table.

—Jon C. Jay

3. Remove the parts shown in photo, clean, and return, using new diaphragm (convex side out). Install new gasket on sight plug, new O-ring on needle/seat, and install in bowl with nut, screw, baffle. Repeat for secondary.

4. Position new gaskets on the metering block. Outside gasket will go on backward, so be careful. Both the float bowls and block are complete, but keep clean until the time comes for installation. Using a clutch-head screwdriver, remove the secondary metering body and plate. The gasket, plate, plate gasket and metering body (in that order) are to be installed later; in the meantime, don't lose those screws.

5. Remove primary metering block by gently prying between air horn and block. Clean block and remove parts shown in photo. Main jets are same size, so don't worry about mixing them. Install new power valve and gasket and other parts, using new O-rings and gaskets. Return mixture screws to same holes, snug down, back out 1½ turns each.

6. Separate the throttle body from the main body, after noting the position of the six screws. Remove plastic choke cover and soak both bodies in cleaner. When clean, install new gasket, assemble and install choke cover.

7. If you disassemble the secondary vacuum assembly for cleaning, you will find a small check ball inside. It's easy to lose, so be careful. Looking down through the top of the carb, there is a Phillips screw directly below the choke plate. It secures the accelerator pump discharge nozzle, two gaskets and a small needle. Clean, replace gaskets, and install with needle point down and nozzles facing toward center of carb. Install the metering block and secondary metering body, plate and gaskets, and tighten screws in a diagonal sequence.

8. Turn each needle and seat until float parallels floor (top) of bowl. Prevent leaks at long transfer tube by coating O-rings with STP. Small transfer tube will seat when bowl is tightened. With car running, adjust floats till fuel appears at the lower threads of the sight plug holes.

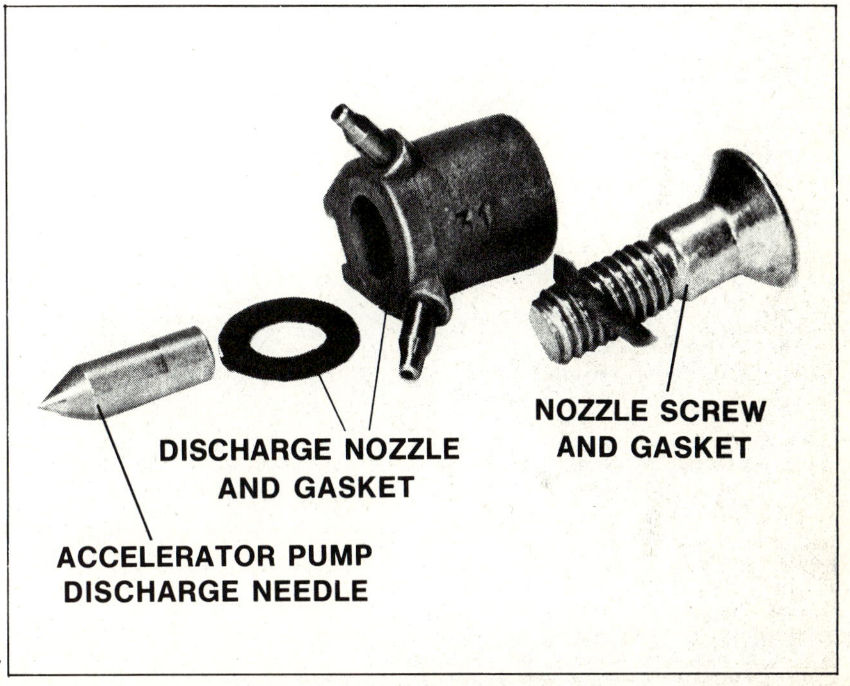

7

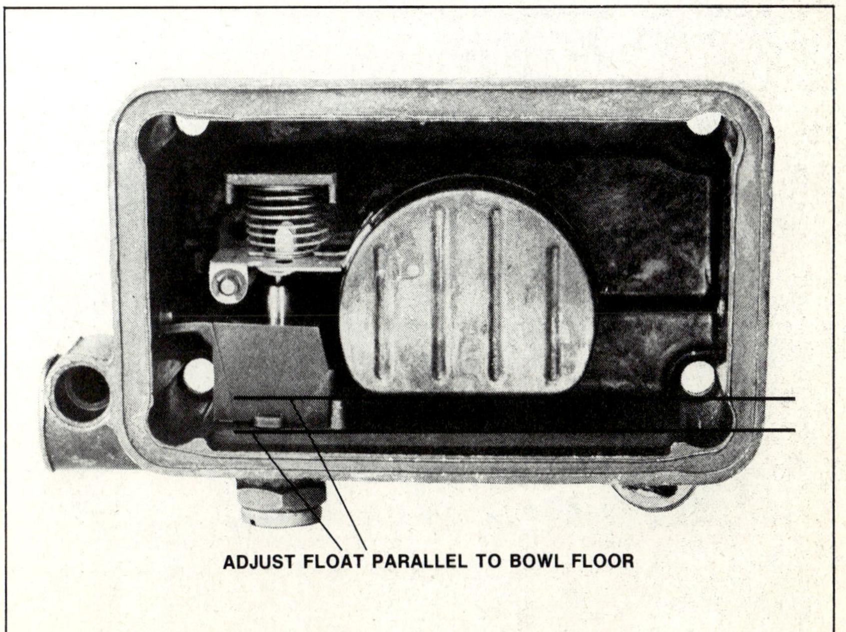

8

The Perfect Match

A bit of backyard science to custom-fit that new Chevy intake manifold.

No two engines are exactly alike. Manufacturing tolerances and slight core shifts during the casting process of such items as blocks, cylinder heads and intake manifolds give each engine its own unique characteristics. Consequently, the manufacturers of aftermarket performance products must design their parts to be compatible with fairly large dimensional tolerances.

This is especially true with aftermarket intake manifolds where correct cylinder head port alignment and sizing may vary by over ⅛-inch. Additionally, due to production costs, one manifold may be designed to work with several different cylinder heads with different port sizes. To further complicate the production of these sophisticated performance components, the manifold manufacturers must also put up with casting variations in their own products, due to core shifts and machining in volume quantities. Therefore, it is common practice to design intake manifolds with runners that are intentionally smaller than the ports of the typical cylinder heads to which they will be mated. This guarantees that the flow of air/fuel mixture will not encounter a protruding edge at the juncture of the manifold and head.

Admittedly, this is a compromise between optimum flow and the realities of the production world, but happily, it also provides adequate material for the enthusiast to go in and precisely match the intake runners to his particular port configuration. The manifold manufacturers encourage such matching, and that's what this article is all about. We'll cover the exact areas that need to be matched, blended or reshaped, how to

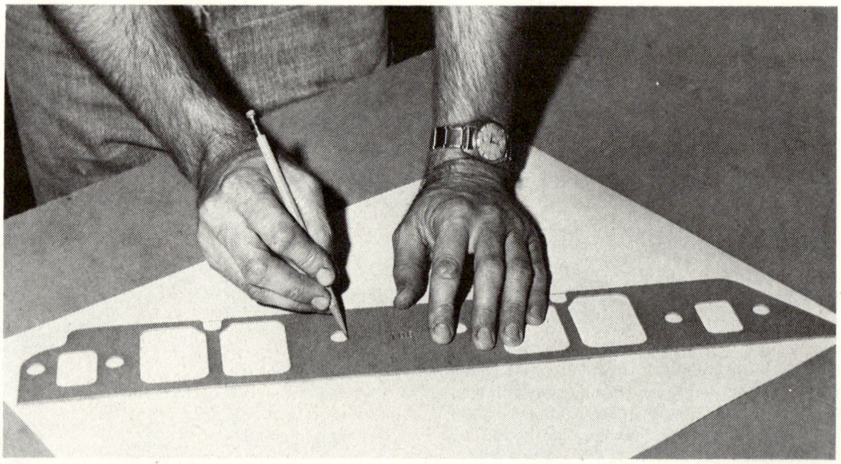

1

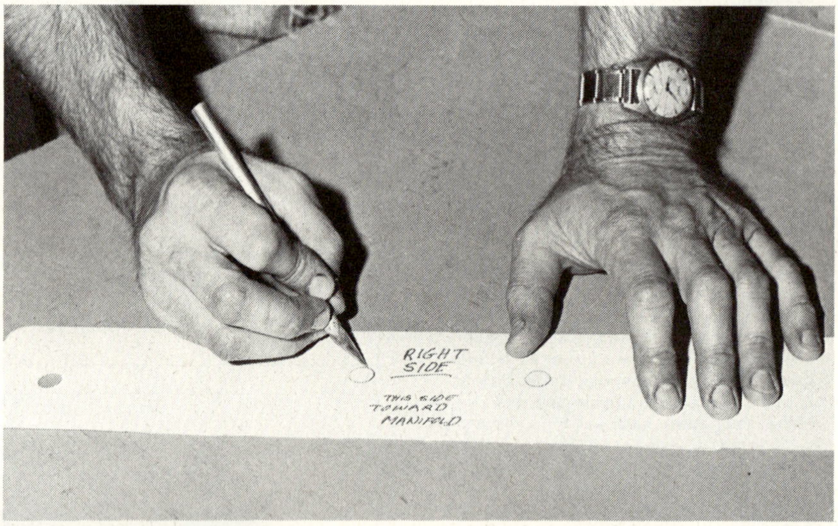

2

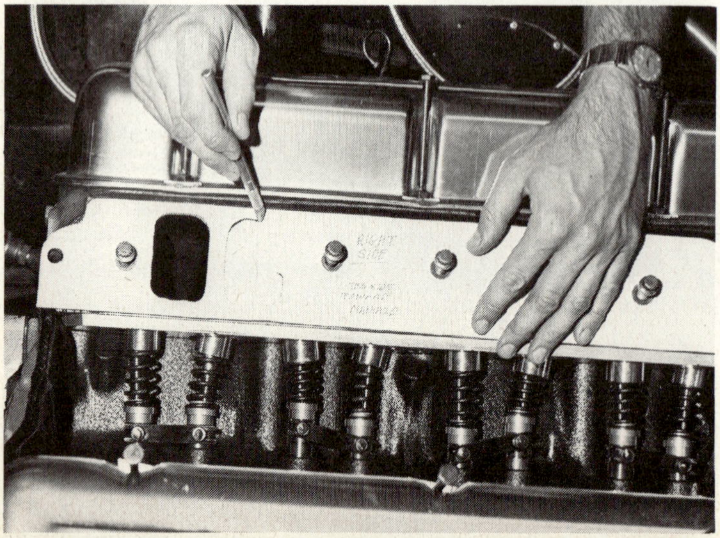

3

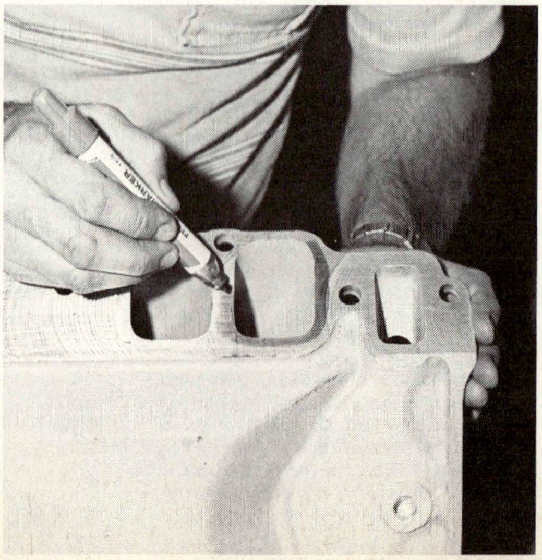

4

go about doing it, and the correct tools for the job. Matching your intake manifold won't produce awesome power gains, but it can give you that extra edge in street and bracket performance. And more important, it's something you can do yourself without spending a lot of money.

Three basic areas of an intake manifold should be considered. The first of these is the matching of the runners to the cylinder head ports. Notice that we said to the ports and not to the intake manifold gasket. In most cases the gasket is larger than the port, to prevent the possibility of the gasket hanging into the port or runner. So, unless both the head ports and the intake manifold runners are matched to the gasket, don't use the gasket as a matching template.

When we polled the four major manifold manufacturers regarding their recommendations on manifold matching, Edelbrock, Holley and Weiand advised matching the runners exactly to the cylinder head ports about the full perimeter of the runner. Offenhauser suggested either matching the runners exactly to the port or leaving the runner about 1/32-inch smaller than the port about its entire perimeter, to form a slight reversion dam to restrict those disruptive reverse flow pulses.

None of the manufacturers recommended a major mismatch at the roof of the runner or along the longest wall, as had been practiced a few years ago. They all feel that improved manifold design technology has precluded the need for such mismatches. Additionally, all the manufacturers feel that when material is removed from the runner opening to achieve matching, such removal should be blended smoothly back into the runner wall so as not to cause an abrupt change in the wall configuration. Depending upon how much material must be removed, Edelbrock, Holley and Offenhauser feel that such blending might extend as much as 1½ inches back into the runner. The Weiand tech advisers cautioned that because runner shape is so important in their designs, they prefer that such blending not extend more than ½-inch back into the runners, for fear of altering flow patterns.

The second area of attention is the receiver or plenum area. Some work should be done in this area to assure that the air/fuel mixture enters the individual runners smoothly, without turbulence or disruption. Because of the way the casting cores fit together, this area often has small ridges or unevenness in it. Generally, what we're talking about doing here is to simply smooth things out and provide proper radiused turns to direct the flow. In effect, we're restoring the receiver area to its "as-designed" configuration. In no case should extensive amounts of metal be removed here, as that might seriously alter the design of the manifold.

The last area of concern is a general

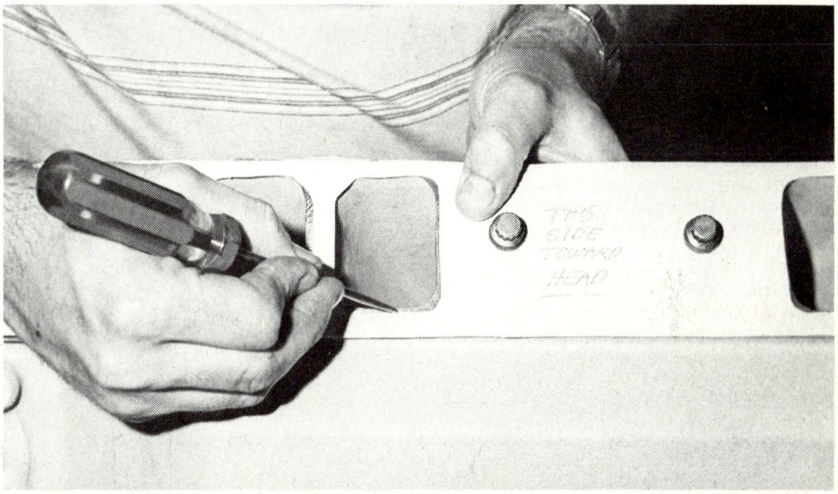

5

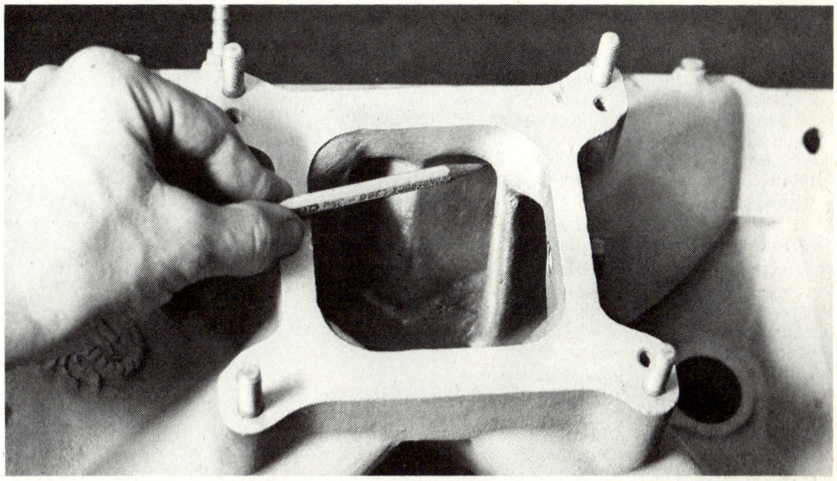

6

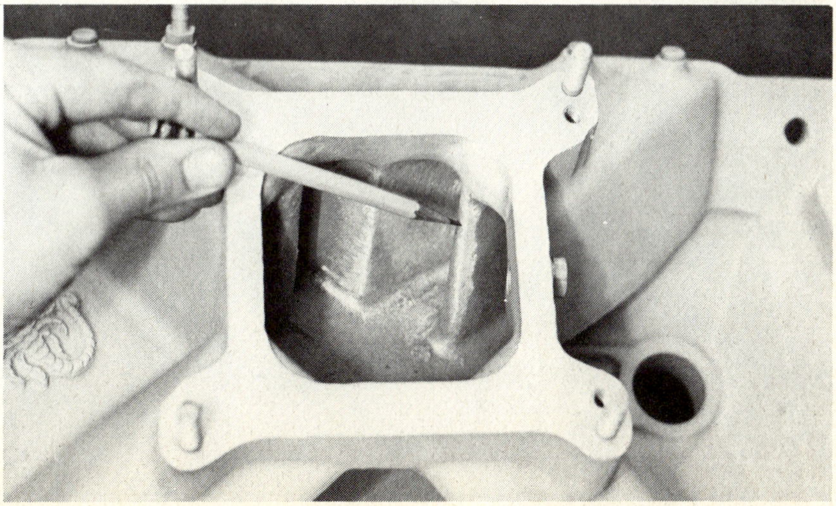

7

1. Before you match the runners, you must have an accurate template of your ports. Using a gasket as a guide, mark the outside shape and the bolt holes on a piece of light cardboard.

2. Cut out the overall shape of the template and carefully cut out the bolt holes.

3. Using manifold bolts as locating pins, position the cardboard on the cylinder head. Mark the cardboard to show which head it was on and which side was against the head. Then, using an X-Acto knife or a sharp pocket knife, cut the port openings in the cardboard to exactly match the head ports.

4. Using either Dye-Kem machinist's bluing or a felt-tip marker, coat the area around each runner opening on the manifold.

5. Again using the manifold bolts as locating pins, position the cardboard template on the manifold, making doubly sure you've got the correct template for that side of the manifold. Then scribe the exact port shape on the inked area around each runner. Now you're ready to begin the actual matching.

6. The area where the roof of the runner enters the receiver or plenum area is frequently rough. It should be smoothed into a gentle radiused turn to direct flow into the runner.

7. The divider wall between two adjacent runners should also be rounded to eliminate sharp edges or roughness.

cleaning up of the runners and receiver area to remove any casting roughness or "dingleberries" that might disrupt flow. Again, the idea here is not to change the shape or configuration of anything, but simply to smooth and polish the air/fuel passages.

Because matching and modification of tunnel ram-type manifolds is a science unto itself, and one best left to professional porting shops, we've restricted our attention to single-plane "spider-type" manifolds that are typical of what most enthusiasts use on the street and strip anyway. Study the accompanying photos and captions and you'll see that such matching really isn't difficult and, in fact, that it is well worth doing.

—C. J. Baker

8. Rotary burrs or special cutters should be used for metal removal. An occasional spray of both the cutter and the manifold surface with a light oil, such as LPS-2 or WD-40, will keep the cutter from loading up with aluminum.

9. After major metal removal has been accomplished with the cutter, the surface should be smoothed and polished with a hard roll on a mandrel. Cutting and polishing can be done with either a high-speed grinder or an ordinary ¼-inch electric drill motor.

10. This cutaway manifold shows a typical "dingleberry" that should be polished out of the runners. Note too the scribe lines showing how the matching of the port should be blended back into the runner.

11. With a polishing stick made from a piece of ¼-inch steel rod with a slot hacksawed into one end and a short length of 60-grit polishing cloth wrapped around it, roughness and dingleberries can be polished out of the runners and receiver area. All polishing in the manifold should be done with 60-grit paper and nothing finer.

12. These are the cutters, hard rolls, mandrels and polishing paper needed to make the job quick and easy, with professional results. Although these tools are frequently hard to find, Cylinder Heads America packages them for distribution at speed shops nationwide. If your local speed shop doesn't handle these items, ask them to get them for you.

8

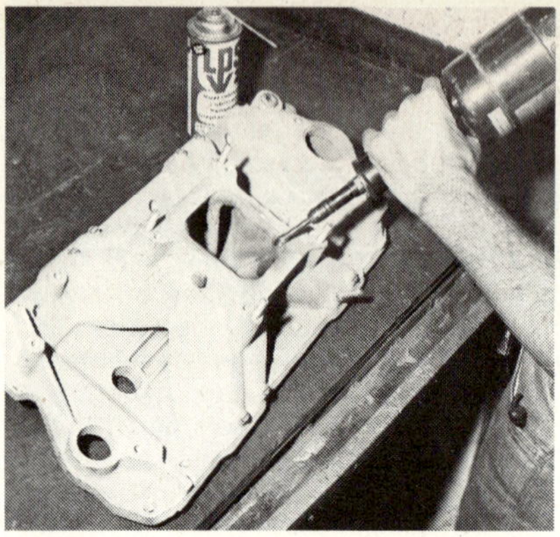

9

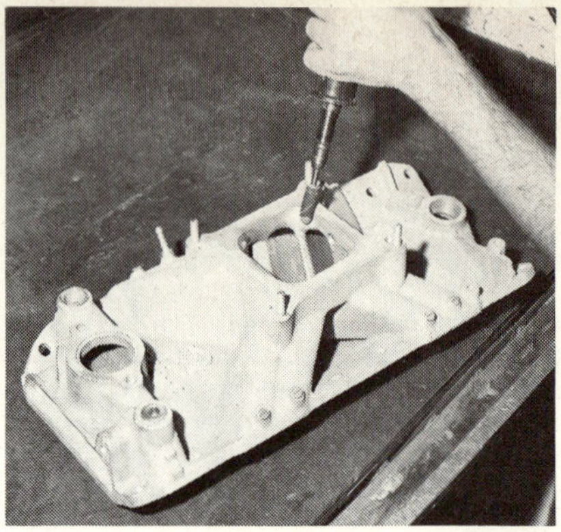

10

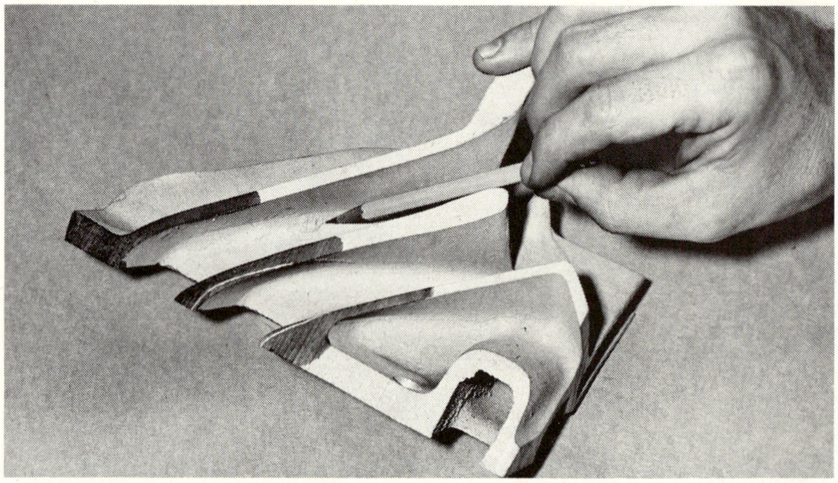

11

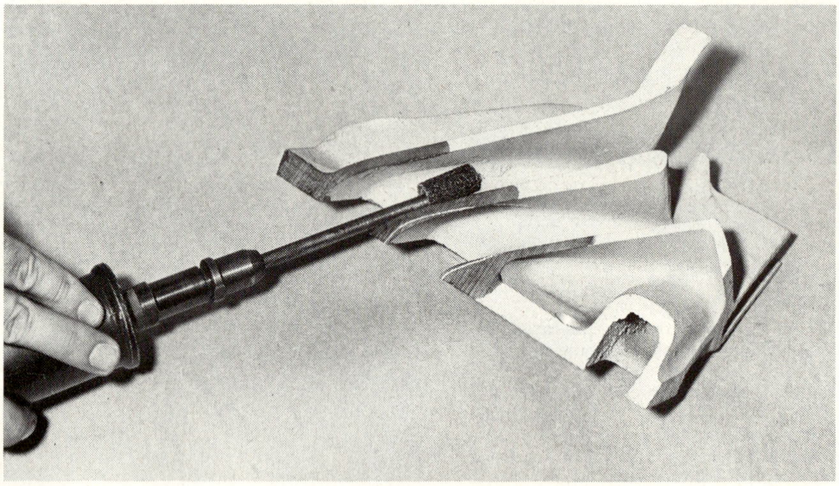

12

Water Power

Water injection—a cure for high-compression engines in a world of low-octane engines.

Since the first energy crisis, when the price of gasoline skyrocketed and the quality nosedived, we have received a steady stream of letters requesting information on what could be done with high-compression engines to make them compatible with poor fuel quality. Until recently, the only advice we could give was to retard the ignition timing, which unfortunately knocks the bottom out of the engine's efficiency, performance and gas mileage, but at least allows the engine to run without destroying itself. The only other alternative was an expensive engine rebuild to install low-compression pistons and/or cylinder heads.

But now there is a reasonably effective cure for preignition in high-compression engines that lets them retain their performance and economy without costly disassembly. It's called water injection, or more correctly, induction charge cooling.

The basic principle of water injection has been in use for many years on aircraft and turbocharged engines, and it works just as well on normally aspirated automobile engines. Water is sprayed directly into the air as it enters the carburetor. The water then vaporizes, substantially cooling the air/fuel mixture before it enters the combustion chamber, thus preventing preignition and detonation. Consequently, ignition timing can be restored for optimum engine efficiency.

However, water injection is required only when the engine is called upon to produce power and combustion chamber pressure is high. So a means of controlling the water injection must be used, and this is where Spearco Performance Products, Inc., in Santa Monica, California, enters the picture. Spearco has developed a complete water injection system, called the Injectronic, which utilizes an electronic control module that monitors both manifold vacuum and engine rpm to start and stop the water flow, as needed, and to regulate the amount of flow to meet the engine's needs.

We installed the complete system, which sells for about $75, on a street-driven 11.25:1-compression 454 Chevelle that had been limited to a total spark advance of only 28 degrees to avoid pinging on today's premium gas. Installation required about an hour, following the detailed instructions. With the unit in operation, it was possible to restore ignition timing to a total of 36 degrees, and this reawakened our dormant musclecar. Gas mileage also increased from 10 mpg to 12 mpg—a 20 percent jump. And the real beauty is that there's nothing else to buy, because water is free. We just add about 1½ quarts of water to the reservoir at each gas fill-up. People who live in cold climates will need to run a 50-50 mix of water and alcohol (ordinary alcohol-base windshield washer antifreeze works just fine).

Water injection isn't a miracle, but for the owners of high-compression engines, or those people who want to get the most out of their machines, it's the next best thing. —C. J. Baker

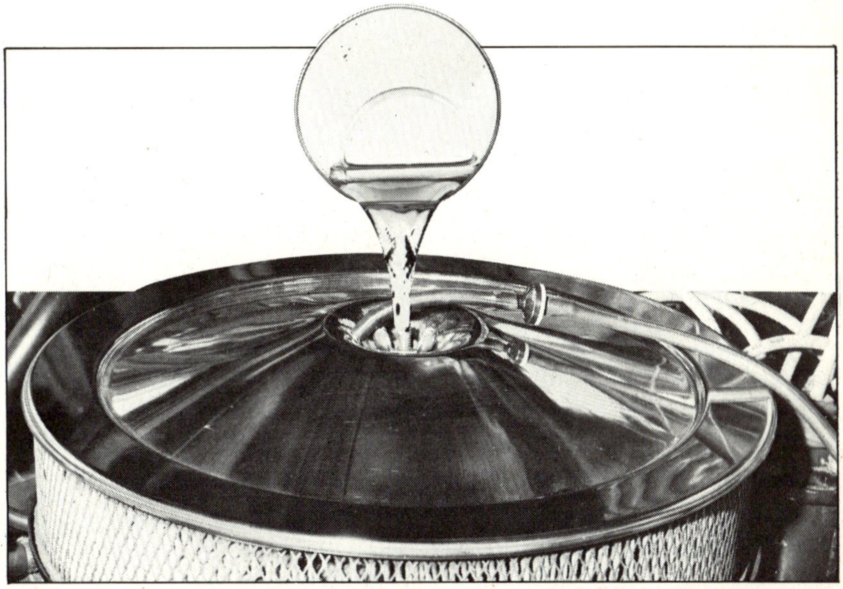
1

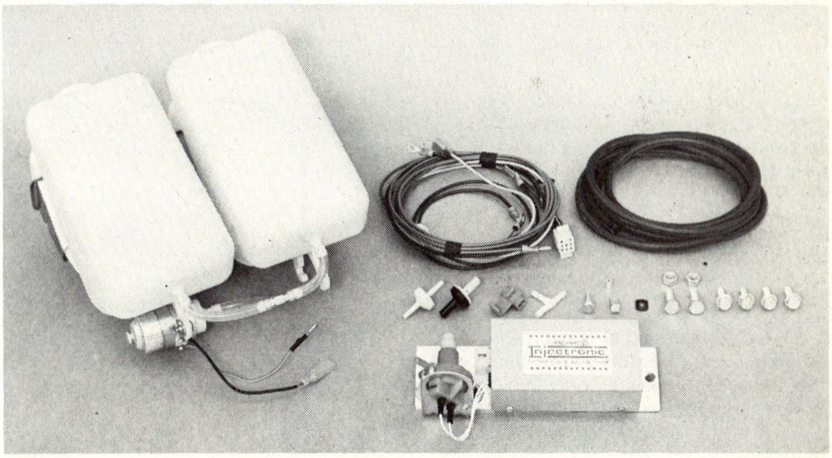
2

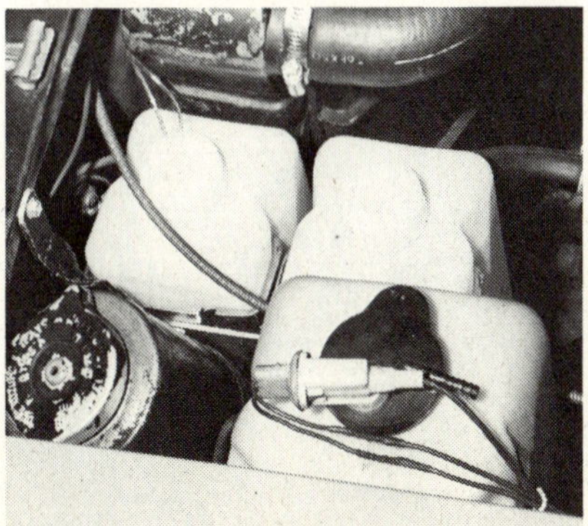
3

1. One small hole drilled in the top of the air cleaner is all that's required to mount the injection nozzle. Water sprays directly into the primary bores of the carb.

2. The complete water injection system consists of a reservoir and pump assembly, the control module and the required hoses, wiring harness and mounting hardware.

3. The reservoir must be mounted below the level of the injection nozzle installed in the air cleaner to prevent siphoning.

Super Tuning

Optimum performance from a Holley carburetor in a street-use application is not all that hard.

Earlier in this book we presented an in-depth feature discussing the relative merits (or lack of them) of various off-road modifications currently being performed on Holley carburetors. But what about the "on-street" use of Holley carbs? Why not offer a similar tech feature to help solve the problems frequently encountered with the street use of Holley carbs? Why not, indeed! And to make such a feature as authoritative as possible, we again sought first-hand information direct from the manufacturer by contacting one of Holley's best-known tech reps, Chuck Gulledge (his remarks are in italics).

Based on his extensive experience in correcting fuel system woes for numerous participants at street rod, street machine and other events across the country, Chuck has developed an orderly procedure for correcting such common complaints as poor idle quality, bogging, lack of power, poor economy, leaks and stalling. We've summarized his procedures and corrective measures. By the time you've finished reading this article, you may be surprised to discover, as we were, that very few of the so-called "carburetion" problems are caused by the carburetor. More often, it is the remainder of the fuel system, lack of maintenance or proper filtering, or the installation that has precipitated the malfunction. Read on. You'll see what we mean.

"Dirt and other foreign material in the fuel probably cause more carburetion problems than any other single thing."

In the vast majority of cases, enthusiasts aren't dealing with new cars, so it cannot be assumed that the gas tank or fuel lines are clean and corrosion-free. Instead, the hot rodder's car is usually several years old or, in some cases, it has been fabricated from scratch. In either case, the probability of dirt, rust, fiberglass particles, metal filings or welding slag and scale in the gas tank or fuel lines is quite high. And, of course, there's no guarantee about the cleanliness of the fuel coming out of the nozzle at the filling station either. Therefore, the installation of a good fuel filter just ahead of the carburetor is essential. Most Holley carbs come with an internal fuel filter positioned just behind the fuel inlet.

Many articles have recommended the removal of such filters to prevent a possible fuel restriction in the event that these relatively small filters should become partially or completely plugged. However, this is not a good idea unless a high-quality in-line filter has been installed ahead of the carburetor fuel inlet, and even then the internal filters provide good insurance for a street-driven vehicle. If the in-line filter is doing its job, the internal filters should remain clean, and in the event that the in-line filter should become plugged in the middle of nowhere, it could then be by-passed temporarily, since the fuel reaching the critical small metering orifices within the carburetor is still filtered.

Some street rodders prefer not to run filters because of space problems or appearance considerations, but this can be a costly mistake. A piece of grit could easily cause an inlet needle to stick open, resulting in fuel overflow from the bowl vents and a probable fire on a running engine. The fact that filters do get dirty and require periodic replacement is testimonial to the fact that dirt and contaminants *do* find their way into gasoline. So if the car in question is exhibiting symptoms indicative of restricted fuel flow, such as a lack of high-speed power, lean surging, popping back through the carburetor or outright fuel starvation, one of the first things to check or replace is the fuel filters(s).

"Any filter, regardless of brand, style or size, is preferable to no filter at all."

Another potential problem involving fuel contamination deals with the possibility of water or other substances in the gasoline. This is quite likely in humid climates on older cars where the gas tank is vented to atmosphere, especially if the vehicle sits outdoors for extended periods. Storage also presents the likelihood of stale or stagnant fuel. Gasoline does not have unlimited storage life. In time, the complex gasoline molecules begin to deteriorate and the light aromatic compounds tend to evaporate away. The result is that the fuel's octane rating, ignitability and even its flammability are reduced, which could cause uneven combustion and a rough running engine—if it'll run at all.

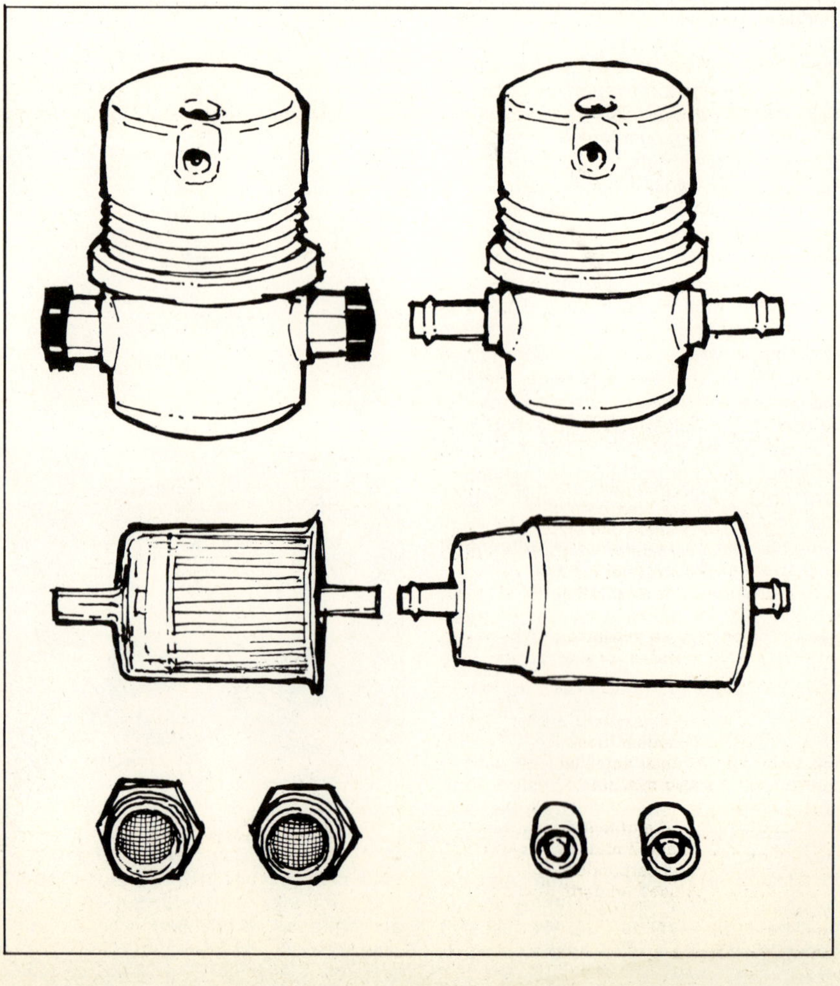

1

"If there's any doubt about the fitness of the fuel, it should be drained from the tank and flushed from the fuel lines and replaced with fresh gasoline."

But there are other properties of gasoline that can manifest themselves as apparent carburetion problems when in fact the carburetor is working properly. Most notable is the Reid Vapor Pressure (RVP). This is the fuel's tendency to evaporate, changing from a liquid to a vapor in relation to changes in temperature and pressure. Unfortunately, this is another property of gasoline that has become extremely inconsistent and unpredictable in recent years. Consequently, many enthusiasts are now faced with severe fuel percolation on high-performance engines that produce high underhood temperatures. The problem becomes critically apparent when a hot engine is restarted after sitting for 5 to 30 minutes, especially if the driver stands on the throttle while the fuel that became superheated in the lines immediately adjacent to the hot engine is still entering the carburetor.

"Heat insulators should be used whenever hood clearance permits."

To reduce the probability of fuel percolation and hot fuel *flash* into the vapor state, heat insulator gaskets that go between the carburetor and the intake manifold are available for most carburetors. The fuel line should also be routed to keep it as cool as possible. In really severe cases, a fresh-air duct can also be run from the grille to vent cool air directly on the carburetor body. A second cool-air duct to supply cool air to the air cleaner also helps reduce inlet air temperatures.

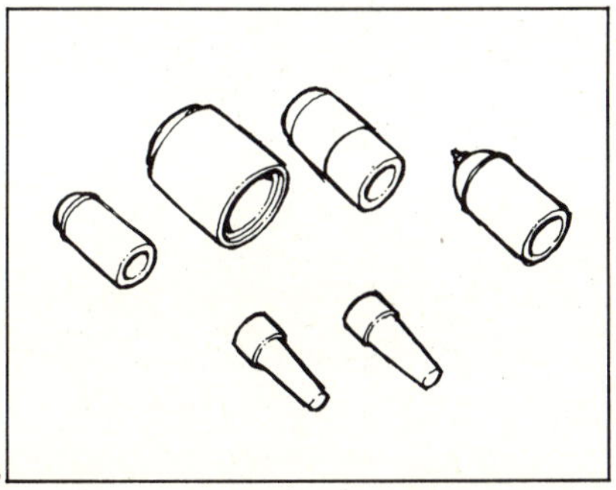

3

1. Shown from top to bottom are in-line filters that use replaceable filter cartridges, disposable in-line filters and the common sintered bronze carburetor filters. Also shown are some special fuel inlet nut strainers that Holley developed for NASCAR.

2. Heat insulators are available in different shapes and thicknesses for most 2- and 4-barrel carburetors.

3. Holley offers assortments of rubber tube plugs to seal off vacuum tubes that may not be used in a particular application. Plastic hose plugs are also available.

4. These high-speed and idle air bleeds (arrows) can be cleared of dirt and debris with a fine wire and carburetor cleaner. The fine wire such as used on parts tags works great.

"Adequate fuel pressure and volume are essential to proper carburetor operation."

Naturally, a good fuel pump and unrestricted fuel lines of a size capable of supplying the engine's needs must be used. Verifying these conditions is another initial inspection step that should be checked before assuming a problem exists within the carburetor.

With the fuel delivery system checked for proper operation, correct installation of the carburetor must be checked. We've already mentioned heat insulator gaskets, but proper linkage attachment and adjustment, choke adjustment and adequate air cleaner clearance must be verified.

"It is imperative to have at least ¾-inch of internal air cleaner clearance

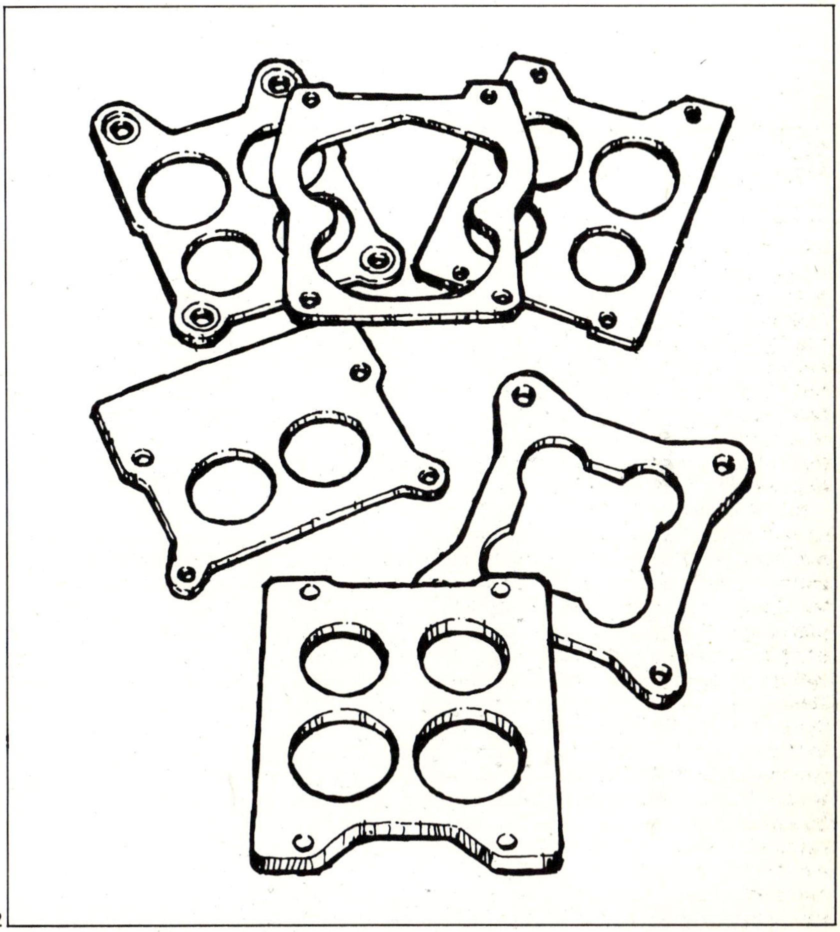

2

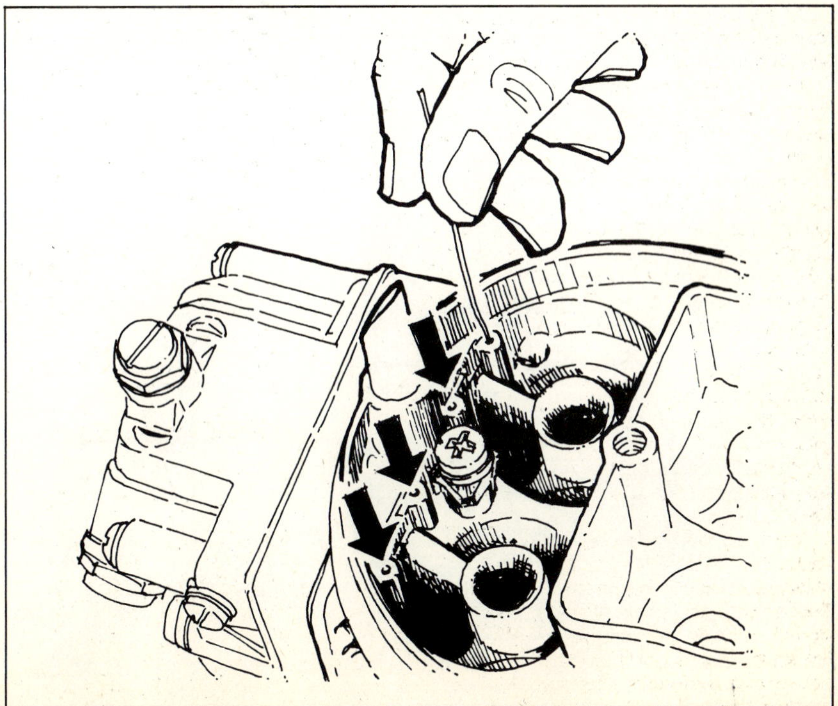

4

above the tops of the bowl vent tubes to prevent normal air flow from disrupting the reference pressure in the float bowls and upsetting the fuel metering. The air also needs this much space to turn smoothly into the carburetor bores. Additionally, the air cleaner element must be reasonably clean and have adequate flow capacity to supply the engine. It must also seal tightly to prevent dirty air from leaking past the element."

While inspecting the installation of the carburetor, make a careful check for vacuum leaks. Such leaks are frequently the cause of poor engine performance that is blamed on the carburetor. Look for disconnected, loose or cracked vacuum lines or tube plugs. Check, too, for leaking intake manifold gaskets.

If everything has checked out up to this point and the problems still exists, *and* you're sure that some other system, such as the ignition or valvetrain, isn't malfunctioning, it is time to begin working on the carburetor.

"The very first thing on a carburetor that must always be checked is the float level."

The fuel level in the float bowls, as determined by the float level (and the fuel pressure and fuel specific gravity), affects the fuel metering of both the idle and main circuits within the carburetor. Fortunately, the float level is easily checked and adjusted on Holley carbs, thanks to the bowl sight plugs and external adjustment. The fuel should just barely trickle out of the sight plug hole with the engine running at idle. Of course the float level must be reset whenever a new inlet needle assembly is installed. And, incidentally, don't be lured into installing oversized inlet needle assemblies. The standard Holley .110-inch Viton needle assembly is more than adequate for any street application. It is also the most trouble-free.

Still other inspections should be made with the carb on the engine. Verify that the idle and high-speed air bleeds, located adjacent to the bowl vent tube(s), are not plugged or restricted. These tiny orifices are especially susceptible to plugging with dirt or fuel residue. A quick squirt of carburetor cleaner and probing with a fine wire is usually sufficient to clear these passages. Disassembly of the carburetor is generally required to permit further inspection, cleaning and adjustment to resolve more serious problems.

"Whenever a carburetor is disassembled, it should be thoroughly inspected for dirt or restrictions in the metering orifices, discharge ports and fuel transfer slots."

This last item, the fuel transfer slots, is especially important, since proper fuel flow from these slots is required for a clean transition from the idle circuit to the main circuit. Many accelerator pump circuits have been erroneously reworked to cover up a bog that occurs because the transfer slots were plugged or the throttle plate stops incorrectly adjusted to get proper action from the transfer slots. As with the previous check of idle and high-speed bleeds, a fine wire and some carb cleaner should be adequate to purge these passages.

Another item to check while the carb is apart is the flatness of the main body metering block mating surfaces. This is rarely a problem on new carbs, but on units that have been in service for a while, verify that these surfaces haven't warped or distorted more than .010-inch. A light dressing with a fine flat file is all that's required to restore it to an acceptable condition, but don't overdo it. Now we're ready to begin tuning the carburetor to the application.

The idle circuit is especially critical on street-driven vehicles. You may have encountered a situation where the idle mixture adjustment screws didn't seem to have any appreciable effect on the idle quality. This is usually caused by the primary throttle plates being too far open at idle. You may have found it necessary to set the idle stop screw on the primaries at such an excessive opening just to get the correct idle speed. However, the cure for the situation is to adjust the *secondary* throttle screw (accessible from the bottom of the carb—see photo) ¼- to ½-turn to provide more secondary idle air flow. This in turn will permit a reduction of the primary throttle opening at idle to the point where the idle mixture screws will once again become functional. It is very seldom necessary to resort to racing modifications on the idle circuits of street-driven carburetors, such as drilling idle air or fuel bleeds, installing restricting wires in the idle fuel metering orifices or drilling idle air holes in the throttle plates.

"A good rule of thumb when adjusting the secondary throttle stop is to position the throttle plates so that the bottoms of the fuel transfer slots are just barely exposed. This setting will also provide a smooth transition into the secondary main circuit."

If, after proper adjustment of the primary and secondary throttle stops, a bog or hesitation occurs when the throttle is opened quickly, it may be necessary to increase the size of the accelerator pump squirter(s). Holley squirters are numbered corresponding to the size of the squirter nozzle. For example, a squirter stamped 27 indicates .027-inch-diameter nozzles. Holley offers squirters in a variety of sizes, but you can also drill them with wire drills and a pin vise. Increase nozzle size one wire drill size at a time until the bog clears up.

On vacuum-secondary Holleys, accelerator pump squirter size must also be coordinated with the secondary vacuum diaphragm spring to prevent bogging and hesitation. If the secondaries open too soon or too quickly, a bog will result. Using the next heavier spring can correct this. If the engine seems down on power or sluggish in response to full throttle, try the next lighter spring. These springs are color-coded: white, yellow,

5. A fine-toothed flat file can be used to restore the metering block surface of the main body to acceptable flatness.

6. The secondary throttle plate opening at idle can be adjusted by turning this concealed stop screw. The idle fuel transfer slot (heavy arrow) should just barely be exposed when viewed from the bottom.

7. Vacuum secondary diaphragm springs come in an assortment of lengths and stiffness to tailor the secondary opening point and the opening rate. See text for details.

8. Shown here are the bay window power valve (left), the four-hole, single-stage power valve (a six-hole version is also offered) and the two-stage power valve (right). Note the gasket shape that must be used with each design. The bay window version is preferable in the majority of applications.

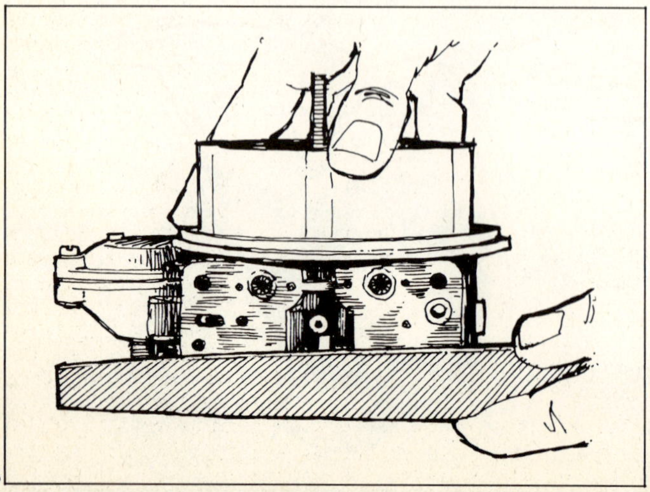

5

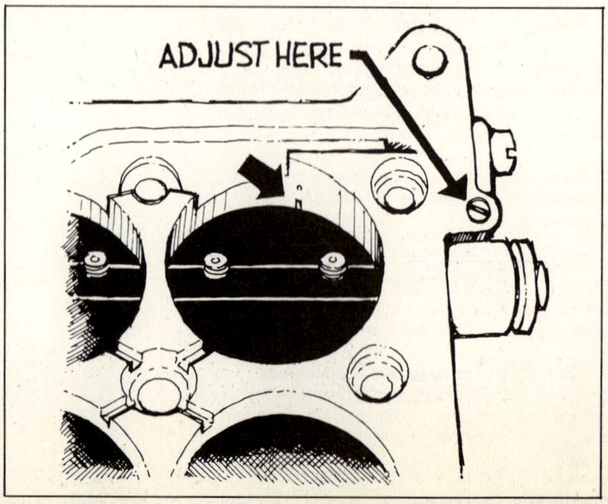

6

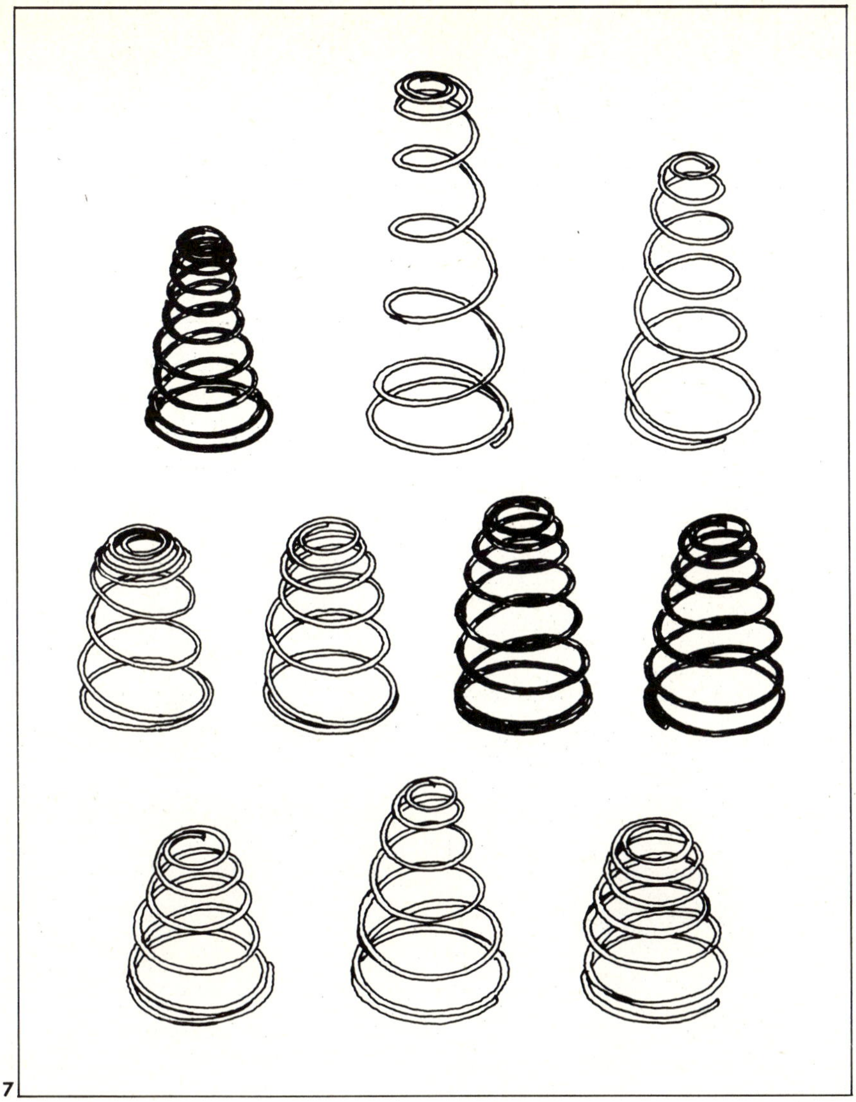

7

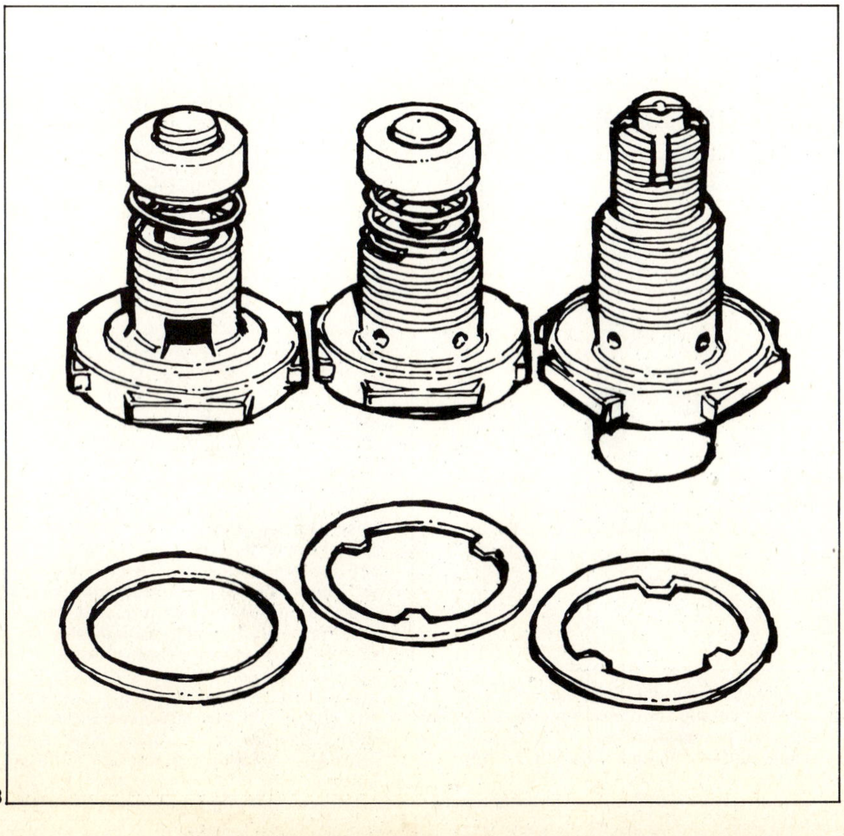

8

red, pink, purple, green, orange, plain (no color), brown and black, from the lightest to the heaviest, respectively.

"The proper spring should be selected before varying nozzle size in an attempt to use as small an accelerator pump squirter as possible. Holley offers an assortment of the most commonly used springs under part No. 85BP-3185."

The only remaining circuits in the carburetor that may need further tuning are the main circuits and the power circuits. Main circuit metering is controlled by jets or metering plates. In the majority of cases, the "out-of-the-box" jetting is very close to perfect, but altitude considerations and engine configuration may require slight adjustment. Holley even offers a high-altitude jetting kit (part No. 85R-4707) for the 6619, 6909, 6919, 6979, 6989 and 7009 RV carbs. In any case, jetting changes should always be done with numbered Holley jets or metering plates. Jets should *never* be drilled. If you find it necessary to rejet the carb more than five numbers either way from the "out-of-the-box" jetting, there is probably something wrong, such as a vacuum leak, restricted passage or inadequate fuel delivery system.

The power circuit is also important to the main circuit. Tailoring the power valve opening point (by manifold vacuum level) will provide smooth off-idle acceleration and proper fuel metering under high-load situations. Power valves are numbered according to the vacuum level (in inches of Mercury, Hg.) below which they are open. For example, a power valve stamped 85 opens at 8.5 inches Hg. and stays open at all vacuum levels lower than 8.5 inches Hg.

To correctly match the power valve(s) to the application, select one(s) that open *at least* 1 to 1½ inches Hg. below the manifold vacuum level that the engine maintains at idle *in gear.* This will prevent unnecessarily rich fuel metering under light loads. For example, if the engine idles at 8 inches Hg., a power valve rating of 65 should be selected. If the power valve(s) that came in the carburetor from the factory is already 1½ inches Hg. or more below the idle vacuum level, there is no need to install power valves with higher ratings to get within 1½ inches Hg.—just stay with the stock rating. Holley also offers two-stage power valves that open partially at one vacuum level and then open fully at a lower vacuum level. Such two-stage power valves can improve gas mileage on some lightweight vehicles with very mild cam timing, but they should be avoided on modified street machines or in any carburetor that has power valve restrictions larger than .076-inch.

These tips for fine tuning and maintaining Holley carbs on street-driven vehicles should be all that's needed to straighten out those problems normally encountered by enthusiasts, because as we pointed out in the beginning, most "carburetor" problems have little or nothing to do with the actual carb. But just in case you do run into something more complex, Holley also sells an excellent technical manual (part No. 36-73) to bail you out.

—C. J Baker

Q-Jet Quickie HOW-TO

The four-barrel Rochester Quadrajet can be stripped and kit-built in less than an hour—and for less than $10.

Just as with almost every other engine component, carburetors do suffer the pangs of age and wear. Consequently, periodic rebuilds can restore a carburetor to its "as-designed" efficiency.

One of the most common carburetors in use today is the Rochester Quadrajet spread-bore, and the following sequence of photos and captions will acquaint you with the basics of an overhaul of such a carb. The small-primary, large-secondary design, using 1.38-inch and 2.25-inch throttle sizes, is well suited for street, economy and high-performance use.

While the carburetor pictured here was originally designed to accommodate a 350 Olds engine, we bought the Delco rebuilding kit for it from a Chevrolet dealer for just over $8. We put both the kit and the carburetor in the capable hands of Edelbrock's Murray Jensen, then listened and watched as he tore it down and put it back together for our cameras.

(1) To begin, remove the five slot and two Phillips screws around the perimeter of the air horn assembly. Then remove the two remaining small screws that are down inside the assembly near the boosters. It's best to lay out the screws roughly in the same pattern in which they were removed from the carburetor and set them aside where they will not pick up any dirt or grit.

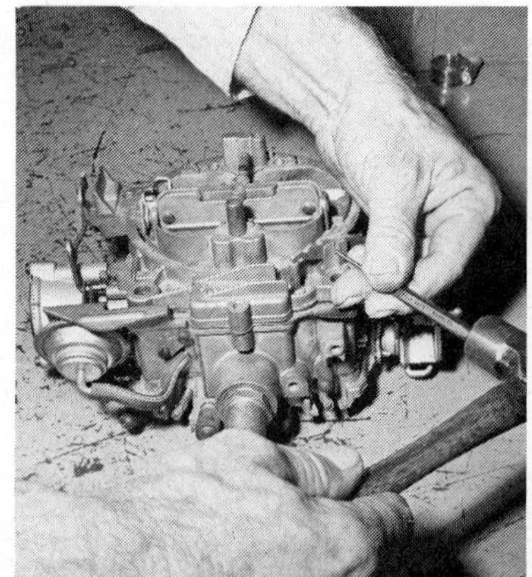

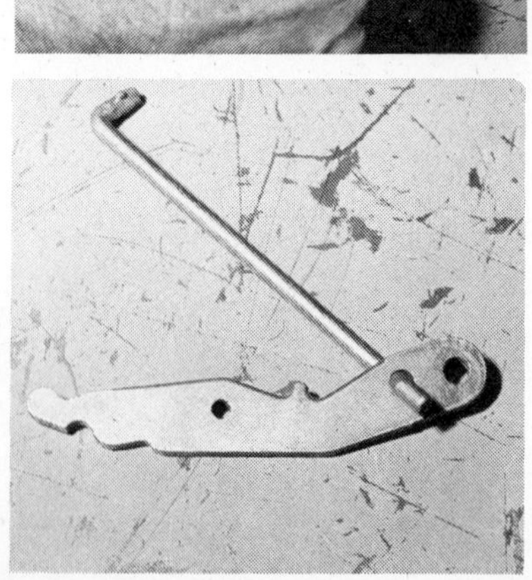

(2) Note the position of the accelerator pump lever arm and link (there is an inner position and outer position) so it can be replaced properly during assembly. Use a drift punch and small hammer to lightly tap the fulcrum pin in toward the air horn body. With the fulcrum pin out, the arm and links can be removed and set aside. The vacuum diaphragm should be removed next, along with any other nonmetallic parts on a Quadrajet, so that they are not affected by cleaner or solvents. Some Quadrajets use two screws and a bracket to hold vacuum diaphragms, and some use a clamp arrangement. If this subassembly is not removed, damage to the rubber diaphragm will result when it contacts solvents.

▲ **(3)** The secondary metering rods are held in the carburetor on a tiny rack, which is retained by a single small screw. Remove the screw and carefully lift the rack and the rods straight up out of the body of the carburetor. Following this, the choke link can be removed by unscrewing it carefully from the side of the air entry housing.

(4) At this point there should be no further obstructions to complete removal of the top, or air horn, section of the Quadrajet carburetor. Lift top off, carefully lift the accelerator pump piston assembly from its well, and remove the air horn gasket a little at a time from the perimeter of the carburetor body until it has come completely away from the carburetor—hopefully in one piece. Check to see that no gasket fragments have fallen into any of the interior wells of the carburetor. ▶

(5) Carefully and safely get rid of any fuel remaining inside the carburetor by tipping it into a container. Then you can lift out the black plastic "filler block" from the fuel bowl area. Use a pair of small pliers to lift out the primary metering rods, and once they are removed, go down in the well with pliers or tweezers and retrieve the spring below in the well. Some Quadrajet models have one tall, thin spring and one shorter, wider spring in this well.
▼

Q-Jet Quickie HOW-TO

(6) The fuel bowl float and its companion needle assembly can now be removed from the main body of the carburetor. The seat that goes with the needle has two or four slots cut in it at the top to accept a very wide screwdriver blade. Use caution here to avoid bending, denting or chipping the brass seat assembly with a screwdriver that is too narrow to do the job properly.

▲ **(7)** With the seat assembly now out of the way, turn out the screw that you will see next to the area where the needle seat was. The removal of the screw will uncover the tiny chrome check-ball. Carefully turn the carburetor upside down over your hand, a rag, or other rough surface to retrieve the ball, in case there is no replacement in the kit you have bought.

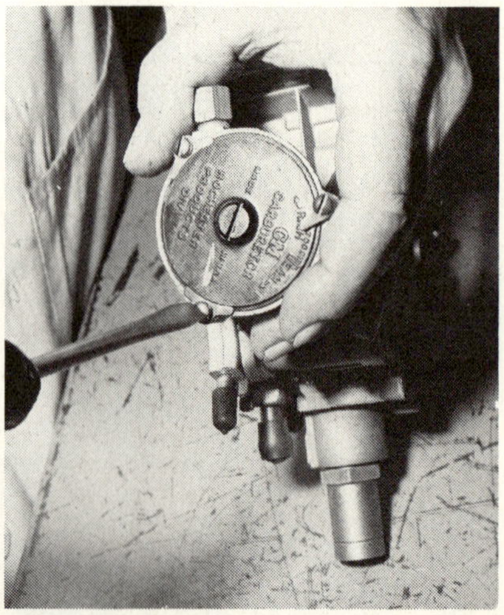

▲ **(8)** Note the exact position of the choke thermostat cover relative to the housing notches, so that the position can be duplicated for reassembly. Loosen the three retaining screws on the cover and remove. Now go to the base of the carburetor and remove the two idle mixture adjusting screws completely from the body of the carburetor with a screwdriver.

(9) Invert the carburetor so that the base or throttle body faces up, and remove the two remaining screws to part the base from the main body of the carburetor. Check the two large and two small channel plugs on the bottom of the carburetor body for evidence of leakage. It's a good idea to use some kind of epoxy compound on all four plugs at this point, to prevent any future leakage, even if you find no evidence that the channel plugs are leaking.

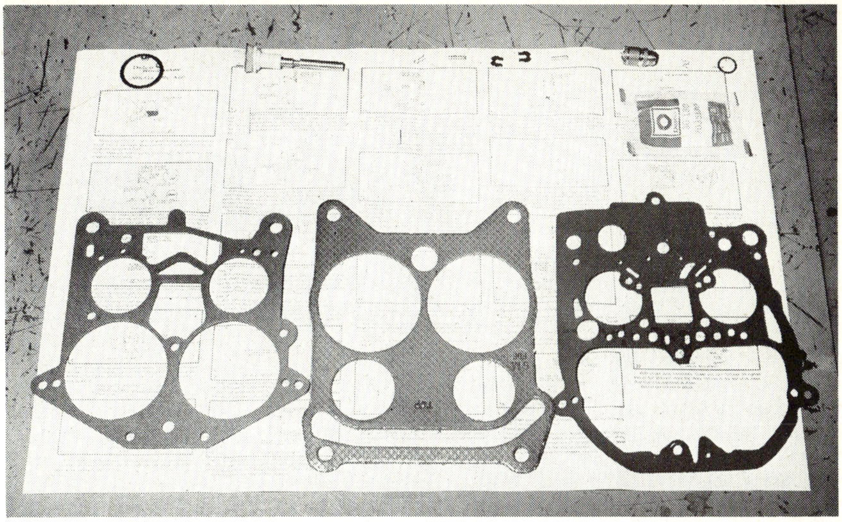

(10) After a complete cleaning, including immersion of all metal parts in solvent and scrubbing with a stiff brush, carburetor subassemblies should be blown out with compressed air, including all passages and wells. Check all new gaskets against old gaskets to see that all holes have been punched through, and match squarely with holes in old gasket. Assemble base to main body of carburetor.

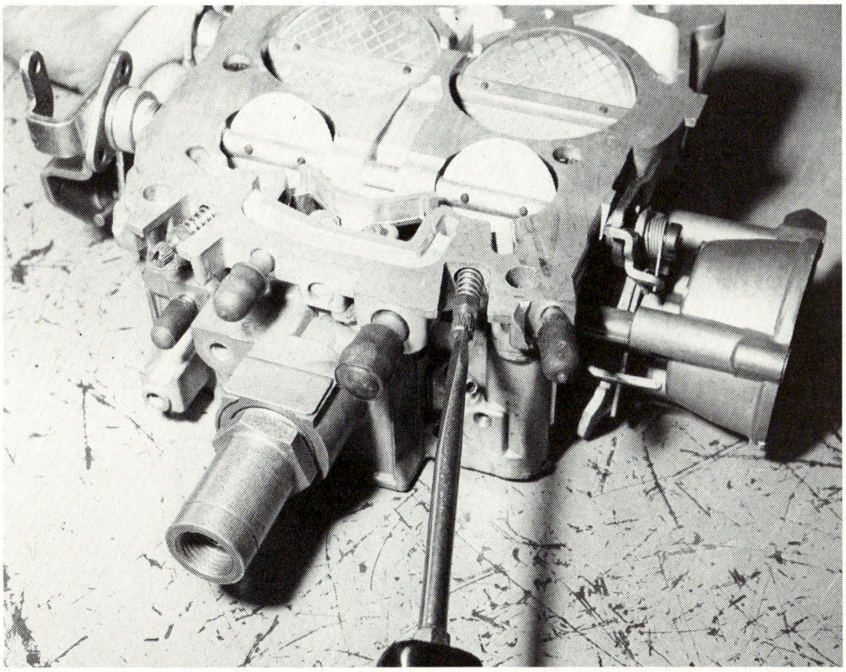

(11) Replace idle mixture adjustment screws by turning them in clockwise, slowly, until each screw bottoms. Then back each screw out three turns. This will be a "ballpark" adjustment for most engines using the Rochester Quadrajet carburetor. Install new check-ball in its well and reinstall cover screw for check-ball assembly. Reinstall float, needle, seat and springs in reverse order and check for proper float pivoting action, properly hung fuel inlet needle and needle action. Install the primary metering rod assembly and check to see that the tiny retainer block is seated snugly and squarely in its seat. Check action of primary metering rods and then replace the large "filler block" in fuel bowl.

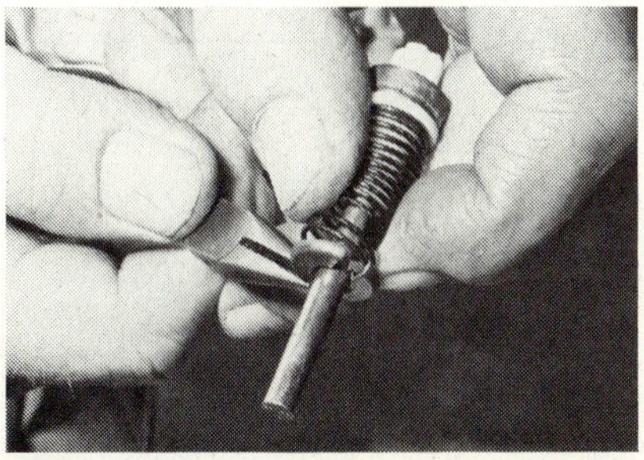

(12) Always use the new accelerator pump piston supplied in Delco rebuilding kit. Swap existing spring and collar from old piston to new. Shoot rubber piston seal with WD-40 or other similar treatment to keep it soft and allow it to act smoothly in the bore. Be sure the piston engages the spring in the well, check the action of the piston several times.

(13) Check the new air horn gasket shape and perforations over the old gasket. Be extremely careful in the area of the primary metering rods to see that the gasket clears rods, springs and hangers in this area. Check and recheck the action of the rods once the gasket is properly in place to avoid trouble later on. Replace air horn assembly atop carburetor body and new gasket, making sure that accelerator pump plunger rod comes through the hole in the air horn body. Recheck the action of the acclerator pump, and then replace all nine inside and perimeter screws in the air horn assembly.

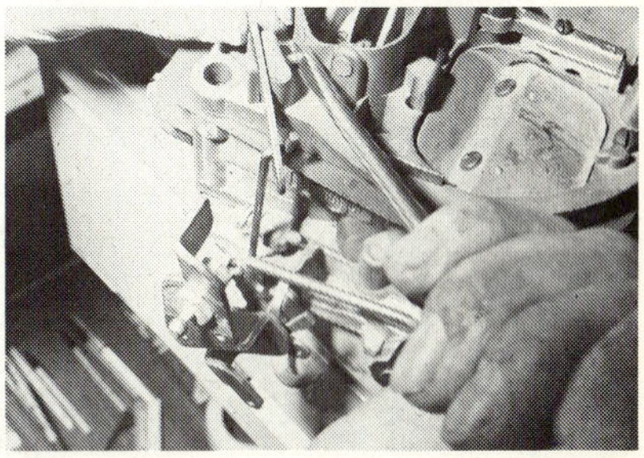

(14) Replace secondary metering rods, bracket and screw; then move to the outside of main body to reinstall vacuum diaphragm links, hoses and brackets. The choke lever link is replaced down inside and behind the thermostat housing. The accelerator pump link goes on bottom-end first, since a tiny tang must clear. Then the link goes upright into the same hole it came out of. Use a wide-blade screwdriver on the back side of the fulcrum pin to push the pin back through the flanges and the arm while your other hand holds the arm down in position to engage the pin.

Taking 5

A close look at two versions of a rugged multipurpose transmission.

On the surface, selecting an optimum transmission would appear to be an easy task. If the car was being built for a single specific class, determining its exact needs and choosing a modified 4-speed with the right ratios would be all that was necessary. But if the car was to be truly versatile, capable of running in a variety of eliminators or brackets, powered by engines ranging from mild to wild, then the transmission must be equally adaptable and versatile. The ability to readily vary first gear multiplication and ratio spreads would be essential. Consequently, the Doug Nash 5-speed is deserving of a look-see.

In actuality, the folks at Doug Nash Engineering and Equipment Corporation in Nashville, Tennessee, call their 5-speed the 4 + 1 Quik Change. This transmission is considerably stronger than any OEM 4-speed. It can be used as either a 4-speed or a 5-speed. Nearly 20 first gear ratios are available, ranging from 2.22:1 all the way to 3.76:1, with a variety of intermediate ratios and ratio spreads. It can even be fitted to function as an overdrive by altering the shift pattern and underdriving fourth gear as much as .86:1. But the versatility doesn't stop there. Both racing and street versions of the 5-speed are offered to suit virtually any need.

In our particular situation, the racing model was selected, but if the car was intended for street use, along with bracket racing activity, the street version would be the logical choice. In either case, both units are so well-designed and tested that no further modification is usually required after delivery. All the racer need do is bolt it in, add the appropriate heavy-duty gear lube, connect the linkage and motor away.

But let's get into the specific features and the differences between the race and street designs. The most important difference between the two versions is that whereas the racing version utilizes straightcut spur gears to eliminate thrust loads, friction and wear, along with "Pro

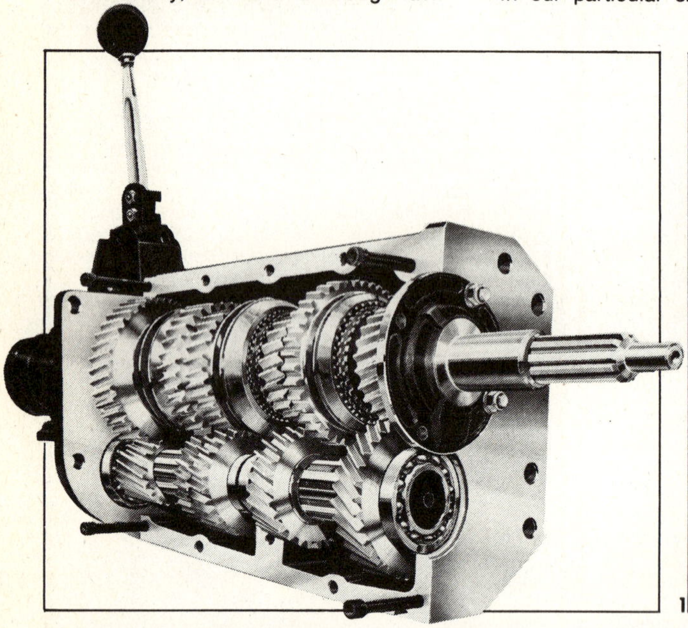

1

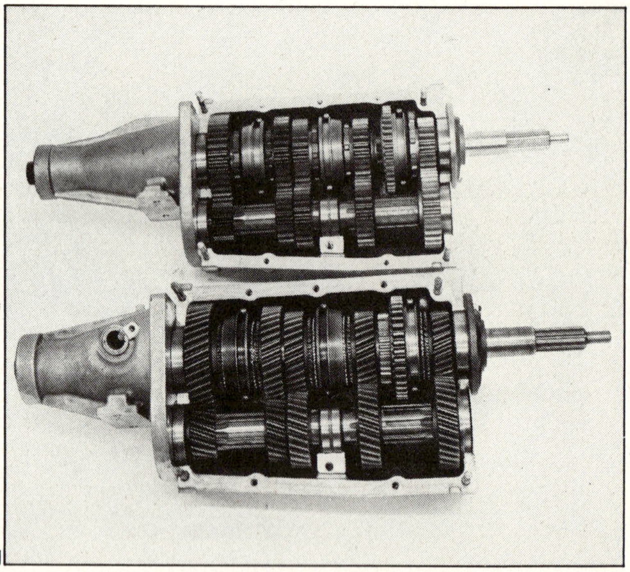

2

1. With the case opened, the street version of the 5-speed can be readily distinguished by the helical-cut gears and synchros. Note the center main bearing on the cluster shaft in both versions to prevent shaft deflection.

2. The DNE street 5-speed utilizes an aluminum case to prevent any corrosion problems. Different bolt patterns permit installation in virtually any American car. Different input shafts also make the swap a simple bolt-in. Note the addition of the speedometer cable provision. The extra fifth gear makes the main case sufficiently longer that the transmission crossmember must be moved 2 inches rearward, but overall trans length is identical to stock.

3. Thanks to a case that splits down the middle, the DNE 5-speed offers quick and convenient access to internal parts for maintenance or ratio changes. RTV sealant prevents lubricant leaks, but care must be taken to use only a 1/16-inch bead of RTV. Otherwise, excess sealant will squeeze into the transmission.

4. This exploded view of the DNE 4 + 1 Quik Change shows the simplicity of the unit. All gears feature bronze bushings, and the shafts are cam-ground to assure adequate lubrication. Premium materials are used throughout.

Shift" sliders to provide quick, positive shifts at high rpm in the "crash box" tradition, the street version incorporates helical-cut gears for relatively quiet operation and smooth-shifting synchronizers and sliders for hassle-free street operation. And to further simplify things, the brass components, hub, slider, keys and springs used in the synchro mechanism are standard Borg-Warner ST-10 pieces, so if they should ever need replacement, they will be readily available.

But that's just the beginning. The racing version uses SAE 9310 steel alloy

3

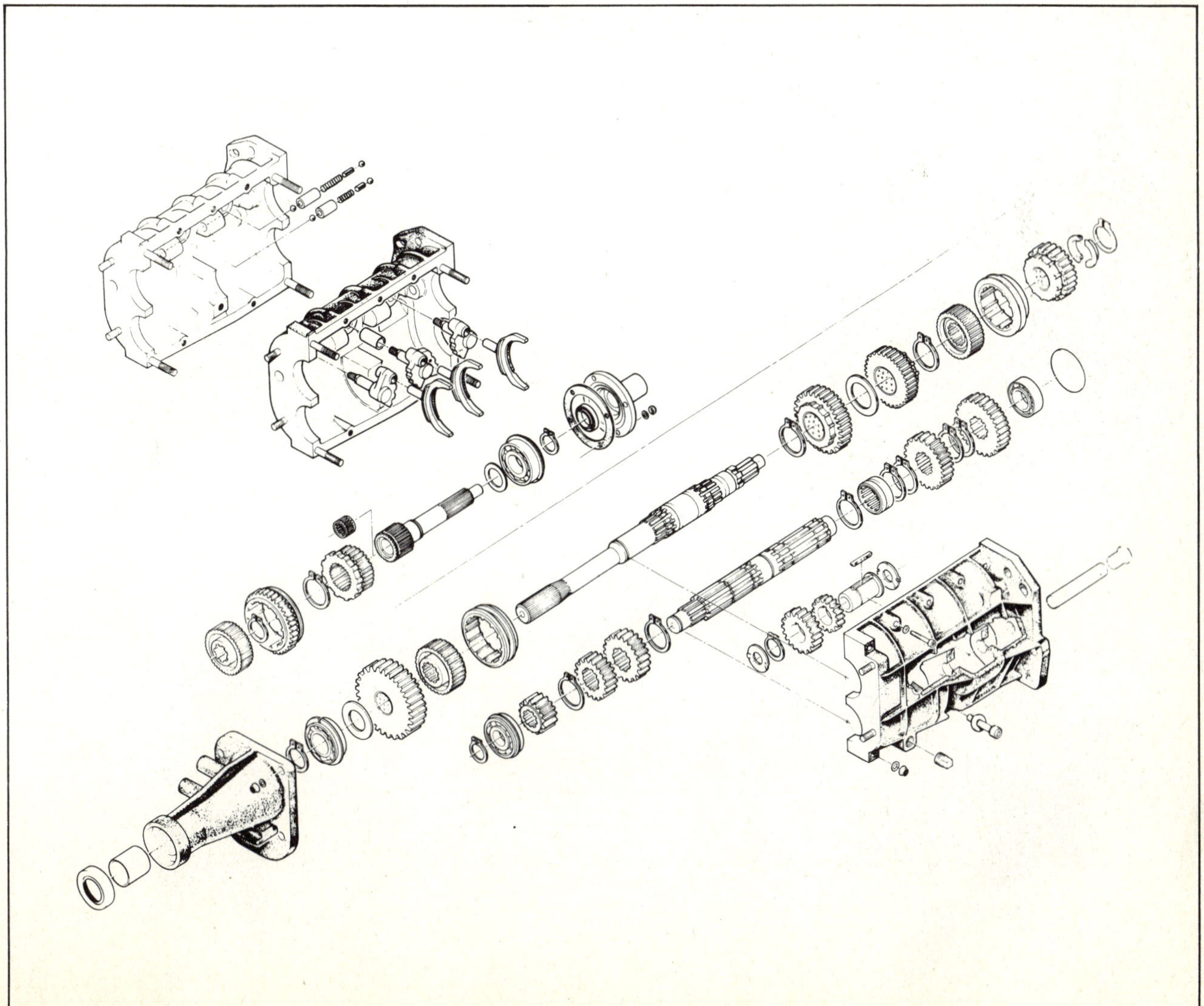

4

gears for ultimate shock resistance, but the street version has SAE 4617 steel alloy gears, selected for superior wear properties. In truth, there is virtually no difference between the relative strengths of the race and street gears, considering the differences in tooth design and angle. In either case, all gears receive bronze bushings to prevent possible galling or seizure to the main or cluster shafts. Both versions utilize the same bearings and SAE 8620 shaft material too, although the street shafts have been redesigned slightly to accommodate the thrust loads imposed by the helical gears.

On the outside, the most notable differences are that the street version uses an aluminum case and tailhousing to forestall any possible corrosion problems that might occur if the magnesium racing case and tailhousing were used. In terms of weight, the aluminum version is about 10 pounds heavier. Additionally, the case has been redesigned to tuck the shifter linkage in as close as possible to the transmission for maximum tunnel clearance. Shifter mounting pads have also been relocated so that the special Hurst 5-speed shifter will coincide with the OEM 4-speed shifter location in as many applications as possible. A provision for a speedometer gear and cable have also been included.

On the street version, tailhousing length and input shaft design have been tailored to provide exact replacement for GM, Ford and MoPar transmissions to simplify installation, although in most cases the Chevy clutch disc must be used.

Gear ratio availability within the street 5-speed line is not as broad as within the racing version line, but the ratios are matched to current-day needs. Low gear ratios of either 3.10:1 or 3.28:1 are offered. Such gearing will really wake up a late-model stocker without resort to high numerical ring and pinion ratios that are hard to live with on the open highway and gasoline consumption.

In this case, we opted for the full racing version, since the car will not be street-driven. In an attempt to select ratios with the maximum variability with a minimum of expense, we chose ratios of 2.80:1, 1.77:1, 1.24:1 and 1:1 in a 4-speed configuration compatible with Super Stock and Super Modified rules. In this setup, the dog teeth are completely removed from the fifth gear slider so that it positively cannot be engaged, in accordance with NHRA's request. However, with this ratio selection, we need change only one cluster shaft gear to get ratios of 2.98:1, 1.77:1, 1.24:1 and 1:1. Or we could change just the input gear for ratios of 2.93:1, 1.85:1, 1.29:1 and 1.04:1. If we used both the new cluster gear and the new input gear, ratios of 3.11:1, 1.85:1, 1.29:1 and 1.04:1 would be obtained. In other words, with only two extra gears, we could alter our basic transmission selection for four different ratio sequences and ratio spreads.

If you're in doubt as to which version, street or race, would be suited to your needs, we would recommend the street version if the car will see *any* street use at all. Relative shift speeds between the racing and street versions is identical to the difference between a "crash box" modified OEM 4-speed and a stock 4-speed, but more important for the racer, the "Pro Shifting" mods on the race version permit shifts to be made at high rpm that might not otherwise be possible. If you select the street version, and at a later date decide to update it for more serious racing endeavors, DNE can Pro Shift the synchros just as they would on any OEM 4-speed.

In closing, we would like to point out that probably the single biggest thing that you can do to a street car for a radical increase in performance, not counting supercharging or turbocharging, would be the installation of a transmission with significantly higher numerical low gear ratio and better ratio spreads. Additionally, such a change is usually a simple bolt-in swap, and there aren't any registration or emission hassles. Five-speeds are a natural for street rods and vans, too. Does anybody want to take 5? *—C.J. Baker*

FIGURING THE ODDS

Calculating transmission gear ratios, combining the ratios of more than one gear set, and then figuring in the gear ratio of the ring and pinion set to get the overall final drive ratio (torque multiplication) may seem like black magic, but it isn't—only simple multiplication and division. It goes like this:

To find the ratio between any two gears in mesh with each other, divide the number of teeth on the driven gear by the number of teeth on the drive gear. This procedure works for any gear combination, whdther it is in the transmission or the rear end. For example, if the driven gear has 32 teeth and the drive gear has 24 teeth, just divide 32 by 24, which equals 1.3333. Rounding it off, the ratio is 1.33:1

To combine more than one gear set, figure the ratio for each pair of gears in mesh, as described above, and then multiply the ratios times each other. For example, if one set of gears in a transmission has 24 teeth on the input shaft driving 32 teeth on the cluster shaft, and that set in turn drives another set with 16 teeth on the cluster and 39 teeth on the output shaft, first figure the ratio of each gear set, which is 1.3333:1 and 2.4375:1 in this case. Then multiply the ratios times each other, 1.3333 times 2.4375, which equals 3.2499. Again rounding it off, the overall gear ratio of the transmission in this mode is 3.25:1. If you wanted to carry this one step further and multiply it times the rear end ratio, such as 4.88:1, for example, you'd get the overall gear ratio, or torque multiplication, from the engine to the rear axle. In this example, the ratio would be 15.86:1.

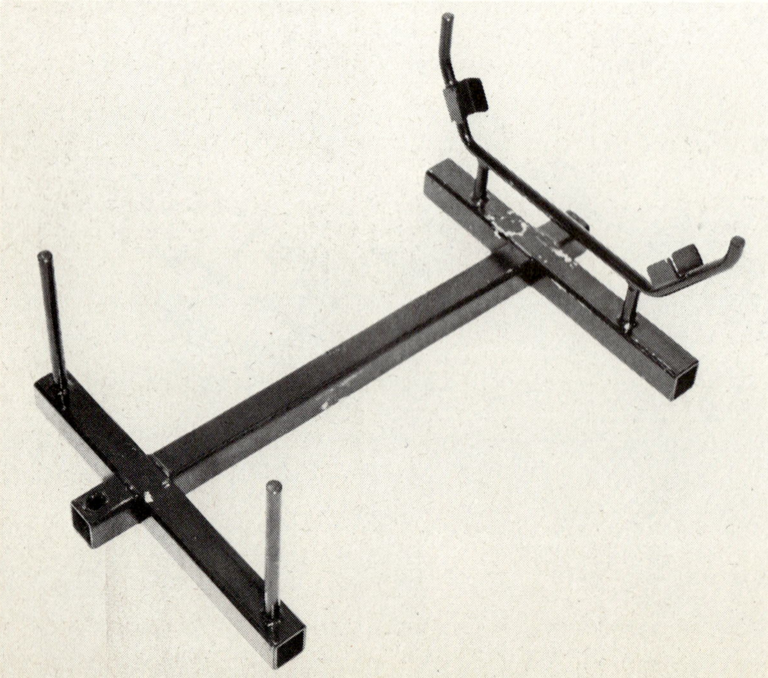

GRAND STAND PLAY

Compared to the conventional 3- and 4-speed manual transmissions, disassembly, overhaul and reassembly of the Nash 5-speed is a piece of cake. The case splits neatly in half to readily expose the innards for easy access and there are no needle bearings to contend with. However, the smoothly rounded case is not especially conducive to lying steady on its side while the work continues. But hurrah, the folks who designed this unique transmission have also come up with a handy stand to rigidly cradle the unit during teardown (after all, the guys at DNE didn't like the transmissions squirming around while they worked on them).

By examining the accompanying photo, you could probably whip up your own version of this helpful holder with a minimum of expense, and if you're like most of us, you'd have it perfect after only five or six tries. But why bother? DNE has these stands available for about $30, and they're guaranteed to fit the first time!

Mix and Match

A high-performance converter doesn't need to be expensive. Just try a little mixing and matching.

Matching torque converter size and stall speed to specific applications—such as high-horsepower engines with relatively high idle speeds, street machines, street/strip usage or even pure racing applications—can provide significant performance gains and increased driveability. And, more important, obtaining such a converter need not cost you an arm and a leg. There are many OEM torque converters that really work slick in performance applications without modification. They're readily available, and factory-reconditioned rebuilts are quite cheap, especially if you pick up an exchangeable core at the salvage yard.

Obviously, these OEM converters don't offer the special features of a high-dollar aftermarket racing converter, such as furnace-brazed fins and Torrington bearings, nor are they as rugged. However, the real beauty of these OEM units is that if one fails, all you have to do is take it back to your local dealer, and for about $40 exchange, he'll hand you a fresh rebuilt and then you'll be back in business.

Our thanks go out to Tony Rossi at Rossi Transmissions in Pomona, California, who shared his years of experience with us, providing the part numbers for the best possible converter choices and combinations.

In some cases, it might be necessary to change flexplates or redrill or elongate the converter bolt holes in your existing flexplate. But in most cases, the recommended replacement converter will drop right into the transmission specified. The only exception to this is when Turbohydro converters are used with Powerglide transmissions, but Rossi offers a special input shaft to permit this increasingly popular swap.

The cost of going fast isn't always as high as you might think!

NOTE: ALL PART NUMBERS ARE SUBJECT TO CHANGE. They have changed through the years and will probably change again in the future. Especially when searching for rebuilt converters or used converters in junkyards, you should ask for the converter desired by the vehicle it was originally installed on, as well as by part number. (Example: "I need an 11-inch '71-'72 Vega Powerglide Torque converter.")

GENERAL MOTORS

Although only Chevrolet applications are listed below, Chevy torque converters will bolt up to Pontiac, Olds, Buick and Cadillac transmissions. Powerglide and Torque Drive converters interchange with each other, but not with any other GM transmissions. Torque Drive transmissions (as opposed to torque converters) are not recommended for high-performance use. Turbohydro 250, 350, 375, 400, CBC Automatic and Super Turbine 300 torque converters interchange with each other, but will not fit Powerglides. The Turbo 250 transmission (as opposed to torque converter) is not recommended for high-performance use. Super Turbine 300s, used by Pontiac, Buick and Olds in the mid-1960s, are basically Powerglides that use a Turbo 350 input shaft, which is why they use Turbohydro converters and not Powerglide converters. The flywheels listed below are applicable only to Chevrolet engines; use with other GM engines requires bolt hole to be elongated or a dual-pattern flywheel to be used, if one is made by the appropriate division.

		TORQUE CONVERTER			ORIGINALLY INSTALLED IN		
USAGE	ENGINE	PART NUMBER	DIAMETER	STALL SPEED (APPROX.)[1]	VEHICLE	TRANSMISSION	USE FLYWHEEL
		TURBOHYDRO TRANSMISSIONS					
Hi-perf street	Big V8 396-427			2000 + rpm	Take any converter from a TH-350 and install in a TH-400		340296
	Big V8 454						3992069[3]
Street/strip	Small V8 262-400	328729	11-inch	3000 rpm	'72-'73 Vega	TH-250, 350	[2]
	Big V8 396-427			3500 rpm			340296
	Big V8 454			3600 rpm			3992069[3]
Strip only [4]	Small V8 262-400	328750	10-inch	3500 rpm	Late '73 to present Vega	TH-250	[2]
	Big V8 396-427			4000 rpm			340296
	Big V8 454			4100 rpm			3992069[3]
		POWERGLIDE TRANSMISSIONS					
High-perf. street	Small V8 262-400	3793984	11-inch	2200 to 2500 rpm	'62-68 Chevy II, Chevelle, except 8-cyl.	Powerglide	Use any P.G. flywheel or 340296
	Big V8 396-427				'69 Chevy II 4-cyl.	Powerglide, Torque Drive	
	Big V8 454				'70 Nova 4-cyl.	Torque Drive	3992069[3]
Street/Strip	Small V8 262-400	3998277	11-inch	3000 rpm	'71-'72 Vega	Powerglide	Use any P.G. flywheel or 340296
	Big V8 396-427			3500 rpm			
	Big V8 454			3600 rpm			3992069[3]

Early Departure

Updating the Super T-10 with first gear ratios compatible with today's drag race tires.

1

In recent years, we have noted an ever-increasing demand for more first gear ratio (torque multiplication) in 4-speed drag racing applications, such as brackets, Stock, Super Stock and Modified classes. This has occurred because of improvements in tire technology and the use of wider slicks. Racers soon discovered that combining more first gear ratio with a compatible flywheel and bigger tires makes the cars launch harder for improved performance. In answer to the demand for more first gear ratio, aftermarket manufacturers began offering modified gear sets, and eventually, Borg-Warner began producing a 2.88:1 first gear version of the Super T-10, called the "extra low" series.

However, there's more to prepping a good 4-speed for the strip than just selecting a super-low gear ratio. To illustrate what we mean, we'll follow through the main steps as the specialists at Trans-End, in Clinton, New Jersey, assemble a Super T-10 and simultaneously update it to either a 2.96:1 or a 2.80:1 first gear ratio, utilizing their specially modified cluster gears. *—C.J. Baker*

2

3

4

1. The Trans-End 2.96:1 gear set utilizes a modified 2.54:1 cluster shaft of SAE 9310 steel. Also used is a mainshaft adapter and spacer. All gears are SAE 9310. Ratios are: 2.96:1, 1.75:1, 1.33:1 and 1:1.

2. A dimple in the face of Borg-Warner gears identifies them as being of the SAE 9310 high-strength steel.

3. Trans-End carefully deburrs all the gears to reduce the likelihood of cracking or chipping.

4. For drag racing applications, the second-design Super T-10 should always be selected, since it uses a larger mainshaft that is considerably stronger. The aluminum tailhousing should also be selected to save weight. The last three digits of the casting number on the case identify its material; 904 indicates nodular cast iron, about 80 percent stronger than conventional cast iron, which is designated by 902.

5. The special 2.80:1 gear set uses the 2.64:1 cluster gear, which is modified. Again, all components are SAE 9310 steel, but no mainshaft adapter or spacer is required with this ratio. Ratios are: 2.80:1, 1.75:1, 1.33:1 and 1:1.

6. The cluster gear should be installed with .009- to .012-inch end clearance. To obtain this clearance, either machine the end of the cluster gear or add shims. Always install shims at the front end of the cluster gear as it is thrust against the rear of the case by the gear tooth helix.

7. Check the mainshaft for straightness. When measured on the ground journal, as shown, there should be no more than .005-inch total runout.

5

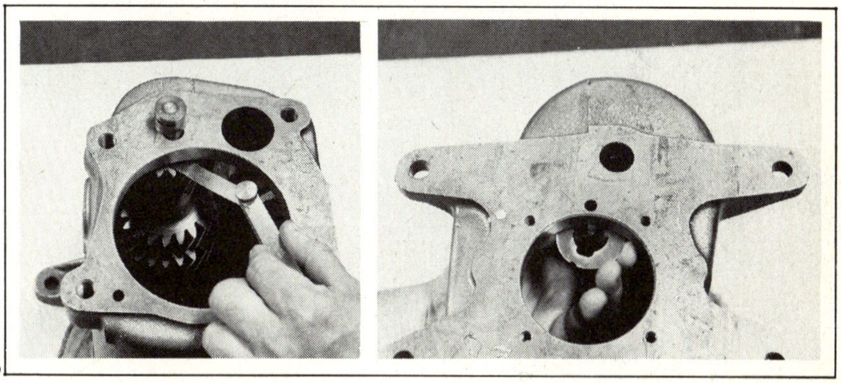

6

7

8

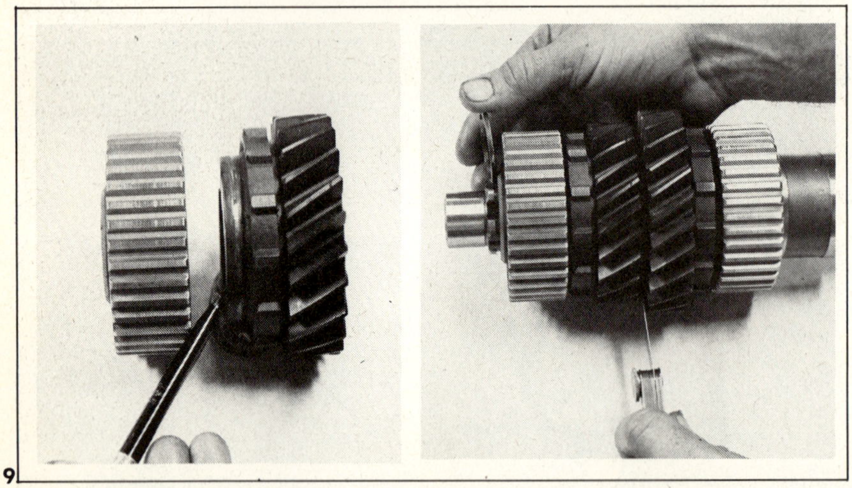

9

10

8. Press second gear onto the mainshaft, along with the adapter (if used). After securing the assembly with a snap ring, critical thrust clearance should be .012- to .015-inch to provide adequate lubrication. If additional clearance is required, it must be undercut into the gear face.

9. Third gear is then installed, again shooting for a thrust clearance of .012- to .015-inch. To obtain the correct clearance, the gear may need to be machined on the indicated face. Once assembled, the clearance is checked behind third gear, as shown.

10. Low gear requires additional work to provide clearance for the reverse idler and to get proper tooth engagement. As much as .080-inch (depending upon clearance) is machined from the front of the gear, and oil reliefs are cut into the back.

11. A spacer is installed on the mainshaft, followed by low gear and a bearing. Thrust clearance on low gear must be .012- to .015-inch, so it is best to begin by taking only a .075-inch cut off the front of low gear, and then check, taking another .005-inch cut if necessary.

12. To gain clearance for the larger low gear, the reverse idler must have up to .080-inch cut from both the gear face and the backside of the gear. Thrust clearance on the reverse idler gear should be .007- to .015-inch.

13. The stock shifter mechanism shafts (right) are replaced with special shafts that do not have an O-ring groove in them. The O-ring

11 

12 

13 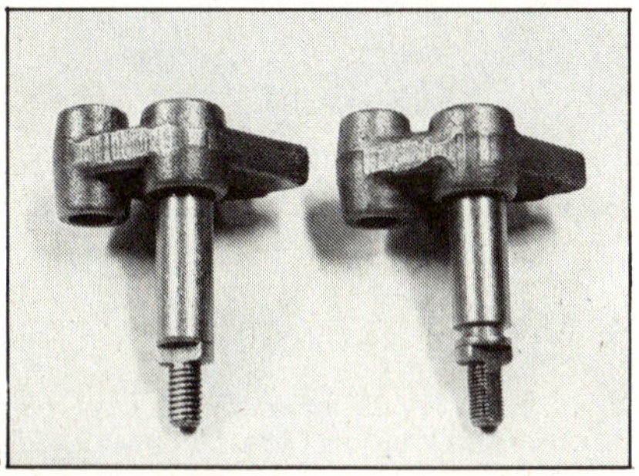

14 

groove has proved to weaken the shafts, often resulting in a failure at that point.

14. With the mainshaft installed, check for perfect mainshaft-to-cluster alignment of the gears, as shown.

15. Many racers have encountered a problem with the tailshaft bushing seizing in the Borg-Warner Super T-10 transmissions, especially when only minimal lubricant is used inside the gearbox. Trans-End has come up with a solution, however. They drill a 5/16-inch hole through the tailhousing and bushing, as shown, while the transmission is disassembled. The hole is tapped for ⅛-inch pipe threads. A common grease fitting (zerk) is then installed. Thus modified, a shot of chassis grease before each meet eliminates the problem.

This particular modification could be made on any transmission, manual or automatic, where tailshaft bushing problems have been encountered. It is particularly applicable on late-model automatics if the driver prefers to shift into neutral, kill the engine and let the car coast to a stop. In such a situation, the pump in the automatic is not turning, and the normal transmission lubrication system is not operating. Consequently, such a driving procedure is not recommended for automatics, but if you must do it, at least the grease fitting will save the tailshaft bushing—if nothing else.

15 

Torque Converter Tips

A selection of "down home" hints for Chevy's automatic transmission.

On a trip to Jacksonville, Florida, we took time out to drop by Turbo-Action and chat with Paul Forte. And although Turbo-Action transmissions and torque convertors have recorded an amazing win record in organized drag racing, we weren't there to do a step-by-step story on how Paul modifies and assembles his well-crafted units. Instead, what we had in mind was to call upon his years of experience for low-buck tips and suggestions regarding automatics. Paul began his discussion with torque convertors.

"The most common question we hear regarding torque convertors deals with how to determine if the convertor is in good shape prior to reinstalling it in the car. Other than obvious signs of physical damage, such as a ballooned housing or a rupture in the outer skin, the only check that can be readily made is for correct operation of the stator sprag or roller clutch. The purpose of the sprag, or clutch, is to allow the stator to rotate, in one direction only, on the stator (reaction) shaft.

"To check for proper action of the sprag, you must have some means of turning the stator hub. The ideal tool for this is the stator shaft from a Turbohydro 350 pump. This shaft is long enough to hold comfortably, and it will fit all '65 and newer GM converters (except the '76 Turbo 200).

"With the shaft inserted in the converter, give the shaft several quick twists in each direction. If everything is working correctly, you should be able to feel the stator hub slip within the stator when twisting in one direction and the heavier mass of the entire stator turning with the shaft in the other direction."

The next item of concern was the need for special tools when disassembling an automatic. Paul pointed out that although most shop manuals call for special (and expensive) tools for working on automatics, in most cases they aren't really necessary. And in fact, even with the large number of transmissions that go through Turbo-Action each year, Paul uses virtually no special tools. Of course snap-ring pliers are required, but they're available at low cost just about everywhere.

"To remove a front pump on most automatics, exotic tools are available, but a couple of 3/8-16x2-inch bolts, some washers and a length of chain will work just as well. First, remove all of the pump bolts. These are 5/16-inch bolts. You will note that two of the bolt holes in the front pump housing are also tapped for ⅜-inch bolts. Insert the two ⅜-inch bolts through the chain and washers and into these tapped holes, as shown. A few good jerks should pull the pump right out.

"On 350 Turbohydros, you will find that a snap ring must be removed from the front of the output shaft to complete disassembly of the transmission. Again, a special tool is specified for removing this snap ring, due to the limited access. However, I've found that a pair of conventional snap-ring pliers, with the tips ground as shown in the accompanying photo, work every bit as good as the expensive special pliers."

These are the words of experience from a man who knows. Now all you need is a little self-confidence!

1
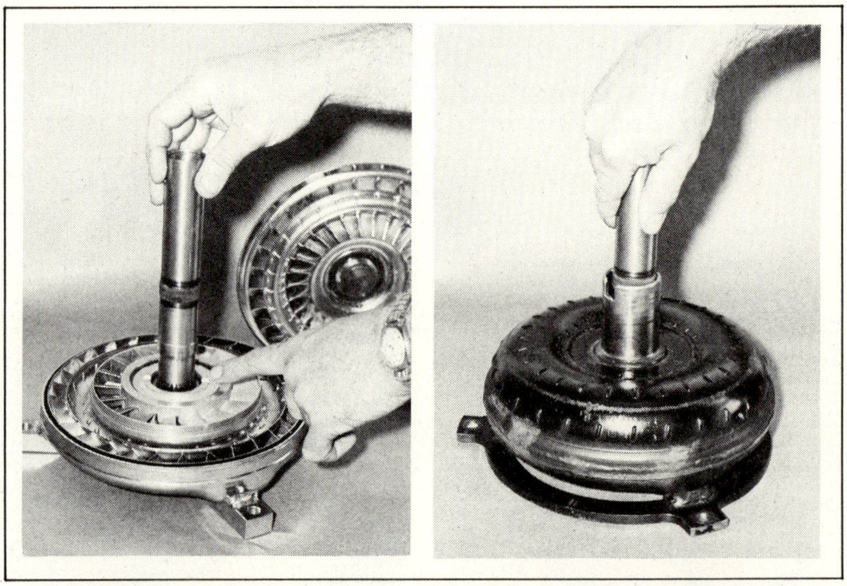

2

3

1. A stator shaft from a 350 Turbohydro fits the stator hub in most convertors. It's a good tool for checking proper one-way action of the sprag. A couple of quick twists, as described in the text, will tell whether the sprag, or roller clutch, is functioning properly.

2. A couple of bolts and washers, a chain and a stout arm are all that's required to remove the front pump on most automatic trans.

3. A pair of conventional snap-ring pliers, with the tips ground as shown, can be used to remove this snap ring from 350 Turbo output shaft.

'Glide Guide

Exhaustive step-by-step procedures for overhauling a Powerglide transmission in your own garage.

The '63 and later aluminum Powerglide transmission is quickly becoming the most popular automatic for bracket racing, class drag racing (where legal), street machines, street rods and mileage applications—and for good reasons. The 'Glide is rugged, lightweight, relatively small, efficient, readily available and reasonably priced, along with being easy and economical to overhaul and modify. Additionally, many transmission experts consider the late-model 'Glide one of the most modern automatics on the market, even though the basic design has been around for more than 30 years. Some even call it "the poor man's Lenco."

Because of its popularity, we are presenting a comprehensive feature on how to disassemble, inspect, overhaul and assemble the 'Glide in your own garage. We gathered this information by following Tom Olson through the complete step-by-step procedure at Rossi Transmissions in Pomona, California. Their years of experience with these transmissions will save you a lot of hassle while simultaneously pinpointing those areas of potential difficulty.

The only special tools required (in addition to common hand tools) will be a pair of snap ring pliers, a brass hammer and a bench vise. Of course you'll need plenty of parts cleaning solvent (available from most auto parts stores or petroleum distributors—see your Yellow Pages under "Solvents"), but under no circumstances should gasoline be used to clean or degrease automatic transmission parts. Besides the obvious fire hazard, gasoline will leave on the parts a film that will dissolve in the transmission fluid and eventually ruin the polypropylene seals in the transmission.

If you're using this feature to guide you through the overhaul or repair of such a trans, this will acquaint you with those areas that should receive special attention during disassembly to verify if they really were malfunctioning. If you're starting with a used transmission about which you know nothing, a few additional checks will be mentioned to determine if the unit is worth repairing. With used Powerglides selling for $15 to $50, depending on the condition and guarantee, it's frequently cheaper to pick up another unit rather than repair any serious damage.

We'll begin with the disassembly, cleaning and inspection, then conclude with the reassembly and clearance checking.

Start by thoroughly cleaning the outside of the transmission with solvent, or by taking it to a coin-operated car wash. A little degreaser sprayed on particularly dirty areas, plus a little scrubbing with a stiff bristle brush, will make the spray wash operation much more effective.

Although we have tried to show each and every step in the reassembly, a question or two may still arise. If so, you can always refer to the Chevrolet Shop Manual at your local Chevy dealership, or in *Petersen's Big Book of Auto Repair.* All of the required gaskets, seals, shims, service components, etc., are available from Chevrolet in a Master Overhaul Kit, or in individual gasket and seal kits, clutch pack kits, and so on. Similar kits are available at many auto parts stores and automatic transmission shops.

Rossi also offers a very complete Master Overhaul Kit that contains all of the required pieces plus special "beef-up" components to improve strength and reliability. In most cases this special kit is a wise investment, although you can get by for less money with the Chevrolet or equivalent kits.

A MATTER OF CONVENIENCE

Disassembly, cleaning and reassembly of many auto components, including automatic transmissions, can be simplified by making a combination parts drain table, cleaning area and assembly stand from an old 55-gallon barrel.

Using a sharp screwdriver or chisel and a hammer, punch a number of drain slots in the top of the barrel. During dissassembly this will allow oil or other fluid to drain into the barrel. Then parts can be cleaned on top of the barrel and, again, the dirty solvent will drain off into the barrel. Prior to component assembly, wash off the top of the barrel and use it as a free-standing workbench, thus allowing you to get at the work from all sides without having to shift everything around.

Here Tom Olson prepares for initial cleaning and disassembly of a Powerglide on just such a barrel.

APPLYING A LITTLE PRESSURE

When overhauling a transmission, if the converter is to be replaced with a rebuilt or a more suitable substitute, simply set the old converter aside before beginning the cleaning process. When the old converter is to be reused, it must be thoroughly cleaned. A converter that hasn't been flushed should never be used with a new or rebuilt transmission, any more than a dirty oil filter should be used with a new or rebuilt engine, and for the same reasons.

If a transmission shop is nearby, take the old converter there and have it flushed. If such a shop isn't available, allow the old converter to drain for several hours by inverting it in a drain pan or similar fixture. Then repeatedly fill the converter with solvent, slosh it around vigorously, and drain the solvent until no trace of dirt can be noted when the solvent is dumped out. The solvent can be siphoned out quickly and efficiently by those who have access to a compressed air nozzle.

Before installing the converter into the transmission, pour about one quart of transmission fluid into the converter neck to lubricate it thoroughly and displace any solvent that might still be inside. This is also good procedure when you're installing any new or rebuilt converter for the first time. Also lubricate the outside of the converter neck with light grease to prevent front seal damage on initial start-up.

'Glide HOW-TO

(1) To remove the valve body, eight bolts must be removed: three at the front, four at the rear and one next to the shifter. After the valve body is out, which may require a little leverage to dislodge the tube shown, set it aside. We'll discuss cleaning it, checking valve operation and gasket replacement during reassembly.

(2) The modulator valve will slip out of the valve body at this time, so remove it and set it aside, too. The exterior of the valve body can be cleaned with solvent at this time, if desired.

(3) To remove the front pump, remove the seven bolts around the perimeter. The two holes indicated here are tapped to accept a ⅜-inch coarse thread. If you have a slide-hammer with this thread on one end, use it to jerk the pump out. If you don't have a slide-hammer, you can improvise with a piece of ⅜-inch threaded rod, some washers and nuts and an old manual transmission cluster gear. Still another approach would be a piece of chain about three feet long and two ⅜-inch bolts.

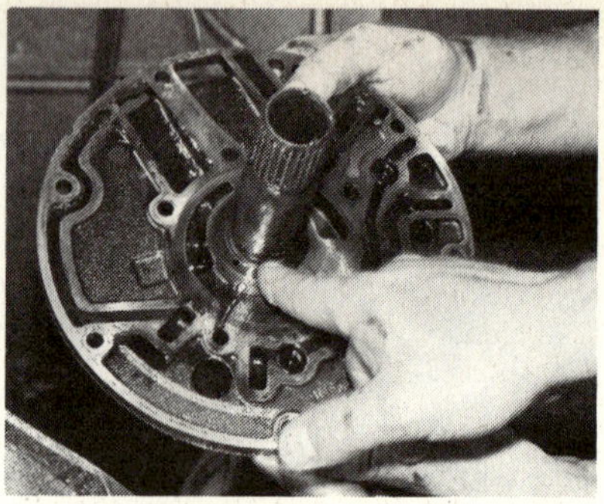

(4) Disassemble the pump. If the gears have scored the face of the pump next to the center tube, you can have it surfaced at a machine shop for a couple of bucks. This is far cheaper than replacing the entire pump.

(5) Whatever amount of material is removed from the pump face can be made up by installing a thicker thrust washer on the front of the pump. Thrust washers are available from Chevrolet in varying thicknesses. Simply add exactly as much thickness as was removed from the back.

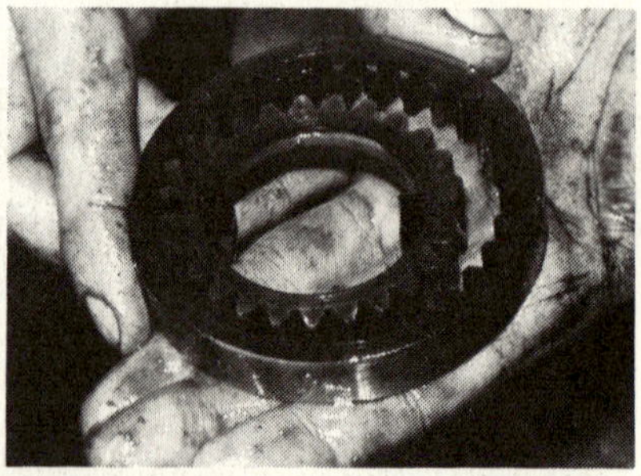

(6) Inspect the pump gears for damage or excessive wear, especially where the inner gear may have been rubbing against the nose of the converter.

(7) Inspect the pump cover for wear in the gear cavity. The half-moon-shaped stator in the cover may have a sharp edge worn on it. If so, remove the sharp edge by lightly dressing it with a fine file.

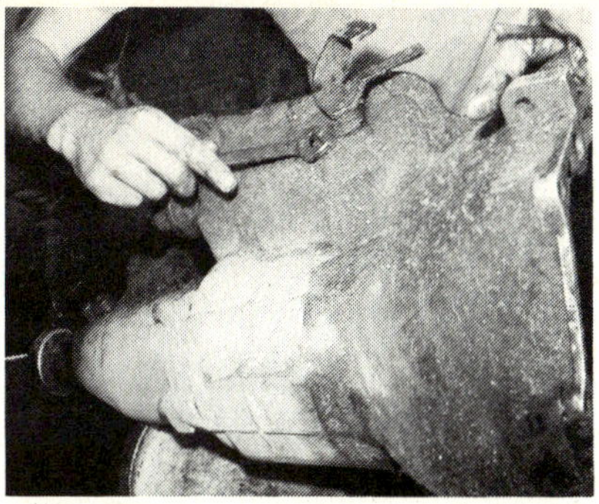

(8) Loosen the band adjustment lock nut on the outside of the case and back the adjustment nut all the way out (about ⅜-inch).

(9) Grasp the input shaft and pull the direct clutch pack out.

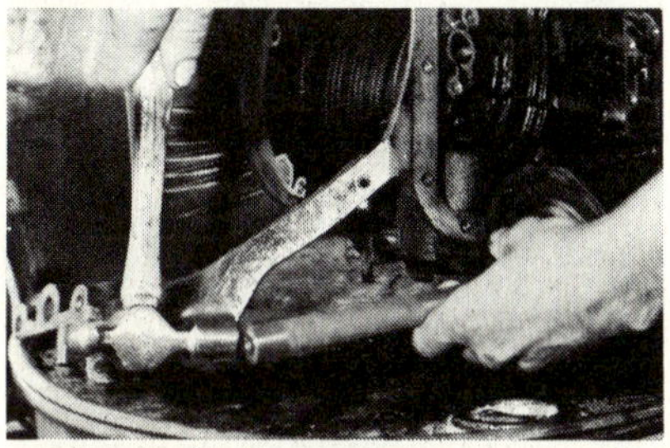

(10) Using a hollow driver to slip over the input shaft pilot, drive the input shaft out of the direct clutch pack from the gear side, as shown. If a hollow driver isn't available, use a brass punch or a wooden dowel. *Do not* hammer directly on the shaft pilot.

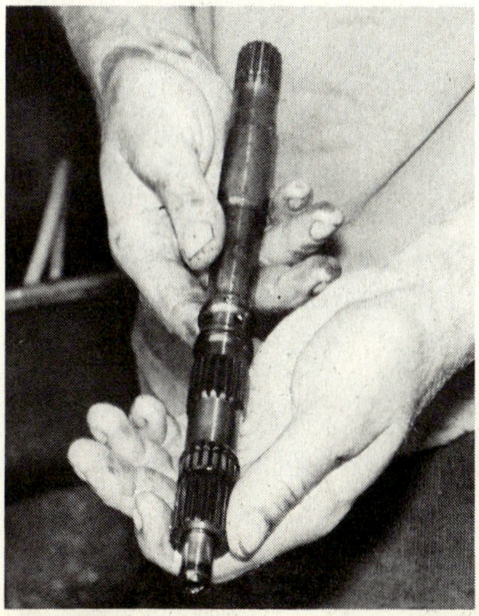

(11) When the input shaft comes out of the clutch pack, it will have a small-toothed brass ring on it. Slip the ring off and set it aside so it won't get lost during cleaning.

(12) Disassemble the clutch pack by removing the large snap ring with a screwdriver or awl.

(13) With the snap ring removed, the front plate, clutches and direct drive hub fall out.

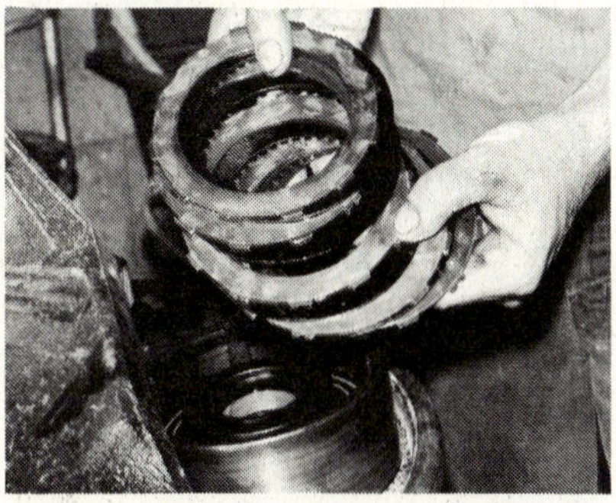

(14) Inspect the clutches for burning or scorching on the steels and worn lining on the friction discs. Check, too, for warpage of the steels, which indicates they are ruined. Some 'Glides will have a wavy cushion spring at the bottom of the stack of clutches bearing directly against the piston and, although it will appear warped, it is supposed to be that way.

(15) To remove the piston from the clutch pack, clamp a 6-inch-long, ⅜-inch bolt in a bench vise (if you used a ⅜-inch threaded rod to remove the pump, that same rod can be used here). Slide the clutch pack over the bolt, and using suitable washers or rings you may have lying around the garage, compress the piston spring retainer against its springs with a ⅜-inch nut.

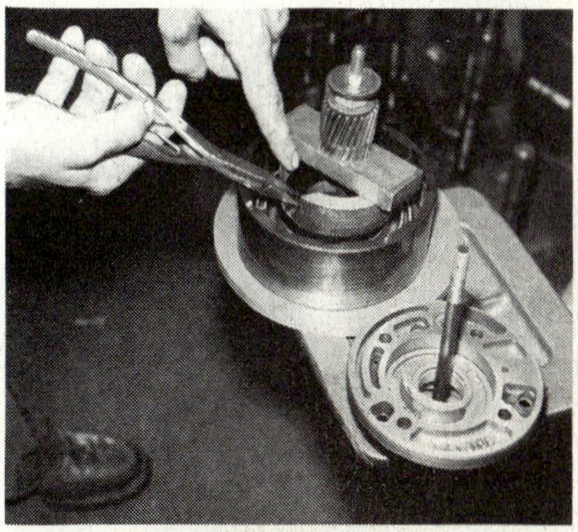

(16) With the retainer compressed, the snap ring that holds the spring retainer can be removed with a pair of snap ring pliers. Then back the nut off and release the pressure on the retainer.

(17) The spring retainer and springs will then fall out of the clutch pack.

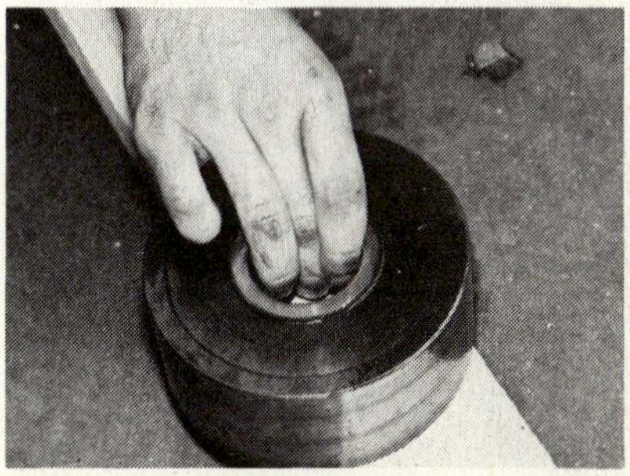

(18) A couple of slaps against a piece of wood will probably be required to get the piston to drop out, since its seal will tend to hold it in place.

'Glide HOW-TO

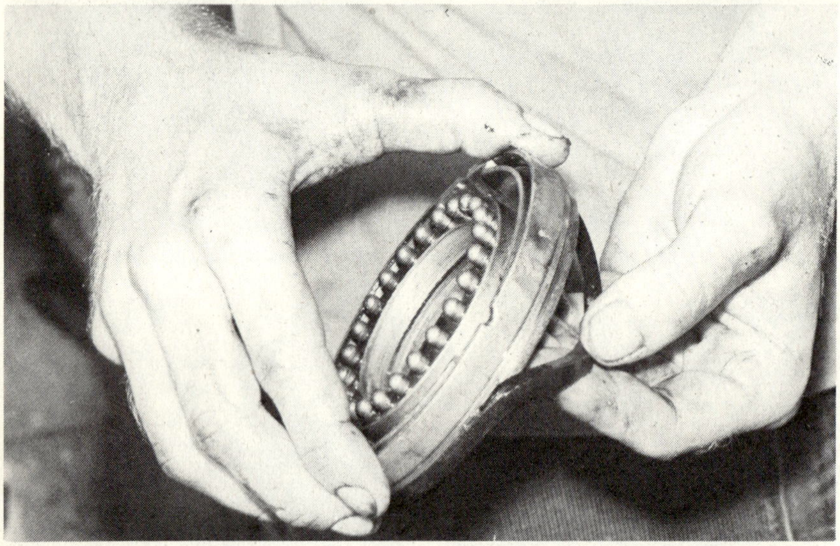

(19) With the piston out of the hub, remove the rubber seal and discard it.

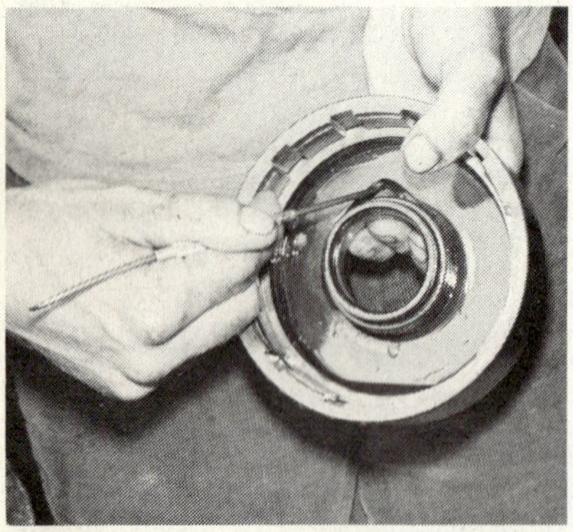

(20) Also remove the inner piston seal in the hub. Discard it, too.

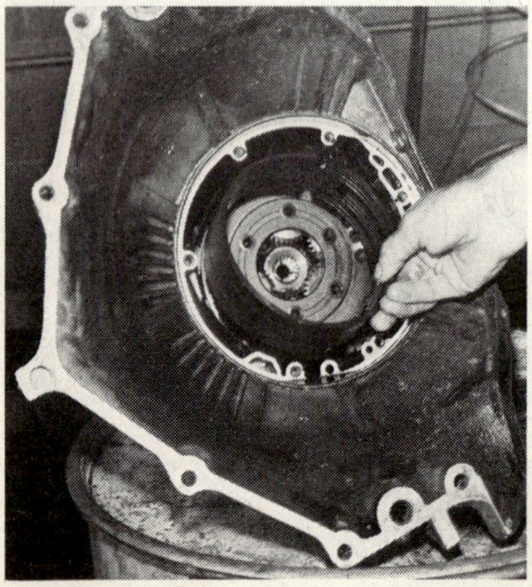

(21) Remove the low gear band and the two band struts, noting how they are positioned in relation to the band, case and servo.

(22) Remove the tailhousing bolts and then slide it off.

(23) Measure the distance from the end of the shaft to the gear and write this dimension down where it won't get lost. Remove the speedometer drive gear by gently tapping it, first on one side and then the other, with a brass punch and hammer. If you damage the gear, most transmission shops have extra ones lying around, or you can get a replacement from Chevrolet. If you're lucky enough to have a gear puller that will reach this gear, use it.

(24) Loosen the single 5/16-inch bolt on the governor and slip it off the output shaft.

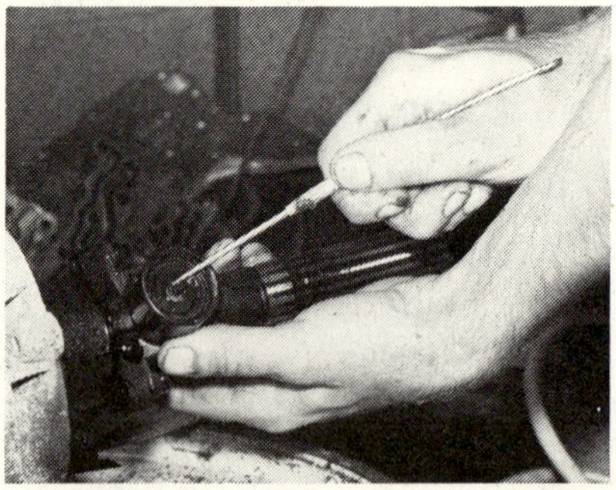

(25) Remove the small E-clip at the end of the governor.

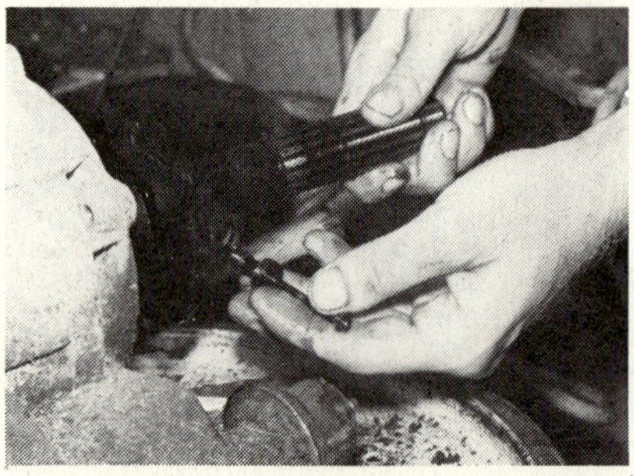

(26) The governor pin will then slip out. Set it aside for cleaning.

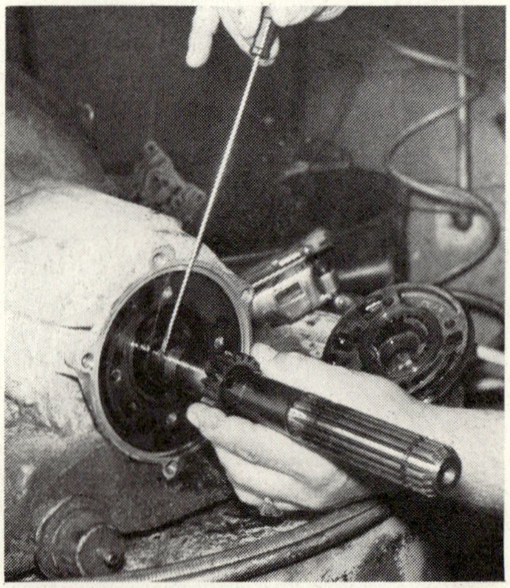

(27) Powerglides built prior to '65½ have a rear pump. All later versions simply have a solid rear support. Whichever it has, remove it. If a rear pump is used, a small pump drive pin, similar to a nail head, will be inserted into the shaft, as shown here. Remove it and set it aside for safekeeping. Inspect the pump gear and case, looking for the same indications described on the front pump.

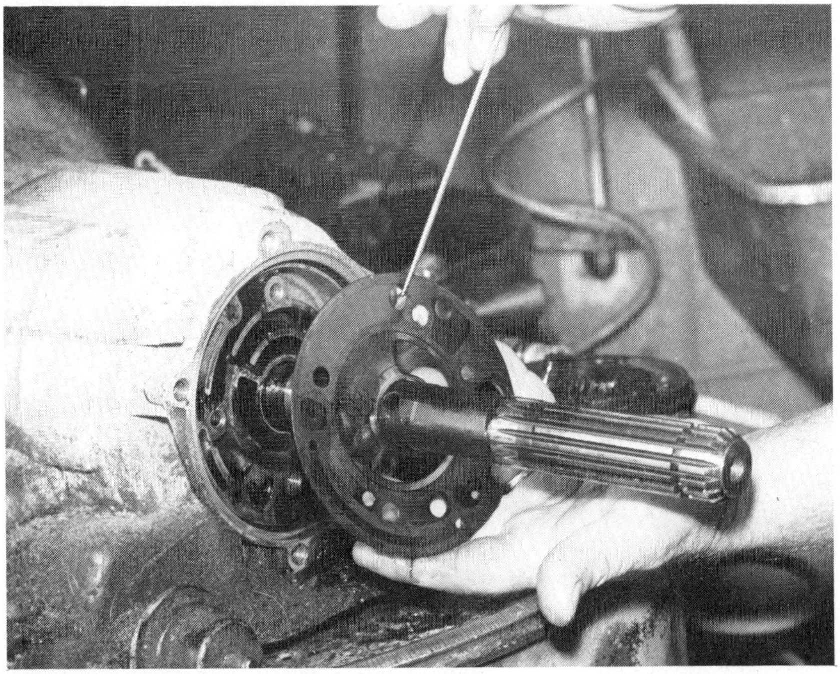

(28) Models with a rear pump will also have this pump wear plate. Note which side faces the pump before removing it.

(29) The entire planetary assembly can now be removed from the front of the case. All that is now left inside the case is the reverse clutch pack and the reverse apply piston assembly.

(30) To remove the reverse clutches, remove the snap ring shown.

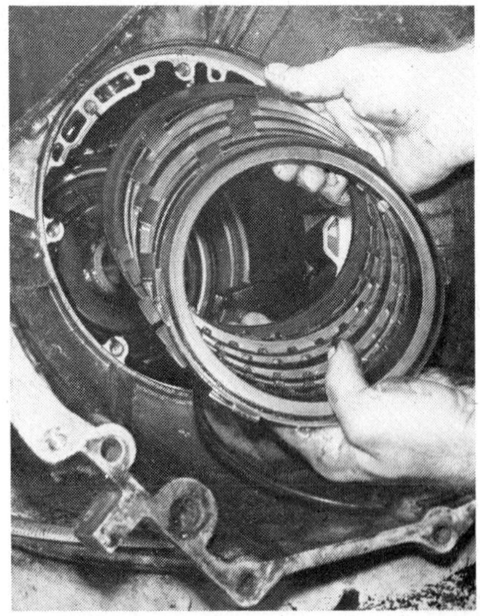

(31) Inspect the reverse clutches just as you did the direct clutches. Again, a bottom cushion spring may or may not have been used. The reverse piston assembly is removed by compressing the spring retainer and removing the snap ring, just as was done for the direct clutch pack. Remove and discard the seals.

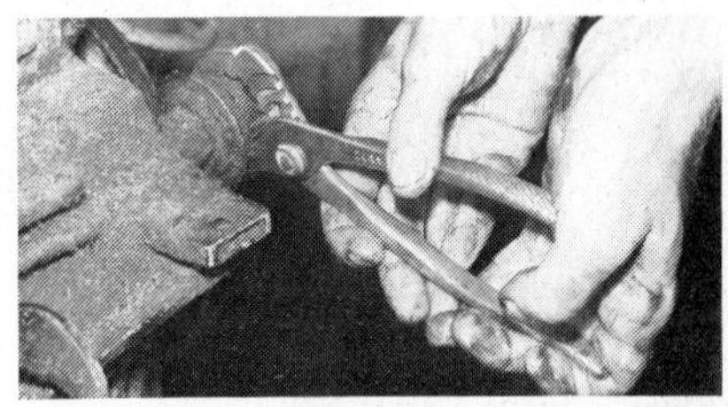

(32) Remove the vacuum modulator. It is usually a good idea to discard the old modulator and replace it with a new one.

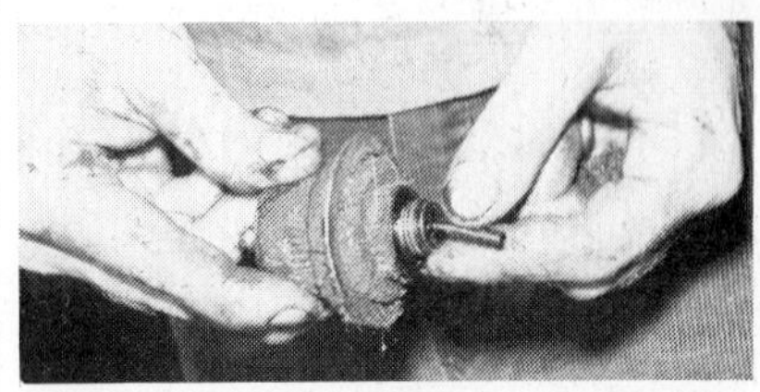

(33) Save the modulator pin when you remove the modulator. Set it aside so it won't get lost.

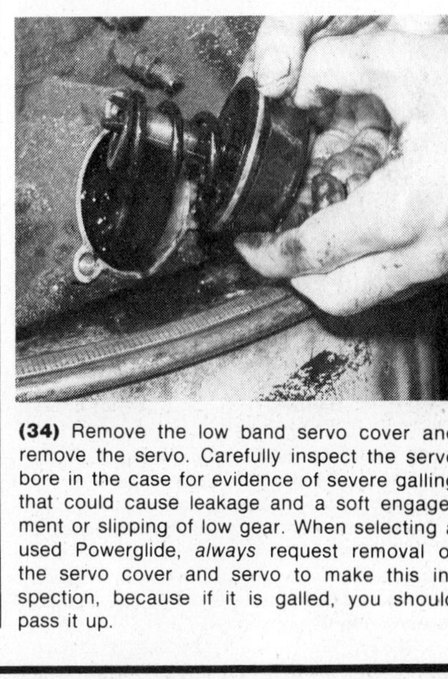

(34) Remove the low band servo cover and remove the servo. Carefully inspect the servo bore in the case for evidence of severe galling that could cause leakage and a soft engagement or slipping of low gear. When selecting a used Powerglide, *always* request removal of the servo cover and servo to make this inspection, because if it is galled, you should pass it up.

'Glide HOW-TO

(35) Now that the 'Glide is all apart, it's time to put it back together. Lay out all the components. The idea is to build each subassembly first, then put them into the case. Friction clutches and low gear band should now be soaked in a pan of clean trans fluid for a minimum of one hour.

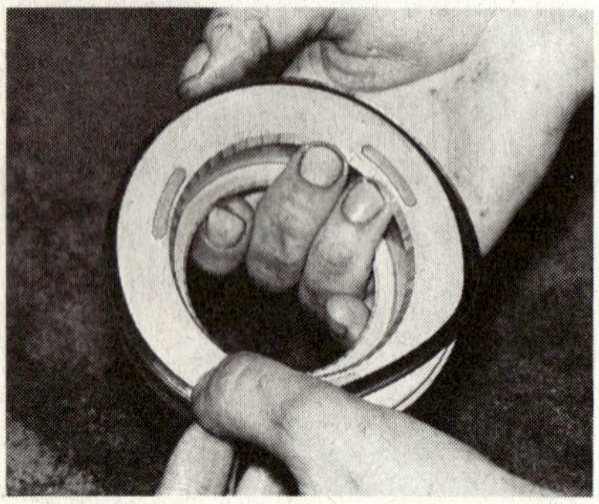

(36) Beginning with the main clutch pack, roll the seal on the piston. The lip on the seal should point away from the piston.

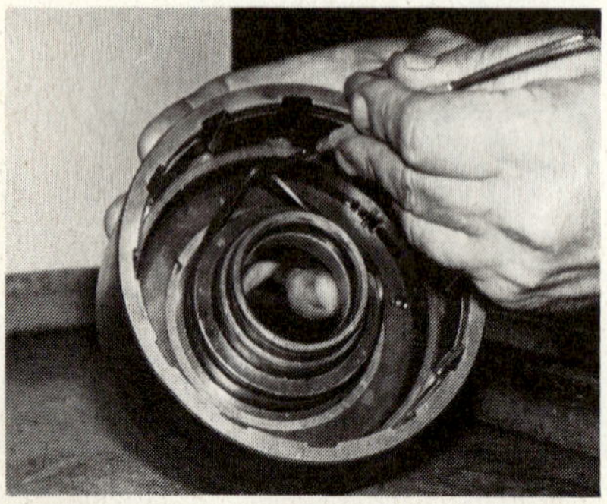

(37) Install the seal in the main clutch drum, with the lip of the seal toward the bottom of the drum. Using a liberal amount of lubricant (Rossi uses a mixture of light grease and STP), reassemble the piston in the drum, along with the springs and retainer. Reassembly is just exactly the reverse of the disassembly procedure.

(38) The ring washer is installed on top of the spring retainer.

(39) Next, the clutch hub is installed, as shown.

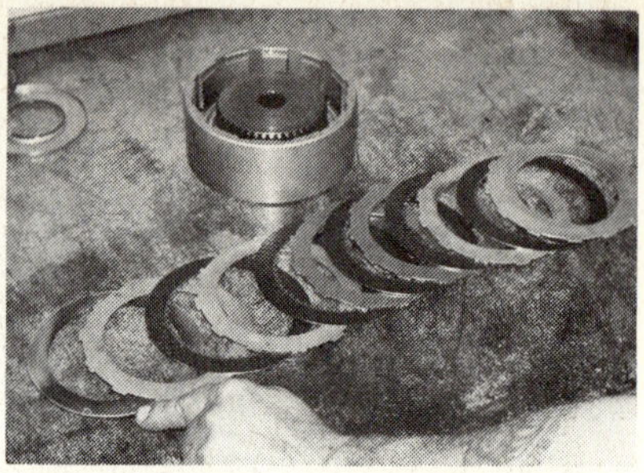

(40) Lay out the main clutch pack steels and clutches (and the cushion washer, if used).

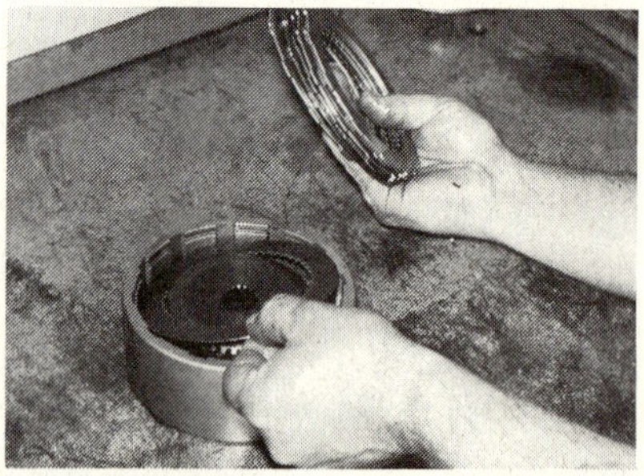

(41) The clutches, steels and cushion spring (if used) are installed in the following order: cushion spring (not all PGs use one), a steel, clutch, steel, clutch, steel, clutch, steel, clutch, steel, clutch, and one or two steels. The Rossi kit uses five clutches.

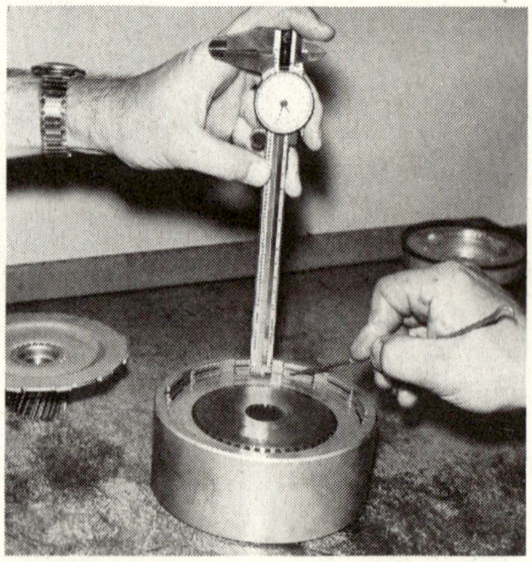

(42) Using a vernier caliper, measure from the retainer lip to the top steel. The correct tolerance here is .055- to .065-inch. Different-thickness steels are available from Chevrolet (.040-, .060- and .070-inch thick) to achieve the proper dimension.

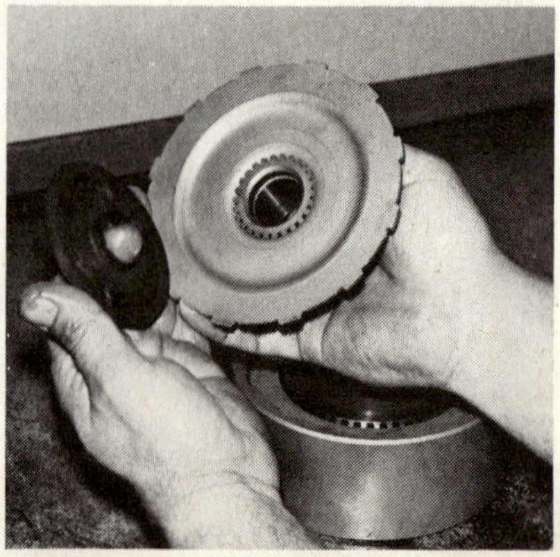

(43) The thrust washer is placed between the hub and the flange, as shown.

(44) Install the flange and the retainer ring.

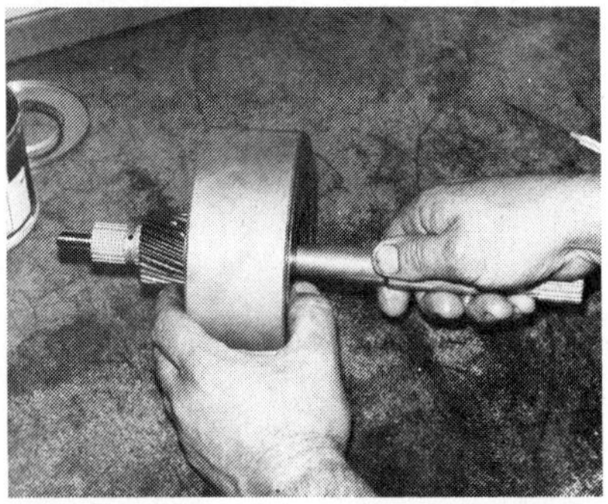

(45) Install the input shaft sealing rings. Insert the input shaft from the side opposite the gear on the main clutch pack. Again, use plenty of lubricant.

(46) Slip the splined brass bushing on the shaft until it is recessed under the gear. This completes the main clutch pack.

(47) The next subassembly will be the pump. Carefully inspect the gears. If they are worn on the converter side, flip both gears over. Normally, the relief on the gears goes toward the front seal. Lubricate everything heavily.

(48) Before proceeding, prime the pump with transmission fluid.

(49) Install the back of the pump, making sure everything lines up, as shown. Start the bolts, but leave them loose.

(50) Align the two halves of the pump with a large hose clamp (or a couple of small clamps put together). Torque the bolts to 20 ft.-lbs. Install a new front pump seal at this time.

(51) Install the pump-to-case O-ring.

(52) Install the pump thrust washer so that the tab lines up with the notch in the pump, as shown. This washer is selective in terms of thickness. It is what will determine the final end-play tolerance at the final step in this part of the 'Glide assembly. Different-thickness washers are available from Chevrolet.

(53) Install the two pump sealing rings. Note that the ends of these rings interlock. *Always* use new sealing rings. This completes the pump.

'Glide HOW-TO

(54) The next item of assembly is the reverse clutch pack in the transmission case. Install the O-ring seals on the reverse piston, just as was done on the main clutch piston. Using plenty of lubricant, install the reverse piston, spring side toward the front of the transmission, with a gentle downward push.

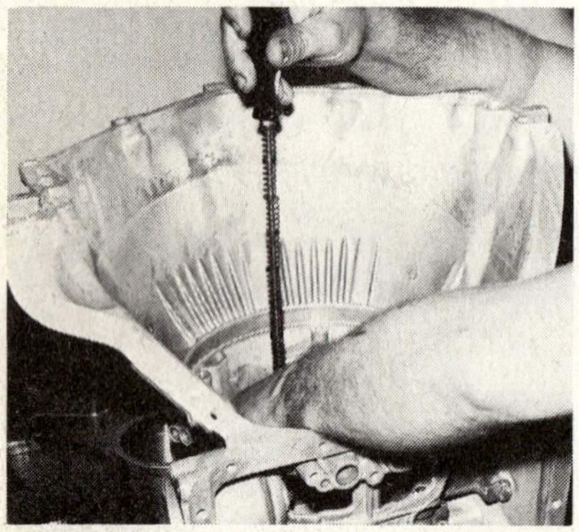

(55) Install the reverse piston springs. A simple method of doing this is to slip the springs over a rod or screwdriver and then slip them onto their bosses.

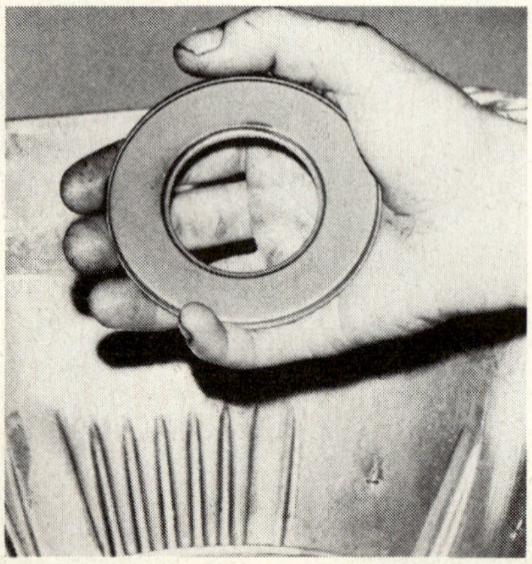

(56) Set the spring retainer in place, compress the spring pack (opposite of disassembly) and install the snap ring.

(57) Install the cushion spring, steels and clutches in the same sequence as was used on the main clutch pack. Each steel has a notch on one tab, which must be installed in the 7 o'clock position, relative to 6 o'clock being at the bottom of the case. The thick reverse pressure plate goes in last (ridge side toward the front of the transmission). It also has a notched tab. Install the pressure ring and snap ring.

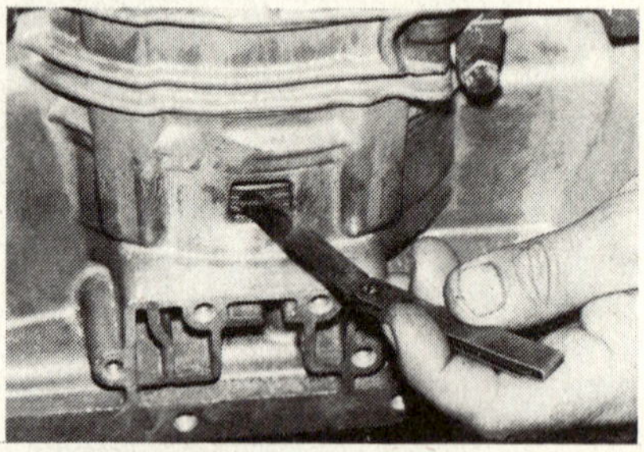

(58) Measure the clearance between a steel and a clutch disc with a feeler gauge. The correct tolerance is .060- to .100-inch. For use with high-idle-speed engines or exceptionally powerful engines, this tolerance should be on the loose side. Chevy has different-thickness steels and snap rings to achieve the desired clearance.

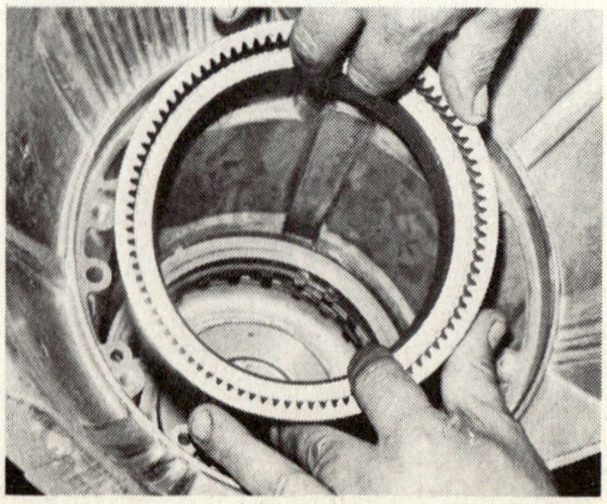

(59) Index the ring gear into the clutches until it fully seats against the pressure ring.

(60) Install the Torrington bearing over the output shaft.

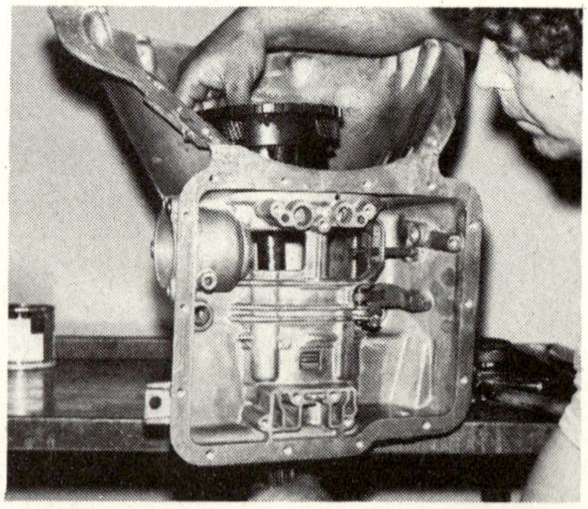

(61) Install the output shaft and planetary assembly in the case. You will probably have to spin and wiggle it to get it to seat fully in the ring gear.

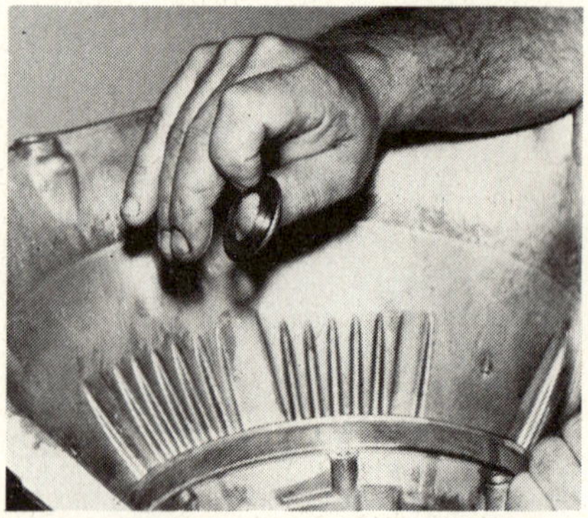

(62) On heavy-duty planetary assemblies, the Torrington, or bronze "hat," washer is installed with the lip toward the front of the transmission. Use plenty of lube.

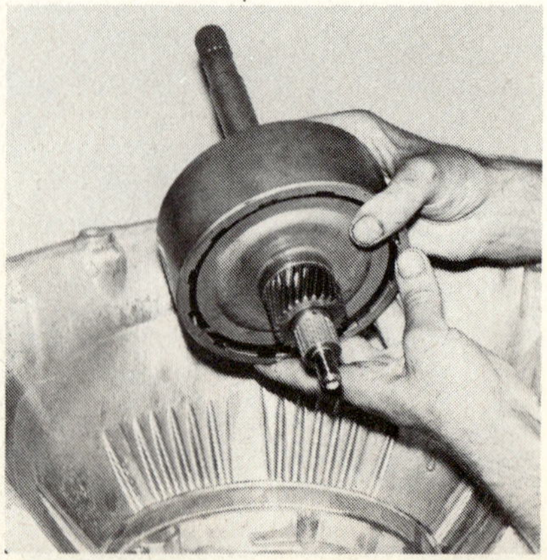

(63) Now install the main clutch pack and shaft assembly. Again, this will take some wobbling and turning to get it to seat fully.

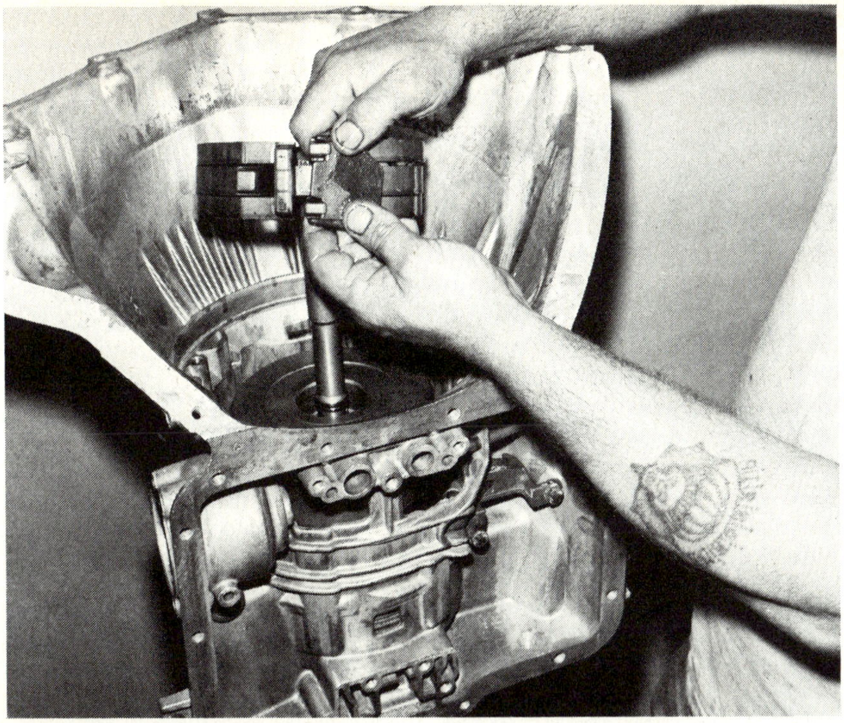

(64) Install the low gear band with the wide end and strut toward the band adjustment side of the case.

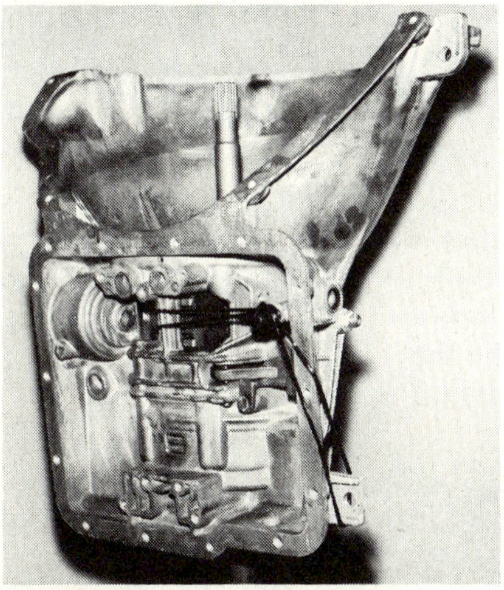

(65) Use an old pump housing O-ring over the servo tab and a bolt in the case, as shown, to position and tension the band during the remaining assembly.

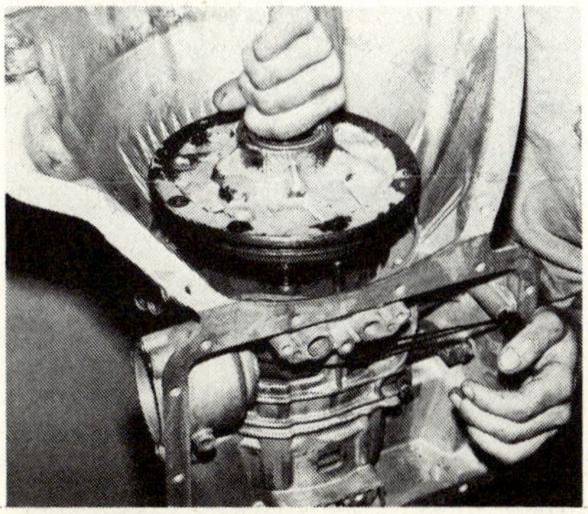

(66) Using two 2 1/2x5/16-inch alignment studs, position the pump to case gasket, lining up all the holes. Lube the pump O-ring, line up the pump and case ports, and slip it over the alignment studs.

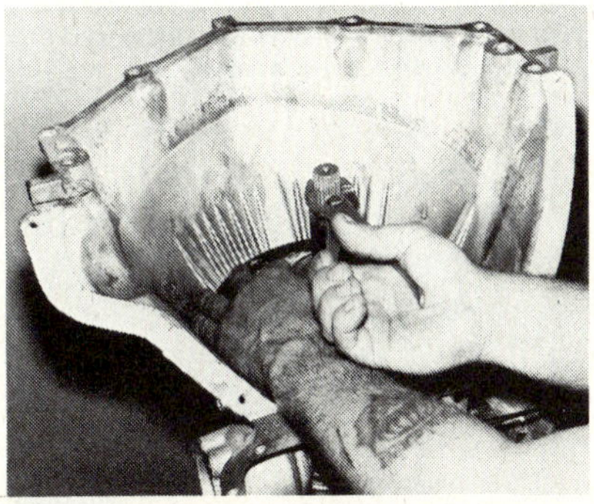

(67) Using a pair of pliers, wiggle the input shaft to line up the shaft sealing rings.

(68) Using two pump bolts directly opposite each other, tighten the bolts alternately (½-turn at a time), using the input shaft alignment in the pump as a guide, to progressively force the pump into place. Wiggle the input while doing this. When the pump is fully seated, remove the guide studs and the two bolts. Install the special rubber-laminated sealing washers under the bolts and tighten them to 15 ft.-lbs.

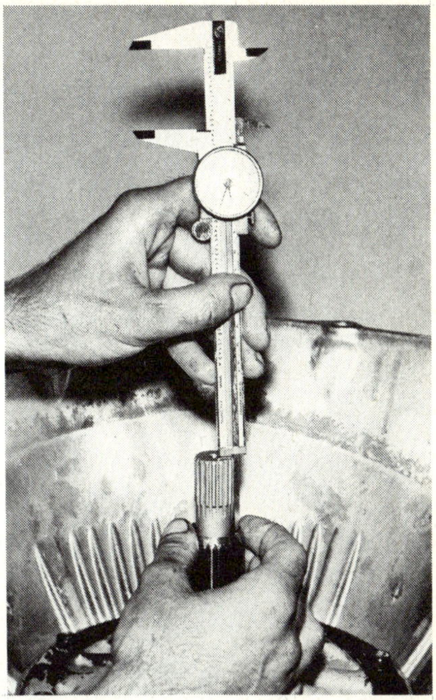

(69) With the transmission supported solely by the back of the case and the output shaft hanging freely straight down, measure the distance from the end of the input shaft to end of stator tube.

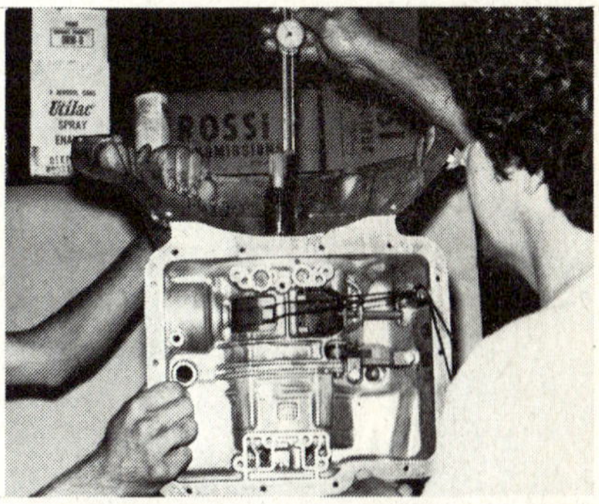

(70) Then reposition the transmission so that it is entirely supported by the output shaft, as shown. The difference between this measurement and the previous one should be .030- to .050-inch. If it isn't within this tolerance, the thrust washer on the back side of the pump will have to be changed for one of a different thickness, as previously discussed.

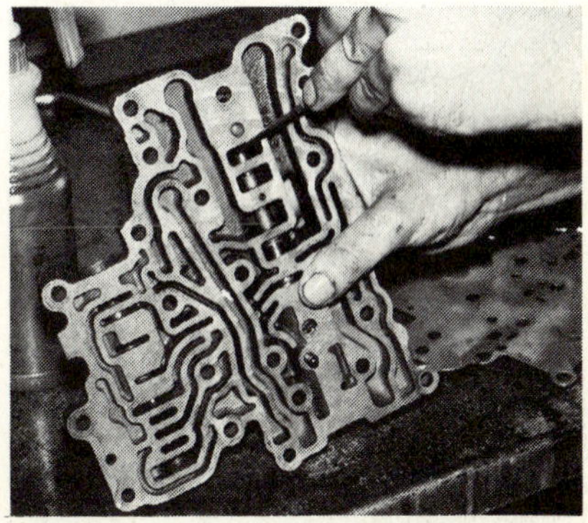

(71) Disassemble the two halves of the valve body. Don't remove the valves. Simply check each valve for freedom of movement, after throughly cleaning the valve body halves with solvent. Use a probe and plenty of light oil, such as WD-40, as shown, when making this check.

'Glide HOW-TO

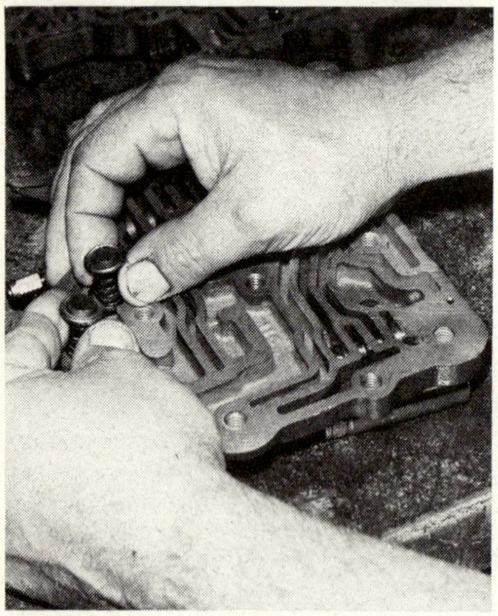

(72) Pre-'66 Powerglides utilize two flat check valves for the front and rear pumps. Later models don't have these valves. The check valves are installed as shown.

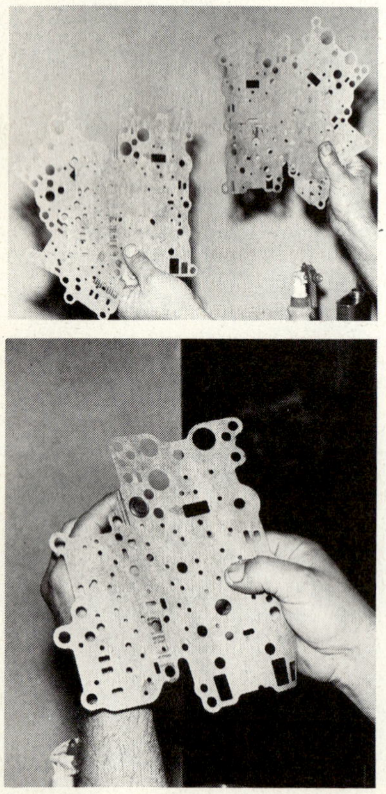

(73) All gasket kits have two sets—one set for the pre-'66 units and one set for the later units. The earlier gaskets have the two extra holes for the check valves, as shown. Be sure to use the correct gaskets for your particular transmission.

(74) To align the gaskets, valve body plate and the two halves of the valve body, install a couple of guide pins in two opposite holes of the valve body top, as shown. Start with the gasket with the biggest holes.

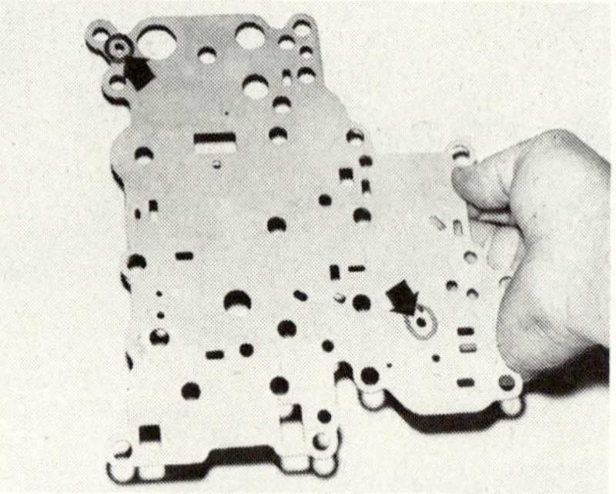

(75) Before installing the valve body plate, drill and deburr the two indicated holes. Opening these holes to a diameter of 3/16-inch improves shift quality and clutch life.

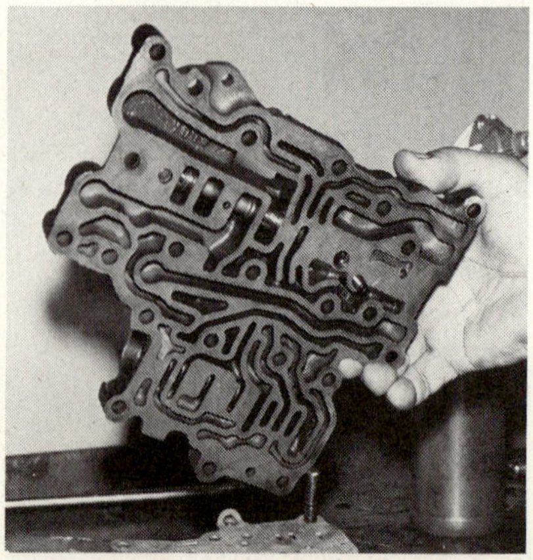

(76) After installing the shift plate, install the second gasket and then the valve body bottom. Torque the bolts holding the two halves together to 15 ft.-lbs.

(77) Lubricate the low band servo bore. Install the new sealing ring on the servo and insert the servo into the bore. It may be necessary to compress the sealing ring with a screwdriver to get it to start into the bore. Don't push the servo all the way in at this time. Leave the cover off.

(78) Remove the rubber O-ring that had been previously installed to tension the low band. Install the band servo strut and gently tap the servo in until it holds the strut.

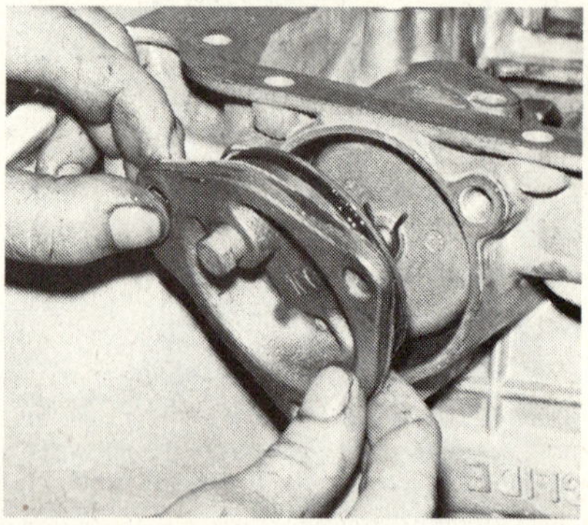

(79) Using a new O-ring and gasket, install the servo cover. Torque the bolts to 20 ft.-lbs.

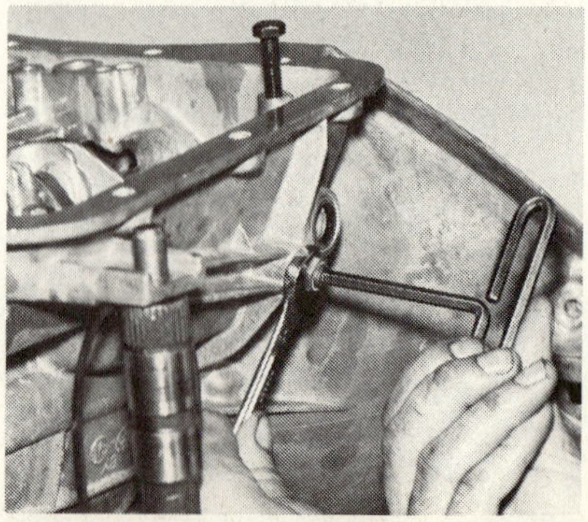

(80) Adjust the low band by loosening the lock nut, turning the Allen screw all the way in and then backing it out 3½ turns. Tighten the lock nut to 15 ft.-lbs.

(81) Install the rear pump or rear support, whichever is applicable.

(82) Early models must have this drive pin installed in the shaft. If the notch in the pump gear that this drive pin slides into is worn, the gear can be turned over. Early models also use a pump wear plate instead of a gasket. Such a wear plate should be installed with the shiny side toward the rear of the transmission.

(83) The fluid baffle is installed under the bottom bolt, as shown.

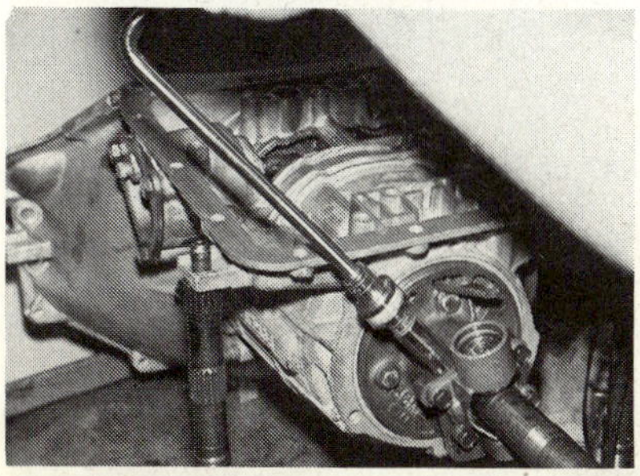

(84) Install the governor so that the retainer screw goes into the hole in the shaft. Tighten the screw to 8 ft.-lbs.

(85) Install the small valve and pin assembly through the governor and install the E-clip. Install the tailhousing O-ring over the rear pump or rear support, as applicable. Also put a new rear seal in the tailhousing.

(86) Install the speedometer gear on the output shaft, using the dimension from the end of the shaft measured during disassembly. Install the tailhousing; tighten the bolts to 25 ft.-lbs.

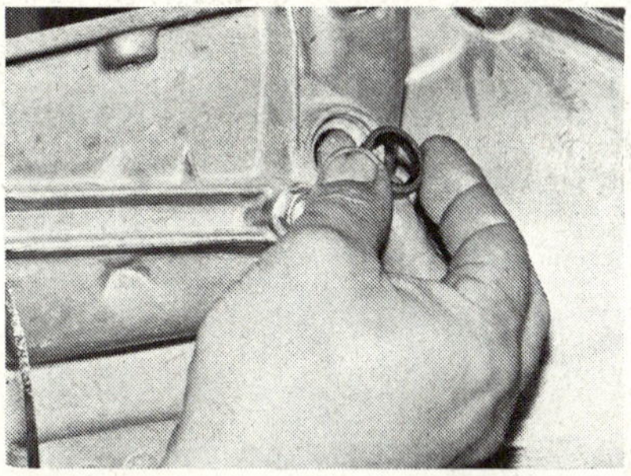

(87) Install the shifter linkage in the case. Use a new seal. Assembly is the opposite of disassembly. It will only go in one way, and the serrations must line up.

(88) Next is the installation of the valve body. Care must be used not to damage the valve-body-to-case gasket, to insert the servo tube into the case, to see that the detent roller slips over the detent plate and that the selector pin engages the manual valve. That's a lot of things to watch, but they are all important to prevent damage to the components and proper operation. Torque the bolts to 15 ft.-lbs., with the guide plate in place. Install the screen, tightening its bolt to 2½ ft.-lbs.

(89) Install the vacuum modulator, pin and valve, as shown. Tighten to 15 ft.-lbs. Install the pan with a new gasket. Tighten the pan bolts to 7½ ft.-lbs. This completes assembly!

12-Bolt Magic

There are few mystical contradictions concerning the Chevy 12-bolt rear end, one the basics are covered.

It shouldn't take reams of statistical data to convince you that the overwhelming majority of Modified, Super Stock and Bracket cars are Chevrolets: A quick look in the pits will uncover legions of early- and late-model Stovebolts in various stages of modification.

Characteristically, one of the most annoying "weak links" in these Chevrolets has been concentrated in that area between the two rear wheels—more specifically, the rearend and axles. Of course, with NHRA, IHRA and AHRA rules permitting the use of any production-type rearend, numerous Chevy racers have been opting for the huskier "Dana 60" assemblies. But Danas are expensive, and they, too, require aftermarket mods to ensure their dependability. Virtually lost in this shuffle has been the stock Chevrolet 12-bolt, which, as it turns out, could quite possibly still be as good a choice, especially for those on a budget.

The 12-bolt is so designated in race lingo because of the number of bolts holding the cover on the housing. It came stock on 1965 and later Chevrolet compacts and passenger cars ('65-'69 Chevy II versions are the narrowest and lightest, while Impala-type units have larger brakes, backing plates, flanges, etc.) and is obviously in plentiful supply and therefore inexpensive. It is also substantially lighter than the Dana 60, and as such offers the racer significant savings in unsprung weight (if you consider that important). Scientific racer types are quick to point out that unsprung weight, as opposed to sprung weight (that which is supported by a vehicle's suspension), cannot be used to any kinetic advantage, i.e., weight transfer. And finally, the Chevrolet 12-bolt has been the subject of much attention by the performance industry, with virtually every component piece now replaceable with a superior-quality aftermarket item.

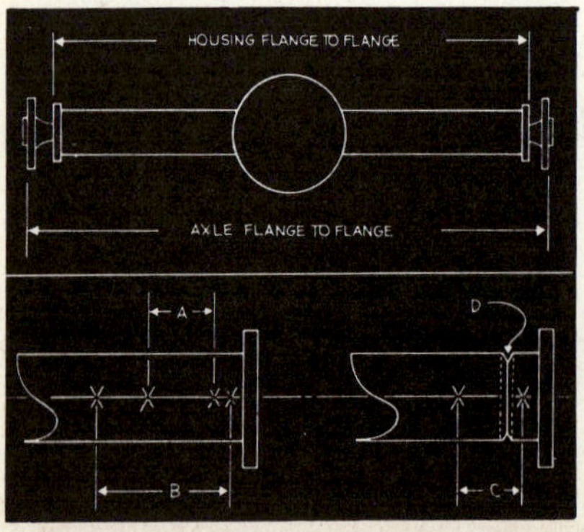

1

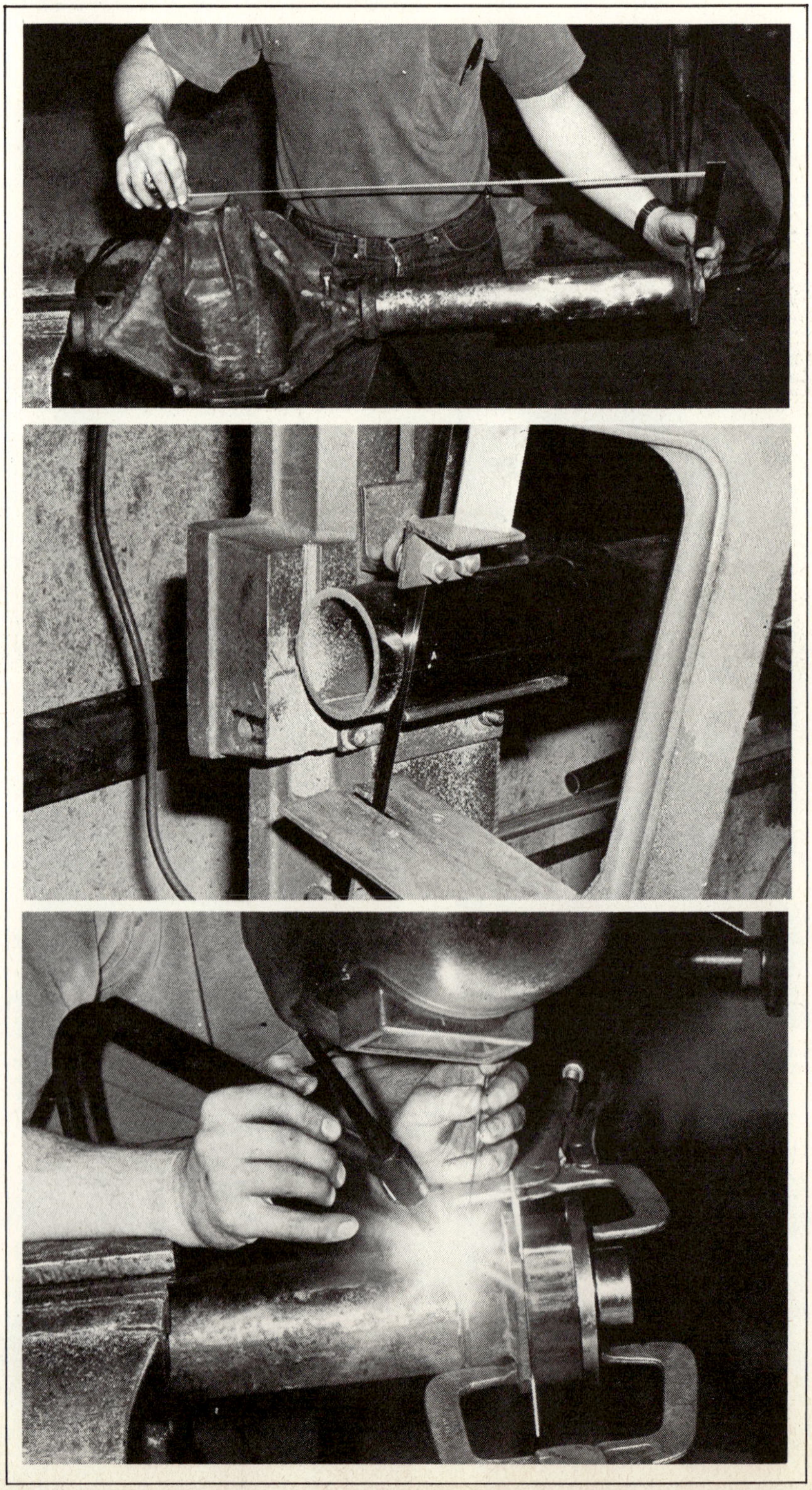

2

One of the people responsible for the development of present-day "bullet-proof" 12-bolt parts is Mark Williams. His Denver, Colorado, based M-W Enterprises, which began manufacturing fully guaranteed forged steel axles over a decade ago, currently offers a vast assortment of highly specialized components for the 12-bolt. Additionally, SEMA chassis builder Williams is constantly called upon to supply racers with 12-bolt units in various stages of completion, and as such has gained valuable insight into the characteristics of drag racing's most popular rearend, in terms of numbers only.

His observations are as follows: "The one part most prone to breakage is the axle. So the racer has the choice of either going to forged steel axles or living with the problem. If the latter course is taken, the most important modification the racer can make is the addition of a 'C-clip eliminator' kit. Stock Chevy axles are prevented from sliding out of the axle housing by means of a C-clip which attaches to a nub on the inboard end of the axle. Unfortunately, any time an axle breaks, there's nothing to prevent the broken axle from unreeling, which generally results in the tire inflicting severe damage to the rear quarter panel/fenderwell area.

"Furthermore, to replace or remove an axle, the racer must unbolt the differential cover and partially disassemble the differential itself to remove the C-clip. This tedious procedure must be reversed for installing an axle. A C-clip kit does away with the stock axle retainer and transforms the burden of axle retention to a large ball bearing with side-thrust loading properties. This bearing is located at the outermost proximity of the axle, and the kit serves to retain broken axles, which usually break in the spline area. To replace a broken axle, the racer need only remove the stock axle flange bolts."

The M-W C-clip eliminator kit is an accredited "safety hub," as required by sanctioning organizations, and the complete kit can be installed with the rearend in place, using only a common hacksaw and appropriate wrenches.

Should a racer choose to replace the stock Chevy axles with forged steel items, compound options are available.

For openers, the racer must decide whether to retain the stock differential (or Positraction unit) or convert to an axle/spool assembly.

The spool, a device which connects both axles directly to the ring gear and which has been accepted for Bracket competition, is now allowed in Super Stock and Modified Eliminators. It's favored by most serious competitors because it provides absolutely equal power distribution to both rear wheels. By nature of its rigidity, the spool is said to improve ring and pinion life, but even here there are differences of opinion. One thing for sure: Its one-piece simplicity eliminates the chance of posi unit breakage.

On the other hand, by its very design, a spool *will not* permit unequal rear wheel travel, which renders spool-equipped cars a tad awkward in day-to-day street maneuvering, and although spool-equipped cars can be driven on the street, a heavy-duty Positraction unit (like Moroso's "Brute Strength" models) is best suited for dual-purpose street/strip punishment.

Forged steel axles are available to match any stock spline configuration, but the forged axles used with spools are generally heftier in size and feature special spline combinations (M-W uses a 35-spline design). Of course any time you consider the expense of forged axles, it's a good idea to consider narrowing the rearend housing to gain additional tire clearance, since forged axles are available in any length you desire.

Oftentimes, due to the complexities of wheel offsets, flange-to-flange measurements and asymmetrical housings, it becomes a real nightmare to determine exactly what it is you need. Perhaps the most foolproof method in determining housing width and the necessary related components is to remove the stock rearend housing, inner panels, etc., and place the car on jackstands in the desired attitude. Roll the new "trick" tires and wheels (minus the rear housing) into position, carefully measuring inner and outer fenderwell clearance. NHRA rules call for the outside of the tire being within 1 inch of the fender in S/S, M/P and Super Mod classes. Make sure that the tires are perpendicular to the imaginery axle line, measuring the front and rear of the tire centerline will assure parallelism. The distance between the bolt flange faces of both wheels will be the correct axle flange-to-axle flange dimension for the shortened housing. By comparing this figure to that of the stock housing (measuring the distance between the faces of both axle flanges), you come up

3

1. The best way to determine the amount to be removed is to measure either from axle flange to axle flange, or from housing flange to housing flange. Marks are made on housing to indicate amount to be removed; the section is then cut off, chamfered and then rewelded.

2. Initial step on any rearend housing narrowing is finding out how much material needs to be removed from housing. After housing is marked, it is cut with a power saw, ends chamfered for good weld penetration, tacked, straightened and welded.

3. Stout, large-diameter, Olds-type bearing (left) replaces stock Chevy 12-bolt bearing. Replacement bearing offers increased side loading capabilities in event of axle breakage.

4. M-W replacement rearend cap is substituted for the stock Chevy unit (left) because stock caps are prone to breakage under strain of competition.

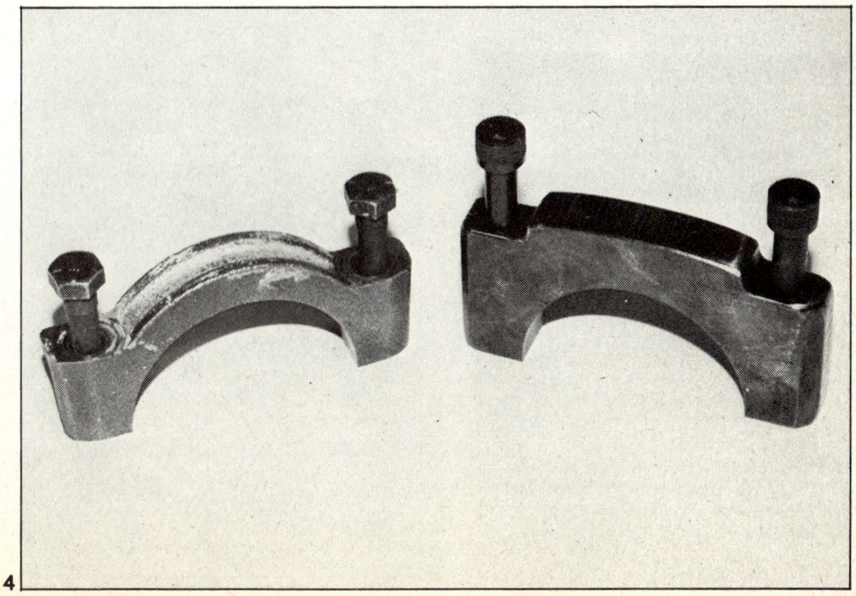

4

with the amount to be sectioned. One thing to remember: When ordering special components, do not confuse axle flange measurements with that of the housing or backing plate bolt-up flange.

Williams recommends that any housing sectioning be done at the outermost extremities in order to retain the structural rigidity of the housing tubes, and that ladder bar brackets, etc., be welded to the housing prior to rewelding the housing ends, to guard against distortion. To assure optimum axle alignment, Mark uses a fixture that is secured by the carrier caps.

This brings us to the final weak link in the 12-bolt assembly, the stock carrier caps. Misalignment causes excessive component damage, as does general fatigue of the caps themselves. To combat this, M-W has developed steel rearend caps which are significantly stronger than the stock units. These steel caps offer greater resistance against carrier walk and generally are capable of holding the carrier assembly in place. This accomplishment alone adds greatly to increased ring gear and pinion life.

With all of the possible modifications made, you can expect to be abusing as reliable a 12-bolt as possible. It should compare to most available rears, Dana 60 included, the only remaining difference being ring and pinion gear sizes. The 12-bolt, 9-inch-diameter ring gear compares quite favorably with the 9-inch Ford and 8¾-inch MoPar but is nominally ¾-inch smaller than the Dana 60 and a full 1½ inches smaller than the Dana 70. Of course with the addition of newly developed 9310 Nickel alloy gears (Schiefer), offered in ratios from 5.14 to 6.14, the Chevy 12-bolt accelerates closer to the imaginary unbreakable status.

In addition to M-W Enterprises (4200 Madison St. Denver, CO 80216), other manufacturers who make related 12-bolt components include: Geis Industries, 1425 E St., San Diego, CA 92101; Henry's Machine Works, 17149 S. Woodruff Ave., Bellflower, CA 90706; Strange Engineering, 739 Howard St., Evanston, IL 60202; and Summers Brothers, 530 S. Mountain Ave., Ontario, CA 91761.

We predict a long life for the Chevrolet 12-bolt—as long as proper precautionary steps are taken!

5

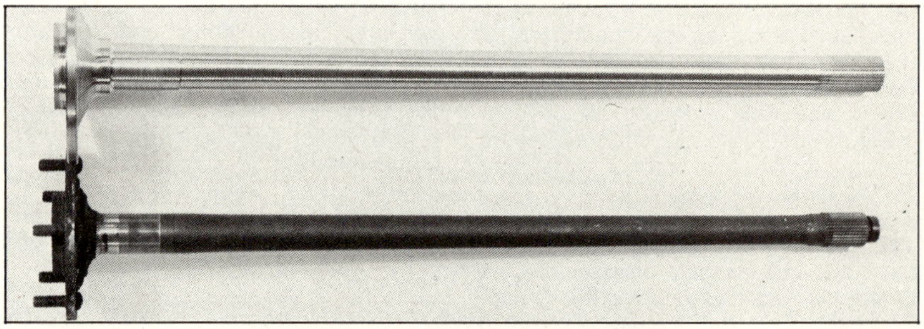

6

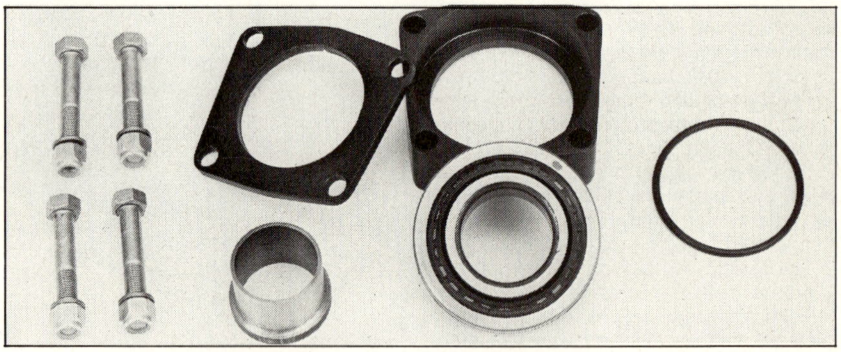

7

8

5. For all-out competition, try to use a spool. It replaces stock differential or posi unit and provides equal thrust to both rear wheels, it's stronger and it's much lighter.

6. Shown is a comparison between a stock Chevy 12-bolt axle (dark finish) and a M-W forged steel unit. M-W's units are forged from snap-resisting alloys.

7. M-W's C-clip eliminator kit facilitates the removal of stock C-clips while functioning as an approved safety hub. The C-clip kit bolts into place.

8. When using C-clip eliminator package with stock axles, bushings must be employed in conjunction with large support bearing. Bushings are not used with forged axles.

Taxing the I.R.S.

Or, how to live with Vette independent rear suspension without going broke . . . or broken.

Anyone who has tried to race a '63 or later Corvette with independent rear suspension (I.R.S.) and stock drivetrain components has probably learned two important facts. First, the Vette I.R.S. hooks up better than most solid "beam-type" rear axle suspension systems. In fact, some competitors who have replaced the stock I.R.S. with a solid axle in their Vettes have had to use slicks up to three or four inches wider than those they were using with the I.R.S. to get equivalent traction. And second, if they are making any power, almost every component in the stock I.R.S. tends to break. It's the axiom of a chain being only as strong as its weakest link; and in the Vette I.R.S. that chain includes the ring and pinion, the differential carrier or positraction unit, the differential pinion and side gears, the inner axles, the inner U-joints, the half-shafts, the outer U-joints and the outer stub axles.

Now, thanks to the combined design and manufacturing efforts of Pepe Estrada, (now acting as a consultant only to Tom's Differentials, 15551 Paramount Blvd., Paramount, CA 90723) and Henry's Machine Works, Inc., (1220 Knollwood Circle, Anaheim CA 92801) each of the components has been redesigned and strengthened to add durability, reliability and safety to the I.R.S. concept.

Working from the inside toward the outside, Tom begins by discarding the stock 10-bolt ring and pinion and posi unit. The case is then enlarged to accept a modified Eaton 9310 nickel-alloy 12-bolt ring and pinion, along with a 12-bolt Eaton posi. A Henry's steel cap and ½-inch bolts replace the stock cast-iron cap and 7/16-inch bolts on the ring gear end of the carrier for additional strength. This eliminates ring and pinion or posi and cap failure. Thus modified, the next components in the chain to fail are the 17-spline inner axles. Therefore, Tom and Henry's have produced special nickel-steel alloy Hy-Tuff inner axles available with either 30 or 31 splines. These inner axles are specially heat-treated and then deep-frozen to increase torsional strength. They are also black-oxided for corrosion resistance. Whether you choose the 30- or 31-spline versions is up to you, since the Eaton posi accepts the Chevy nickel-alloy 30-spline side gears (part No. 3995839) and pinion gears (part No. 3995843) or the Pontiac

Shown from left to right are the stock 17-spline inner axle and side gear and small U-joint, the 30-spline Hy-Tuff inner axle and the 31-spline large-flange inner axle, side gear and U-joint.

31-spline side gears (part No. 9778342) and pinion gears (part No. 9778343) without alteration. Naturally, it will also accept the Corvette nickel-alloy 17-spline side gears (part No. 6260528) and pinion gears (part No. 3995843) as well. Pepe heat-treats all side and pinion gears, just as he does the ring and pinion and inner axles, for greater durability.

Proceeding outward, the next failure area in highly stressed applications is the inner U-joints, which are 3⅝ inches wide. To beef up this weak point, Dana-Spicer 5-160X U-joints are installed. These U-joints are 4 3/16 inches wide and offer an approximate 55 percent increase in strength over the stock pieces. Of course, to accept these larger U-joints, larger companion flanges are required on the inner axles, and Tom and Henry's offer the inner axles, again in both 30- and 31-spline versions, with the larger flange. These particular inner axles have the companion flange both splined *and* welded to the inner axles to withstand the most severe loads.

But the companion flange is only half of what's required for the bigger U-joints. The half-shaft yoke also has to be larger. You guessed it—Tom has those, too. However, the 2½-inch-o.d., .090-inch-wall thickness, stock half-shafts have been known to twist up and fail in high-stress situations, so Tom's big-yoke half-shafts are 3 inches in diameter with .151-inch wall thickness. The outer ends are also fitted with the big yokes and big U-joints. Big companion flanges for the outer stub axles are offered, too—another potential failure point eliminated.

That brings us to the last, the most crucial and the most bothersome component of the I.R.S. system, the outer stub axle. These outer stub axles are prone to two types of failure. First, the bearings fail, due to improper or infrequent lubrication, and weld themselves to the outer axle. And second, the outer axle, which is only 1⅛ inches in diameter, tends to break at the juncture of the wheel flange. When this happens, the wheel comes off the car, damaging the body in the process, and subsequently throws the car out of control, which could damage *your* body. All of that is bad enough, but disassembly of the outer stub axle from the trailing arm to repair a failure or to service the disc brake rotor or emergency brake shoes and drum is an equally big hassle. Everything is pressed into the trailing arm. The outer axle bearings, rotor and emergency brake drum, outer axle, hub and suspension bracket, flange and U-joint yoke are all bolted and pressed together. It is an engineering fiasco. To disassemble it for repair requires special tools and a lot of labor, and it almost always results in a destroyed trailing arm and/or outer axle.

To solve these breakage, service and safety problems, Tom and Henry's have come up with a special two-piece outer axle with a removable flange. You could call it a "free-flange" or a "floater-flange" design. This permits access to the complete braking system by simply unbolting the wheel and flange retainer nut. The flange, which is splined to the axle shaft, can then be removed. Incidentally, this new flange is 4130 steel alloy for extra strength to tolerate extreme side loads for those units that find their way into road racing applications. The axle itself is 1½ inches in diameter at the flange mating point to eliminate the possiblity of breakage in this critical area. The remainder of the outer axle has been increased to 1⅜ inches in diameter, and the inner end has 31 splines in place of the stock 17 splines for increased strength. Material and heat-treat are the same as used on the inner axles. Additional features include a special slip-fit Timken inner bearing, available only from Henry's, while all other components, such as the outer bearing, are stock Corvette items. What's more, all of this can be installed in the stock hub without any modification or machining.

Yet another approach is to completely replace the Chevrolet center section of the I.R.S., be it the standard 10-bolt arrangement or a 12-bolt modification, with a Dana 60 center section. Such a conversion has been developed by Strange Engineering (739 W. Howard St., Evanston, IL 60202). Although this is one of the most comprehensive and expensive of the I.R.S. strengthening alternatives, it

2

3

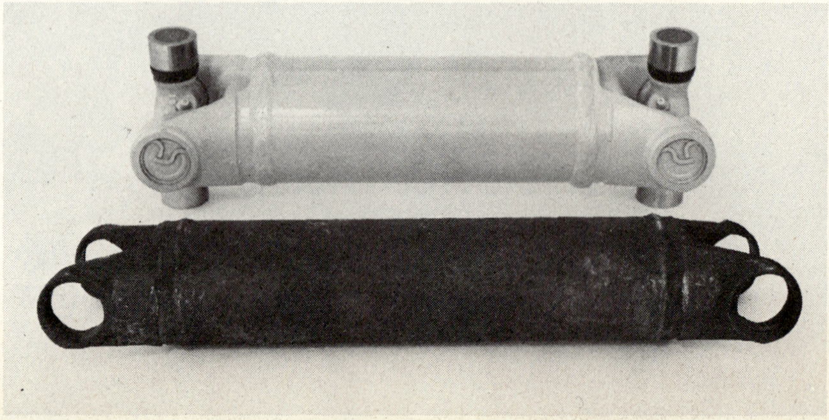
4

2. Here is a size comparison between the stock 3⅝-Inch U-joint and the heavy-duty 4 3/16-Inch Dana-Spicer unit.

3. Shown is a complete Vette case modified with 12-bolt gears and posi. This particular one belongs to John Lingenfelter, who won Super Stock at the '75 NHRA Winternationals with his '73 Vette.

4. The stout 3-inch-diameter big yoke half-shaft (rear) offers far more strength than the stock 2½-inch shaft.

5. This illustration, supplied by Strange Engineering, shows the relative location of the I.R.S. components. Shown, too, is the use of the Dana center section and the Strange Engineering spool.

6. Gary Trautloff, manager of Henry's Machine Works, Inc., and Pepe Estrada examine the array of Vette I.R.S. high-strength components that they developed.

7. Note the increased diameter of the 31-spline outer axle, shown with the special Timken bearing, compared to the stock 17-spline outer axle.

8. This is the trick two-piece outer axle with removable flange.

has to be considered the ultimate, last-ditch approach to solving the breakage problem. There is simply nothing stronger currently available.

The Strange Engineering solution is to replace the entire Corvette drivetrain with Dana/Spicer components, with a modified "60" series case and the Strange lightweight spool. This is basically the same combination that has been so successful in the Pro Stock cars. For the half-shafts, Spicer 1480 series U-joints and components are used (rated at 3330 ft.-lbs. of torque!), while 3½-inch-o.d., .083-inch-wall tube replaces the stock 2½-inch-o.d., .090-inch-wall half-shafts. The stock stub axles are replaced with Hy-Tuff steel forgings using 1½-inch-diameter, 35-tooth splines, and either ½-inch or ⅝-inch wheel studs, depending upon the application. For ease of maintenance, the stock rear wheel bearings are replaced by sealed ball bearings; and for weight reduction, the cast-iron bearing carrier is replaced by 7075-T6 aluminum.

The rear cover of the Corvette center section is actually a structural member of the car's frame. In addition to mounting the case securely to the frame and absorbing the engine's torque reaction, it mounts the single transverse rear leaf spring. To keep costs to a minimum, Strange retains the stock cover, bolting it to the Dana case with a ⅝-inch-thick 7075-T6 aluminum adapter plate. This plate also adds strength to the cut-down Dana case (which is 12 pounds lighter than stock) and provides mounts for the lower control arms.

Corvette rear wheel geometry is unique in that the half-shafts double as upper control arms in locating the rear wheels. This is a very good design because there are no slip yokes. Under extreme power, slip yokes tend to lock up, limiting suspension movement. The Strange design uses the spool bearings to control axial half-shaft movement. In the unlikely event of a U-joint failure, the wheel no longer has a positive camber location (as is the case with a stock Corvette). Strange provides a mount for a "safety strut" to eliminate this problem. This strut will be available in the near future, and it will fit stock Corvette rears as well.

Nothing has been left to chance. Depending upon your application, you 'may not need all of these components, such as the bigger half-shafts and U-joints, but at least now you have the option. Naturally, these pieces aren't cheap, and a complete conversion will cost you several hundred dollars for the Henry's set-up and well over $1000 for the Strange Dana conversion. But then again, consider the cost of a broken ring and pinion, posi and case, or a lost wheel and a destroyed rear quarter panel. Consider, too, the possible cost of a lost victory. We think it's cheap insurance. Don't you?

Of course, depending upon the application, not all Corvettes need the all-out bulletproof components mentioned above. For example, street-driven versions seldom have big enough tires and ample traction to shatter the drivetrain pieces as drag strip machines do, but street machines can still benefit from less ambitious (and less expensive) modifications. Such a modification is the replacement of the stock Corvette 10-bolt posi unit with a special "Brute Strength" 10-bolt posi sold by Moroso Performance Sales, Inc., 737 Canal St., Bldg. 23, Stamford, CT 06902. Dubbed model No. 8305, the Moroso unit features a thicker, webbed flange on the beefy, reinforced case to reduce ring gear deflection. The pinion gears are forged from H-11 aircraft steel bar stock, making them approximately 50 percent stronger than the Chevrolet "service package" gears.

The posi unit contains 22 discs instead of the stock 18 discs, to improve the bias ratio by over 100 percent for more evenly distributed loads to the side gears, thus improving durability. The spring plates are extra-thick, and special chrome silicone springs yield over 800 pounds of preload on the disc packs for maximum traction. This particular unit is designed to accept gear ratios from 3.90:1 through 6.14:1. Stock-type inner axles are still used, and costly spacers or installation kits are not required. And what's more, the unit is guaranteed against breakage for a full year.

—C.J. Baker

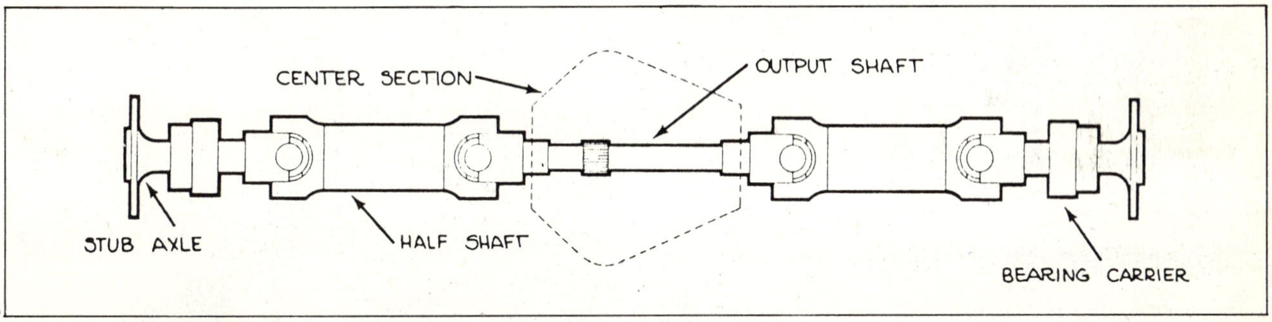

5

6

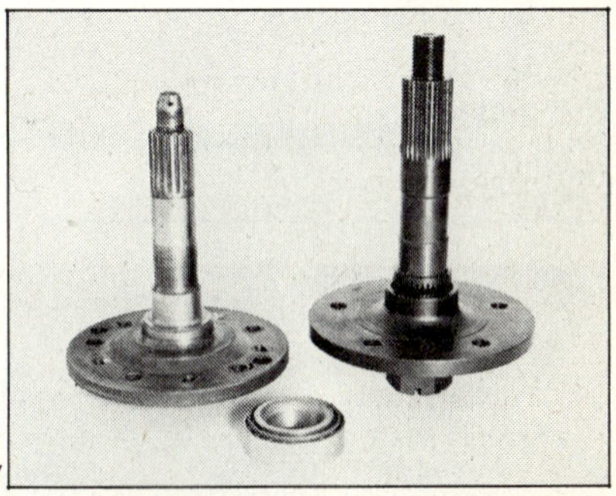

7

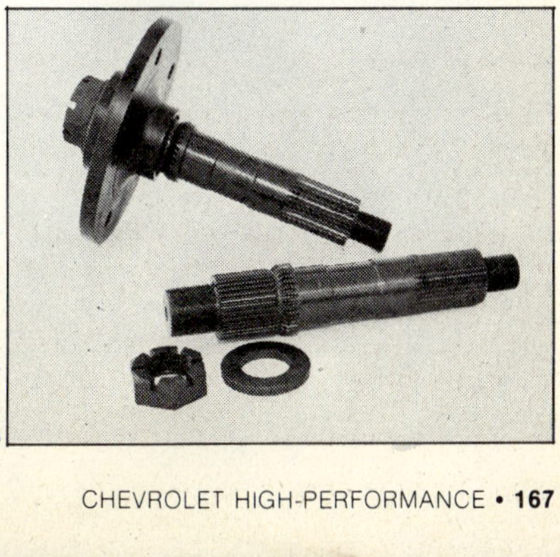

8

Cheverra: The Poor Man's Porsche

How to build a killer Camaro in 52 hours.

Company presidents are quick to promise miracles when it comes to the "easy," "no-hassle," "do-it-yourself" installation of their particular products, right? For example Herb Adams, the former Pontiac engineer and current president of Herb Adams VSE (Very Special Equipment), made the comment a while back that two guys could turn a bone-stock Firebird or Camaro into one of his company's mean-street Fire-Am/Cheverra road-racing rigs in 26 hours flat. Having spent two glorious weeks with a shop-built Fire-Am, we were skeptical that such a drastic change in handling and braking could be accomplished in such a short amount of time, and we told him so. Adams re-

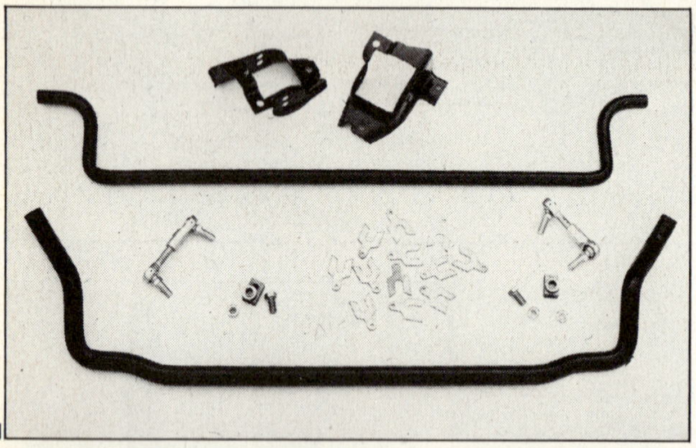
1

2

3

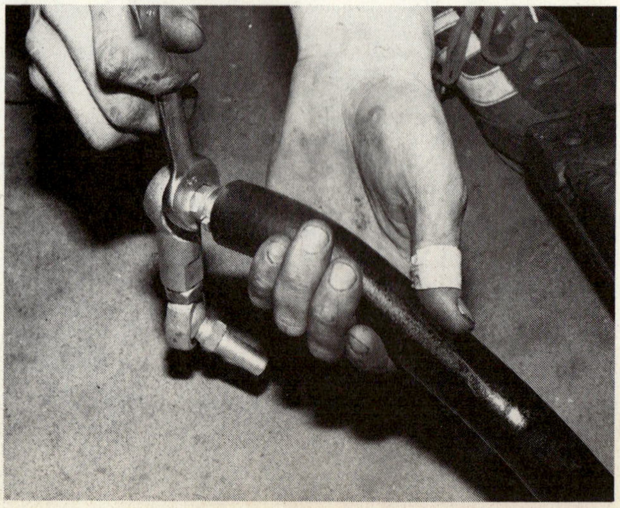
4

sponded with an open invitation to make him prove the point, and we were only too happy to oblige.

HOT ROD Magazine accepted an invitation to watch two guys bolt one of these babies together, without special equipment or modifications, in somebody's garage at home. In this case, the garage was well-equipped, to be sure, and the same two-man crew had some experience at turning Firebirds into VSE Fire-Ams. But this time the car was a Chevrolet Camaro, their very first non-Firebird, and this time the tools would all be your basic "backyard" items: hand wrenches, jackstands, a small drill motor, some blocks of wood, crowbars, hammers—nothing exotic. The only "outside" work required at all, according to Adams, would be the final alignment of the front wheels and suspension.

But he was wrong about that: Chief mechanic Bill Fisher went so far as to execute an unscheduled garage-floor alignment job (using nothing fancier than a carpenter's level, two 10-foot lengths of channel iron and one ball of string),

1. Recommended starting point in the step-by-step transition from Chevy to VSE Cheverra is the hardware shown here: replacement leaf-spring brackets and big anti-sway bars. VSE "wrench" Jeff Odell handled the back half of the project, a true bolt-on operation, in less than two hours.

2. After wedging apart the leaf spring and rear subframe with a common two-by-four, Jeff swapped in the VSE spring bracket. Simple device relocates the front eye of the Hochkiss-type spring ¾-inch higher than stock, effectively lowering the rear of the car. In reducing "roll understeer," modified GM bracket increases "anti-squat" of stock suspension. The result is improved traction for cornering and dead-stop launching.

3. Because Z28-model Camaros come equipped with a rear sway bar, mounting hardware and brackets were already in place for the larger (1-inch-diameter) VSE replacement. Proper mounting holes are predrilled in all Camaros, and VSE provides the rest.

4. The front sway bar kit does away with rubber-mounted links connecting a stock Camaro's stabilizer bar and control arms. Rubber allows for "lost motion" in front suspension, so VSE provides replacement links made from steel.

5. The kit shortens "swing-arm length" of the sway bar 10 percent and allows adjustment, so the Cheverra owner can fine-tune under- or oversteer.

6. Original Camaro frame holes must be drilled out slightly (to ½-inch) to accommodate the larger VSE mounting brackets, then fitted with replacement "rivnuts" (arrow). VSE and some Pontiac dealers own a trick rivnut-collapsing tool, as shown, but rivnuts can also be tack-welded in place. (No other welding is required in a Stage I Cheverra conversion.)

7. All the preceding pieces combine to produce a markedly different Camaro, especially when installed in conjunction with the recommended VSE battery kit and wheel/tire combination (see "boxes"). However, what really puts the Cheverra into road-racing territory is the VSE Front-End Lowering and Structure Kit, which consists primarily of four subframe-to-body spacers and a pair of subframe-to-firewall struts. Installation of the solid spacers effectively lowers the car around its chassis, while eliminating the deflection of the stock rubber body mounts. VSE's frame struts connect the upper control arm frame mounts to the firewall, where as much as ¾-inch of deflection occurs in a stock Camaro.

8. Production Camaros are built with rubber biscuits between the front subframe, body and radiator baffle.

9. VSE removes the mounting bolt and rubber mount, both of which are retained and reinstalled along with a solid spacer. Spacers of varying thickness are supplied with kits.

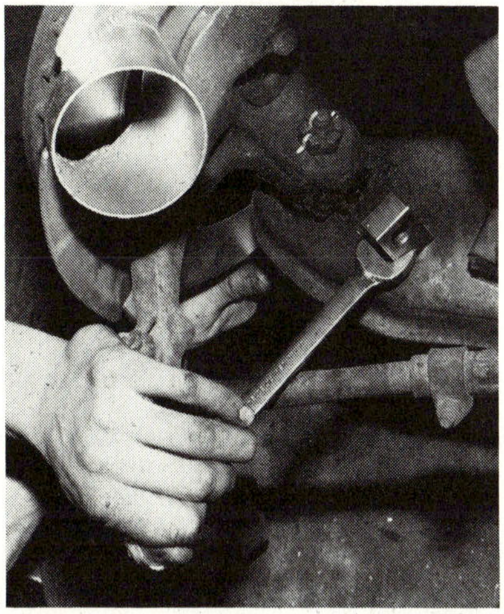

5

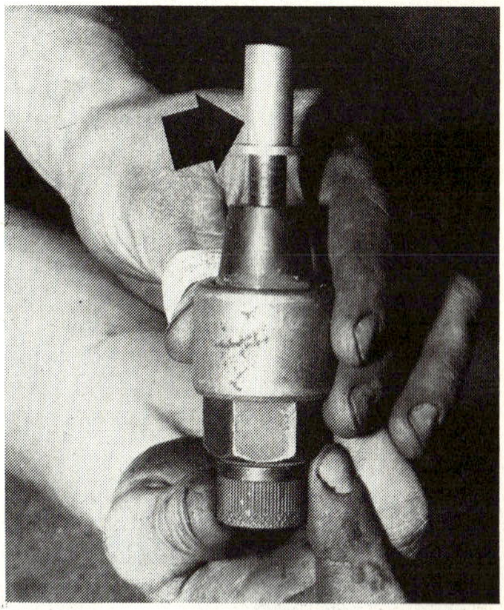

6

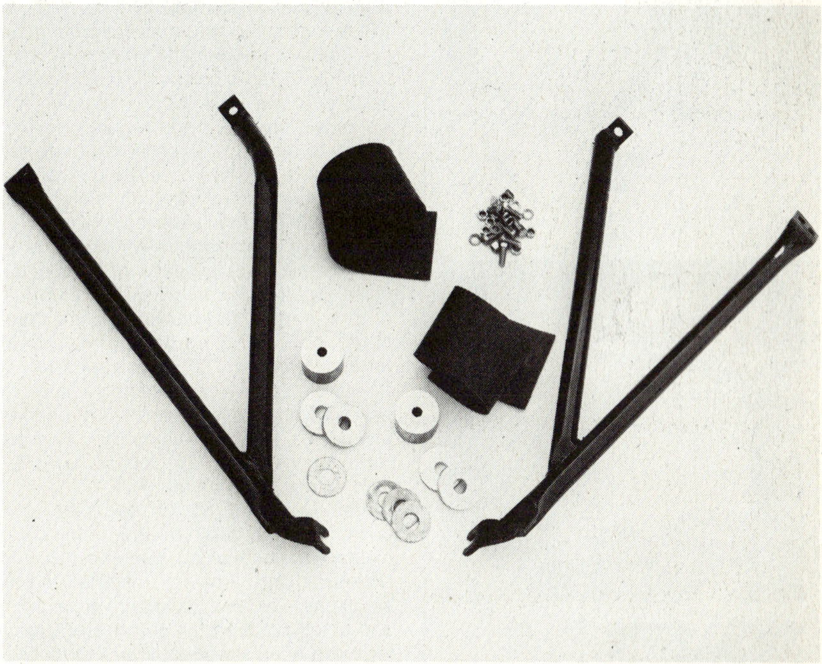

7

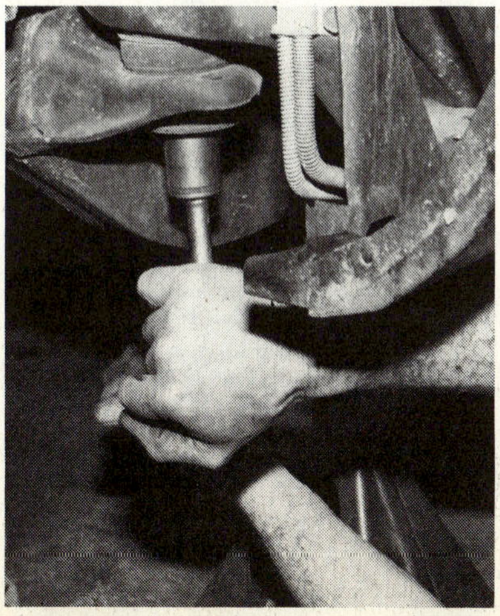

8

9

just to emphasize the do-it-yourself nature of the Fire-Am/Cheverra conversion process. Right after that, he handed me back the keys to a 1978 Camaro Z28 that I had personally driven into that garage exactly 39 man-hours earlier (two men working 19.5 hours each), quicker than advertised.

We should note that the front coil springs had both been removed, shortened (by ¾-coil), and reinstalled by the time I arrived to roadtest the stock Z-car, so the entire operation came a little closer to the company's own 52-hour conversion estimate for a Stage 1 Cheverra. (Two more preinstalled VSE items were the chrome rollbar and trick steering wheel, but neither is considered a Stage 1 requirement.) But a Cheverra doesn't have to be built, or paid for, all at once; Herb Adams designed the whole program in stages, with due consideration for everyday Camaro owners who want to build their Cheverras in stages, weekend by weekend, and keep driving their cars in the meantime.

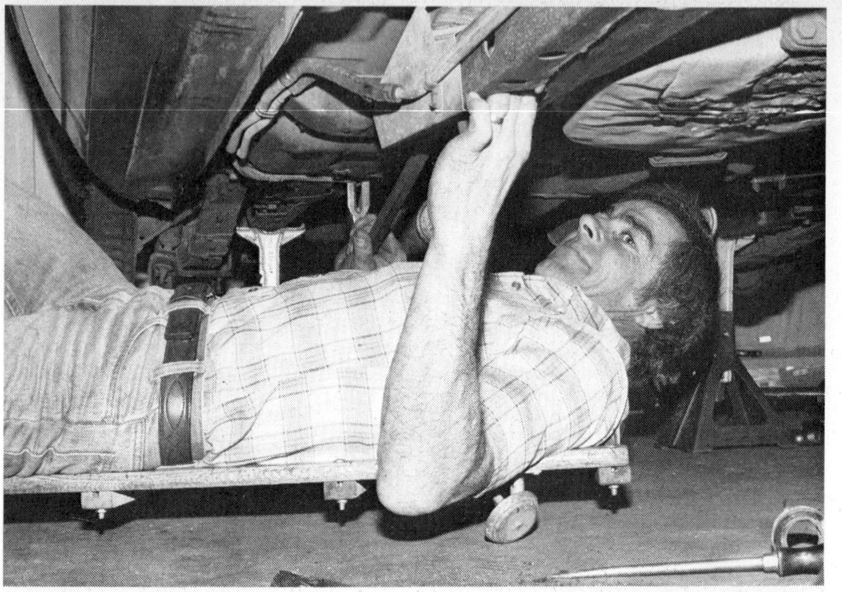

13

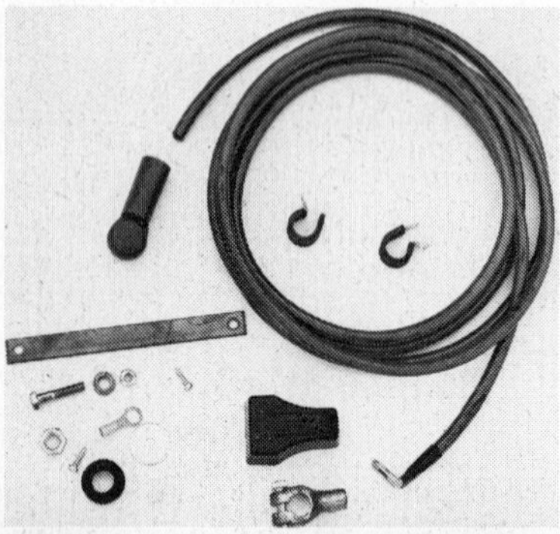

1

11

12

Anybody in a particular hurry, on the other hand, to get out and start mixing it up with the Porsches can save time by turning the whole job over to one of VSE's four authorized installation centers (in California, Maryland, Michigan and New York), and be back on the road in a couple of days. Another deal in the works will possibly bring running, rolling, complete State 1 Cheverras right into the showrooms of Chevy dealers by the end of the summer. (Details on either plan are available upon request.)

But most of these cars will continue to fall together over a series of weekends, going together in the driveways and garages of people dissatisfied with the handling and/or braking characteristics of their late-model ('70½-up) Camaros. A full-color "cutaway" Cheverra poster, printed complete with parts descriptions and prices, is now available for $2 from Herb Adams VSE (125 Ocean View, Pacific Grove, CA 93950, 408 372-2767). Better yet, send $5 for the Cheverra poster *plus* a copy of the company's master catalog, which features detailed Cheverra and Fire-Am installation instructions, valuable handling information and a full rundown on the ever-expanding Herb Adams product line. Meanwhile, don't try to stay too close to another Camaro with "Cheverra" decals—and that goes for you Porsche drivers, too!

—Dave Wallace

10., 11., 12. Any Camaro/Firebird can benefit from VSE's battery relocation kit, which makes use of the stock GM mounting tray. This simple operation decreases front-end weight on the nose-heavy GM F-body cars by one full percent, which Herb Adams equates to 10 inches of engine setback! Benefits include reduced understeer, improved braking and better traction. (Spillfree, high-output batteries like the Z-car's Delco Freedom unit are recommended for trunk installations.)

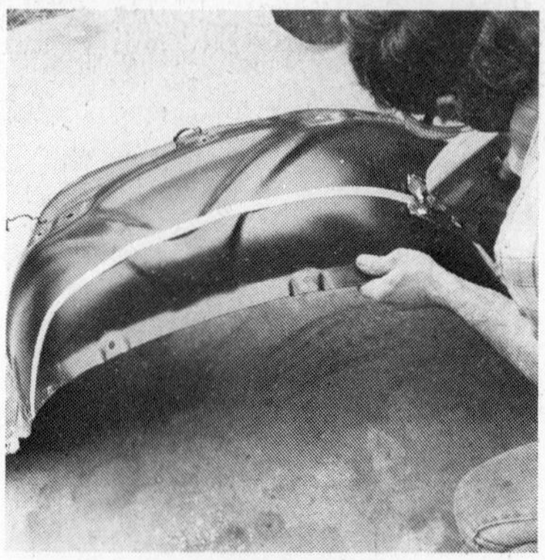

14

15

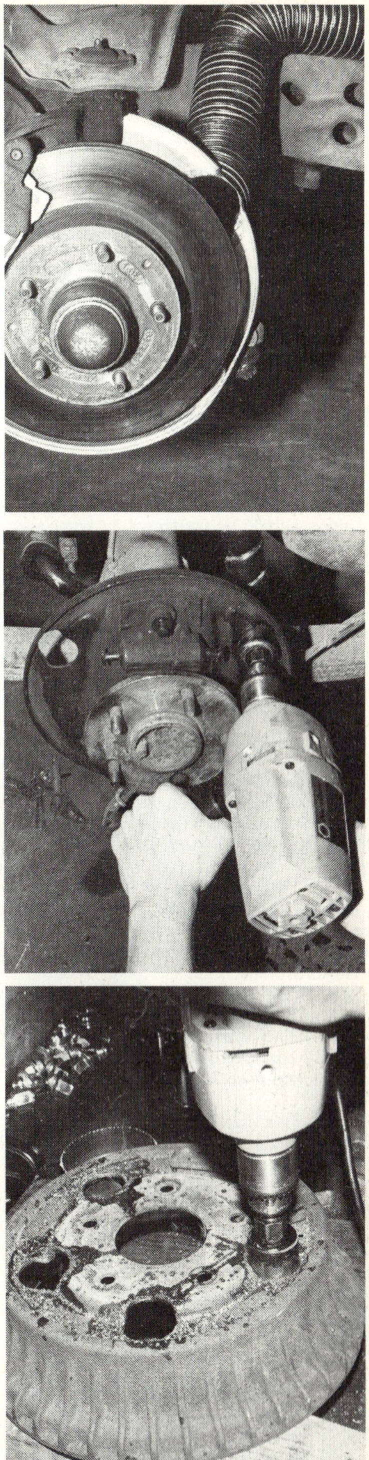

16

13. At the rear of the subframe is a bolt and rubber body mount. After removal of the Grade 5 mounting bolt, VSE's Bill Fisher pries the frame and body apart with a crowbar, then fishes out the rubber mount. A solid VSE spacer is installed in its place. Once again the stock rubber biscuit is retained and reinstalled with the same bolt, this time beneath the subframe. A rubber strip is provided to isolate the body from subframe.

14. VSE sells a special 15x8-inch, zero-offset steel wheel and recommends HR60 Goodyear rubber, but first you've got to make room. Catalog provides guidelines for trimming the front wheel tubs, as demonstrated here by Bill Fisher, and snipping the outer fender lip. (Tubs receive three cuts each for frame clearance, tire clearance and brake ducts.) Late-model Camaros also require some heavy blows to flatten out this section of the rear fenderwell, but no cutting; the fender lip can be folded over with channel locks for outside clearance.

15. The new, closer relationship between body and frame is reflected in the rubber fan shroud, which must be trimmed as shown to clear the fan blades. Installation of the passenger-side strut requires relocation of the coolant recovery tank to the area formerly occupied by the battery.

16. VSE Brake Cooling Kit brings ducted backing plates for the front discs, a section of flexible ducting and instructions for properly ventilating stock rear drum webs and backing plates, as demonstrated by Jeff Odell. Full metallic linings are also available for use with power-assisted brake systems.

Getting It Handled

Are snaky S-turns or freeway transition sweepers "turning" your Vette off? Well, relax. With a little help from Dick Guldstrand, you can get it handled.

1. In order for all the big pieces to work at full efficiency, the basic suspension components must be in good shape. All bushings should be inspected for wear and replaced as neccessary. Lower control arm ball joints are especially prone to wear, usually needing replacement by 30,000 miles. With 70,000 miles on my chassis, they had long gone south. Guldstrand replaced them with his heavy-duty ball joints, which can be adjusted to compensate for wear. Upper ball joints are less subject to strain and usually last upward of 100,000 miles.

2. Things were easy with no engine in the chassis; if it were still there, we might have had to remove front upper control arm studs because engine interference doesn't allow arm to be pulled off stud. They must then be spot-welded back in.

3. The idler arm is usually the first steering part to wear and cause a loss in steering response. It should be replaced with this heavy-duty piece.

4. Solid eccentric upper control arm bushings were installed to eliminate unwanted geometry changes. Alignments are made easier, since fewer shims are required.

5. Front wheel bearings should be inspected and repacked regularly; replace if brinelled or discolored. Valvoline Val-Lith No. 2 EP Wheel Bearing Grease (part No. 608) with lithium is recommended for repacking; Dick has found it outlasts any other competitive brand. Caution: The wheel bearings must be super-clean to use this grease, because it simply will not mix. To correctly adjust the wheel bearings, turn nut in until it bottoms; then back out to first notch; readjust after alignment.

In the two years that my '69 Corvette languished engineless, I had ample opportunity to save pennies, and—more important—to plan the ultimate street machine: a dual turbocharged big-block Corvette (bets are already being taken concerning my longevity after its completion).

Construction was seriously begun when the body and chassis were trailered down to Corvette expert Dick Guldstrand (11924 W. Jefferson Blvd., Culver City, CA 90230) for "a little" suspension work. Without the engine interfering with the front suspension, I thought the car would be much easier to work on. The only drawback was that the final ride height couldn't be set until the engine was installed.

Dick has developed a wide variety of suspension packages which range from a relatively soft street-type touring setup to all-out, ultra-stiff racing kits. In his catalog, a 1-10 scale has been devised

1

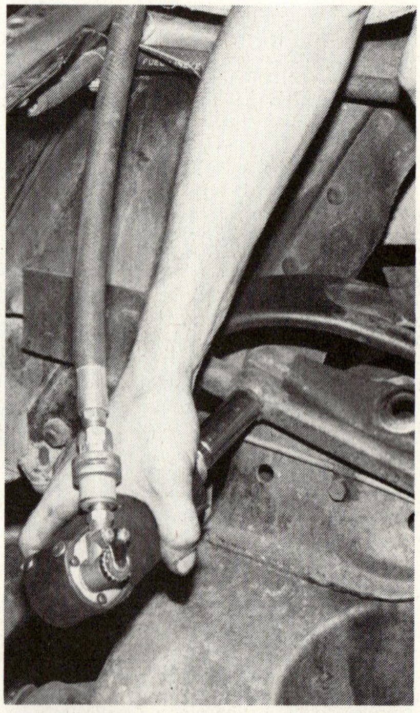

2

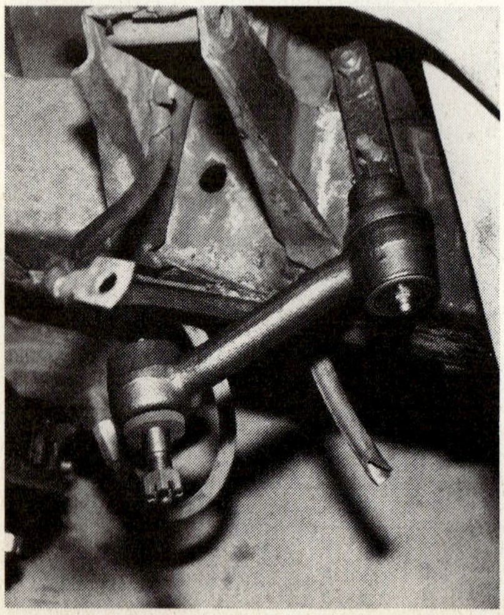

3

4

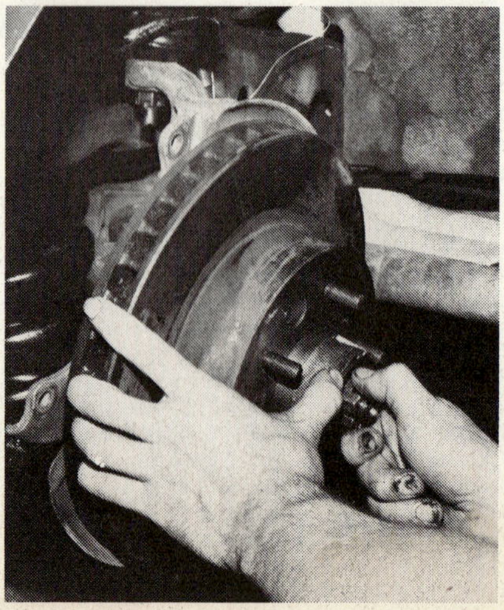

5

to rate the various components, "1" being the most streetable and "10" indicating an all-out racing component. Most parts range somewhere in between. The main differences are the amount of metal substituted for rubber components in the suspension, and shock and spring rates. By the time the car has gone all the way to a full-race configuration, it is totally unsuited for the street; in fact, under highway conditions it probably would be inferior to a less-stiff slalom setup, since the street is not as smooth as a race course but is full of undulations, dips and bumps, in which a suspension having more compliance or "give" would stick to the "hollows" better than an unyielding race setup.

Nevertheless, you must face the fact that the better you want the car to handle, the harsher the ride will be, at least to the point of "skitter"; that is, the point where the suspension is so stiff that its oscillation becomes completely out of sync with the road surface. Obviously, this varies from car to car, and between different types of road surface, so compromises have to be made. The factory makes too many compromises, since smoothness and comfort are major factors in their criteria, in order to please the average person. If you don't want to compromise comfort and luxury, don't be changing springs!

Speaking of springs, Dick prefers to tune the suspension with them, rather then emulating the usual factory practice of relying on ever-larger sway bars. While this leads to some very high spring rates, Dick feels that sway bars put excessive load on the outside-loaded wheel and tire, and that their effect is not steady-state like springs, but rather all or nothing—obviously not a desirable characteristic for smooth transition into and out of corners!

Despite all these sophisticated suspension components, how well the car will ultimately handle comes down to the size and type of wheel and tire chosen (having the car's center of gravity as low as possible is also vitally important). No street tire can match a race tire for ultimate adhesion characteristics; but race

6. Heavy-duty steering rods replace the stock stamped rods. They screw up to the existing tie rod ends and eliminate flex caused by the heavy loads high-performance tires generate. Ground wrench flats are provided for easy adjustment. The steering rods should be used with a heavy-duty relay rod; some Vettes, such as '60s big-block-equipped ones, already have the good relay rod, so I was in luck.

7. Anyone can have quick steering by merely moving the tie rod attachment to the front hole in the steering knuckle. Just don't plan to do any paralled parking, since you do lose some leverage with this modification.

8. Dick believes in stiff springs. His big-block slalom spring (left) is rated at nearly 1000 in.-lbs., compared to 550 pounds for the factory heavy-duty F41 spring and a wimpy 280 pounds for the standard production spring (right).

9. With the heavy front and rear springs, the stock front sway bar is adequate; the rear one can be removed entirely. However, it pays to install these heim-jointed rollbar links in place of the stockers, to reduce body roll and improve cornering response even further.

10. At present, Dick feels Bilstein gas-filled shock absorbers—which adjust themselves automatically to compensate for changing conditions—are the best for most all-around use. Three different settings are available for Corvettes: grand touring, salom and racing. The last are really too harsh for the street, and even many racing applications, so the slalom units were chosen (two 319945 front and two 328032 rear). Buy them from Dick or the Bilstein Corp. of America (11760 Sorrento Valley Rd., Ste. P, San Diego, CA 92121).

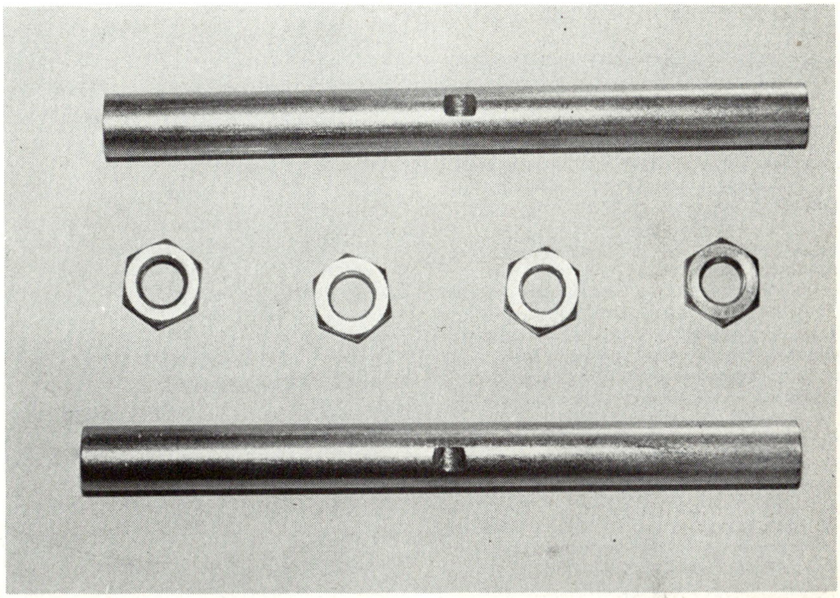
6

7

8

9

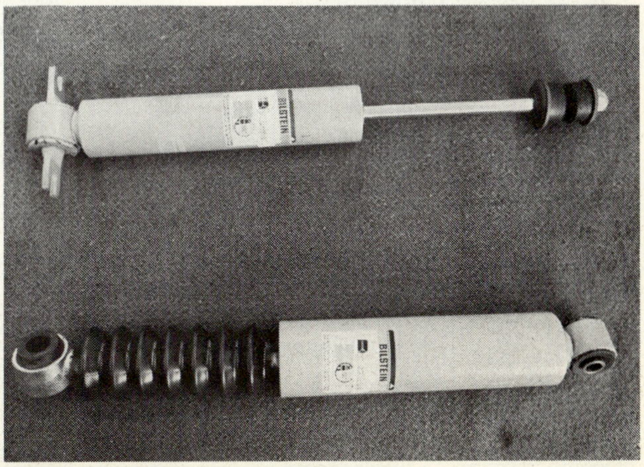

10

tires wear out quickly, have thin sidewalls and poor resistance to road hazards, and are illegal on the street anyway. So again, compromises have to be made. The next best solution is a good, wide, high-performance street radial, like the Goodrich Radial T/A. Put the biggest, widest tires on the biggest, widest wheels you can fit under the car (preferably with flared fenders), with as little negative offset as possible, to prolong wheel bearing life. And if you can afford to buy tires frequently, it helps immensely to shave half the tread depth off to reduce "squirm."

With the correct wheels and tires, springs and shocks, bushings and links, and few mods to the rear axle and brakes as detailed herein, you're ready for the fiercest roads. *—Marlan Davis*

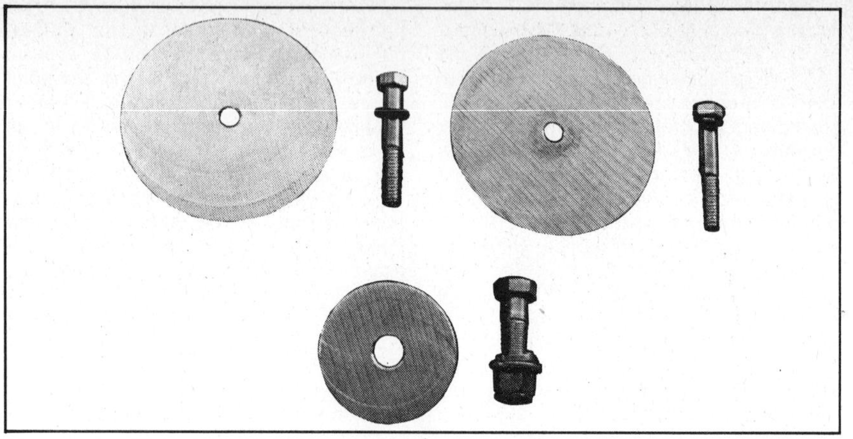

11

11. One of the most important modifications is this rear suspension locating kit. The aluminum discs eliminate rear crossmember flex and prevent separation of the differential from the frame.

12. Another weak link is the camber adjusters. They often get bent, especially when a mechanic takes the easy way out and bends the rod to adjust camber instead of using the eccentric adjuster, which is harder and more time-consuming. This is what probably happened to the stock piece at top. Dick's rod eliminates the lower bushing and the eccentric adjuster that often slips; instead, adjustment is provided by left- and right-hand-threaded heim joints. They're also stronger and better able to handle wide tires.

13. There is no rear slalom spring offered *per se* for the Vette, other than a single-leaf add-on for the stock spring. Instead, a five-leaf, 350-pound/inch-rate small-block full-race spring (bottom) is used for this application. 190-pound/inch stock spring (not shown) has nine leaves, while 305-pound F41 (top) has seven. Guldstrand gets by with fewer leaves because he increases the thickness of each leaf. Fewer leaves, in combination with special adjustable long spring bolts, are required for getting the car "down in the weeds."

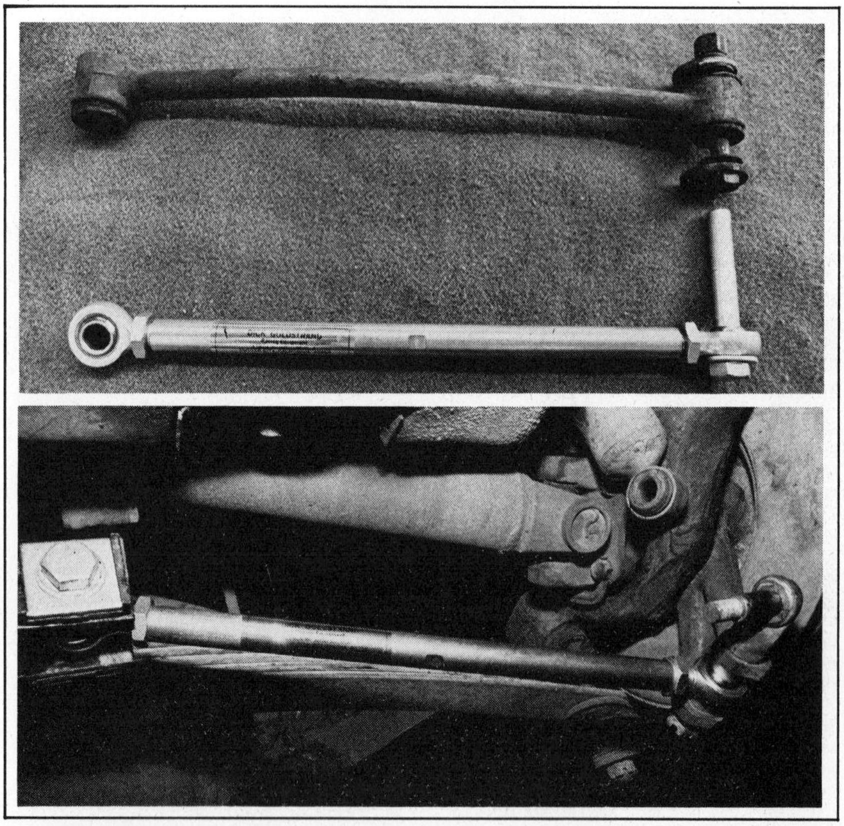

12

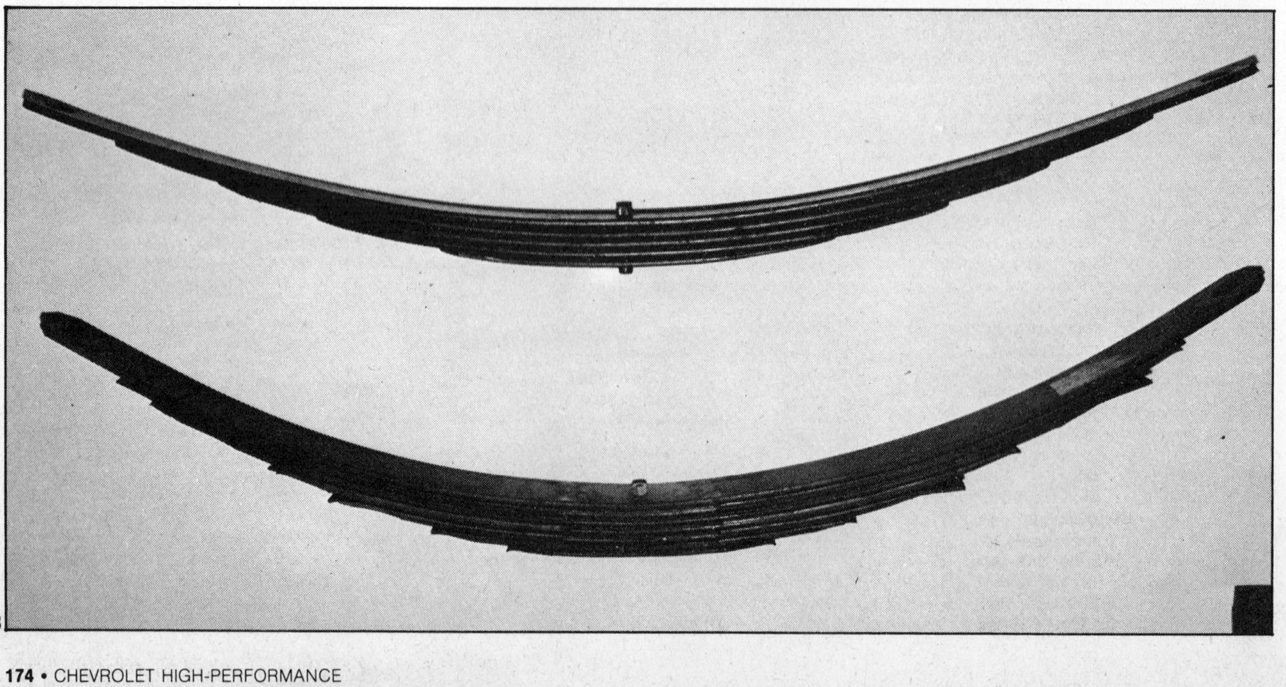

13

14

15

16

17

14. About the worst aspect of the stock independent rear suspension is the factory press-on method of wheel bearing retention. Since the Corvette went to disc brakes, there's no longer any reason for it, because disc spindle failure will not result in the wheel falling off, as would have happened with drums if not for the press-on bearing. You need plenty of expertise and a hydraulic press to remove them for lubrication, so many people don't bother; or sometimes, if removed, they are reinstalled incorrectly. Dick converts them to a slip-on configuration easily removed for lubrication. After they are removed, the stub axle is turned on a lathe (A), so bearing can be slipped on by hand. The bearings are then shimmed to adjust freeplay, which should be .001-.003-inch. After sizing them on the car with the seal in place (B), remove and repack, then reinstall, tightening spindle nut to 150 ft.-lbs. (C).

15. Stock 7/16-20x1¼ wheel studs were replaced by ½-20x3-inch Moroso No. 4618 Mopar/Ford stud kit with .615-inch-diameter knurl. These heavy-duty studs are a must to keep wheel on the hub during hard cornering.

16. I suggested using L-88 pads, but Dick pointed out that they're useless until they get hot, which could present problems on the street. Instead he has developed an improved lining material which increases fade resistance while avoiding hard pedal feel.

17. Wheel and tire choice is important to achieve optimum handling. Back-spacing is critical to avoid excessive negative offset, which kills wheel bearings. No stock aftermarket wheel has correct back-spacing, so you must go to a modular wheel, like the three-piece bolt-together lightweight high-strength mags made by Jongbloed Racing Enterprises (1521 E. McFadden Ave., Ste. G, Santa Ana, CA 92705). His manufacturing process enables virtually any possible back-spacing to be made. For a Vette, 3½-4 inches, measured from the hub mounting face, is preferred. G50-15 Goodrich Radial T/As will be used up front on 15x9-inch wheels, with L50-15s on 15x10 wheels out back. With this suspension setup and these wheels and tires recommended alignment should be: 0-¼° negative camber in front, with 1-1½° positive caster without power steering, or 2-3° positive caster with power steering. Out back, camber should range from ½° negative for stock tires to ¼° negative for very wide tires. Toe-in should be measured separately for each side; 1/16-inch or less works well in front, and aim for 3/32 out back. For maximum performance, inflate the front tires to 36 psi and the rears to 33. The fenders will have to be flared for this combination, so next it's off to the bodyshop.

The Naked Launch

Suspension and chassis ideas for bracket cars and other racers.

The naked launch is that frozen moment when everyone can see a race car that is about to leave the starting line. It is that fraction of a second when the car either gets the needed bite or it doesn't. It is that most agonizing of times which certainly means the difference between winning and losing.

We've all seen the car that goes up but not too far forward or spins the tires or plants the tires and *then* moves or comes off the line in a heavy body roll with front tires splayed. Even the rankest amateur could tell you the difference between a car that leaves and a car that doesn't. Edward Quay knows, too. He's seen it for 15 years, and he's learned about it by experience and a great deal of time spent with one of Bill Jenkins' early "groups," S&W Race Cars and Scott Shafiroff. He feels strongly enough about it to make his livelihood building race car chassis, and although he can construct a better way for any brand race car, the Camaro is his forte.

He holds two factors to be paramount in the correct preparation of race car suspension. The first is to make it possible, through proper construction, for the rear wheels to have complete and total contact with the race track. To ensure this, the spring rates should be as soft as possible. A heavily sprung car will have a tendency to bounce, and with each change in the car's attitude (gear-shifting, etc.), the weight tends to transfer from front to rear and back again. This means loss of time and, most likely, loss of the race.

Quay reiterates this when discussing shock absorber valving. He wants a shock with bump (compression) capabilities that are three times as soft as the jounce (bounce or rebound). What, then, would be the criteria for a properly built race car? Quay concurs with others in his field that the chassis should be built as rigid as possible. Hence, all forms of frame connectors, full frames in unibody cars and full roll cages, which can all be used as stress members and stiffeners.

In the rear, both springs, whether they be leaf or coil, should have an even rate. Quay relates to this as part of an ideal condition in which the front tires could be carried one inch off the ground all the way through low gear. That way, all the weight will be "transferred" to the rear wheels and the car will not realize any lost time or motion—it will be traveling only in a *forward* direction. To put it another way, the car should be neutral, without body lift or roll, with the suspension and tires doing all the work.

Wheelie bars should not be expected to do this job. According to Quay, they can be viewed as a safety valve and not as a crutch to make the car work. He says that all rear axle housings will rotate on initial shock of the tires and housing-mounted bars will lift the rear wheels upward from the pavement enough to lose a significant amount of traction throughout the entire gear range. An ideal situation would find the car with spring-loaded wheelie bars for the first 15 feet and then taken over by a body-mounted bar for the rest of the trip.

Quay refers to a ladder bar or four-link traction setup in the above instance. It would be detrimental with a Lakewood type, because that device tends to negate all oscillation from the rear half of the leaf spring and make the front of the car rise and the rear tires plant very hard, which provides the ideal wheelstand situation.

In a ladder bar car, there is no way to kill *all* reaction, and the first thing the car will try to do is raise its nose. The torque reaction from the axle housing is taken by the ladder bars, thus transferring the force from one part of the car to another. In their attempt to push the body and frame up, the ladder bars are actually using a point on the ground (where the tires rest) from which to push, so the tires are pushed downward, planting them against the ground.

The amount of body lift realized by a ladder-bar car is relative. It is not as severe as that of the Lakewood type, but it can be improved by installing yet another traction device. Even though these two devices we've talked about work well, the four-link works even better. Quay favors it, as long as it is mounted properly. Positioning the upper and lower links through the vertical centerline of the axle is not fully correct. The two lower links may be attached this way, but the upper bars should be positioned at a point *forward* of the axle centerline. This way, rotation of the housing will be minimal and the car will want to go ahead instead of upward. Again, it is important to maintain even weight distri-

bution across the rear of the car, making certain the driver or his equivalent weight is in the driver's seat at the time.

At the rear axle, be it Dana or any other, Quay stresses the advantage to be gained by making the unsprung mass ever lighter. The wider the axle housing, the heavier it will be. Quay is fond of fitting his Camaros with 38.5-inch Danas and then using a 14x32 tire on a wheel with 3 inches of positive offset. But you don't need to be as extreme; instead, try to keep the total weight of the axle, wheels and tires at 400 pounds, or less. You can also use the largest possible tire (in width and diameter) that will go under the car.

Detailing the suspension will be basically the same in any vehicle, as long as the car is equipped with unequal-length A-arms and coil springs. Ed suggests that the front be lowered as much as possible, with specially treated springs or relocated upper control arms. Spring rates will be different in the front, because the left side will need a V8 rate coil to compensate for driver weight, while the right can use a 6-cylinder version from the same type automobile. Shock absorbers should be valved for a great deal of jounce and very little bump (this means about a three-to-one ratio).

Ed advocates running the suspension as close to the bump stop as possible, but NHRA allows only a 1-inch minimum here, so beware of the technicalities. If you have the coil springs cut, then cut the suspension snubbers a commensurate amount to retain equal suspension travel without interference. Lowering the body will cut down on the amount of air under the car at speed. Even so, at full-ride the car's height will be an inch higher than it is at full-rest.

Front-end geometry boils down to two specifics on a race car. One is the amount of toe-in, which ideally should be zero throughout the suspension's travel. But because the most rigid frame will bend under acceleration, it is best to preload the suspension by setting static toe-in at 1/16- to 1/8-inch. The best way to accomplish this is by lengthening or shortening the tie rod's threaded ends. But most stock suspensions can't retain a usable toe-in pattern throughout the suspension's travel. Bending the steering arms and shortening the suspension members by trial and error is the only way to perfect such a pattern. It's time-consuming but necessary if you want to be right on the money.

Camber should also be at zero on any drag race vehicle. This is necessary because, if the tire is not straight up and down in its movement, drag will result and will scrub off precious tenths of a second. Cars such as the Camaro use shims between the A-arm and frame to change the caster as well as camber. They are usually available in thicknesses from 1/32- to 1/8-inch. These shims are open at one end and may remove themselves if the frame flexes too much. Supplant them with washers of the correct width after completing the final wheel alignment.

Quay further advises that front-end travel is to be determined by track conditions. If the track is "slippery," then the travel limiters should be loosened. Leave the shocks alone. On a good track, tighten down on the adjustable shock before fretting about the limiters. This talk of track surface prompted another piece of insight: If you're running a track overpopulated with Vegas and/or Corvettes, remember that these cars have a wheelbase that is 10 inches shorter than any Camaro. This means you could wind up staging on the shiny spot and not on the succulent rubber build-up. Just remember to use everything to advantage, and don't spin out.

—Ro McGonegal

1. Eddie Quay lives two blocks from his shop in Pottstown, Pennsylvania. Suffice it to say he spends too much time there, but someone who has thrashed for Jenkins can get downright used to it. His specialty is building chassis for Modified and heavy bracket cars.

2. Quay favors the four-link traction idea and recommends its use whenever money and time allow. Most important is the positioning of the upper links. They should be mounted ahead of the axle's vertical centerline so that rotation of the housing will be at a minimum.

3. The ladder bar is an intermediate step from Lakewood-type to four-link. It actually uses a point on the ground from which to push. It can be used with leaf or coil rear suspension.

1

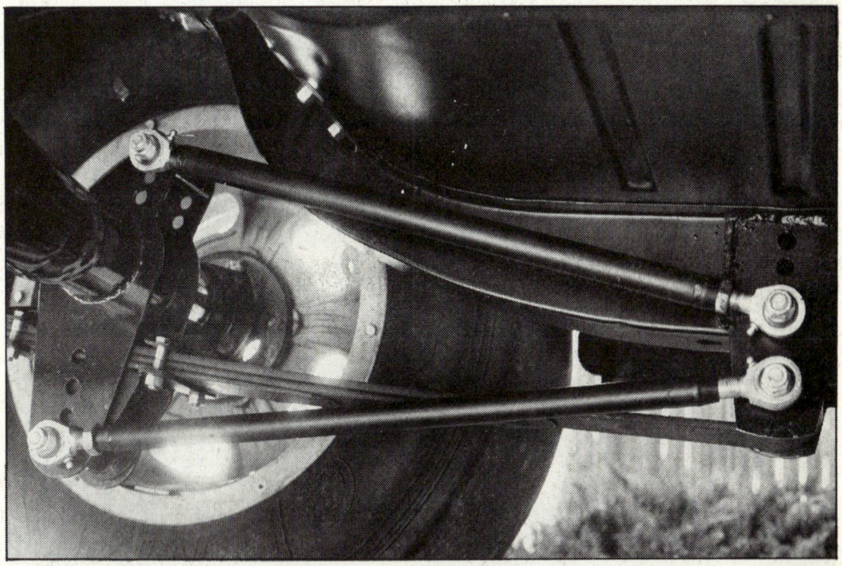
2

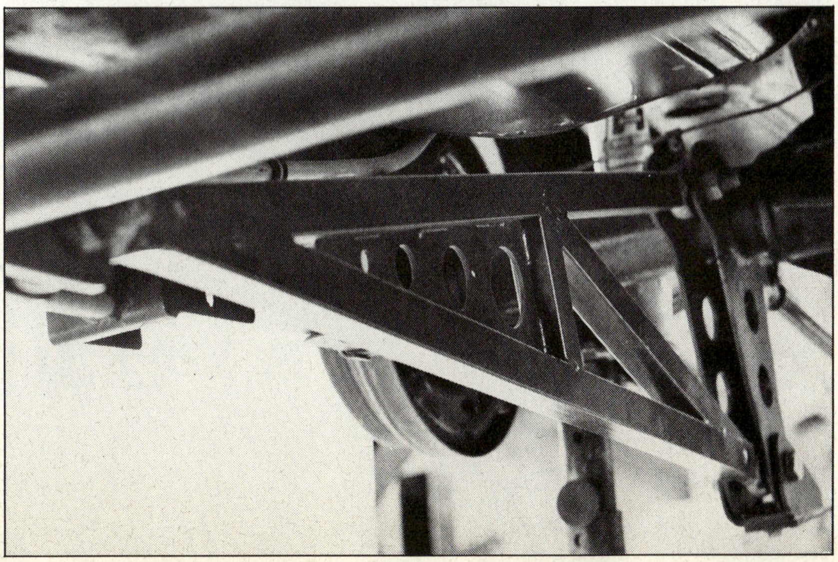
3

V8 Vega Swap Kits

V8 Vega Kits, swapdom's most popular conversion aids, make this hands-plant a virtual shoo-in.

Studellacs, V8s and Fordmobiles . . . engine swapping is as old as hot rodding itself. While the turbochargerists may disagree, it remains a well-known fact that replacing a poor four, or a sick six, with a healthy injection of V8 juice is the cheapest way to increase your car's get up and go.

One of the earliest conversions was putting a flathead Ford V8 into the spot originally occupied by a four-cylinder "banger." The famed '53 Studebakers, fitted with stock OHV Caddy V8s, became instant winners, both on and off the drag strip. Even the old faithful flatmotor gave way to the Rocket 88 during times past. And with today's interest in V8 Vegas, it's still happening. According to Andy Herbert (of Herbert Automotive), a second-generation engine swapper, "The latest swapping favorite is the V8 Vega." We agree. No less than seven separate companies either sell or manufacture kits for this, the most popular of conversions.

While the kits are simple, the fit is tight. Most swappers recommend that you attach the exhaust system (or custom headers) to the engine before you lower it into the engine compartment. But that's after you remove the four-banger, steam the engine bay and attach the front and rear motor mounts. Some conversion kits feature top motor mounts that attach to the front of the block. Others pick up the attachment points on the

Don Hardy Race Cars, Inc.
202 W. Missouri St.
Floydada, TX 79235,
(806) 983-3774

Herbert Automotive
8792 Fruitridge Rd.
Sacramento, CA 95826,
(916) 381-0667

Hooker Headers
1032 W. Brooks St.
Ontario, CA 91762,
(714) 983-5871

Hurst Performance, Inc.
50 West Street Rd.
Warminster, PA 18974,
(215) 672-5000

Doug Thorley Headers
7403 Telegraph Rd.
Los Angeles, CA 90040,
(213) 722-3774

Motion Performance Parts
598 Sunrise Hwy.
Baldwin, NY 11510,
(516) 223-3172

Trans-Dapt of California
Box 4157
Compton, CA 90024,
(213) 432-5461

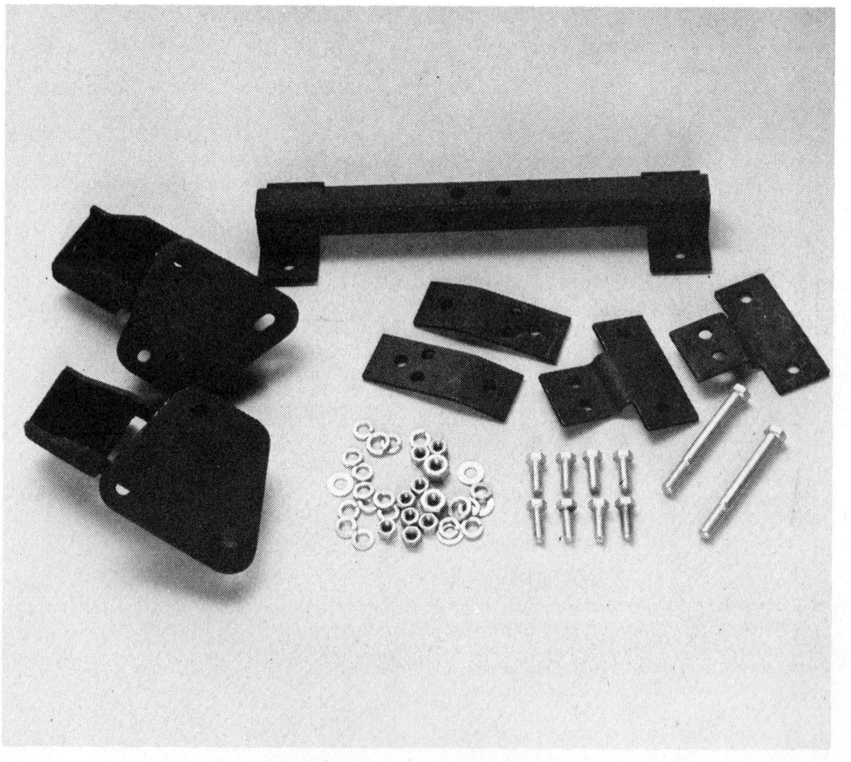

1

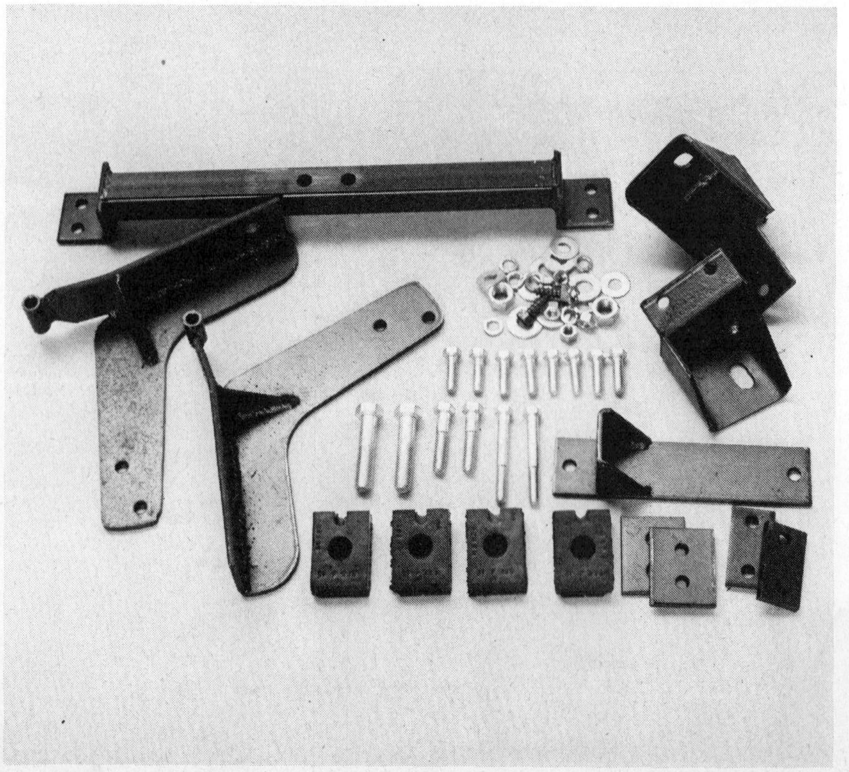

2

3

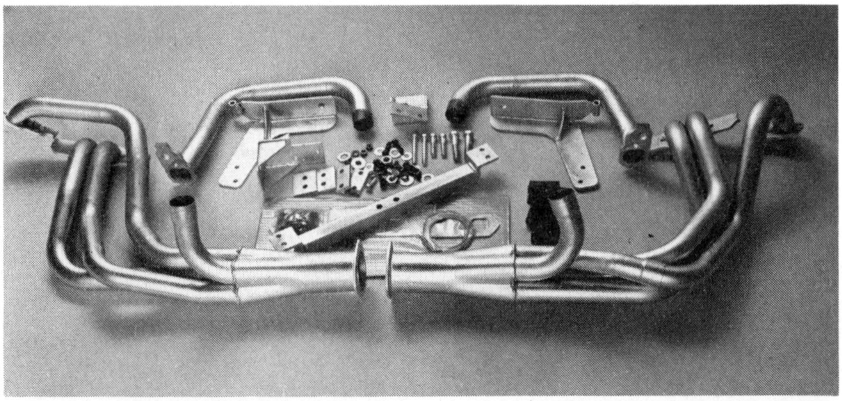
4

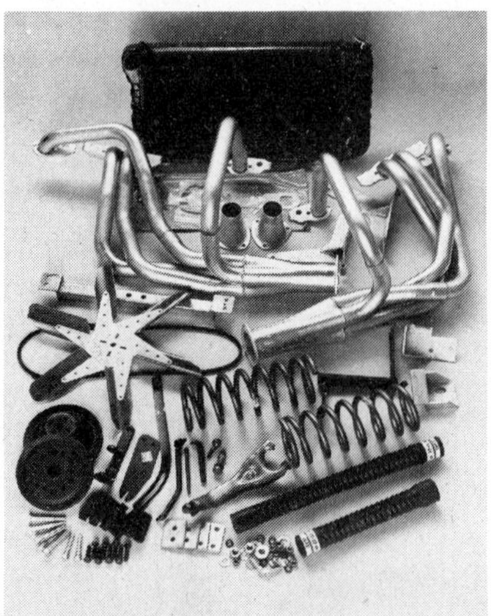
5

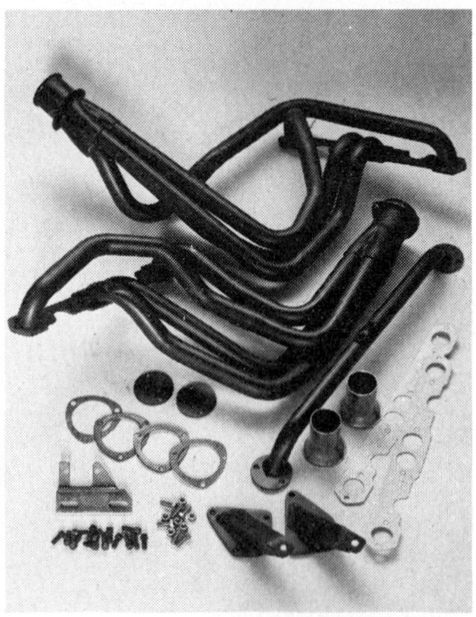
6

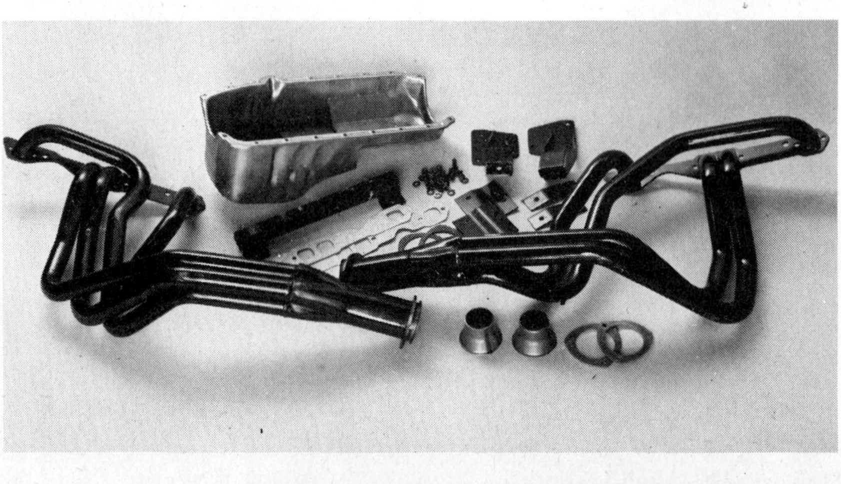
7

side of the block. However, all will be using pre-'68 units, because today's "safety" mounts won't work. As we said, it's a tight fit, so chances are you'll have to dimple the trans hump in the floor, grind a portion off the front of the starter housing to clear the frame, or take a half-inch off the right side of the bell-housing to get the necessary clearance. The '69 or later Chevy oil pan will have to be notched to clear the tie rod. However, there are alternatives. Herbert Automotive, Motion and Milodon offer pans specifically fabricated for the V8 Vega swap.

Now you've got to realize that when you double the inches, you have to double the amount of heat that must be thrown off by the Vega's cooling system. Fortunately a recored (like four inches thick), heavy-duty Vega radiator with switched outlets, a shorter '68 water pump, the necessary top and bottom shrouding, and 1½x15½- and 13½-inch flexible water hose will keep the water temp from climbing in heavy traffic. Hookups include a '70-'71 Chevelle throttle cable and new alternator bracket with '64-'67 V-belts. In addition, you'll need to reverse the heater core outlets to clear the right-hand cylinder head by cutting two new holes in the plastic core. If you are Balloon Foot Bodine, you can retain the stock Vega rearend with 2.93 gears, but you'll need to install new springs from a '63 Corvair van. These units are the same heights as the Vegas, but are wound with heavier wire.

As you might have guessed, each kit manufacturer has his own pet way of making the swap that much easier. But like Rome, a swap isn't made in a day. Take your time, read their installation directions again and again, and then proceed with caution. Because if the swap is successful, there will be nothing "vega" about this hair-raising ride.

—Gray Baskerville

1. HURST: This six-piece mounting kit uses mostly GM parts and bolts into place without the need for welding.

2. TRANS-DAPT: Their weld-in kit features front motor mounts that attach to the front of the block and to the water pump.

3. HARDY RACE CARS, INC.: Super Twister Vega kit is another weld-in that includes many of the support pieces needed for the swap.

4. HOOKER HEADERS: Not only does Hooker offer a complete mounting kit, they have headers specifically designed for the conversion.

5. MOTION: Their Phase III Super Vega package is probably the most complete V8 Vega swap kit currently on the market.

6. DOUG THORLEY HEADERS: Like the Hooker kit, Thorley's nicely detailed package includes specifically designed headers for the conversion.

7. HERBERT AUTOMOTIVE: Along with mounts and headers, Herbert offers a modified pan and radiator. A V6 conversion is also available.

Get Down and Boogie

Parts and tips make a Vette handle like a vacuum cleaner.

The Corvette has enjoyed a reputation as one of America's better handling cars, but even so, it can be vastly improved upon. Because Chevrolet has no way of knowing who the eventual owner of each Corvette will be, production suspensions are an acceptable compromise between ultimate cornering and a smooth boulevard ride. In fact, the OEM standard suspension leans more toward the concept of smoother riding; and the optional heavy-duty suspensions, such as the F-41 package, achieve the compromise between handling and ride. But for uncompromised handling capability, a good deal more work must still be done, and that's what this article is all about.

Before we get into the actual nuts-and-bolts specifics, we must point out that we'll be working primarily with the '63 and newer models fitted with independent rear suspension (IRS). Although '62 and earlier Vettes can also benefit from improved handling, the structural strength of the early chassis leaves a lot to be desired. Consequently, the substitution of significantly stiffer springs, increased shock absorber control and stiffness, along with tires of more than 7 inches of tread width, will create considerable chassis deflection and distortion unless the chassis is adequately reinforced by a knowledgeable chassis designer. Without such modification, the best approach is simply to utilize the factory heavy-duty pieces in conjunction with 7-inch-wide tires mounted on 8-inch rims that maintain the stock wheel offset on all '62 and earlier Vettes.

Since we mentioned tires and wheels on the early models, that's the logical place to start with modifications on the '63 and newer Vettes. The proper tire-and-wheel combination has a greater effect on the handling of a Corvette than any single thing you can change. Now that sounds a little strong, but in actuality, all of the suspension components merely support and supplement the interaction of the tire with the pavement. For street use, late-model Corvettes can effectively use tire tread widths up to a maximum of 10 inches. In many cases, the wheelwell openings will have to be flared to accommodate tires of this width, but it's better to spend the money

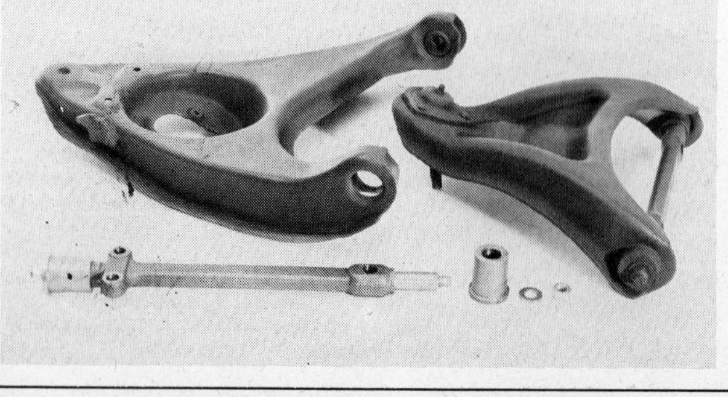

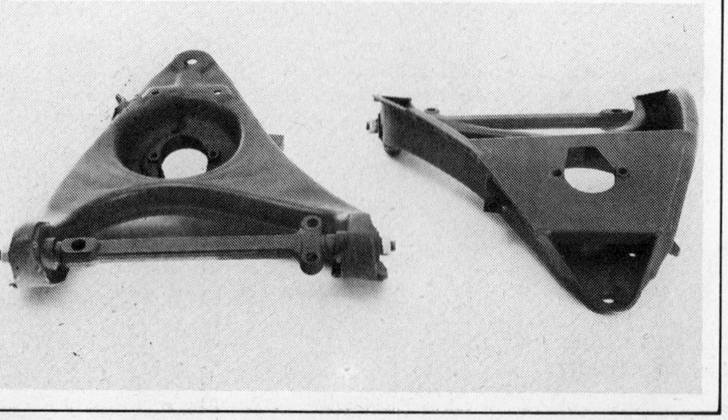

1

for the bodywork now than to have to spend it later for fiberglass repairs. And for off-road and racing applications, tires of the maximum width allowed by the sanctioning body rules should be used. In either case, both front and rear tires should be of the same size to retain good, balanced handling.

In addition to tires, the wheels that they're mounted on are equally important in terms of handling. At a minimum, the wheels should be at least as wide as the tread face of the tire, but ideally, wheel width should exceed tire tread width by about 1 inch. This width provides a stable trapezoidal tire/wheel cross section that resists "rollunder" of the tire and keeps the tread on the pavement. The wheel offset must also be kept as close as possible to stock. A lot of engineering time went into calculating wheel offset, in terms of a lever arm working against the shocks, springs, bushings and wheel bearings; and contrary to popular opinion, wheels with large amounts of negative offset will *not* improve handling by virtue of a wider stance on the Corvette. In fact, in some situations, negative offset can disrupt handling to such an extent as to make the car dangerous to drive, due to a loss of body roll control, tire sidebite and altered steering geometry. Excessive negative offset also wipes out the wheel bearings on Corvettes in short order.

Once the tires and wheels have been selected to fit the application, further improvements become a matter of suspension tuning, using various components to achieve the desired result. There are two primary suppliers of such components: Dick Guldstrand Enterprises (11924 W. Jefferson Blvd., Culver City, CA 90230), and John Greenwood Sales, Inc. (25700 Princeton, Dearborn Heights, MI 48125). Both firms offer complete lines of special suspension components for both street and racing use, including some very exotic pieces that totally replace the OEM suspension with a redesigned suspension featuring different principles of operation and revised geometry. Regardless of whether improvements are made to the stock suspension or whether it is replaced with something more exotic, the tuning process is essentially the same. It is a matter of matching spring rates, sway bar sizes and shock control, along with eliminating suspension slop through the replacement of rubber bushings with bearings or metal equivalents, and correcting bothersome alignment changes through the entire range of suspension travel inherent to the stock design. The extent of the modification and the severity of control is dependent upon the application and desired ride comfort. In other words, there are different levels of compromise between a harsh, all-out performance ride and comfort.

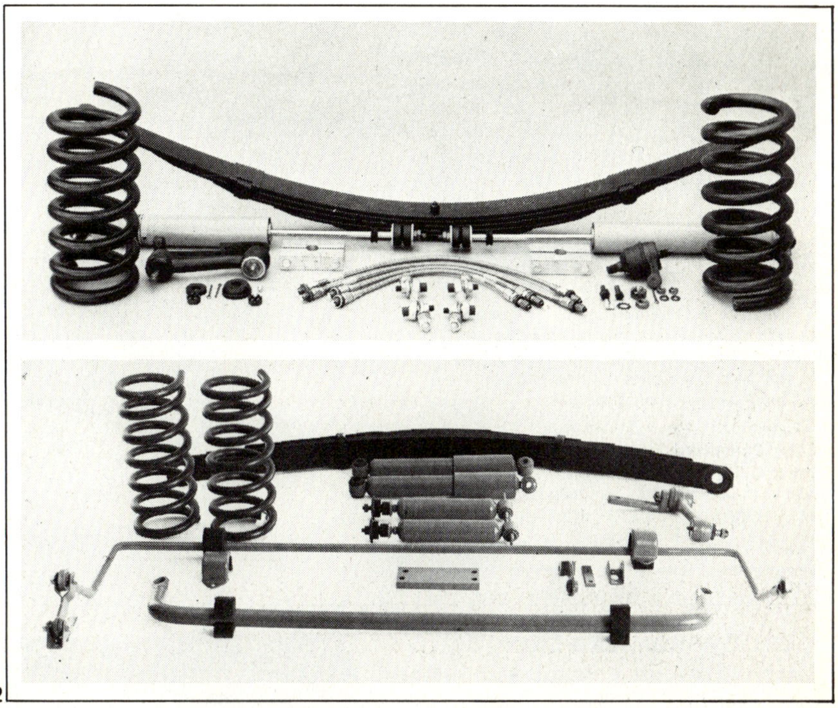

2

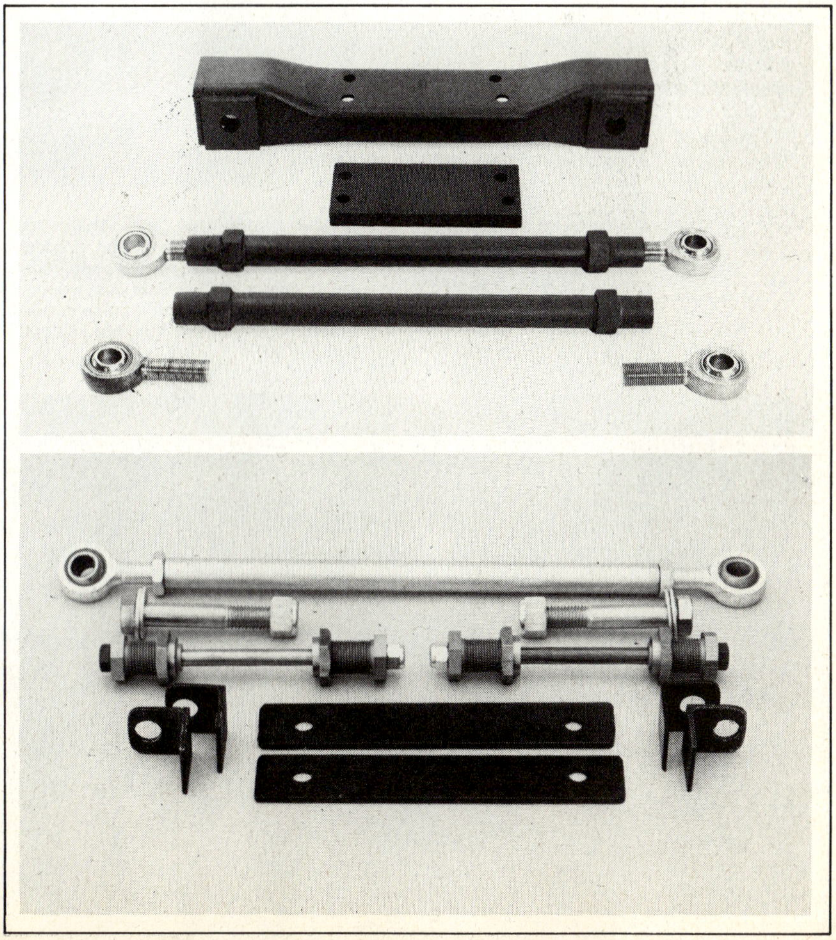

3

1. Shown here are front A-arms from Greenwood Sales that have been specially reinforced for competition use. Note, too, the replacement of the stock rubber bushings with metal bushings to eliminate geometry changes under load. Both Greenwood and Guldstrand offer such bushings. The upper bushings are eccentric to permit camber adjustments.

2. These groups of components vastly improve Vette street handling with wheels up to 10 inches wide and high-performance street tires. These packages are even suitable for weekend slalom or tight road course racing. Top collection is from Guldstrand for the big-block, including adjustable ball joints, steel-braided brake lines, adjustable shocks, etc. The lower kit from Greenwood Sales fits small-block Vettes and includes special anti-sway bars.

3. At the top is the Greenwood strut rod carrier kit. Below is the Guldstrand strut rod, spreader bar for front camber and screw-type adjusters for setting rear toe-in, thus eliminating shims.

The accompanying photos and illustrations depict the variety of components and packages available from Guldstrand and Greenwood. The differences are in spring rate, shock and sway bar control. Whenever possible, both the front and rear of the car should be done at the same time, to achieve balance; but if you're short on cash, start with the front first. Specific details and suggested alignment settings are supplied with the components, based on data supplied by the customer, so we really can't give specific settings here. One thing we can tell you though: After you've experienced a Vette that has been properly set up for ultimate handling, you'll probably realize that what you've been accepting from your road machine was merely the very beginning. —*C.J. Baker*

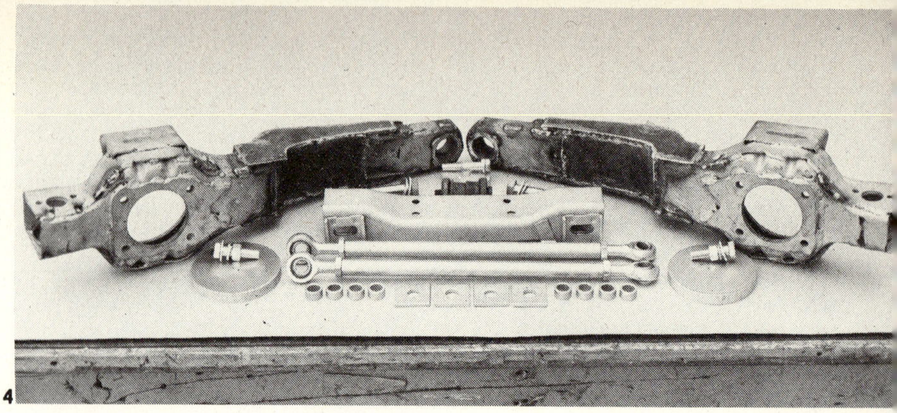

4

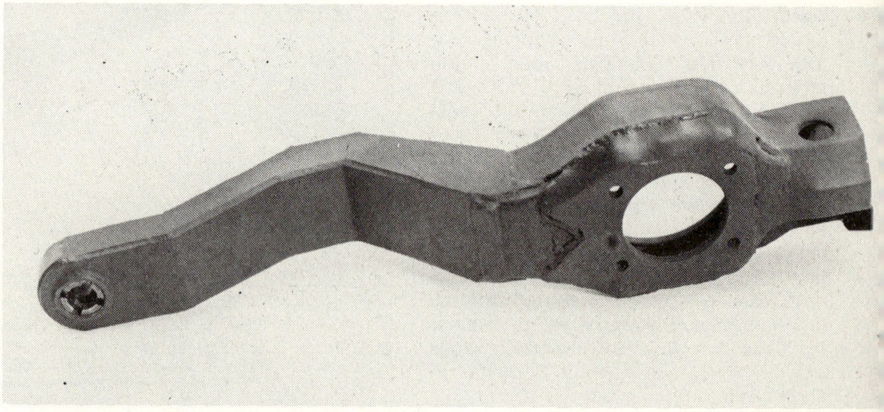

5

4. These are the Guldstrand pieces required for drag strip use: notched rear trailing arms, strut rod carrier kit and differential locating kit. Not shown are a rear spring and shocks that must be matched to the car.

5. The Greenwood kit eliminates toe steer throughout the full range of front suspension travel and Greenwood's recontoured rear trailing arm (shown) permits the use of very wide rear tires.

UNCOMPROMISED EXOTICA

This is it—the end of the line, where nothing but the best will do. Shown here is an illustration of the Greenwood rear suspension system, which, unlike the stock Vette suspension, eliminates any rearend squat on acceleration or rise during braking. The rear suspension virtually controls the attitude of the entire car, keeping it flat in the corners and stable as a rock. The design is so good that only two different spring rates are required to tailor the suspension to virtually any race track. It's not a low-dollar conversion, but it works.

The accompanying photo shows the actual rear suspension pieces, along with a reinforced front A-arm that includes sealed ball bearings instead of the stock rubber bushings. The sway bar is used at the rear.

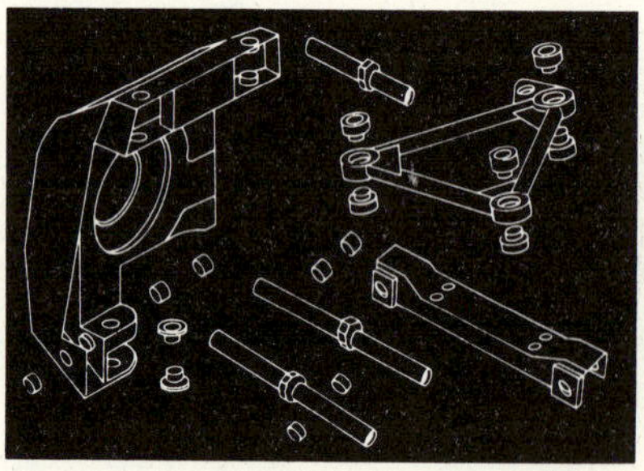

Serious racers, take note. This is the trick Greenwood package, featuring a reddsigned rear suspension and related components, along with reinforced front suspension pieces with ball bearings used in place of the A-arm bushings. Expensive? Yes, but you get what you pay for.

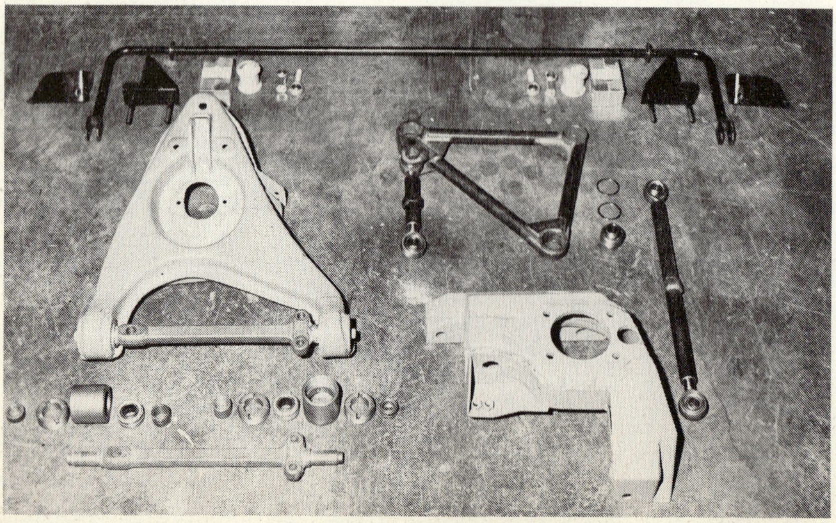

HANDLING—¼-MILE AT A TIME

Drag racing the Corvette presents speci problems in terms of suspension. The fro of the car must be modified to elimina bump steer (change in toe setting througho travel) problems inherent to the stock su pension. Kits to correct this problem a available from both Guldstrand and Gree wood. And at the rear of the car, provisior must be made to accommodate wide dra slicks and to prevent breakage under load.

Dick Guldstrand pointed out that a larg percentage of Corvette rear suspension ar rearend drivetrain component failure is th result of the rear suspension bottoming u der acceleration and thereby imparting binding action to those components. He co tends that the heavy-duty stock componen are capable of handling up to 600 horsepov er without failure if steps are taken to i crease the spring rate and prevent botton ing. Additionally, if the camber contr brackets are dropped 1½ inches, cambe change can be largely eliminated through th range of suspension travel, thus keeping th full tread face in contact with the pavemen Positively locating the differential within i rubber mounts also reduces rear suspensio geometry variations. Kits are available t accomplish all these changes.

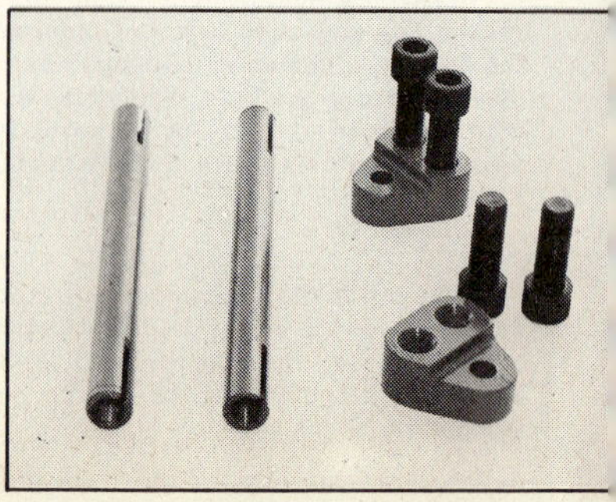

Swap Meat

Choice cuts of muscle for homebuilt haulers.

Contemporary engine swaps no longer emphasize the traditional high power-to-weight aims of an engine exchange. The brute acceleration requirements of yesteryear's swapping rodders have given way to the more practical parameters of our times—namely, economy, handling and everyday functionability. There's a whole new generation of enthusiasts out there who are going to have to approach the subject of swapping with plenty of fresh perspectives, and not a whole lot in the way of hands-on experience. Hopefully, this review will shed some guiding lights.

Basically, engine swaps can be divided into three categories: common (small-block V8s into Vegas), oddball (turbo'd big-block into an early Chevy pickup), and strictly OE (original equipment). It really wouldn't be practical to look into all of the permutations and combinations that fall under any of these headings, but there is some common ground to cover. As unlikely as it may seem, all the above groups have their roots in OE—and that's the area most often ignored. In view of the veritable volumes of published swapping material, OE has always drawn the short end of the stick. We're going to do our best to remedy that right now.

OE is actually the easy way out (or the easy way in, depending on just which side of the chain hoist you're standing). Today's hot swappers have an entire generation of formal factory engineering to learn from, and that fact alone makes a whole slew of exchanges infinitely more practical. In fact, there's a good chance that nearly any practical swap that an enthusiast can think of was once executed by a new car factory in one form or another. That's not to say you're likely to come across *every* powertrain/chassis combination as a one-time assembly-line option, but chances are that Motown has already mass-produced something similar. What this boils down to in practical terms is a wealth of almost untapped information and a fair share of hardware that's already available for everyday applications.

Even today, the factories are on our side. General Motors' current practice of "one-size-fits-all" car building, where "corporate" engines from each division are shared among the five sister divisions, makes late-model exchanges infinitely easier. That's not to say that too many folks can afford to tinker with brand-new high-dollar iron, but if they do so desire, the parts and pieces are just waiting to be screwed together.

This broad look at swapping theory and practices should help to make mechanical matters a little clearer for first-time transplanters, and hopefully it will provide fresh information for experienced exchange students as well. With the details available on our identification charts, we hope to get everyone off to a proper start. And once the factual foundations are laid down, our specific guides to actual swapping practice can be put to work.

ENGINE IDENTIFICATION

Locating suitable swapping material has always been a somewhat confusing proposition for engine exchange enthusiasts, mostly because there's so much to choose from. Even when a potential powerplant has been found, the problem of determining exactly what lies under its valve covers still remains. The following information should help to clarify matters a little, but considering the investment involved in a powerplant purchase, there's no better way to thoroughly investigate an engine on hand than by doing a complete teardown and inspection.

The key to this guide is knowing the car that the prospective motor came from. Most of our clues are based on having a complete vehicle to inspect rather than just the engine itself. But as you'll see, there are instances where information applies to separated assemblies as well. In some isolated cases, the same ID code may appear on different engines from a particular manufacturer. The same may apply to a particular engine used in different model year applications by the same auto maker.

Availability and desirability are the main parameters for our listings, but certain combinations are considered so rare and hard to find that we've passed over them for that very reason. If any doubt remains as to an engine's ID after its manufacturer's codes are deciphered, check out other reference sources. Repair manuals (like *Motors* or *Chiltons)* or even the original factory service texts can contain the confirming clue as to a powerplant's true identity. The key codes to most popular swap material are listed as follows:

All Chevy V8s can be identified by a two- or three-letter code which can be found following the engine's serial number stamped into the front right-hand side of the block below the head.

CLASSIC COMBINATIONS

Although 1955-'57 Chevrolets don't exactly fall within our late-model musclecar parameters, their mass popularity as swapping material at least merits a mention. Essentially, all three variations on what was basically the same body/chassis platform are easily adaptable to both OE-style small-block power as well as to later-model rat power. The process of mating either motor with an early car has actually been refined to the point where certain combinations make for a better fit than the factory original.

As can be expected, both pure-stock assembly-line power teams—a small-block backed by either a manual trans or the Powerglide automatic—require the simplest arrangements. With the standard-shift OE bellhousing bolted to the back of a mouse motor, virtually any 3- or 4-speed Saginaw, Muncie or Borg-Warner GM gearbox will fit. Early cast-iron Powerglide installations are equally easy, because the original V8 cars were attached to their 'Glides by a pair of bellhousing mounting ears. Later-model automatic-backed powertrains must be "side-mounted" with an aftermarket kit attached at the side, rather than at the front of the engine block, and a late-model rear crossmember must be custom-mounted to match. Big-blocks will fit in underhood too, but valve-cover-to-firewall interference problems may require some sheetmetal tweaking. Kits containing all the necessary hardware for any of these combinations are available from a number of the swap shops listed here.

HELPFUL PUBLICATIONS

Quite possibly the ultimate source for original equipment interchange information is the Hollander manual. It's referred to in the trade as "The Bible," and for many a good reason. Cross-referenced listings identify components and systems, as well as the cars they came in and the other models they'll fit. It takes a bit of practice to become proficient in its use, but like any other source of specifics, it's well worth the effort to learn. Boneyards nationwide depend on Hollanders for what-fits-what guidelines, and the serious swapper can reap many benefits. Editions are available covering current and past production vehicles and foreign cars as well. Further information can be obtained directly from Hollander Publishing (12320 Wayzata Blvd., Minnetonka, MN 55343, 612-544-4111).

The detail and diversity of engine swapping is delved into in this fourth in

CAMARO

Year	Engine	Codes
1967	327/275	MK,ML,MM,MN
	350/295	MS, MT, MU, MV
	396/325	MY, MZ
	396/350	EL, EQ, EY
	396/375	MQ, MR
1968	302/290	MO
	327/275	EA, EE
	350/295	MS, MU
	396/325	MW, MY
	396/350	MR, MX
	396/375	MQ, MT
1969	302/290	DZ
	350/295	HQ, HR, HS
	350/300	HA, HE, HB, HP
	396/325*	JB, JG, JU, CJG, CJK, CJU
	396/350*	JF, JI, KA, CJF, CJI
	396/375*	JH, JL, KC, CJH, CJL JJ**, JM**, KE**
1970	350/300	CNJ, CNK, CRE
	350/360	CTB, CTC
	396/350*	CTW, CTX
	396/375*	CKD, CTY
1971	350/270	CGK, CGL, CJG, CJD
	350/330	CGP, CGR
	396/300*	CLC, CLD
1972	350/200	CDG, CKD, CKK
	350/255	CKS, CKT
	396/240*	CLA, CLB, CTA, CTB
1973	350/175	CKB, CKU
	350/245	CLJ, CLK, CLL, CLM

*Early 1969 big-blocks are indicated by a two-digit code and displace 396 cubic inches. Late 1969 and later big-blocks are indicated by a three-digit code and displace 402 cubic inches.
**Aluminum cylinder heads

CORVETTE

Year	Engine	Codes
1967	327/300	HE, HO, HH, HR
	327/350	HP, HT, HD, KH
	427/390	IL, IQ, IM, IR
	427/400***	JC, JD, JF, JG
	427/435***	IU, IT, JE, JA, JH
1968	327/300	HE, HO
	327/350	HP, HT
	427/390	IL, IQ
	427/400***	IM, IO
	427/435***	IR, IT, IU
1969	350/300	HY, HZ
	350/350	HW, HX
	427/390	LL, LM
	427/400***	LN, LQ
	427/430	LO**, LV**
	427/435***	LR, LT, LX, LP** LU**, LW**
1970	350/300	CTL, CTM
	350/350	CTN, CTO, CTP, CTQ
	350/370(LT-1)	CTR, CTU, CTV
	454/390	CGW, CRI, CZU
	454/465	CZL**, CZN**
1971	350/270	CGT, CJL
	350/330(LT-1)	CGY, CGZ, CJK
	454/365	CPH, CPJ
	454/425	CPW**, CPX**
1972	350/200	CDH, CDJ, CKW, CKX
	350/255(LT-1)	CKY, CKZ, CRS, CRT
	454/270	CPH, CPJ, CSR, CSS
1973	350/190	CLA, CLB, CLC
	350/250	CLD, CLH, CLR, CLS
	454/275	CWM, CWR, CWS, CWT
1974	350/250	CLD, CLR
	454/270	CWM, CWR, CWS

***Three 2-barrel carburetors
**Aluminum cylinder heads

CHEVELLE

Year	Engine	Codes
1967	327/275	EA, EB, EC, EQ, EE
	327/325	EP, ER, ES, ET
	396/325	ED, EK, EM
	396/350	EF, EJ, EL, EN, EU, EV, EW
	396/375	EG, EX
1968	327/275	EA, EE
	327/325	EP, ES, ET
	396/325	ED, EH, EK
	396/350	EF, EL, EU
	396/375	EG
1969	350/300	HB, HE, HP
	396/325*	JA, JK, JV, CJA, CJB, CJV
	396/350*	JC, JE, KB, CJE
	396/375*	CJD, CJF, JD, KD, KF, KG**, KI**, KL**, KH**
1970	350/300	CNJ, CNK, CRE
	350/360	CRT
	396/330*	CKR, CKS
	396/350*	CTW, CTX, CTZ
	396/375*	CKO, CKQ, CKT, CKU, CTY, CKP**
	454/360	CRN, CRQ
	454/450	CRR, CRV, CRS**
1971	350/270	CGK, CGL
	396/300*	CLB, CLL, CLP, CLR, CLS
	454/365	CPA, CPD, CPG
	454/425	CPP**, CPR**
1972	396/240*	CLA, CLB, CLS, CTA, CTB, CTH, CTJ
	454/270	CPD, CRW, CRX, CPA
1973	454/245	CWA, CWB, CWC, CWD

*Early 1969 big-blocks are indicated by a two-digit code and displace 396 cubic inches. Late 1969 and later big-blocks are indicated by a three-digit code and displace 402 cubic inches.
**Aluminum cylinder heads

NOVA

Year	Engine	Codes
1968	350/295	MN, MS, MV
	396/350	MR, MX
	396/375	MQ
1969	350/300	HA, HB, HE
	396/350	JF, JI, KA
	396/375	JH, JL, KC
1970	350/300	CNJ, CNK, CRE
	350/360	CTB, CTC
	396/350*	CTW, CTZ
	396/375*	CKQ, CKT, CKU, CKW, CKP**
1971	350/270	CJD, CJG
1972	350/200	CKD, CRL, CDG

*Early 1969 big-blocks are indicated by a two-digit code and displace 396 cubic inches. Late 1969 and later big-blocks are indicated by a three-digit code and displace 402 cubic inches.
**Aluminum cylinder heads

V8 INSTALLATION TIPS

	Engine Tabs		Mounts		Bellhousing Mounts		Bell-housing	Clutch Fork	Oil Pan	Oil Pump Pickup
	left	right	left	right	left	right				
SMALL-BLOCK										
Manual Trans	3836325	3836326	3768180	3768180	3728507	3728508	3755343	3737454	3735640	3754942
Cast-iron Powerglide	3836325	3836326	3768180	3768180	3728507	3728508	-	-	3735640	3754942
Aluminum Powerglide & all THM	Use sidemount kit		3990916	3990916	**	**	-	-	3735640	3754942
BIG-BLOCK* **										
Manual Trans	*	*	3768180	3768180	**	**	3755343	3737454	3902371	3860366
All THM	*	*	3990916	3990916	**	**	-	-	3902371	3860366

*Use aftermarket tabs (Hurst, Cyclone, Trans-Dapt) in six-cylinder frame holes

**Rear crossmember installation required. Use mount number 3870182 for PG and THM 350 and number 3872248 for THM 400.

***Tube headers required

NOTE: All part numbers listed are Chevrolet's, but compatible are also available from aftermarket sources.

a series of all-time best-selling Petersen publications. The *Complete Book of Engine Swapping No. 4* digs for the specifics, with plenty of photos to show you the way. It's a look at the practical applications of many of this story's particulars, and you actually get to see the rolling results of the kind of backyard thrashing that we try so hard to inspire.

Further references to most of the powertrains reviewed here can also be found in another massively illustrated Petersen guide book, the *Big Book of Auto Repair.* This hefty volume calls the shots for both the beginning and the experienced do-it-yourselfer, with literally loads of step-by-step how-tos. Every approach to an engine swap should start with an open book rather than a hastily opened hood. *—Al Kirschenbaum*

ENGINE FITTING GUIDE I

	CRITICAL DIMENSIONS								
SIZE	A	B	C	D	E	F	Oil Sump	Starter	Weight
Small-blocks	26.50	27.00	19.50	26.00	20.50	25.00	rear	left	575
Big-blocks	30.50	30.50	22.00	27.00	23.50	29.50	rear	left	685

KEYS TO ENGINE DIMENSIONS

All figures are given in inches and pounds. Starter list refers to the side of the engine that the starter motor bolts to. Oil sump list refers to location of sump.

DIMENSION

A Front of water pump pulley to the point of firewall interference at either the distributor or a valve cover
B Front of water pump pulley to rear face of block
C Overall width without exhaust manifolds
D Overall width with exhaust manifolds (OE)
E Top of valve covers to bottom of oil pan sump
F Top of carb air horn to bottom of oil pan sump

SWAP SHOPS

The following are some of the more popular sources of supply for Chevy-based engine swapping parts and information.

Advance Adapters, Inc.
12120 Woodruff Ave.
Downey, CA 90241
(adapters, hardware)

Ansen Automotive Engineering
13720 S. Western Ave.
Gardena, CA 90249
(adapters, hardware)

Chassis Engineering & Speed
2584 E. Fender, Apt. E
Fullerton, CA 92631
(adapters, hardware)

Cragar Industries
19007 S. Reyes Ave.
Compton, CA 90221
(headers)

Crown Manufacturing Co.
858 Production Pl., Box 2860
Newport Beach, CA 92663
(adapters, hardware)

Cyclone Automotive
7040-D Lankershim Blvd.
N. Hollywood, CA 91605
(headers)

Don Hardy Race Cars
202 W. Missouri St.
Floydada, TX 79235
(adapters, kits and hardware)

Doug Thorley Headers
7403 Telegraph Rd.
Los Angeles, CA 90040
(kits, headers)

Herbert Automotive & Marine
(formerly Herbert & Meek)
8792 Fruitridge Rd.
Sacramento, CA 95826
(adapters, kits and hardware)

Hooker Headers
1032 W. Brooks St.
Ontario, CA 91761
(kits and headers)

Horsepower Engineering
2489 E. Washington Blvd.
Pasadena, CA 91104
(custom headers)

Hurst Performance, Inc.
50 W. Street Rd.
Warminster, PA 18974
(mounts and shifters)

JR Headers
P.O. Box 504, Lemon Grove
North Adams, MI 49262
(headers)

Kelmark Engineering, Inc.
Box 120
Okemos, MI 48864
(adapters and kits)

Kustom Headers
1331 W. Cedar, Box 328
Standish, MI 48658
(headers)

Milodon Engineering
7711 Ventura Canyon
Van Nuys, CA 91402
(oiling systems)

Moon Equipment Co.
10820 S. Norwalk Blvd.
Santa Fe Springs, CA 90670
(adapters and custom fabrication)

Moroso Performance
Carter Drive
Guilford, CT 06437
(motor mounts, crossmembers, oil pans)

Motion Performance
598 Sunrise Hwy.
Baldwin, NY 11510
(adapters and kits)

Offenhauser Sales Corp.
5300 Alhambra Ave.
Los Angeles, CA 90032
(adapters)

Positive Performance
2970 E. LaPalma Ave.
Anaheim, CA 92806
(custom headers)

Sanderson Headers
1379 N. Carolan
Burlingame, CA 94010
(headers)

Speed Research & Development
147 Pennsylvania Ave.
Malvern, PA 19355
(kits and chassis components)

Trans-Dapt Co.
Box 4157
Compton, CA 90224
(adapters)

Tubular Automotive
248 Weymouth St.
Rockland, MA 02370
(headers)

Wilcap Co.
2930 Sepulveda Blvd.
Torrance, CA 90510
(adapters and custom fabrication)

Bracket Attack

A complete step-by-step journey through the construction and preparation of a bracket/econo altered.

It isn't often that HOT ROD magazine has the space to present a detailed account of the construction of a modern drag racing vehicle, but with the ever-growing popularity of bracket racing, and the interest in Competition Eliminator racing with econo altered-type cars, we deemed it an appropriate time to initiate such a project. We begin with the preparation of a dual-purpose car to fit both econo altered classes and bracket racing criteria. We'll start off with the construction of a modern, state-of-the-art chassis by Warren Brogie at Brogie Race Cars (1838 E. Ball Rd., Anaheim, CA 92805). Second, we'll detail the preparation of a bullet-proof small-block Chevrolet engine intended for consistent, trouble-free, low-maintenance bracket racing. We'll detail the aluminum work, body installation, painting and assembly of the complete car. We'll upgrade the engine to an all-out Comp Eliminator degree of sophistication, and we'll even take you along as the staff spends a day at the track sorting out the car and offering their driving impressions.

Third, we'll continue the construction of our dual-purpose car with the preparation of a quick and durable small-block Chevrolet engine that will power our ground pounder. Component selection and assembly of this bracket engine, which will also serve as the foundation for the all-out version of the engine to be tested at a later date, was handled by

1

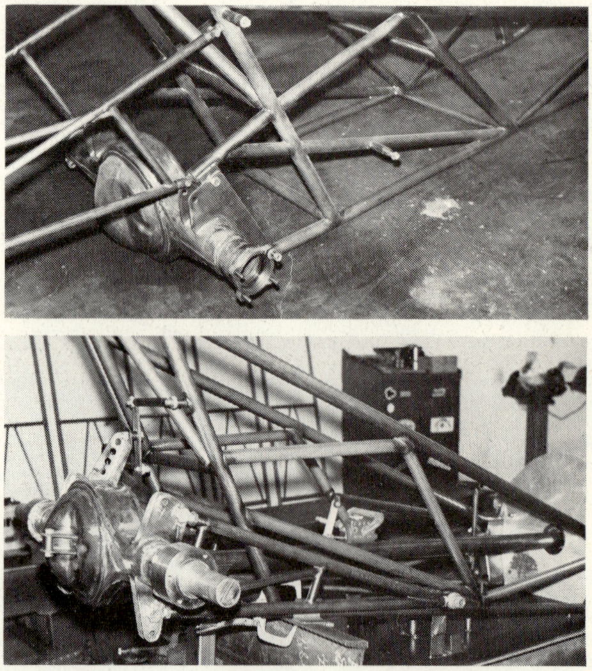

2

3

4

5

1. Brogle's basic chassis is assembled in a jig to assure perfect fit and alignment. He has all of the pieces precut for immediate assembly and delivery. A block is mocked up with a driveline tube to assure alignment; 4130 chrome moly tubing is used in accordance with the SEMA 10-1 Spec. Our car has a 110-inch wheelbase.

2. Brogie offers both solid rear axle mount cars and optional suspended rear axle cars. There are many pros and cons of the two mounting systems, although Brogie feels the suspended rear offers superior traction with everything but all-out big-block engines. We opted for the suspended rear, since a small-block engine will power the car.

3. The 36-inch lift bars are constructed of 4130 chrome moly and utilize ¾-inch Teflon-lined 4130 rod ends. Axle and chassis attachment points are fully gusseted for maximum strength and to reduce flex. Rear track bar upright reduces flexing and provides a more rigid attachment of the rod-end bung. At the front end of the lift bar, both tubes extend all the way forward to capture the rod-end bung. Extra lower holes on the axle mount are for wheelie-bar attachments.

4. The upper rail ahead of the motor plate slip fits into the bigger tube past the uprights for increased strength in this area. The uprights are 1½x3-inch rectangular mild steel tubing to prevent the cracking problem in this high-stress area on some higher-powered cars. Positioning of the upright is such that welding heat concentration on the bottom rail is reduced. Similar positioning theory is used throughout the car.

5. Front engine mounts are attached to the chassis with hose clamps. Brogie offsets these brackets precisely the right amount so that they can simply be reversed when swapping big-block and small-block engines. This allows the mount to sit in exactly the same location on the chassis tube without scratching new marks in the paint. Lots of thought went into this car.

6. The lower frame rail is fitted with a tube to pass the battery cable through it for hassle-free, sanitary installation.

7. The 4130 chrome moly tubular front axle is fitted with Stiletto spindles.

8. In back, the narrowed Ford 9-inch station wagon axle housing is fitted with Summers Brothers housing ends, axles and wheel studs. Edco 6-puck magnesium disc brakes are used, along with Edco rotors, hats and brackets.

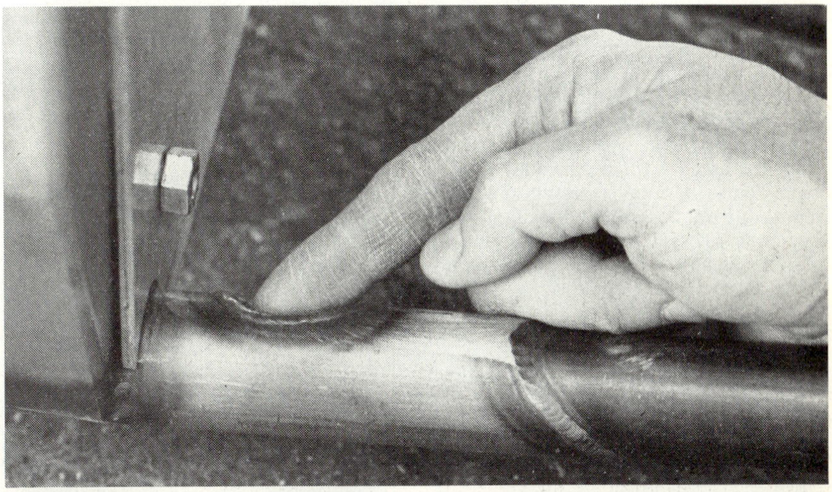

6

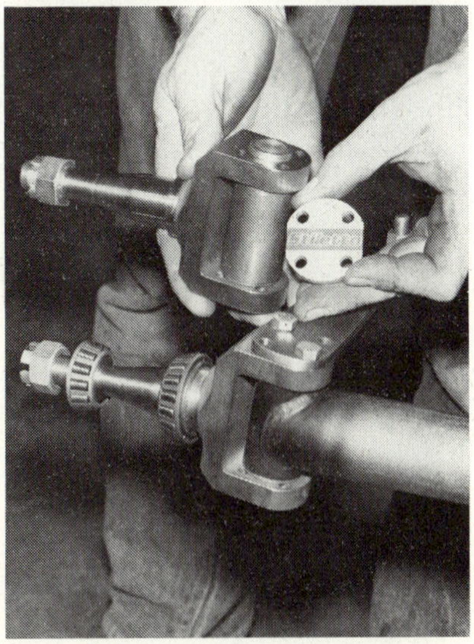

7

8

John McClintic at McClintic Industries (9701 Trumbull S.E., Albuquerque, NM 87123). McClintic Industries has one of the most complete engine-building facilities to be found anywhere for both high-performance street and full competition engines. Additionally, John McClintic is always on top of the latest winning combinations. So stay with us as we describe the what, how and why of our 305-cubic-incher.

Finally, we'll put the engine and transmission in place, add chassis sheetmetal, fiberglass body, paint and a host of support equipment. You'll note, too, that our engine does not yet have headers, a cooling, induction or an ignition system. Those will be added as well when Project Bracket Attack nears completion as it is readied for its first trip to the strip.

It'll be a fun project filled with loads of information and tips for anyone who is thinking of building a similar car—and even if you're not thinking about it now, you might after you read this. So follow along as our bracket/altered concept is transformed into a rolling, fire-breathing race car.

In the end, completion of the bracket car took longer than anticipated, but that was the only bad news connected with the project. The good news is that the resultant car exceeded our every expectation in both performance and appearance. One look at the detail work, the glistening chrome, the polished aluminum, and the dazzling paint and you'll see what we mean about appearance. Less obvious, but equally impressive, was the initial performance. After a few initial outings, Warren Brogie, who built the chassis and actually assembled the car, clicked of an incredible 8.95 elapsed time at 148.27 mph! That kind of performance is creditable in its own right; but in view of the NHRA index for A/EA of 8.85, and if you consider that the car is currently 65 pounds heavy for the class, that the combination really hasn't been dialed in as of this writing and that this is our relatively mild bracket version of the engine rather than an engine that takes full advantage of A/EA rules, you'll realize that we have a winner on our hands.

Although the 305-cubic-inch engine in the car is completely legal for A/EA, engine builder John McClintic points out that the addition of lightweight valves, better valve springs, better porting and increasing carb size to the legal maximum of 850 cfm (the car here has a 750-cfm Holley on it) would allow the engine to rev freely to 9500+ rpm for greater power output. The increased power and rpm, in conjunction with more rearend gear ratio (5.14:1 at present), along with paring the car down to minimum legal weight, should put the car in the 8.60s. That would be fast enough to grab the NHRA National Record for A/EA and make the car a serious threat during eliminations at major meets—and

9

10

9. Special hand-built Monroe coil-over rear shocks (coils not in place during this trial fit) control the rear axle. These are state-of-the-art shocks featuring all the latest technology in weight transfer control, and they are double-adjustable. Very trick pieces. Note, too, the Watts linkage attached to the top of the housing for a high roll center. This is done to encourage the car to slide rather than dig in if it ever gets sideways. The small bracket in the center of the housing is for the parachute shroud attachment.

10. A foot brake pedal was installed in our car, since we will be using an automatic transmission. A ⅞-inch bore Edco master cylinder is used in conjunction with Brogie's linkage for ratios from 8:1 to 11:1.

11. Here Warren carefully fits and aligns the roll cage to the chassis. Cage height can be specified by the customer to fit various driver heights.

12. Engine struts are added to each side of the engine to control fore-and-aft movement. These are a $65 option on any of Brogie's cars.

13. A ⅛-inch-thick steel driveshaft tube with ⅞-inch tubing loops was added. Note that the tube has had wedged shaped pieces welded in to enlarge the rear of the tube for up-and-down movement of the driveshaft with the suspension.

14. A Russell Race Car Products transmission heat sink cooler is mounted under the driver.

11

12

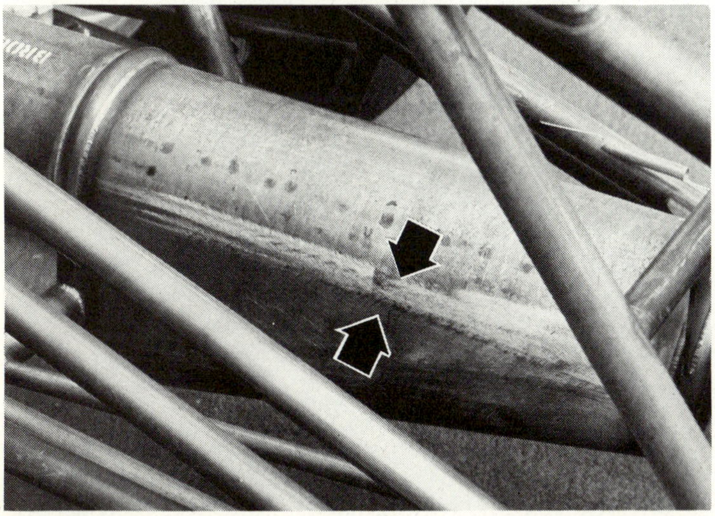

13

14

BROGIE ECONO ALTERED CHASSIS AND OPTIONS

Basic chassis $2985
Meets SEMA 10-1 specs; 106-inch, 110-inch or 118-inch wheelbases available; torsion-bar front suspension; solid rear suspension; tubular front axle; spindles; steering box; steering linkage; steering wheel; brake linkage; pedal and master cylinder; throttle pedal; narrowed Ford 9-inch rear axle housing; housing ends; aluminum engine plate; front mounts; trans mount and grade-8 fasteners used throughout

Basic chassis with mounted '23-T turtle deck body $3785
Built-in tonneau cover; extended cowl to meet firewall; raised flip-open rear deck and bolt-on body tree with battery mounts

Body only $585

Options:

Sprung rear suspension $900
36-inch lift bars with ¾-inch 4130 rod ends; Watts linkage track bars; coil-over shocks; springs and rear driveshaft loop

60-inch wheelie bars $300

Chassis sheetmetal $525
Two-piece aluminum seat; bulkhead under seat; belly pan and T-shell-shaped fuel tank

Summers Brothers axles $294

Summers Brothers spool $220

Axle bearing (pair) $42

Axle housing brace $125

Edco disc brakes $463
6-puck magnesium calipers with lining, hats, rotors, caliper brackets and fasteners

Brake lines and fittings $85

Parachute mount $150
Includes shroud-line mount; release lever; pack mount and body modification to recess pack

Engine strut rods (pair) $65

MANUFACTURERS OF OUR CHASSIS PRODUCTS

Russell Performance Products
22628 S. Normandie
Torrance, CA 90502

Stiletto Racing Products
Box 925
Rialto, CA 92376

P&S Machine Products
12256 S. Woodruff Ave.
Downey, CA 90241

Summers Brothers, Inc.
530 S. Mountain Ave.
Ontario, CA 91761

Edco Specialty Products, Inc.
1525-15 W. MacArthur Blvd.
Costa Mesa, CA 92626

Monroe Auto Equipment Co.
1 International Dr.
Monroe, MI 48161

Cragar Industries
19007 S. Reyes Ave.
Compton, CA 90221

Goodyear Tire & Rubber Co.
1144 E. Market St.
Akron, OH 44316

that's just exactly what we intend to do with it. But enough for the future plans. Let's take a close look at the car as it is shown here.

The car was essentially complete, going through trial assembly before paint and chrome were applied. One problem that cropped up was interference between the kickout on the side of the Competition Engineering oil pan and the Weber Ultra-Tork starter. There was simply no way the starter would fit. Competition Engineering suggested grinding off the side of the starter case—a questionable procedure at best, but that still wouldn't have alleviated the problem. Consequently, we contacted Moroso for one of their new No. 2101 Ultimate II oil pans, which solved the starter problem, but, unfortunately, the total depth of 8½ inches created ground clearance problems in the sump area on our super-low chassis. Brogie was faced with modifying either the kickout on the Competition Engineering pan or the sump on the Moroso pan, although it really was a toss-of-the-coin decision. The Competition Engineering pan won the toss, and Brogie took a full ½-inch out of the side kickout in the starter area.

Also added were the Brogie wheelie bars; an Edelbrock Victor Jr. manifold; a Holley 750-cfm R-4779 double-pumper carb; a Yother Performance Products air scoop; an Autotronic Controls crank trigger pickup for the MSD-7A ignition; a fully equipped Lenco aluminum rearend center section housing a Summers Bros. spool and Richmond Gear Division (formerly Eaton) 5.14:1 pro gears, a Delco breakerless tach-drive distributor; Accel plug wires; NGK B9FS spark plugs, Russell braided stainless steel lines; a Moroso 4-inch-face tach and seat upholstery. Rounding out the safety equipment for the package is a Simpson drag chute, Simpson 5-point harness, firesuit, gloves and the latest Simpson Super RX-1 full-face helmet.

The combination of the McClintic-prepped small-block and the A-1 Transmissions Powerglide, complete with trans-brake and an 8-inch 5500-rpm converter, offers consistent, maintenance-

15

16

MANUFACTURERS OF OUR ENGINE PRODUCTS

BRC Racing Components
7091 Belgrave Ave.
Garden Grove, CA 92641

Chevrolet Motor Division
General Motors Corp.
General Motors Bldg.
Detroit, MI 48202

Competition Engineering
25 Edwards St.
Box 310
North Madison, OH 44057

Manley Performance Products
13 Race St.
Bloomfield, NJ 07003

Moroso Performance Products, Inc.
Carter Dr.
Guilford, CT 06437

Perfect Circle
Dana Corp.
Box 321
Toledo, OH 43691

Speed-Pro
Sealed Power Corp.
2001 Sanford St.
Muskegon, MI 49443

TRW Replacement Division
8001 E. Pleasant Valley Rd.
Cleveland, OH 44131

Winona Tool Manufacturing Co.
4730 W. Highway 61
Winona, MN 55987

MANUFACTURERS OF MISCELLANEOUS PARTS AND COMPONENTS

A-1 Automatic Transmissions
7239½ Woodley Ave.
Van Nuys, CA 91406

Autotronic Controls Corp.
6908 Commerce St.
El Paso, TX 79915

Holley Carburetor Division
Colt Industries
11955 E. Nine Mile Rd.
Warren, MI 48090

Hooker Headers
1032 W. Brooks St.
Ontario, CA 91781

Moroso Performance Products, Inc.
Carter Dr.
Guilford, CT 06437

Sears, Roebuck and Co.
Sears Tower
Chicago, IL 60684

Simpson Safety Equipment, Inc.
22630 S. Normandie Ave.
Torrance, CA 90502

Unique Metal Products
8745 Magnolia
Santee, CA 92071

Weber Performance Products
Box 1328
4840 E. 12th St.
Kansas City, MO 64141

17

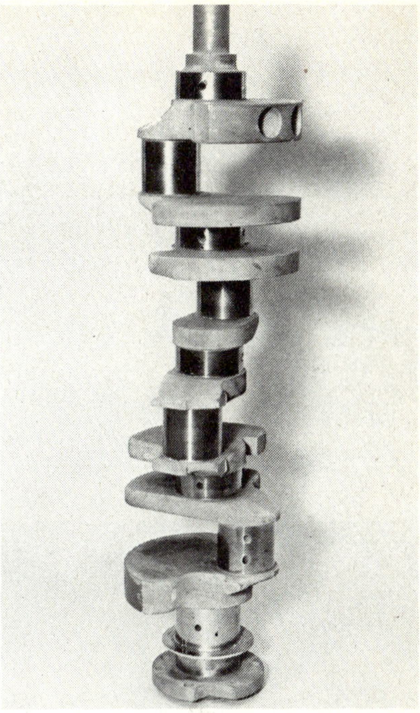

18

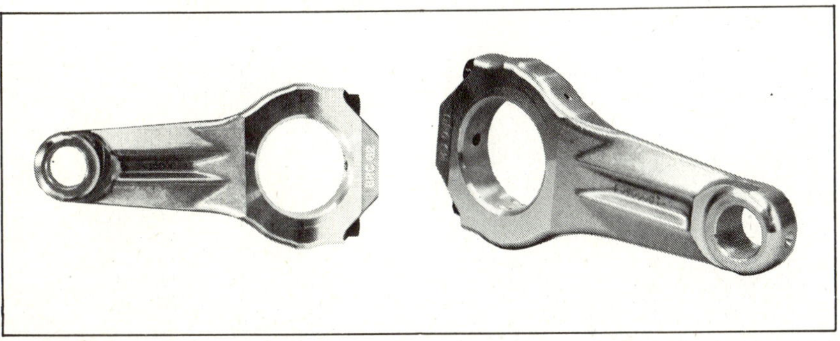

19

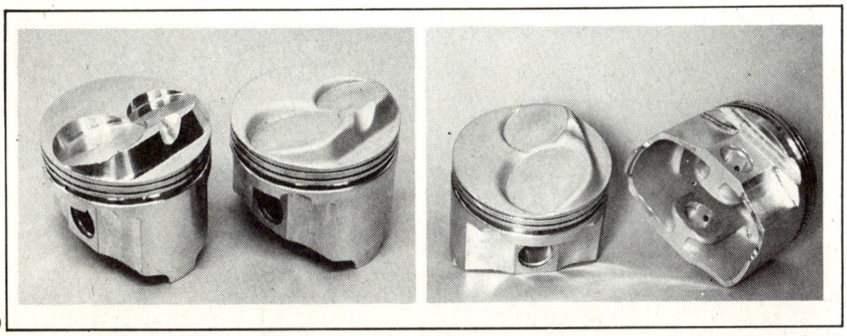

20

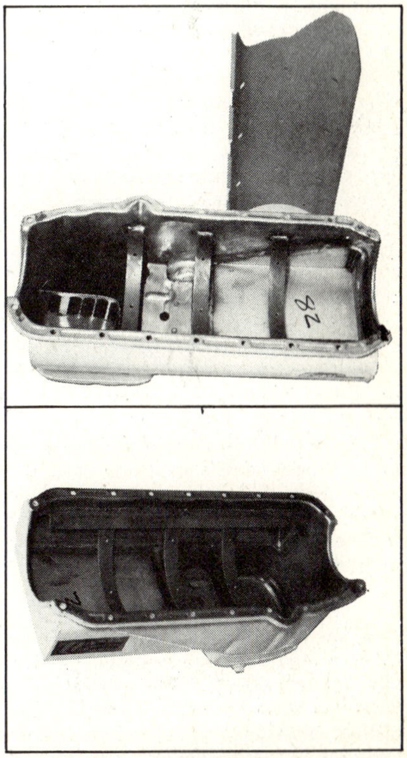

21

15. S&M Tire polished the spindle mount Cragar Super Tricks mounting the Goodyear front drag tires. The aluminum catch tank (not yet polished) is a $50 option, and the tow hoop welded in the lower left frame rail is a $20 option.

16. A Russell Race Car Products steering wheel is mounted on the P&S Machine Products steering gear. Brogle's placement of the instruments (oil pressure and water temp) allows the gauges to be read through the steering wheel openings. Engine and transmission shown in the car are strictly for mock-up and fitting purposes. They aren't the ones that will be used in our car.

17. The foundation of our 305-incher is the high-tin-content No. 366246 block, which has a 4-inch bore and four-bolt caps. McClintic Industries fully machine the block, including squaring the decks, align bore and hone, and boring and honing .020-inch oversize to true the cylinder bores. Also shown here is the Moroso small-block oil pump with bottom pickup. A 1-inch Moroso spacer properly lowers the pump into the pan sump.

18. The crankshaft is the '69 large-journal, 3-inch-stroke version. McClintic prefers to use this crank, cutting down the rod journals to the early 2-inch diameter. This also allows precise indexing of the throws and slight stroke variations, if so desired. The finished crank is cross-drilled, and the throws are streamlined to reduce windage, as shown in the photo of the assembled short-block. Clevite 77 bearings are used throughout the engine.

19. Because the same short-block will be used for our all-out altered engine, BRC aluminum rods with a 5.600-inch center-to-center length were selected. This is .100-inch shorter than stock to increase low-rpm torque with the automatic transmission. The BRC piston pins are free-floating and retained in the pistons with double Tru-Arcs.

20. The pistons used are the BRC No. 53, silicone alloy versions with a side shoulder for stability in the bore. McClintic smooths the dome, lays back the intake relief, and glass-peens the dome (as compared to the as-received configuration) to enhance flow and flame propagation. Speed-Pro .043-inch compression rings are used with 1/16-inch cast-iron second rings and 3/16-inch standard tension oil rings. Although not shown, the pistons will be gas-ported to improve top ring sealing. McClintic finds oil ring tension isn't a factor in power output—oil in the combustion chamber is.

21. Buttoning up the bottom end is a modified Competition Engineering pan and Teflon-coated windage tray. The pan features a deepened sump, windage scrapers and windage relief kickouts, and special metal and foam baffles around the oil pump pickup.

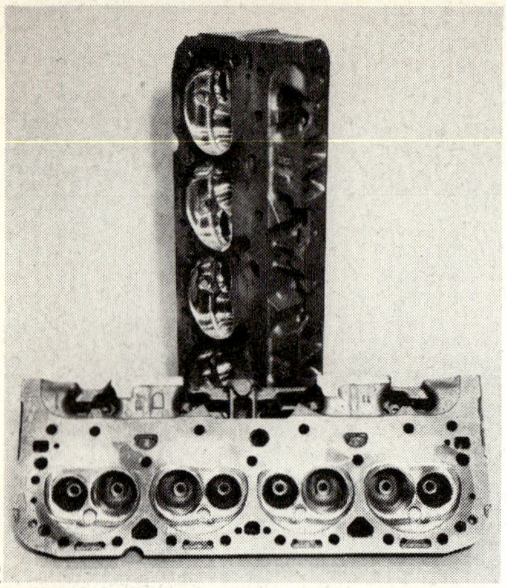

22

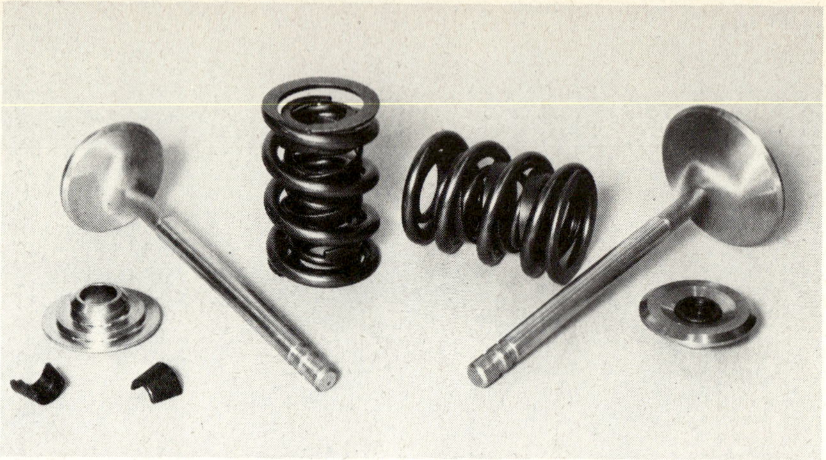

23

24

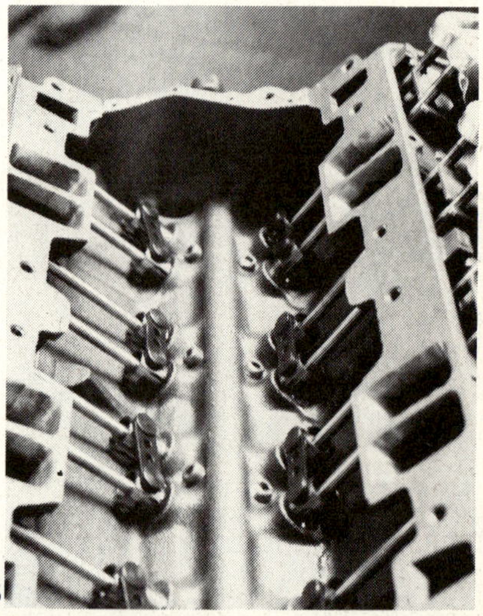

25

free power to the beefy Goodyear 14.5x32-15 slicks. Rollout on these D-1 compound gumballs is 103 inches. Each pass has been straight as an arrow—a tribute to the chassis engineering. So far, the best performance has been obtained with an ignition lead of 44 degrees, .032-inch valve lash and stalling the converter to 5200 rpm against the internal trans-brake. The 1-2 shift is made at 7800 rpm, and the car goes through the lights at about 8400 rpm.

As we mentioned earlier, we plan to optimize the A/EA combination in the near future. Beyond that, well there is a 900+ hp, twin-turbocharged, intercooled, 305-cubic-inch small-block going together in Gale Banks' research and development facility in San Gabriel, California, and rumors persist about an all-aluminum big-block for the little altered, so.... It just could be that the best is yet to come. Believe it!

—C.J. Baker

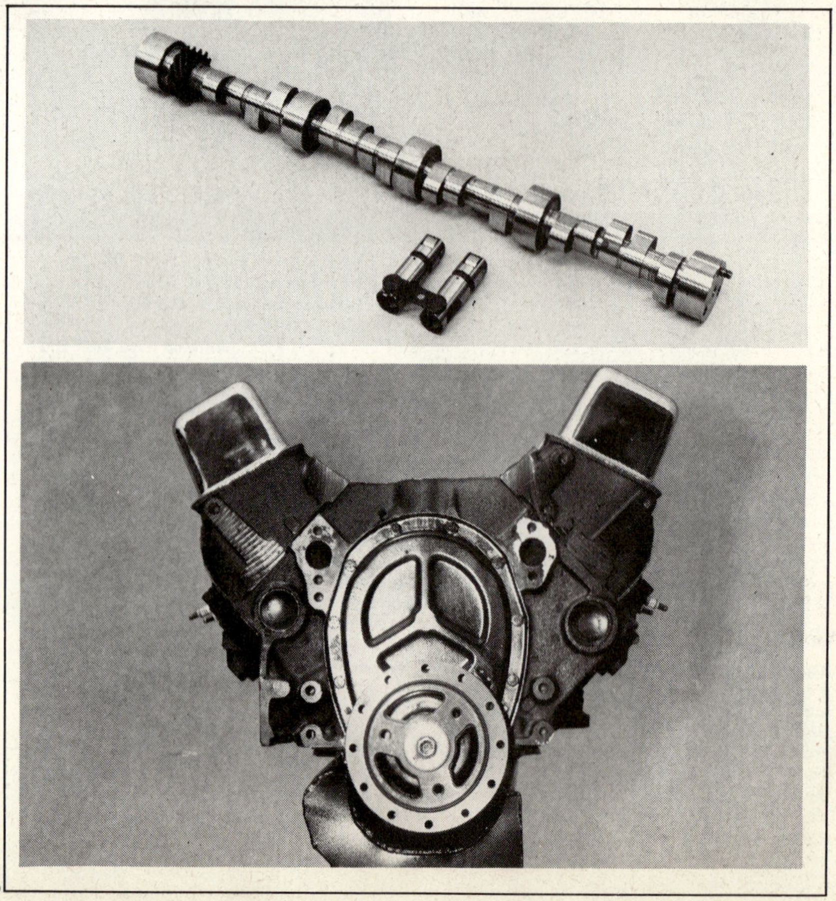

26

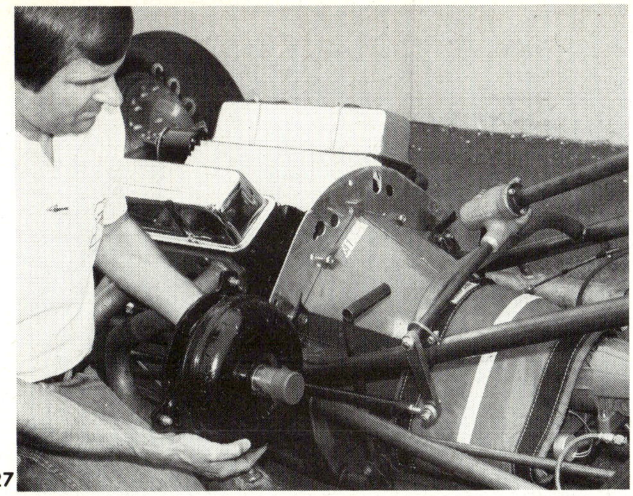

27

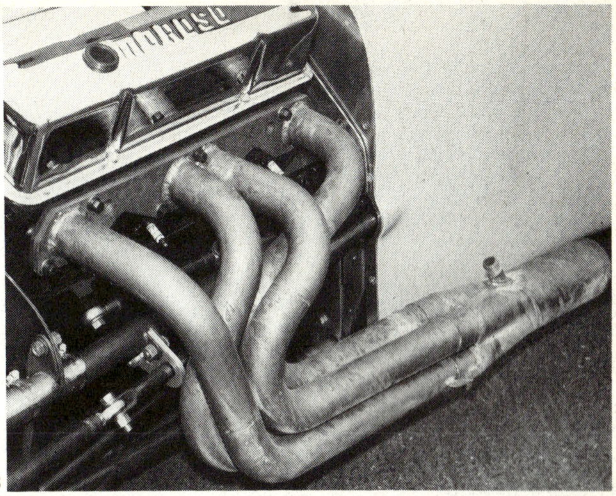

28

29

22. The cylinder heads, which are based on the No. 3965784 angle-plug castings, are prepped as a set of Econo Altered heads would be. The chambers are cleaned up and polished, with a 56cc-to-57cc volume achieved by angle-milling. Not much material is removed, so durability will remain good. The intake ports are cleaned up a maximum of $\frac{1}{2}$-inch back into the port. The exhaust ports are left untouched. A three-angle valve job is done, and the exhaust seats are fully radiused. Winona bronze guide inserts are used, and the heads are installed with No. 3830711 Chevy steel-shim head gaskets and socket-head head bolts—studs are not used.

23. Completing the heads are Manley stainless steel 2.05-inch intake and 1.60-inch exhaust valves, used in conjunction with Competition Cams' 10-degree retainers and keepers. The springs are also from Competition Cams—the No. 913 outer and No. 914 inner springs are installed at 1.75 inches with 180 pounds seat pressure and 475 pounds pressure at .650-inch lift.

24. Atop the heads, 7/16 chrome moly rocker studs are used with BRC aluminum rocker arms. Perfect Circle valve seals are also used. The pushrods and guide plates are Chevrolet items. Note the cleanup work on the intake ports.

25. Oil flow in the engine is restricted with .040-inch orifices in the lifter galleries. McClintic also plugs all valley oil drain holes except at the front. During a run, as much as 2 quarts *may* accumulate in the valley, but at least it doesn't drain back on the spinning crankshaft. Moroso chromed valve covers top the engine, and a degreed harmonic balancer seals the front. Note the configuration of the oil pan for windage and scraper relief.

26. Actuating the valvetrain is a CS308B-R6 roller tappet cam from Competition Cams. This cam features .678-inch and .600-inch lift and 308 degrees and 319 degrees duration at .020-inch lift on the intake and exhaust lobes, respectively. Lobe centers are 106 degrees. This cam works from 3500 to 4000 rpm up, and it has 10,000 rpm stability with lightweight valves. Consequently, this same cam will be used in both the bracket and altered versions of our engine. The cam is installed 2 degrees advanced to help low-rpm power with the automatic transmission.

27. The final trial assembly began with the installation of the McClintic 305-cubic-inch Chevy V8 and the trick A-1 Automatic Transmissions Powerglide. This trans includes an internal brake. Also used is an A-1 8-inch torque converter. The transmission assembly is wrapped with a SEMA-approved Simpson safety blanket.

28. With the engine in the car, the Hooker headers were bolted into place. These headers feature $1\frac{7}{8}$-inch-diameter primaries 28 inches long with $3\frac{1}{2}$x12-inch collectors. The headers will be chromed for final assembly. Note the provisions for the pan evacuation system.

29. Also bolted to the engine during the trial assembly was the Weber No. 397101 Ultra-Tork starter. This unit features heavier windings to spin a high-compression racing engine fast enough for sure starts.

30. To keep front-end weight to a minimum and to reduce power drains on the engine, a Moroso aluminum water pump and electric water pump drive are used. This also permits water circulation while the engine is off. An oversight on the Moroso system requires milling the water pump neck supports to clear the special toothed-belt drive pulley, as shown.

31. The aluminum sheetmetal work on our altered was handled by Larry Storch at Unique Metal Products. The two-piece seat is held in place with Dzus buttons, as is the kick panel.

32. The bottom of the body is sealed and streamlined with an aluminum bellypan. The rectangular opening permits air flow to the Russell Performance Products heat-sink transmission cooler.

33. Unique Metal Products also fabricated the water tank immediately ahead of the engine, the water overflow catch can just behind the front axle, and the fuel tank ahead of the axle. Note that both fuel tank and water tank have all the necessary fittings for fuel injection in case that setup is ever desired in the car. Note the quick-release saddle mount for the dual Holley electric fuel pumps. Such a mount permits quick service of the pumps, should it become necessary.

34. The body used on our bracket/altered is Brogie's own fiberglass '23 T turtledeck. Unlike most '23 T roadster bodies, Brogie's body incorporates a raised turtledeck, an extended cowl and a built-in tonneau cover to reduce the amount of modification normally required with the more traditional '23 T body. Here the turtledeck is held in place to approximate its eventual mounting position.

35. The super-hot Autotronics MSD-7A racing ignition system is mounted just ahead of the steering box, with the MSD capacitive-discharge coil mounted just above the steering box. Note, too, the Brogie-made shifter, the bulkhead coupling of the transmission cooler lines to permit simple, drip-free engine/transmission removal and the simple gauge mounting system.

36. This high-angle frontal view shows how tightly the massive slicks are tucked under the body for optimum streamlining. Front tread width is 46 inches from center to center.

37. This rear view of the car shows the incredible amount of rubber on the ground to assure traction in even minimal conditions. The rear tread is 40 inches from center to center of the Goodyear 32x14.5x15 slicks. Those skins have an 18-inch section width! The slicks are mounted on Crager 14x15 polished Super Tricks with 6-inch backspacing.

30

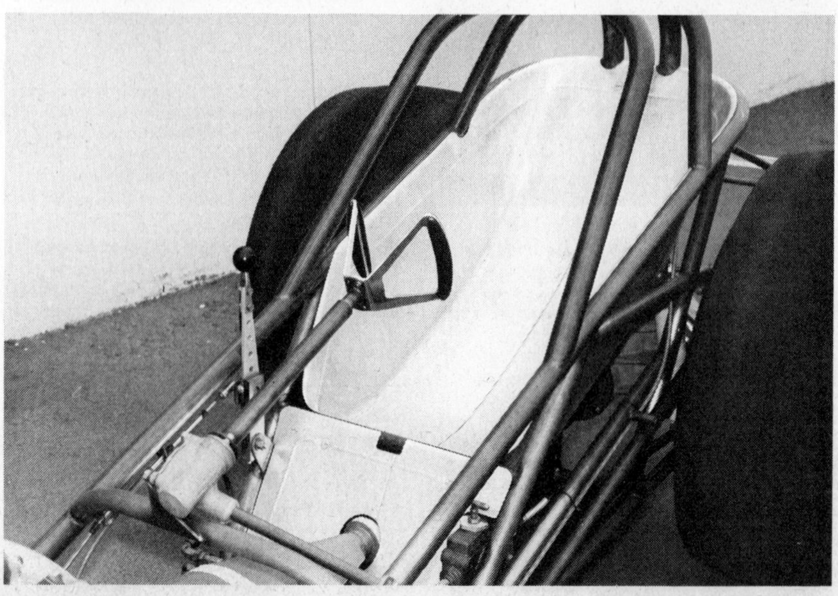
31

32

33

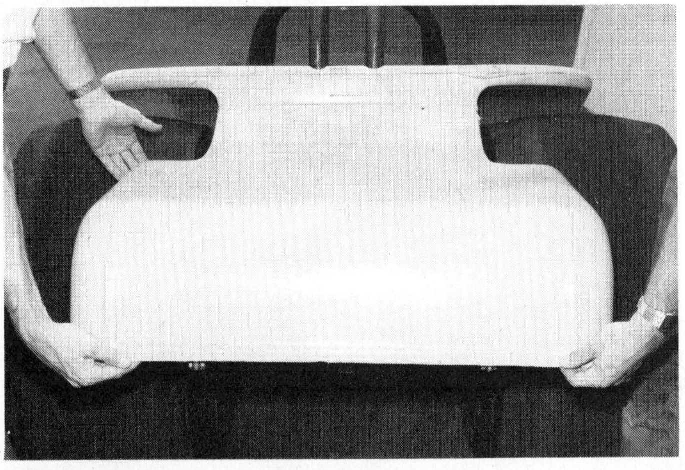

34

35

36

37

POWER PER DOLLAR

We have been able in the past to list the exact prices to purchase an altered chassis from Brogie Race Cars. Unfortunately, because the range of aftermarket engine components is so wide, and because many pieces can be purchased used, engine costs cannot be so exactly listed. We can tell you though, that if reliability and power are primary concerns, the cost of building a competitive engine is never cheap. For example, John McClintic commented that most of the fully prepared. race-ready, small-block Chevrolet *short-blocks* that he builds go out the door at something around $2800. Of course, such a short-block would feature all-new parts, the high tin block, streamlined and cross-drilled crank, fully prepped Z28 or aluminum rods, aftermarket pistons and a roller cam. If you were on a tight budget, shopping carefully for good used parts, a decent racing short-block with a flat-tappet cam might be possible for as little as $1000; but that would be cutting a lot of corners. Machining and labor costs have a way of running even the most frugal estimates up to four figures.

Total engine cost depends on the component selection and the amount of modification (such as cylinder head porting) you desire. A reliable engine with components to rev 9000+ rpm, a few alloy parts, and professional assembly typically falls into the $6000-to-$8000 range nowadays—and that's a drag racing engine. If you want to get serious about a 500-mile roundy-round engine, such as used in NASCAR, figure a minimum of $10,000 per engine.

Racing is expensive. Don't ever let anyone deceive you into believing otherwise. Cheap engines can be built, but they usually lack durability; so the cost of frequent repair or replacement often equals the cost of the top-line engine. For anything other than bracket racing, it takes the best of everything just to be competitive. Power costs money. *—C. J. Baker*

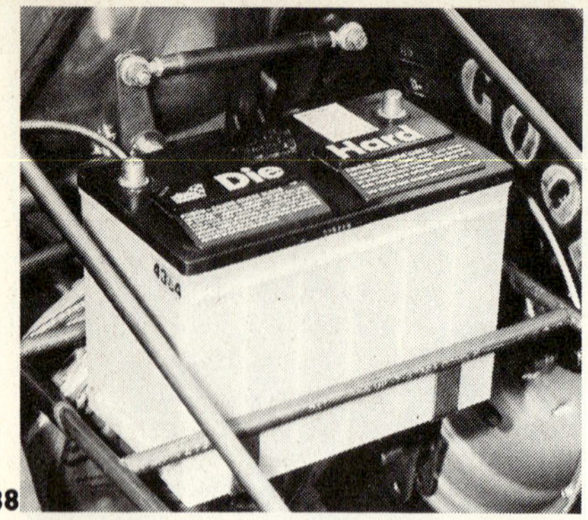

38

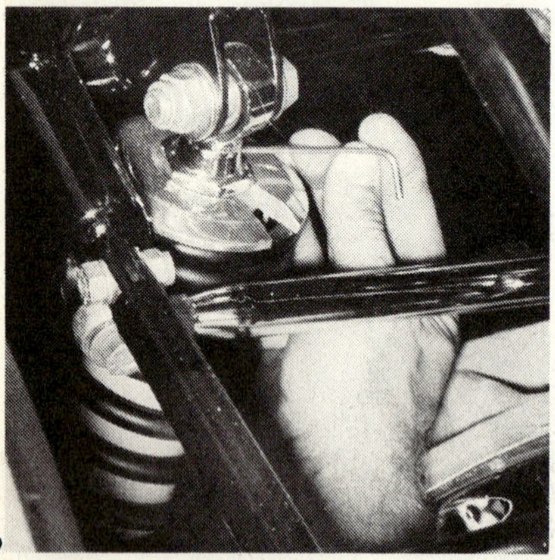

39

40

41

42

43

38. Mounted under the turtledeck is the Sears No. 4384 Die-Hard battery. Brogie's chassis and body are designed to accept two such batteries, one ahead of each slick inside the body, or a single setup, such as shown here. Initially we'll try the single battery, but for the all-out altered configuration, the second battery may be added.

39. Here Warren Brogie adjusts the special coil-over Monroe rear shocks to fine-tune the suspension to varying track conditions. Adjustments are quick and simple.

40. This engine close-up shows the attention to detail. Virtually everything is polished or chromed, including the Hooker headers, Moroso valve covers and engine-mounting hardware. Dual Holley pumps feed dual pressure regulators into the 750-cfm double-pumper. The manifold is an Edelbrock Victor Jr. Russell Race Car Products supplied the braided stainless hose and anodized fittings.

41. Shown here is the Autotronic Controls crank trigger for the MSD-7A ignition. Shown, too, is the Moroso electric water pump drive and the Moroso aluminum water pump. Note the Grade 8 bolts everywhere.

42. With the body removed, the cockpit is clearly visible. Creative Auto Interiors of Placentia, California, did the upholstery. The safety harness and trans blanket are from Simpson. Note the trans shifter, which was custom-made by Brogie. The button atop the shifter is for the internal trans brake in the A-1 Powerglide. Everywhere you look, the detailing on this car is perfect.

43. Warren Brogie and Don Vandermee remove the body from the bracket altered (a two-minute job) while Lee Kelley looks on. It's maintenance features like this that make an altered such a desirable bracket car, as opposed to a stocker.

One of the original ideas behind our bracket altered was to give the staff members an opportunity to actually strap into a purebred open race car and make a blistering pass down the quarter-mile. In concept, it was a great idea. In reality, it didn't work out well.

Getting the entire HOT ROD staff together in one place at the same time for anything other than free beer or paychecks is next to impossible. However, Lee Kelley, C.J. Baker, Gray Baskerville and Bruce Caldwell did manage to meet Warren Brogie and Don Vandermee at Orange County Raceway in an attempt to find out what it was like to drive a genuine race car.

Unfortunately for the staff, an altered chassis bears no resemblance to a passenger car that's designed to fit anybody and everybody. An altered is actually tailored to fit the driver's size, which determines steering wheel placement, pedal and seat position. The driver's height also determines the roll cage height. Since Warren Brogie was to be the primary driver of the car, he's the one our altered fits. Of the staff members present that day, only Lee Kelley fit the car comfortably. C.J. Baker's legs were a bit too long, preventing him from being able to properly actuate the brake pedal, and his helmet was jammed up against the roll cage. Gray Baskerville decided that he fit his camera better than he fit the altered, and Bruce Caldwell . . . well, the accompanying photo sequence says it all.

C.J. did manage to make a restrained pass in the car (it's hard to tell your right foot to go when you can't get your left foot on the brake), as did Caldwell, but it was Kelley who really got with the program. On his first pass, Lee staged the car, accidentally rolled out of the starting beams before nailing the throttle, and still ran a 10.34 at over 137 mph!

It was a day of fun, and everyone who was there came away believing that if you want to go drag racing, want to go fast and want to have fun, a car built specifically for those goals is the only way to fly.

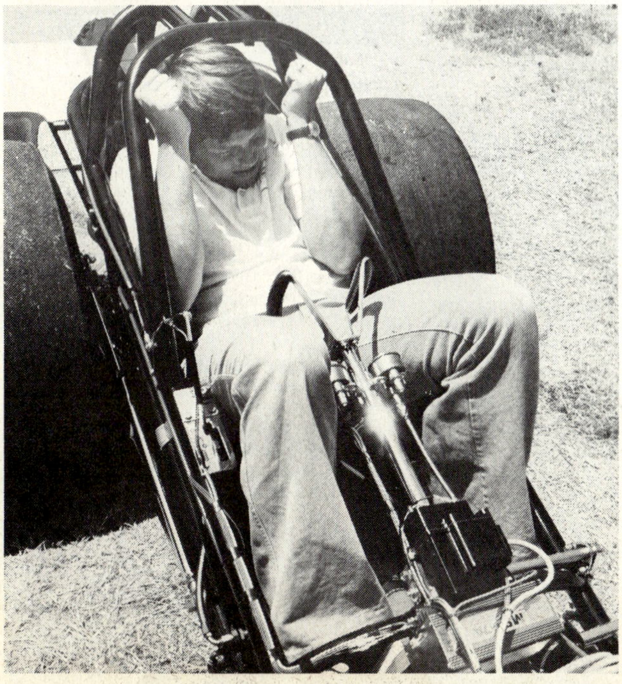

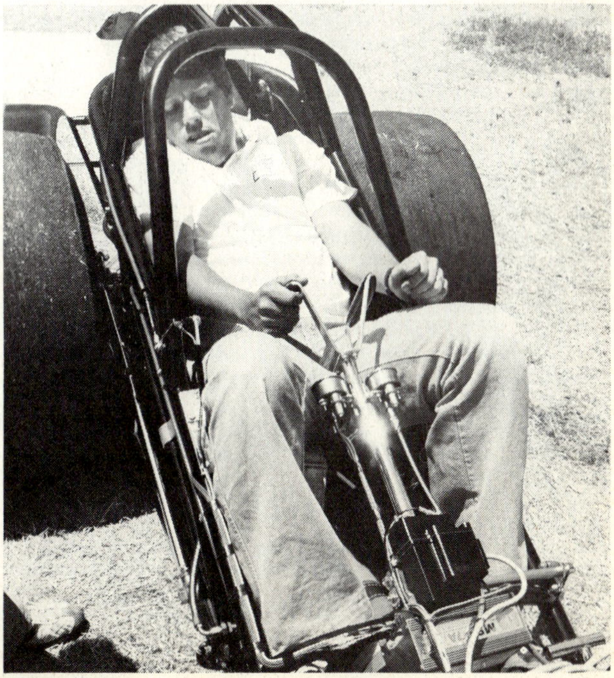

Getting into a snug-fitting altered isn't easy. Here, Bruce Caldwell demonstrates how to cram 10 pounds of editor into a 5-pound car. Note his changes in expression!

How to Build a Street Machine

"The best way for a beginner is to unscrew it, then carefully screw it back together."

(Li'l John Buttera)

Don't tackle any project without adequate preparation. Ask for help or advice from someone who has had experience with your kind of project. A lot of guys who don't know what they're doing point with pride at a botched-up job because they did it themselves—and it looks like it! You must take one step at a time to realize the full monetary and commonsense rewards from that first experience. Remember, if you do fail, don't be afraid to fix it, because we all foul up."

Ah, yes, the wisdom gained from building 200-plus cars in the last 20 years is something Little John Buttera can impart to our readers in general, and to his son Chris in particular, for that's what this yarn is all about: how a 19-year-old kid built his first street machine.

"You know, kids are crazy and do dumb things," mused the mustachioed Li'l John the other day. "Take my kid Chris, for instance. He had this perfect '69 Z28 Camaro and wrapped it around a telephone pole one night while showing off in front of his pals. He's your typical kid who likes cars and the type who can identify with late models because that's what he and his pals mess around with. About a year ago he bought his Camaro from an old guy who liked fast cars and who had it equipped with all the factory hot rod options. Chris got it for $2500, put $1000 worth of wheels and tires on it, and then promptly bent it in half. He couldn't afford to turn his back on his investment, so he went out to the garage and began to fix it.

"You might say that Chris sort of got forced into it. I wasn't going to let him walk away from his bent Camaro, even though he had no real automotive background—except from hanging around me or what he had learned from bench racing with his pals at the local Jack-In-The-Box. That crash turned out to be a blessing in disguise, because instead of taking a $3500 bath, Chris learned how to take a car apart, clean up the various components, detail them, then put the entire package back together again by himself. He got himself a six-year education in six months by becoming aware of how the wiring was run, how the upholstery went in, how the front and rear suspension worked, how the window glass was attached, how the fenders bolted on and how an entire car was constructed."

1. The second Camaro was completely disassembled (including wiring harnesses, upholstery, sheetmetal and exterior brightwork) by Chris Buttera before the "parts" car was primered and painted by Terry Finch.

2. The 302-inch Z28 engine retains all the high-performance factory-supplied hardware, augmented by a Milodon oiling system, an Accel ignition package and this somewhat unique Crower-built induction assembly.

3. The two-quart Moon receiver tank was fitted with a removable Holley float bowl (and float assembly) to control the fuel level. Fuel is pumped into the top of the tank so gravity can be used to help prime the injector pump. Bottom line (arrow) goes to the pump, while the top lines are bypass returns.

4. However, Chris decided to spend some money dressing up the engine compartment by polishing or plating the water pump, fuel pump and alternator mounting brackets, pulley and timing cover.

1

2

3
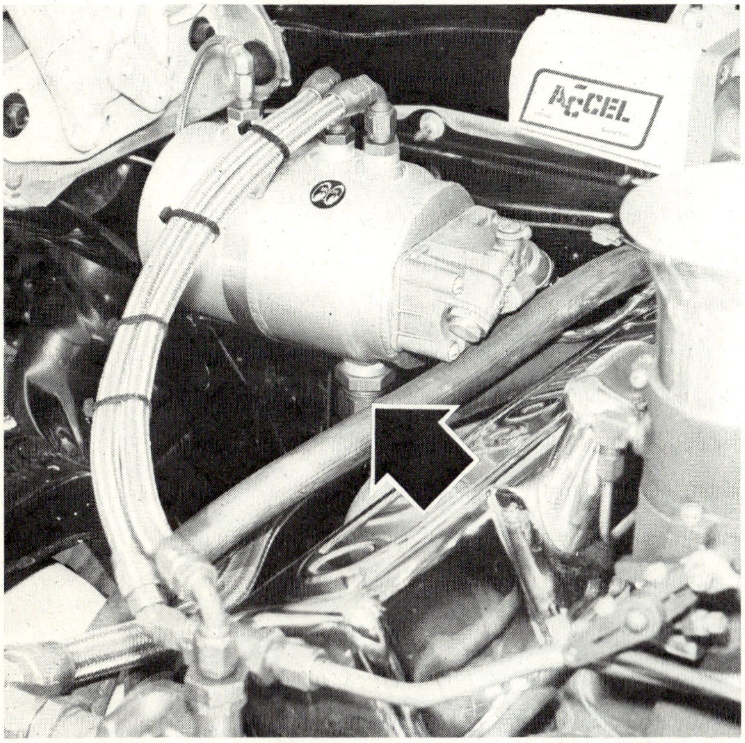

4

In a body shell, that is what this story is all about. We're simply going to take a car apart, rebuild or refinish the various assemblies, then reassemble it in the age-old tradition of building a street rod. We know that many street machiners will simply add some aftermarket wheels, tires, headers and/or chrome goodies to their otherwise "stocker," and consider it done. We also realize that others will go the full blower route, using the eye-catching appeal of a polished 6-71 Jimmy huffer, wheezing through the hood of a flamed and fuming shagnasty, to build a car that looks more at home in the staging lanes at the local drag strip than huffing down "burger lane." However, Chris' Camaro is what we like to call the middle ground. His Z28 combines the best of the bolt-on world with some of the most eye-catching performance tricks—he never forgot, however, that he must not adversely affect the car's potential resale value.

In essence, Chris' six-month learning process involved nothing more than finding a '69 Camaro "parts" car that had no more than a near-perfect body. He could then strip it down to the bare shell, sell off what he didn't need, take the high-performance factory components from his bent original, detail them, and carefully reassemble the total package—and eventually wind up with what you see on these pages.

UNSCREWING IT

Because Chris unscrewed his original '69 Z28 with a telephone pole, all that was left for him to do was to cannibalize it for its high-performance options and store them away for future use. "The secret here," according to his fidgety father, "is to remember where everything goes. Now this isn't easy, so Chris took a Polaroid of each assembly and made notes on how it came apart; then he bought a '69 Camaro Service Manual, so he could see and read how those components went back together again. He then took everything out of the wrecked Z28; and that included the headliner, upholstery, wiring harnesses, window glass, trim—everything that could be used to turn his parts car into another Z28."

SOCKING IT AWAY

It's all well and good to remember how something comes apart and goes back together again, but if you can't find that fastener or rod end, you're in trouble. "If Chris learned one thing," Li'l John, "it was how to keep track of all the pieces. He used grease pencils (or China markers), labeled boxes and coffee cans and stored them in such a way as to make it easy for him when it came time to massage each component. By doing it that way, he didn't run into any memory problems.

MOTOR RESPONSE

"It was both Chris' and my thought that, from a psychological standpoint, his Z28 should have a distinctive motor. The idea, therefore, was to build a performance engine to fit the Z28's image of high performance. The engine out of the original car was good; it had all the right components; and when we made a leak-down test, we found the 302-inch small-block Chevy V8 was in great condition. I guess Chris could have taken the engine apart and had it balanced and blue-printed, but why spend the money when you don't have it?"

Consequently, the little 302-incher was simply steam-cleaned, repainted a glossy, spray-can black and fitted with four basic elements: an Accel distributor, a Milodon oiling system, Kenny Moores headers (Positive Performance Headers, Fullerton, California) and one streetified-type set of Crower port injectors.

"The reason we decided to equip the 302 with an Accel distributor is simple.

1969 Z28 CAMARO
- FULLY DETAILED FRONT SUBFRAME
- CAD-PLATED UPPER AND LOWER A-ARMS
- CAD-PLATED MONROE SHOCKS
- CAD-PLATED STABILIZER BAR
- 1969 302-INCH CHEVROLET V8
- ACCEL DISTRIBUTOR
- MILODON PAN AND ROCKER COVERS
- CROWER INJECTOR
- MOON HOLDING TANK
- 1969 Z28 REAREND FITTED WITH STRANGE ENGINEERING BEARING RETAINERS
- CENTER LINE WHEELS 7x14 FRONT, 9x15 REAR
- FIRESTONE TIRES P205/7OR14 FRONT, SS RADIALS LR60-15 REAR
- SIDE MARKER LIGHTS REMOVED
- REAR FENDER LOUVERS REMOVED
- STOCK 1969 Z28 INTERIOR
- OPTIONAL HOOD
- LIFT BARS
- KENNY MOORES HEADERS AND EXHAUST SYSTEM

We wanted to fire all the fuel from the injectors, and under compression that could be difficult with a less powerful ignition system. As far as I'm concerned, a good ignition is the most important part of a performance engine.

"As for the 'Dooner' oiling package, I felt that a high-capacity pump with a big oil pan was a plus in the engine-cooling department. Oil is a coolant, too, and by lowering the oil temperature, you can lower the temperature of the engine.

"The reason behind the exhaust headers is this: While your out-of-the-box-type headers are good, Chris wanted something distinctive, something that didn't allow the rest of the system to fall below the rocker panel line and kill the Camaro's profile by having anything hanging down along the bottom of the car. We wanted to make the exhaust system as unobtrusive as possible and still keep its performance image. The balance of the custom exhaust system is made up of Corvair-type turbo mufflers which combine a nice shape with a mellow tone (Why tee off your neighbors with a set of raspy pipes?) and two 2½-inch-i.d. tailpipes shaped to keep the exhaust system hidden from view."

But the big deal was, and is, the fuel injectors. "I guess Chris could have bolted a blower on his engine, but that would mean he would have had to do away with the original hood line, because you gotta hack and slash the hood so the blower will clear the sheetmetal. We decided to add a set of port injectors and still retain the clean performance image.

"The only problem we ran into was finding a cheap and simple way to prime the injector pump. The way we solved that item was to take a two-quart Moon aluminum tank and bolt a Holley removable float bowl to one side of the tank. It was simply a matter of hooking up a Carter competition-type electric fuel pump near the stock gas tank and running a fuel line from it to the firewall-mounted 'receiver' tank. What happens here is that the electric pump fills the receiver tank until the float shuts it off. A No. 12 steel-braided line is run from the bottom of the tank to the injector pump, which always keeps the injector pump primed. Two return lines (one a high-pressure relief valve on the pump and the other the bypass jet) are rerouted back to the top of the receiver tank. It's very simple and works just as easily.

"The injector itself is strictly a bolt-on item. The only thing you have to do when ordering a system like this is to tell the Crower people how their injector is going to be used, and they'll program it for your particular application. They'll flow it for you, send you the right port nozzles and bypass jets. Chris' unusual induction system cost $750 for the injectors, pump, belts, pulleys and jets, plus an additional $170 for the receiver tank and about $50 for the steel-braided fuel lines, although he could have saved himself $40 if he had used rubber hoses. However, he's got something different, with a heck of a lot of visual clout per dollar."

DRIVELINE DISCUSSION

The only other thing Chris did to the engine compartment was to use a Deuce Factory thermostatically controlled electric fan (because the injector pump pulleys and belts interfered with the stock radiator shroud) and to fully detail and paint the inner fender panels and firewall—but we'll get to that later. Meanwhile, the stock Z28 clutch and flywheel package was replaced by a Schiefer "street/strip" single-disc model used in

5

6

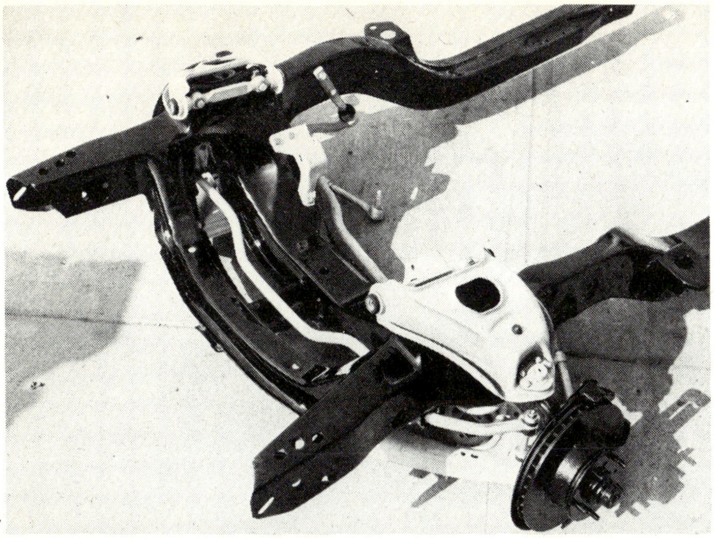

7

8

5. Injector stacks have been cut down for underhood clearance. Other than some basic detailing and the addition of a pair of Milodon rocker covers, the underhood package is basically factory stock.

6. Now any novice street rodder can do this to his front suspension. Here Chris (left) begins to assemble the Camaro's underpinnings.

7. Note that the welds on the front subframe have been ground and filled with "bodyman in the can" and then painted with high-gloss black acrylic enamel.

8. You can see that the upper and lower A-arms, shocks, springs (cut down one inch) and stabilizer bar have been cad-plated to contrast with the gloss black subframe and floor pan.

9. See how the Kenny Moores-massaged headers are tucked up inside the subframe so the entire exhaust system doesn't hang down below the rocker panel line. Again, note how the trans mount has been cad-plated and the floor pan painted a contrasting black.

10. As we've stated before, this is a '69 Z28 Camaro that any novice could build. The wheel/tire combination is appropriate, the paint layout is contemporary, and there are just enough body modifications (like the removal of the side marker lights and rear fender louvers) to visually clean up the exterior of the car.

9

10

conjunction with a steel flywheel and a pair of new U-joints, because "you can't touch one part of a car without touching other parts." To round out the balance of the driveline, Chris retained the stock axles (although he checked them for twist) and added a set of Strange Engineering bearing retainers to keep the wheels in the axle housing in case he snapped an axle. The stock 4.11 "posi" was taken apart to make sure all the bearings were good; and the brakes were completely rebuilt using good, aftermarket, NAPA-type linings and rebuilding kits.

As Li'l John points out, "Essentially, all the changes Chris made to his car were done for the sake of improvement, not for the sake of change. By upgrading the Camaro with longevity and safety in mind, you can compensate for a power increase without having to worry about undue breakage."

SUSPENSION EVALUATION

Except for one coil cut out of the front springs and the addition of a pair of high-performance Monroe tube shocks, the front suspension on Chris' Camaro is box-stock. "We could have made the front end lower, but Chris didn't want to use small front tires, and he didn't want his new Firestone P205/70R14 radials rubbing on the inside of the fenders." However, all the various suspension components were either cad-plated or spray bomb-painted, and the front-end assembly was equipped with all-new tie rod ends and ball joints.

Like the front suspension, the rear suspension was simply taken apart, stripped of old paint, road grease and dirt buildup, then either cad-plated or re-painted. Chris bought a traction bar kit from a friend and used it, along with a set of new Monroe Road Handler shocks and a chrome-plated Milodon differential cover. The axle housings were sanded, then painted with a combination of Rustoleum and out-of-the-bomb semi-gloss black enamel paint. Again, all the suspension bushings were replaced with new factory units.

WHEELING AND DEALING

To ensure that distinctive Z28 performance-oriented look, Chris opted for Center Line "Indy" rims which measure 7x14 inches on the front, 9x15 inches on the rear, with a 3½-inch offset on the front and the rear. The tires temselves are Firestone P205/70R14s on the front and LR60-15 SS radials on the back. The front skins are 25 inches tall, with a 6-inch tread width, while the rear meats are 27½ inches tall, with a 9-inch tread. According to Li'l John, "I figured I had one crazy kid, so if he went out and cut a tire in the middle of Resume Speed, Iowa, he could find a replacement without too much trouble."

FRAME UP

As you might recall, Camaros, like most late models, rely upon combining a unibody with a front frame section that

11

12

13

14

15

bolts to the firewall. Well, Chris unbolted the front frame section and had it sandblasted. Then, with a lot of help from Terry Finch, he detailed the welds with an air grinder, a file or two and some plastic filler, before having painter Terry cover it with black acrylic enamel laced with a hardening agent that promotes a high gloss and durable finish. The rear frame section was also wire-brushed, cleaned, detailed and painted a semigloss black with a spray bomb.

BODY, PAINT, UPHOLSTERY AND SOUL

Since Chris bought a '69 Camaro parts car for its body and not its used-up engine or running gear, little was needed in the way of straightening out the body, but it did need a thorough stripping. It's harder to do it this way, but the only means to ensure a first-class paint job is to do what Chris did: remove all the trim, tear it apart and take it down to bare metal. It's hard, dirty, no-good, son-of-a-gun work; but if you want it straight, you've got to play it straight.

To make sure his Camaro would be clean, Chris had Steve Davis fill in the side marker lights, as well as remove the louvers in the rear fenders, because they would interfere with the paint layout. Every salvageable part of the "bent" original was retained, because Chris didn't know what he could or could not buy. For instance, the Z28 had its own interior, and securing genuine replacement could have been a problem; but it wasn't, and all the OEM (Original Equipment Manufacturer) rubber is still available from Chevrolet.

The paint layout is based on the stock Camaro silver, set off by Dennis Ricklefs graphics, encompassing red, yellow and blue panels separated by red, yellow and blue pinstriping. "The thought behind this Terry Finch-squirted paint scheme," Buttera points out, "was to accent the Camaro's body lines without destroying the performance look of the car. A paint job like this could have cost Chris $1500, but he saved himself a lot of money by doing all the dirty work."

The interior of the Z28 is bone-stock and in keeping with the philosophy that this particular car is a learning process, not a radical rod. It's a first step, a getting your feet wet and eventually ending up with a brand-new old car. The rear kick panels, or inside trim panels, are brand-new, and one seat has a new cover. Other than that, what you see here is the original '69 Z28 upholstery. The interior also sports new rugs, the best of the window glass from both cars, and although the stock wiring looms were retained, Chris had to rewire both the ignition and fuel pump circuits.

The old man is proud of this particular car, because his son persevered, put in a lot of time, and can now see the value of his work. "Everyone likes it. It's a tribute to showing people you can have something nice without having to spend a lot of money or buy a lot of equipment. It's a nice car to his peers, but Chris hasn't destroyed its resale value. It's a resto-machine, and he's gained the knowledge that he can do it. There are times when car building is boring. It's a dirty job. It's removing 10 years of oil, lube and road grime. But he proved to himself he could do it."

Yes, Chris Buttera certainly did prove it to himself—and how!

—Gray Baskerville

11. Rearend housing has been detailed and equipped with Strange Engineering bearing retainers to keep the wheels in place, should Chris break an axle. Buttera also checked both the "posi" bearings for wear and the axles for unwanted twist.

12. This is how Chris' Camaro appeared before the body was rubbed out. Check out super body fit and subtle rake.

13. As we've said, half of Chris' Camaro was painted with a spray can. Here he squirts the power steering box hanging on a backyard-style support system.

14. While the all-black interior is completely stock, it does feature a Hurst shifter and linkage, as well as an aftermarket wheel.

15. The traction bars are of undetermined origin, but the chrome-plated rearend inspection cover is a Milodon item. Rear sway bar and gas tank retaining straps were also cad-plated.

Novamotion

Joe Sherman's '77 Nova is so stock you won't believe it—and it runs low 10s at over 130!

Building a successful bracket racer doesn't have to be an expensive proposition. Frequently, it's just a matter of doing the best you can with what you already have, buying only what is absolutely essential to meet the safety requirements or to achieve the desired performance. Innovation and doing things for yourself, rather than buying special components or hiring someone to prepare the car for you, is part of the allure of e.t. handicap racing, since everyone can be equally competitive regardless of mechanical ability or ability to invest in a race car.

Perhaps this whole philosophy behind building a bracket racer is best exemplified by Joe Sherman, a 38-year-old tool-and-die maker at Edelbrock who resides in Huntington Beach, California. Here's how Joe sums it up: "Keep it simple, keep it cheap, do everything for yourself that you can, and don't spend any money that isn't absolutely necessary." Does Joe Sherman's approach work? Well, you may remember that several years ago Joe had a '70 Chevelle big-block that ran 9.60s at over 143. Now he has the small-block-powered '77 Nova that you see in the accompanying photos. Its best run to date is 10.35 at 131.20, and Joe estimates that he's only invested about $3000, including the engine and transmission, to achieve this kind of performance from what started out as the family grocery-getter.

Sure, Joe Sherman is an accomplished machinist, and he has a lot of friends and contacts in the speed equipment industry, but it's really nothing that most guys with a little initiative couldn't duplicate in terms of product sponsorship and careful do-it-yourself work on a similar car. In today's inflationary times, building such a fast and competitive race car is still possible. If you follow Joe's procedures, you could duplicate this hard-charging Nova, or apply most of the ideas to whatever kind of bracket racing doorslammer you choose. Pay close attention to the accompanying photos and you'll see what we mean.

—C.J. Baker

1

2

4

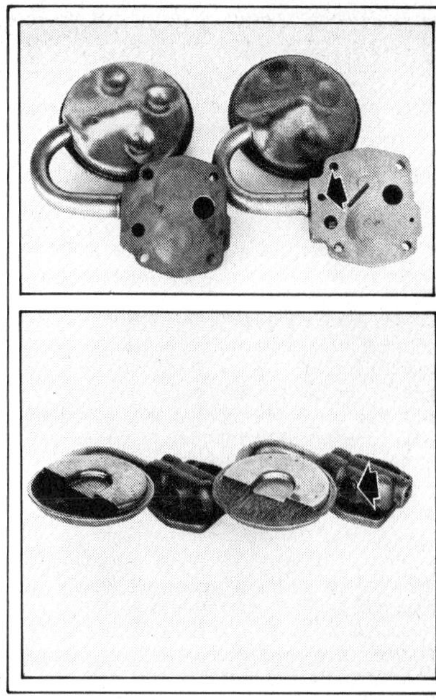
7

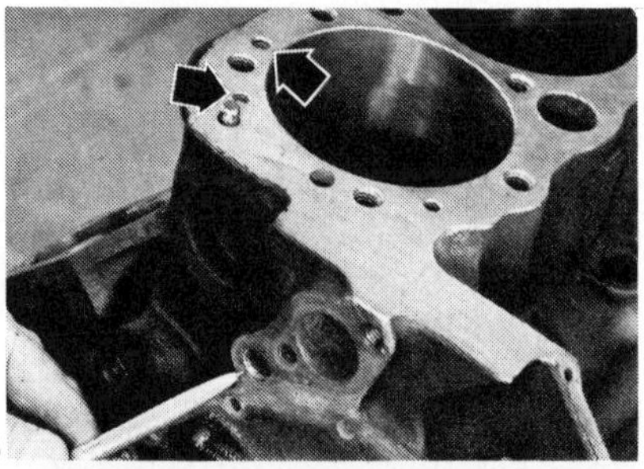
3

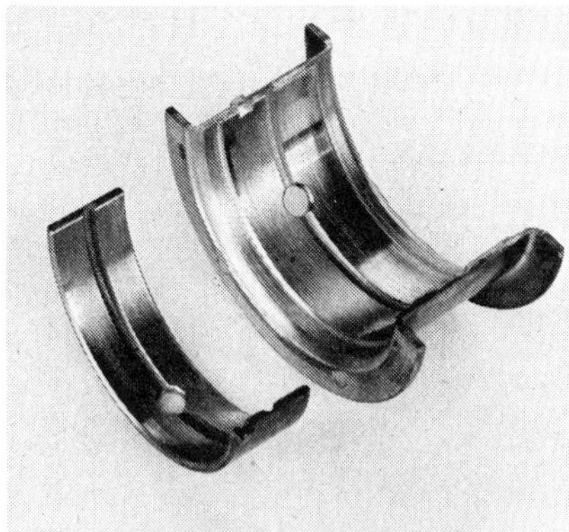
5

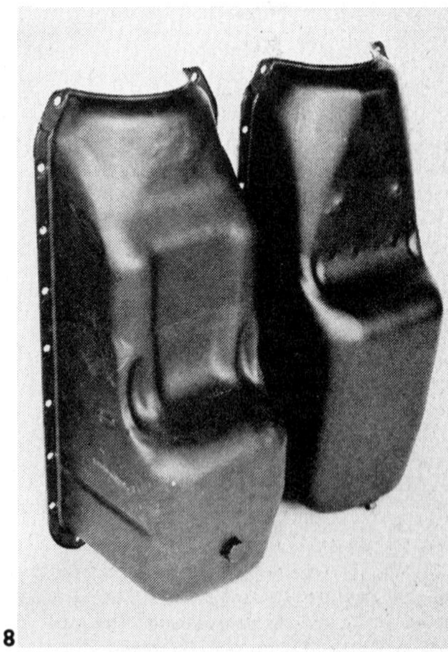
8

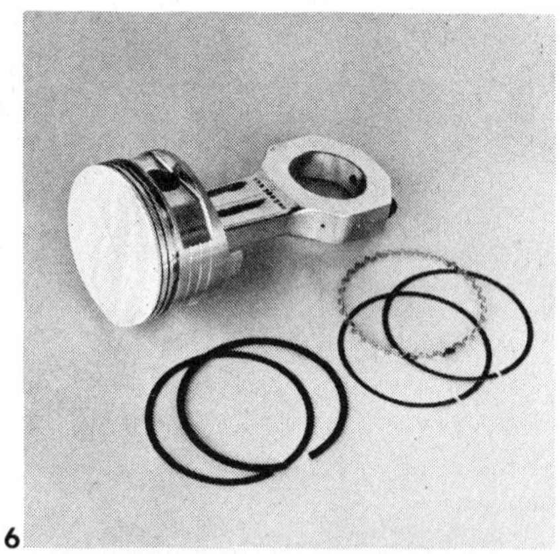
6

1. This is the normal pre-race dance. First, Sherman boils the 11.5/28.5-15W Firestone 9-Compound slicks in a water burnout. These slicks are mounted on 15x10 Center Line wheels with 4⅞-inch backspacing. Next, Joe does a short dry burnout to assure that the slicks are free of water, usually pulling the 4/28-15 Goodyear Front Runners, mounted on 15x3½ Center Lines, right off the ground. Then it's to the starting line for another arrow-straight low-10-second pass.

2. In keeping with the philosophy of saving money, a used '69 Corvette block was purchased in a salvage yard for $75. The cylinders have been bored .030-inch and honed to a very smooth finish compatible with the moly compression rings. The block has not been align-bored or decked. On both decks, the lower water passages between the front two and rear two cylinders have been blocked (arrows), and the center cooling hole and spark plug cooling holes (circled) have been drilled on the driver's side only. These modifications are intended to provide more uniform cooling of the engine, based on water circulation patterns within the block.

3. Another water circulation modification to the block includes blocking the water bypass hole on the front of the passenger side of the block. Also plugged are the front two cooling holes (arrows) on the passenger side deck. Note that the pressed plugs normally used at the front of the oil galleries have been replaced with screwed-in plugs.

4. A close inspection of the lower end of the four-bolt block reveals that the pan rails have been notched to clear the aluminum connecting rods. A stock 3.48-inch-stroke crankshaft is used, which, in conjunction with the 4.030-inch bore, yields a displacement of 354 cubic inches. The counterweights on the crankshaft have been turned down ¼-inch to help compensate for the light piston/rod assembly, which saves excessive drilling of the counterweights when balancing the final assembly. This reduces crankcase windage. The oil holes have been chamfered and cross-drilled. The crank work and final balancing were done by Reath Automotive.

5. Joe Sherman uses the GM Morraine 400 main bearings, but he drills out the oil hole to the same diameter as the oil passage in the block that supplies the main bearing. The enlarged oil hole in the bearing is also chamfered. Bearing clearances are held to about .003-inch on both the rods and mains.

6. Because this engine is intended to run at speeds as high as 9500 rpm, lightweight aluminum rods and pistons have been used. Both are from Venolia. The aluminum rods are .300-inch longer (6.00 inches center-to-center). The big ends are 2.1-inch diameter, and the caps are pinned to hold the Federal Mogul 3200AP bearings. Rod weight is 560 grams, which is about the same as the Chevy rod, but the aluminum material offers elasticity to absorb some of the energy of starting and stopping the piston at 9000+ rpm. The No. 284PS pistons are held on the rods with Venolia lightweight pins (.072-inch wall, 2½ inches long) and single .050-inch Tru-Arc pin locks. The pistons, which weigh 415 grams, have been internally and externally lightened and feature extra-deep valve reliefs. The domes have a ¼-inch radius, and the intake pocket is radiused slightly. They are cut for a .043-inch Speed-Pro top ring located .090-inch down from the deck. Ten .050-inch gas ports help load the ring. The second ring is a 1/16-inch cast-iron version, and the oil ring width is 3/16-inch. Both are Speed-Pro rings, and the oil ring features super-low tension.

7. Engine oiling and oil control have been given considerable attention. The Z28 oil pump, which features a pressure balance slot in the cover, is used. Note that the pickup is also secured by a setscrew (arrow). Since this pump was never intended for 9000+ rpm use, modification to the pressure relief circuit is needed. The bypass hole is enlarged to .45-inch (stock is .3-inch) and a .045-inch shim placed under the relief spring. The spring retainer pin diameter is also increased from 3/32-inch to ⅛-inch. Access for enlarging the relief passage is indicated by the arrow. Larger hole is then tapped and plugged.

8. The oil pan used is the standard '63-'65 pickup or passenger car pan that is ¾-inch deeper than the Corvette pan, and it holds six quarts, which is perfect. An expanded sheetmetal windage screen is also used.

9. To properly position the oil pump in the deeper pan, a ¾-inch homemade aluminum spacer is used with a Moroso anti-slosh baffle. The big-block oil pump driveshaft is ¾-inch longer than the small-block version, so it works perfectly.

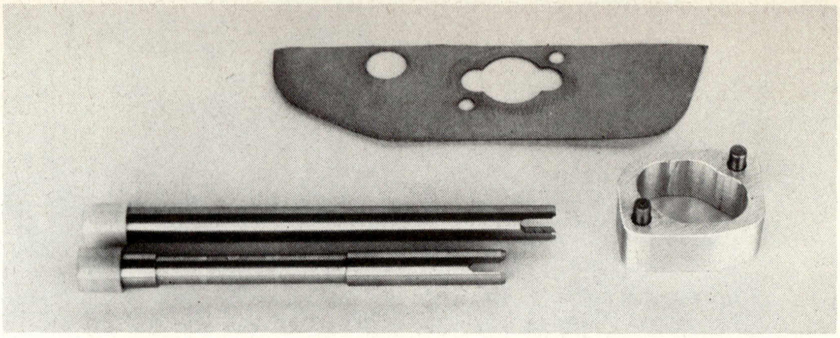

9

10. The cylinder heads used are the latest turbo slant-plug versions. The heads were originally given a basic street porting job at Joehnck Racing Heads in Santa Barbara, California. This included opening up the intake and exhaust ports, polishing the entire length of the ports, and a nice valve job. Since that time, Sherman has enlarged the intake ports another .100-inch in height and .070-inch in width at the flange area to complement the Victor Ram manifold that is now in use on the car. The heads have been milled to provide 62.5cc combustion chambers. The valves are Manley stainless steel versions that feature necked-down stems just behind the valve heads. The intakes are 2.05-inch and the exhausts are 1.60-inch. Both have stems that are .100-inch longer to allow installed spring heights of 1.900-inch with special offset retainers, which are also from Manley.

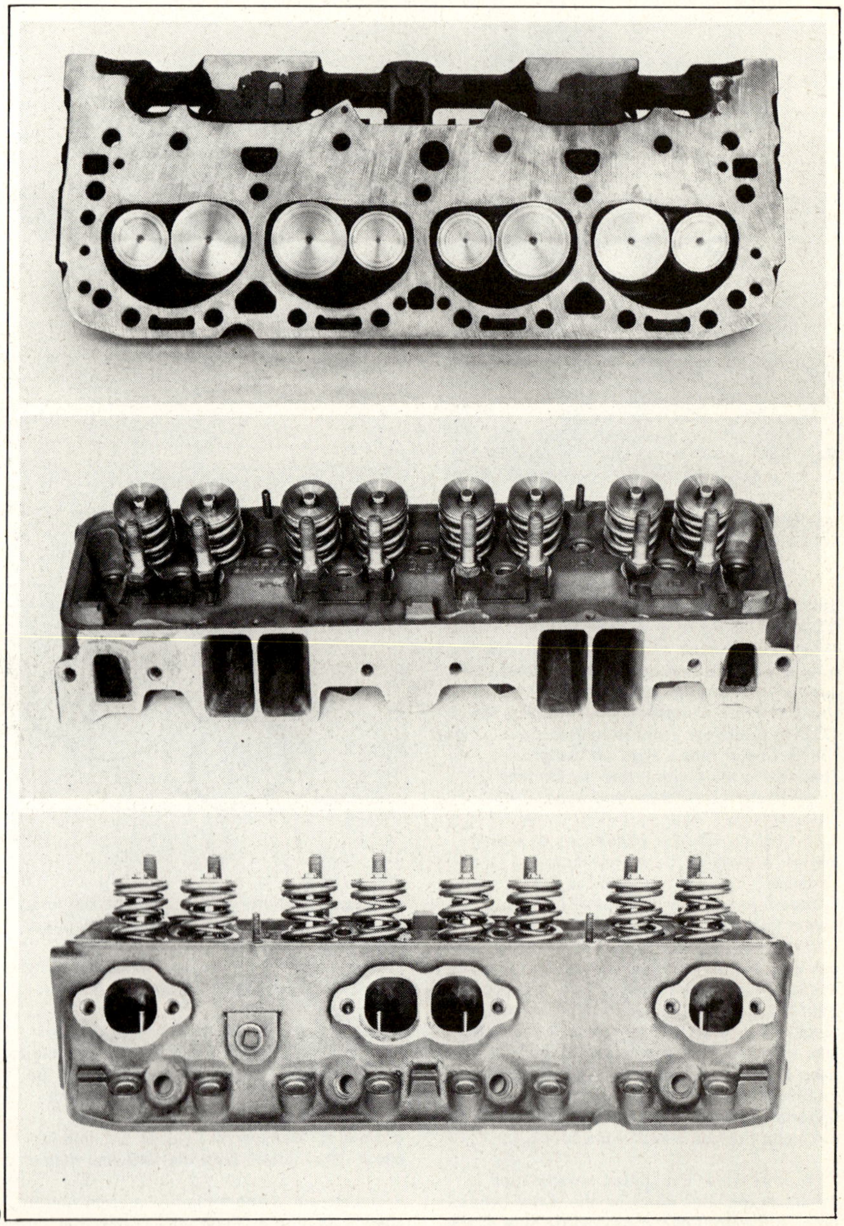

10

11. Completing the heads are Sig Erson guide plates (the holes had to be elongated slightly to achieve proper valve stem alignment) retained by big-block 7/16-inch rocker studs. The base of these studs had to be shortened .125-inch to prevent bottoming in the stud bosses. Erson heavy-wall ⅜-inch pushrods actuate Crane roller bearing rocker arms. The intakes are 1.6:1 and the exhausts are 1.5:1. Since Sig Erson triple stainless steel big-block valve springs are used, the rocker arms had to be machined to clear the larger diameter (1.560-inch) springs and retainers. A Jomar double-bar Pro Stock-type stud girdle (not shown) is used to help stabilize the valvetrain.

12. The camshaft is driven by a stock Chevrolet silent timing chain and nylon-covered cam sprocket. The crank sprocket is a steel Borg-Warner unit. Although not shown, the cam used is a Sig Erson 675AB roller cam that has 274 and 286 degrees duration at .050-inch valve lift on the intakes and exhausts, respectively. The lobe lift is .447-and .440-inch on the intake and exhaust, which equates to .715- and .660-inch gross valve lift with the 1.6:1 and 1.5:1 rockers. Valve lash is set at .024- and .026-inch, respectively. Completing the front of the engine is the big 350 harmonic balancer and a Moroso 3-inch crank pulley.

13. The top of the engine is buttoned up with an Edelbrock Victor Ram manifold supporting twin Holley R-4224 660-cfm carbs. Number 75 jets are used all the way around.

14. A ½-inch-i.d. fuel line runs from the back of the car, where a single Holley high-performance electric fuel pump supplies the fuel, to a single pressure regulator that supplies both carbs through ⅜-inch-i.d. lines. Pressure is set at 6½ psi. The gasoline used is Daeco Super Premium racing gas from Daeco Engineering in Wilmington, California. The valve covers are Edelbrock's deep-drawn aluminum versions that provide clearance for stud girdles and high-lift cams. The spark is supplied by the latest Pulsar System from Stinger Ignitions. Stinger plug wire is also used. Joe estimates the power output of this 354-cubic-inch small-block at 650 hp.

15. Here Joe Sherman carefully checks the ignition timing, which is held to a maximum of 39 degrees. The spark plugs are Champion BL-9Ys gapped at .040-inch. Note that no fan or fan shroud is used, and that the stock grille is still in place behind the aluminum sheeting shown in other photos.

16. The headers are 1⅞-inch diameter with primary tubes 31 inches long. They are manufactured by Hooker. Sherman added the pan evacuation system that is plumbed into the collector. The converter dust shield is left off to provide some air cooling to the A-1 Transmissions C-Model 8-inch converter, which is rated as having a 5300-rpm stall speed. The modified Turbo-Hydro 350 transmission is also from A-1, and Simpson Safety Equipment provided the transmission safety blanket.

17. Joe's ideas about not spending any unnecessary money certainly hold true to form at the rearend of his '77 Nova. That's right, it's stock Chevy 10-bolt rearend, using a stock Chevy 3-series posi unit and cast caps. Even the stock axles are used. The gears are Zoom 5.57:1 units set up with .005-inch backlash. Joe feels that no more than this is needed, since the automatic transmission is so easy on the drivetrain. As a tip, Joe suggests using straight Valvoline 90W gear oil in the rearend, which provides less posi slippage than high-pressure positraction lubricant. Remember, keep it cheap and keep it simple.

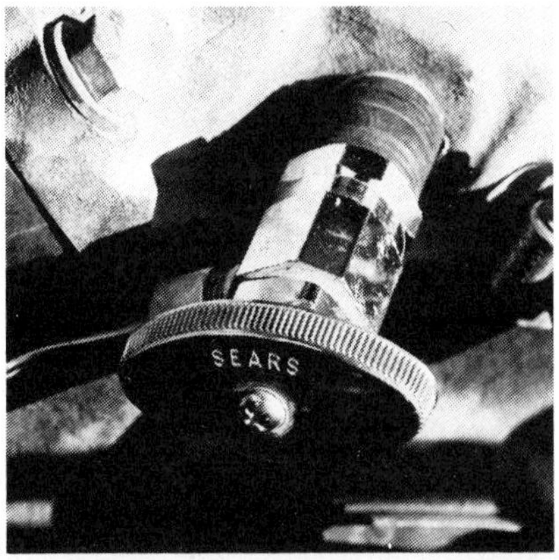

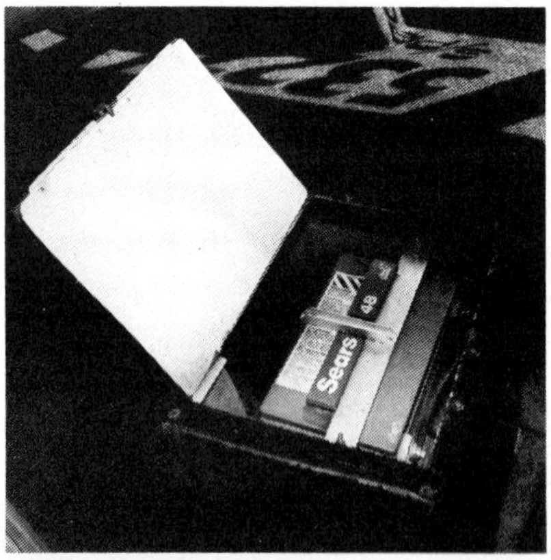

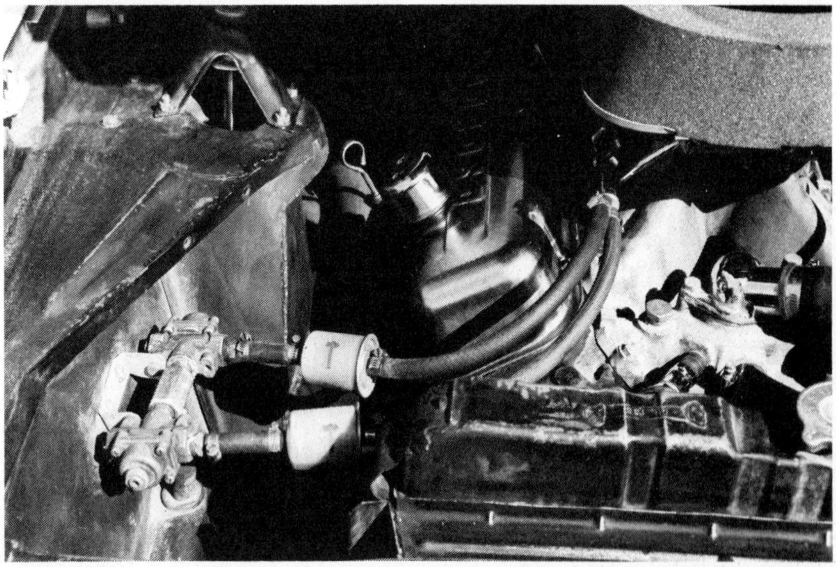

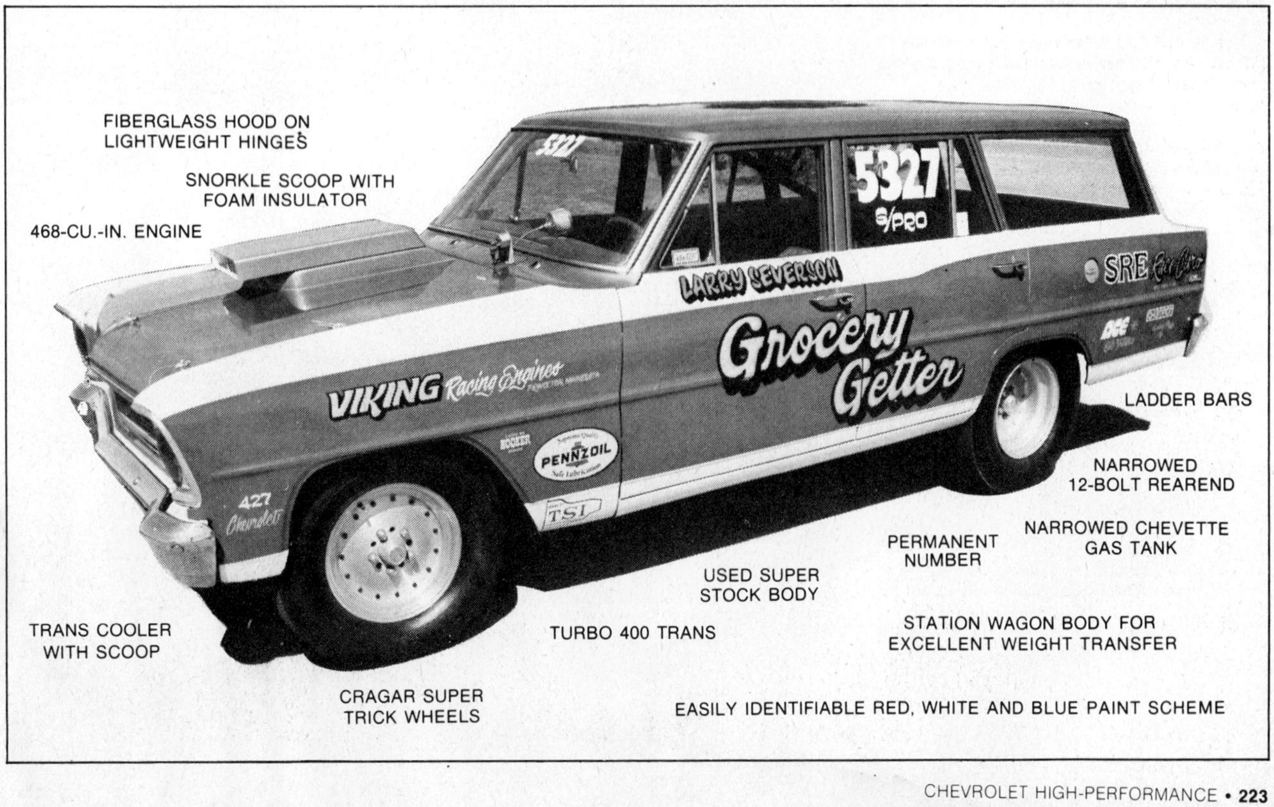

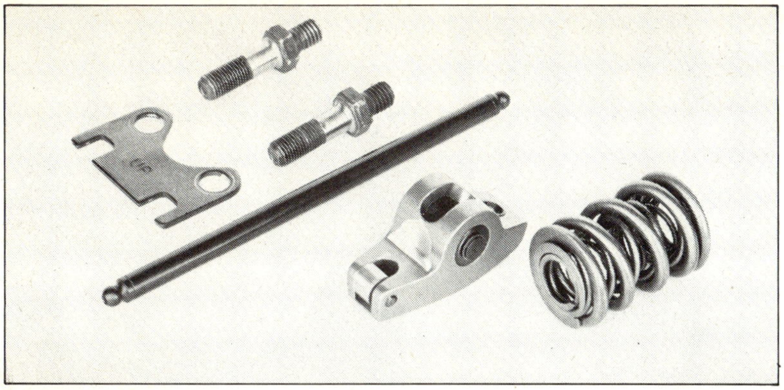
11

12

13

14

15

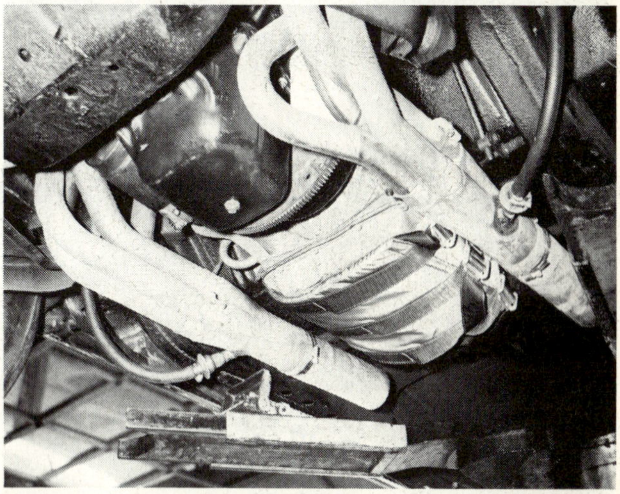
16

17

18. Sherman's '77 Nova was ordered with the police suspension package. He has removed the front sway bar and front disc brakes. A '65 Chevy master cylinder now replaces the stock master cylinder. An old set of front hubs was salvaged from the scrap bin at a local Chevy dealership, and the rotors were cut off. The front shocks are Cure Ride 90/10 versions, which Sherman swears by. The drilled beam just ahead of the upper A-arm is a homemade aluminum torque strap for the engine. Although originally equipped with power steering, the pump has been removed.

19. Suspension modifications at the rear are minimal. Genuine Suspension traction bars are used with the stock police option springs. The shock absorbers in use here are actually a mistake. Sherman erroneously bought Gabriel Pro Shockers for a '77 Camaro, which utilize a spike-type bottom attachment, so he made up a couple of aluminum adapter blocks to mount them on the Nova. These adjustable shocks are set in the soft position.

20. To gain extra wheelwell clearance while simultaneously improving the rear U-joint angle, Sherman lowered the front leaf spring mount 1¾-inches on each side by welding a couple of small tabs on the existing mounts. He also extended the Genuine Suspension traction bars with bar stock so that they would contact the leaf spring directly under the spring eye.

21. The rear spring shackles were replaced with homemade shackles that are considerably heavier than stock. Additionally, Sherman made the driver's side shackle ¾-inch shorter than stock to preload the chassis under acceleration.

22. Sometimes Joe Sherman's ability with a milling machine becomes evident. Most guys would have welded a simple D-ring on the frame for attaching a tow rope, but Joe whittled his own from aluminum stock and bolted it in place.

23. In place of the front sway bar, Sherman has installed a piece of strap aluminum to support two block drains to facilitate cooldowns between runs. In actuality, the small-block rarely exceeds 180 degrees, so he is considering elimination of the cooldown hassle altogether.

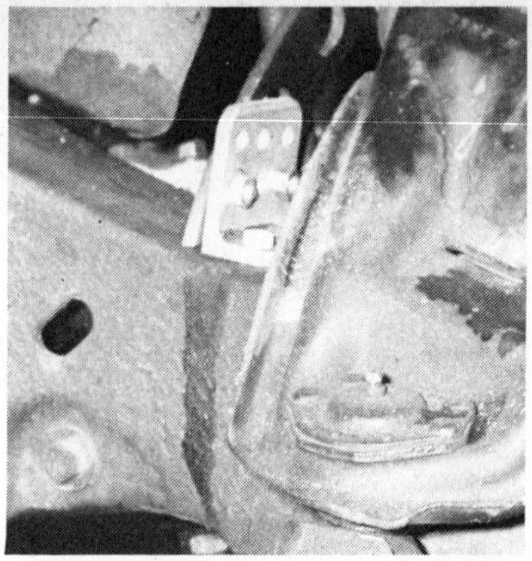
18

19

20

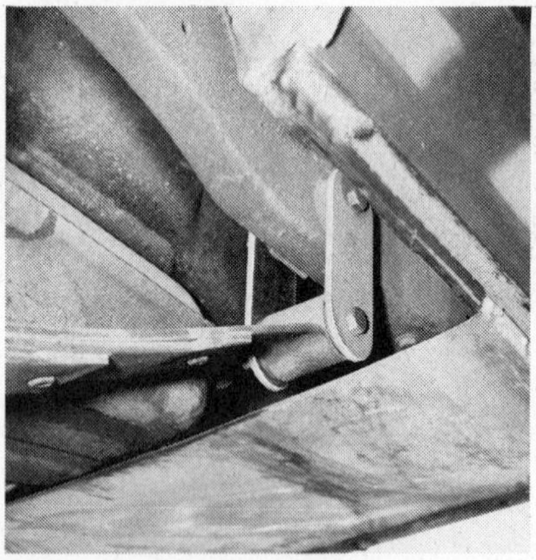
21

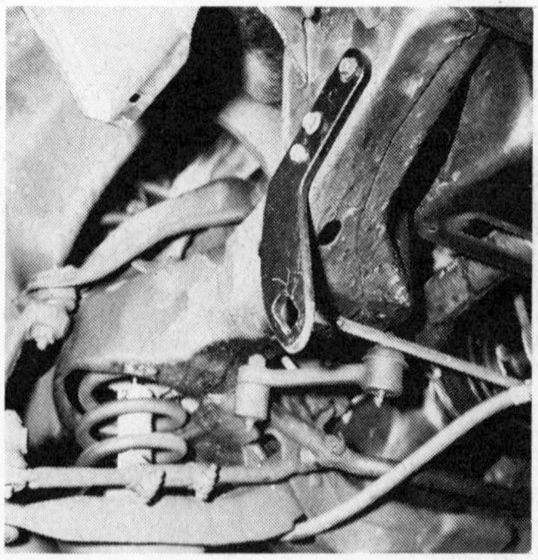
22

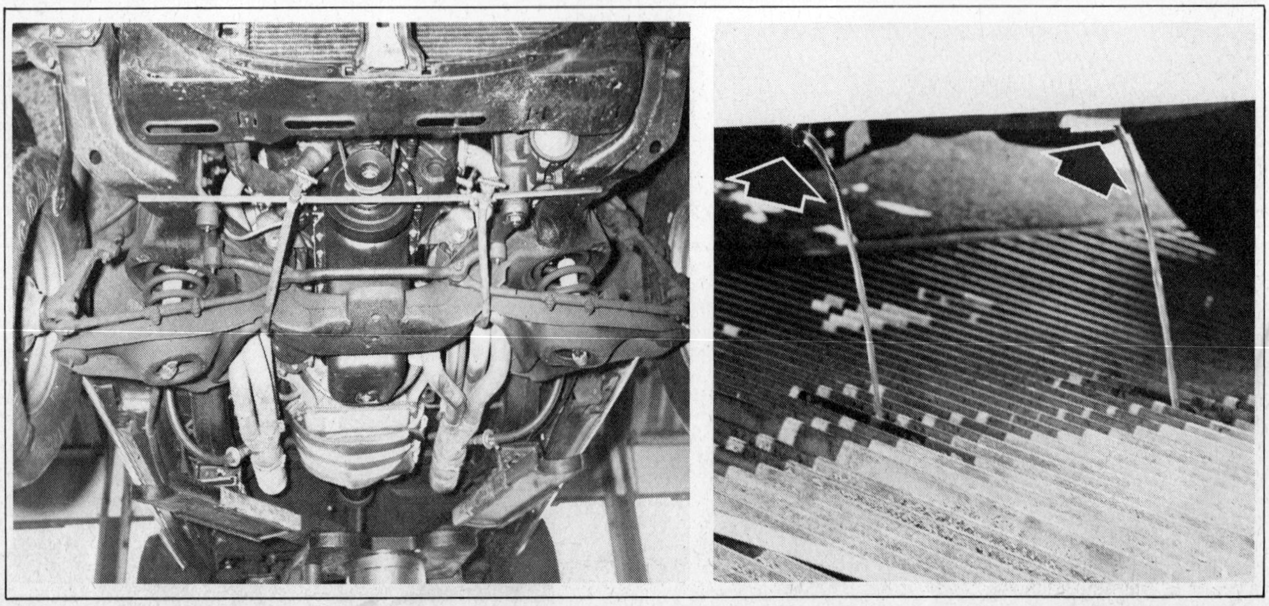
23

24

25

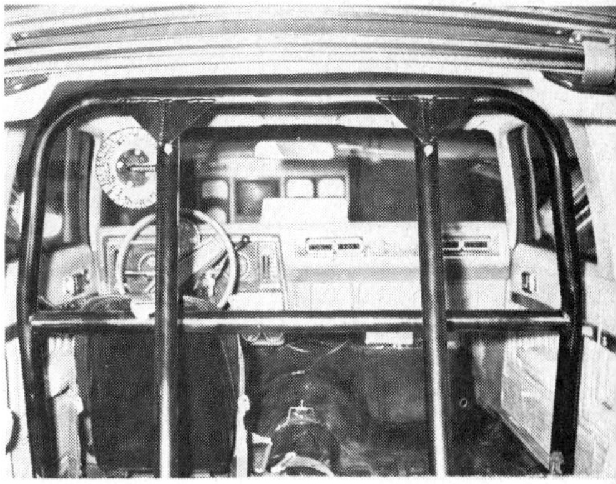

26

27

28

29

30

24. The interior of the Nova isn't quite as plush as it used to be, but a few memories linger on. Note the electric door locks and the power windows, which still work. The tach and the water temp and oil pressure gauges are from Auto Meter; and yes, fans, that's the stock column shifter that still controls the Turbo 350. As Joe said, "Don't spend any money that you don't have to."

25. The stock seats, front and rear, have been removed, as has the carpeting and padding, in the interest of reducing weight. The car weighs in at 3270 pounds, less driver. A fiberglass racing seat has been installed. All the safety harness is from Simpson.

26. Additional safety equipment includes a very neat roll cage fashined from 2-inch tubing by Duke's Roll Bars in Fullerton, California. The switch atop the driveshaft tunnel is an ignition kill switch to avoid the possibility of unintentionally locking the steering in an emergency if the column switch was used.

27. Another of Joe Sherman's low-buck speed secrets is streamlining. A sheet of very thin aluminum sheetmetal covers the stock plastic grille. Another sheet of .100-inch thick aluminum replaces the front bumper to keep the front of the car very smooth, while eliminating 75 pounds of bumper at the same time. Cutting the hole in the hood to mount the fiberglass hood scoop that was required with the Victor Ram manifold is the only modification Sherman made to the otherwise stock sheetmetal. Even the vinyl top was retained!

28. More streamlining is achieved by blanking off the cowl air inlets and by putting a simple strip of duct tape over the upper windshield molding. Low-buck? You bet!

29. Streamlining is also a consideration under the rear of the car, where a thin sheet of aluminum blanks off the gap between the stock gas tank and the rear bumper. The stock fuel pickup, mounted in the top of the tank, has been turned 180 degrees toward the rear of the tank, necessitating the fuel line going over the back of the tank and then forward, as shown. This simple, no-buck trick allows Sherman to make a pass with as little as three gallons of gas in the car.

30. Whenever Joe parks the car, a crowd immediately gathers, looking at this, inspecting that, sliding underneath in search of trick suspension or some other speed secret that simply doesn't exist. "People just don't want to believe that you can go fast without spending a bunch of money, but you can—and this car proves it." That says it all.

Double Vision

A look at two trick Chevy Suburbans used as dual-purpose utility/race support vehicles.

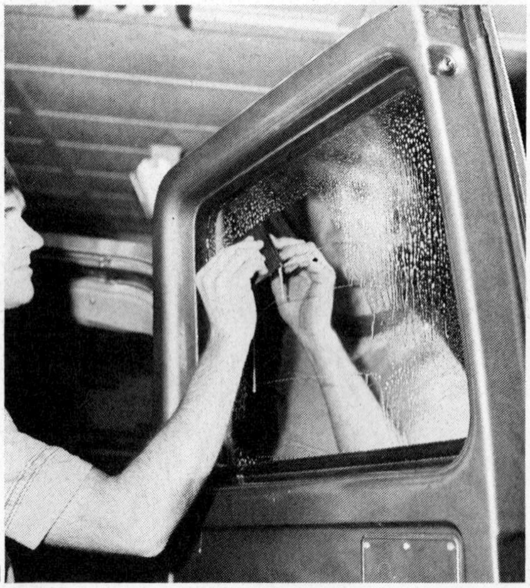

1

2

It takes only one quick walk through the pits at any drag race to discover that the Suburban, distributed by both Chevrolet and GMC, has become one of the more popular tow and race car support vehicles; and when you stop and think about it, a Suburban does make a lot of sense. It's an excellent trailer-towing vehicle (GM had that in mind when they designed it) that offers adequate weight, handling and stability, engines up to 454 cubic inches, a large secure cargo area, and comfort for six or more passengers. When it isn't being used to pull a trailer, it offers exceptional versatility as a multipurpose vehicle. The Suburban has a better ride and handling than vans with equivalent GVW (vehicle plus cargo) ratings, and it's less obtrusive than a van for normal driving.

No sooner had we started this project, based on an older '76 Chevrolet Silverado Suburban C-10, than we discovered that Top Fuel racer and custom upholstery king Tony Nancy was undertaking a similar project based on a '75 Custom Deluxe Suburban C-20. This provided us with an opportunity to present a double-barreled feature with some interesting comparisons for those of you who might be contemplating the selection of a Suburban for similar purposes.

For example, our test vehicle is a C-10 (½-ton model), whereas Nancy's is a heavier C-20 version (¾-ton). Additionally, ours has panel rear doors, while Tony's is fitted with a tailgate. There were factory engine changes made between the '75 and '76 models (both trucks are equipped with the 454s) to improve gas mileage, and we'll get into those as well. (Later-model Suburbans also incorporate subtle differences between each model year through '80, but since all versions are by and large the same, our discussion applies to an any year Suburban.)

The selection of either rear panel doors or the fold-down tailgate is really a simple matter of personal preference. There's little question that the tailgate version is more visually appealing, and it also provides the capability of opening the rear window for increased ventilation. On the negative side, the window must be completely lowered before the tailgate can be opened. And during daily usage, having to lean over the tailgate to reach packages or groceries can be a little inconvenient. To accommodate frequent access to the cargo, we selected the rear doors.

Since we anticipated doing a lot of trailer towing with our vehicle, we opted for the 454 engine—there's just no substitute for brute torque. It's interesting to note that about 25% of all '75 and '76 Suburbans were equipped with the 454 and Turbohydro 400. Another 64 percent got the 350 engine and Turbohydro 350 transmission.

We also elected to retain the standard suspension, as opposed to the heavy-duty option, to make the ride as comfortable as possible. However, to facilitate handling, we ordered the heavy-duty shock absorbers and the stout 1-inch-diameter front stabilizer bar. This blend of equipment makes a very stable, yet pleasant, vehicle, regardless of whether a trailer is being towed.

We took delivery of our project vehicle at the GM Proving Grounds in Mesa, Arizona. The 400-mile drive back to Los Angeles provided us with a good baseline and overall feel of the vehicle. We were told by Chevrolet Engineering that the California emission-legal 454 was the most powerful combination in the entire '76 Chevrolet line. The performance of our Suburban seemed to bear this out, as it has good full-throttle response in the 30 -to-70-mph range. Standing-start performance with the stock 3.07:1 rear-end ratio and a curb weight of 5600 pounds isn't too exciting, but it would be in a lightweight C-10 pickup fitted with a 4.10:1 gear. Incidentally, Chevy says that the 454 in a C-10 pickup was the quickest vehicle they made in '76, although the Vette still gets top speed honors due to its superior aerodynamics. For some reason, for '76 the California emissions package added 10 hp and 15 lb.-ft. of torque to the net power ratings of the 454 over non-California engines.

We've heard complaints of 6 to 8 mpg for 454-equipped Suburbans, but the '76 454 engines were improved. Sure enough, our 454 recorded 12.2 mpg, with and without the air conditioning, but subsequent city driving produced figures of 10 mpg, as did open highway driving with a 5000-pound trailer load.

To achieve these mileage gains, Chevrolet changed the cam and eliminated the controlled vacuum spark system which had restricted vacuum distributor advance to high gear only. The '76 454s have full-time vacuum distributor advance. If you're interested in backfitting an earlier 454 with the '76 cam, it is Chevrolet part No. 3963544 and is the same cam used in the '70 454 engines.

Among other observations on the drive to Los Angeles, we noted that the Suburban was very comfortable. The large glass area provides unobscured visibility but also lets in a lot of sunshine. The seating position and angle, coupled with the tilt steering wheel, is excellent, and interior noise levels are low with the windows closed. No doubt the extra insulation included with the Silverado option contributes to this. About the only negative item we noticed is that the electric tach is sensitive to current fluctuations in the vehicle's electrical system.

Once the basic road test was out of the way, we began to modify and add to our Suburban's appearance, utility and comfort, and to ready it for the job of trailer towing. The first step was to re-

3

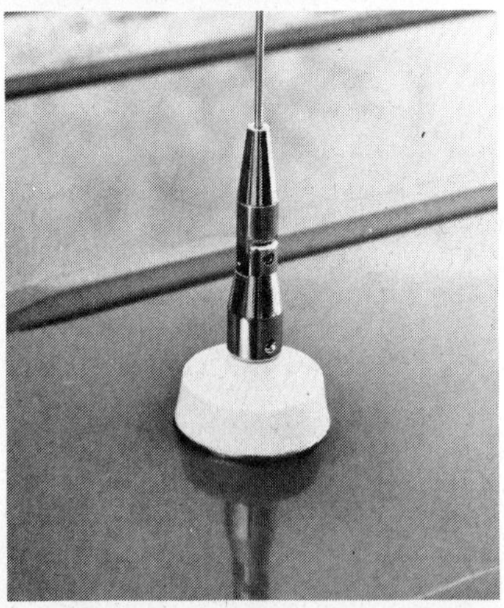

4

1. Care, patience and plenty of Windex is needed to apply Trimbrite sunscreening, but it isn't difficult. The mirror-like material adds both privacy and cooling.

2. The Silverado option includes woodgrain trim, door pockets, carpeting, full instrumentation and much more. The Delco Cadillac AM/FM Stereo Weather Band signal-seeking radio fits the stock opening. A center console was added to hold the Hy-Range III CB within convenient reach.

3. Mounted above the mirror is the Super Snooper radar detector from Autotronics.

4. The Hellcat 7 CB roof-mount antenna has a fold-over base for low clearance protection.

5. Demountable trailer mirrors are a combination of a demountable (no longer available) set and stainless steel heads from a swingaway set, both from Sears.

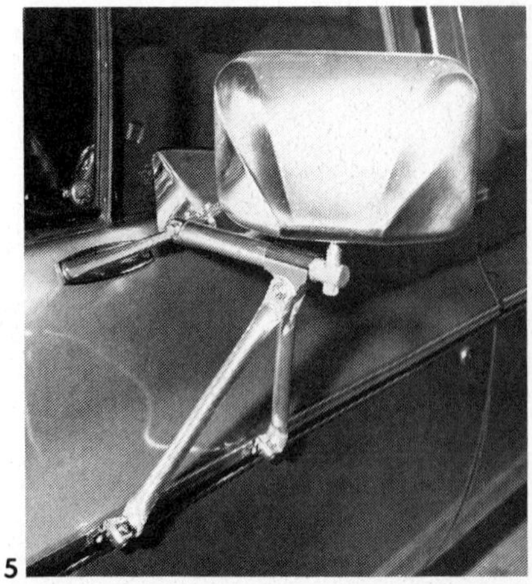

5

Tony Nancy was among the first to realize the practicality of the Suburban as a race car tow vehicle and as a dual-purpose utility vehicle for his custom upholstery business. However, when Tony took delivery of his '75 C-20, he was distressed by the "buslike" appearance of his plain beige truck. He quickly set about altering the appearance through the use of Western 16½x10 aluminum wheels fitted with 32-inch-diameter Formula Desert Dog PCV 12.00-16.5 tires and a paint scheme designed and applied by Carter that gives the truck a taller, shorter look. The entire vehicle was repainted with acrylic enamel, using light and dark brown and orange accent stripes. Acrylic enamel was specified to prevent accidental spotting from fuel spray landing on the finish while the vehicle is at the races.

Other exterior additions include a set of Per-lux driving lights, an American Carrier luggage roof rack and a sunroof by Martin. Completing the overall color scheme of the Suburban is dark bronze mylar sunscreening applied to the rear side door and rear windows by SSC Solar Control. The front door side windows are also covered, but there a light bronze sunscreen material has been used. The overall effect is striking.

Of course, one would expect the interior of Tony Nancy's Suburban to be a real eye-catcher, and indeed it is. The entire floor area has been carpeted with a rich dark brown high-pile curl material, as have the kick panels and the lower portions of the doors. The stock bucket seats have been redone in a natural leather-colored Naugahyde and a golden brown crushed velour to complement the carpeting. The overall appearance, accented by the simulated woodgrained trim on the instrument panel and door panels, gives a luxurious look to the vehicle, but then what did you expect from Tony Nancy—vinyl seats and rubber mats?

place the stock wheels, tires and hubcaps with a set of Goodyear Custom Polysteel LR78-15 Radials mounted on 15x8½ Chrome Mojock, eight-spoked wheels from Appliance Industries. The LR78-15 size provides a full 9-inch tread face with deep water relief grooves to reduce the likelihood of hydroplaning on wet roadways.

The next modification was to apply Trimbrite "Solar Shield" chrome sunscreening to the rear side and back windows. The chrome sunscreening reflects a great deal of solar heat and keeps the vehicle noticeably cooler. Consequently, it takes the dual air conditioners far less time to cool off the interior when it is hot. An additional benefit is the mirror-like sunscreening that provides a certain degree of privacy inside the vehicle.

The next area of concern dealt with those items that make driving a little more comfortable and trouble-free. We installed the Delco AM/FM Stereo Weather Band signal-seeking radio with four 4x6 speakers mounted in the front kick panels and in the rear air conditioning shroud. This particular radio was an option on '76 Cadillacs and features push-button tuning (the buttons can be set for five AM stations and five different FM stations), plus a signal-seeking bar which, when depressed, automatically scans and selects strong station signals for crystal-clear listening. Of course, the radio can also be tuned manually. Another feature is that it has a separate button that selects either 162.40 or 162.55 MHz for continuous local weather information and forecasts from the nationwide network of government weather stations. A small "W" illuminates on the radio dial whenever it is within range of one of these stations. As you might expect, a radio that does all this isn't cheap; in fact, the over-the-counter price was $445 without speakers.

Of course weather data is only part of the essential information you need to survive on today's double-nickel highways; hence the reason for a Hy-Gain, Hy-Range III, 23-channel CB radio. Among the many features on this quality transceiver are: an automatic noise limiter and noise blanker, transmit and receive indicator lights, adjustable antenna tuning, automatic modulation control and a built-in microphone pre-amp. We also used a Hy-Gain "Hellcat 7" fold-over roof mount antenna. The fold-over feature prevents impact damage to the 52-inch whip.

Mounting the CB radio within convenient reach was a problem, due to the configuration of the Suburban dash, but the problem was quickly solved by a commercially available console that fits both Suburbans and pickups. We purchased the console from Mother Trucker (5955 Van Nuys Blvd., Van Nuys, CA 91401), which sells both retail and mail order; contact them if you're unable to locate a similar console locally. We simply knocked out the tray bottom in the console lid and enlarged the opening to fit the radio and mike cord. The CB was then fastened to the inside of the console so that the top could still be opened.

In our past experience with cross-country trailering, we have learned that even a good CB radio isn't always enough to keep you out of trouble. We have repeatedly been "saved" by a small box mounted just above the rear-view mirror. That black box was a Snooper radar detector manufactured by Autotronics, Inc., in Richardson, Texas. This same company also has an improved version, called the Super Snooper, which features improved filtering for rejecting spurious signals not actually generated by a speed monitoring radar unit, such as nearby CB transmissions. It has an adjustable volume control for the warning beeper and a flashing light that cautions to "check speed" when it detects a radar signal. There is also a sensitivity switch to adjust the unit for optimum range.

Although we haven't actually had a chance to field-test the new Super Snooper, we can readily attest to the fact that the earlier, less sophisticated model would give us a clear warning of an impending radar speed check about ¼-mile *before* the radar unit was able to get a reading on *our* vehicle. Of course, terrain, trees, building and strength of the radar signal can reduce this warning distance, however.

Prepping the Suburban for trailering really didn't take much work, since it was already equipped with a hitch platform and trailer wiring harness. All we had to do was buy an appropriate hitch head and ball to fit the platform and then attach a wiring connector plug compatible with our trailer.

Although the Suburban came equipped with dual outside mirrors, they are marginal for trailer towing. So to avoid the ugliness of large, permanently attached trailer mirrors, we purchased a pair of

6

SILVER BULLET SPECS

BODY

Model	C-10 Panel Door
Length	218.7 inches
Width	79.6 inches
Height	71.5 inches
Wheelbase	129.5 inches
Curb Weight	5580 pounds
Gross Vehicle Weight	6800 pounds
Cargo Capacity	143.5 cubic feet
Load Floor Area	40 square feet
Tire Size	LR78-15 radials

ENGINE

Displacement	454 cubic inches
Bore & Stroke	4.25x4.00 inches
Compression Ratio	8.25:1
Carburetion	Rochester 4-bbl spread-bore
Net Horsepower	250 @ 3800 rpm*
Net Torque	385 lb.-ft. @ 2800 rpm*

*These specifications are only for vehicles sold in California, which are equipped with the mandatory California emissions package. Similar vehicles sold outside California have a net horsepower of 245 @ 3800 rpm and a net torque of 365 lb.-ft. @ 2800 rpm for the 454-cubic-inch engine.

demountable mirrors from Sears. We discovered that these mirrors and mirror housings left a lot to be desired in terms of rugged durability, although the attachment method and hardware was really trick. A second trip to the Sears automotive department confirmed our findings, as the demountable mirrors were being discontinued in favor of a more rugged "swing-away" design. The new model features stainless steel mirror housings and weathertight rubber moldings, but unfortunately, they must be permanently mounted. So we compromised. We got a set of the swing-away mirrors and installed them on the demountable hardware from our first set. The result is a really nice (but somewhat expensive) set of durable and demountable trailer mirrors. If you want to duplicate our setup, you might be able to locate a set of the demountable mirrors that a Sears store has in stock. If you can't find the demountable units, the swing-away models are still a good choice, even if you can't remove them when they're not needed.

The most notable change in our Project Suburban is its exterior appearance. We wanted to transform the dark silver, bus-like truck into something more contemporary. To fill our needs, Kenny Youngblood designed an accent treatment with sweeping curves of black and earth colors to set off the bleak dark silver. A bright silver overlay along the sides and rear of the vehicle further complements the overall treatment. A semi-gloss black was added to the hood, headlight shrouds and bumper inserts. This unusual yet tasteful design was adeptly executed by Jeff Scozzaro, at Jeff's Place, in Anaheim, California. But Jeff is his own man, too, and he added a final touch by covering the side and rear stripes with a light mixture of blue pearl and clear acrylic enamel to add color variations as the lighting changes.

The next step was to add those items that would contribute to the convenience of race car trailering and support, such as a set of Sears twin air horns. These horns give off a powerful blast and are invaluable for preventing emergency situations and gaining the attention of slowpoke drivers who unthinkingly obstruct the normal flow of traffic. Another item from Sears that we added was a 200 VA Inverter/10 Amp battery charger. This unit can be used to convert the 12-volt battery current into the equivalent of 110-120-volt AC to power most small hand tools. It can even be used to power 110-volt lights for night work, and can also be used as a conventional battery charger.

Carrying adequate racing gas is a problem for most racers, since the quality of available pump gas has become so nconsistent in different parts of the country. Consequently, we arranged with Hooker Headers to install one of their auxiliary fuel tanks to carry our racing gas. Unfortunately, at the time, Hooker did not have a tank designed to fit the Suburban, but they modified an existing tank to fit. The tank holds roughly 15 gallons, mounts under the left side of the vehicle and vents into the stock evaporative emissions system. We installed a separate electric fuel pump to transfer the race gas to the race car, but also plumbed the auxiliary tank into the main fuel system (through a selector valve) so that it can be used as emergency fuel.

The remainder of the support modifications are at the rear of the vehicle and include a 2¾-pound dry chemical fire extinguisher, a cigarette lighter receptacle wired to the auxiliary battery for powering various 12-volt lights and accessories, a rocker switch to turn on the dome and cargo area lights and a pair of Hella quartz-iodine interior lights (available from J.C. Whitney & Co.) mounted on the insides of the rear doors. When the doors are opened fully, they provide illumination for working on the race car and to light the trailer during loading and unloading.

With the race support modifications completed, we began to concentrate on improving gas mileage. In stock trim, we managed 12.2 mpg on the highway, and felt we could only go up. The first step was to examine available rearend ratios.

Suburban C-10s fitted with a 454 are available with three different gear ratios: 3.07:1, 3.40:1 and 3.73:1. In the interest of better mileage, we tried each ratio and observed virtually no difference. When we queried Chevy Engineering about this, they confirmed that this vehicle/engine combination is relatively insensitive to gearing changes unless the vehicle is being used to haul heavy loads. Because of this input, coupled with our own testing, we can say that rearend ratio choice should be based solely on application rather than mileage. We elected to stay with the 3.07:1 ratio.

Next we removed the stock cast-iron intake manifold and Rochester Quadrajet carb to install a Holley Street Dominator manifold and the recommended R-6979-AAA 600-cfm carb. Connecting the numerous hoses, lines, throttle and cruise control linkages, throttle cable and dashpot brackets took some fabrication, but the result was worth it. With the emissions systems intact and operating, mileage climbed from 12.2 to 13.5 mpg. But there's more good news. With the stock manifold and carb, the engine had noticeable hesitation just off idle, especially if it was cold. With the Holley system, all hesitation vanished. As far as we are concerned, the improved driveability

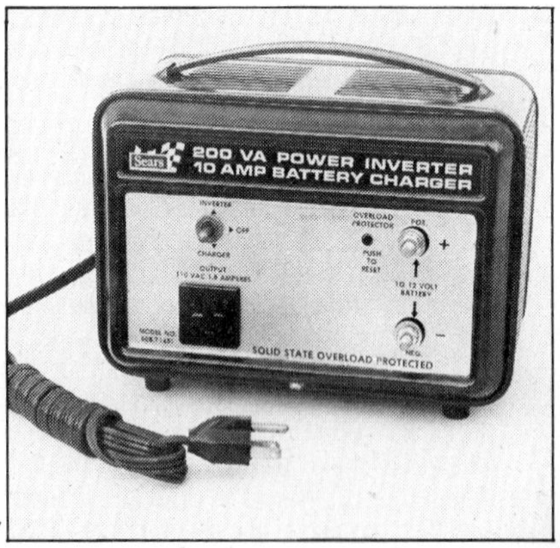

7

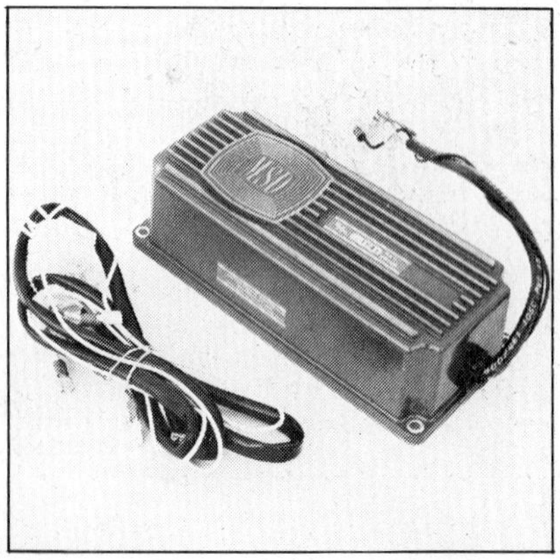
8

9

6. Suburbans are offered with either a tailgate or double door rear openings. Choice depends upon personal preference and application. Cargo area is huge.

7. Trackside 110-volt power is supplied by a Sears power inverter/battery charger combo.

8. A Moroso externally adjustable cam timer and the latest MSD-2 multiple spark ignition were used to squeeze gas mileage gains.

9. Sears dual air horns, complete with compressor and air tank, were added as a safety measure for trailer towing operation.

alone was worth the trouble of installing the system.

The next thing we did was to install a Moroso No. 6625 quick-change cam timer in place of the stock camshaft sprocket. This cam timer is externally adjustable for easy cam timing changes. Since it is designed for the early ¾-inch-wide heavy-duty timing chain instead of the ⅝-inch chain, we also had to install the heavy-duty chain (part No. 3860036) and the associated wider crankshaft sprocket (part No. 3860035). Cam timing is really a fairly haphazard thing, due to manufacturing tolerance "stack-up," so it is impossible to say exactly what changes will work best. It's simply a trial-and-error procedure, using 4-degree changes

10

11

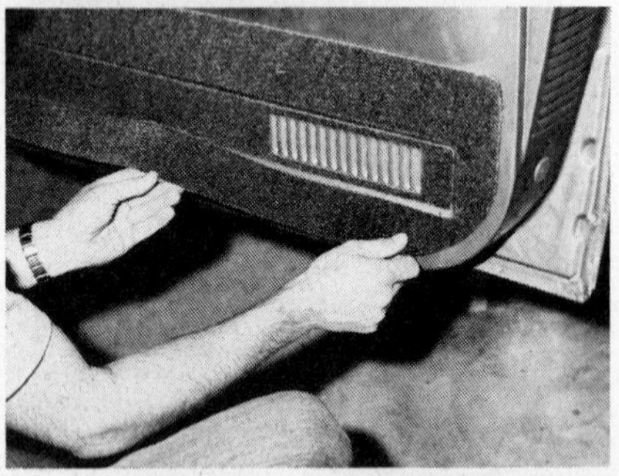

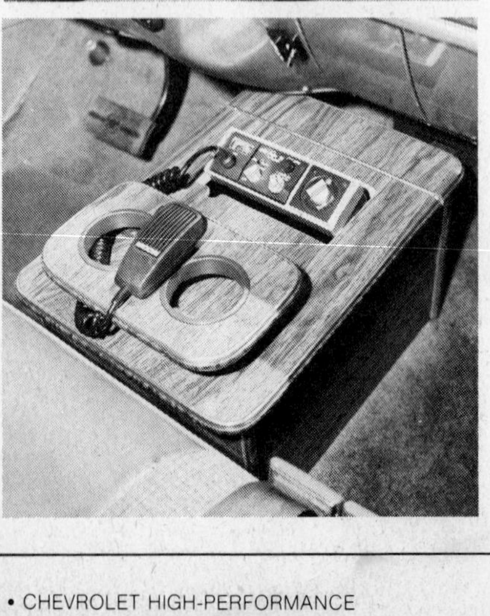

CHEAP FRILLS

One look at the interior of Tony Nancy's Suburban was all it took to make us realize just how elegant and comfortable a Suburban could be. Although Tony had started with austere rubber floor mats and vinyl-covered seats which he redid with deep pile carpeting, crushed velour and Naugahyde, the Silverado option on our '76 model already included medium-blue pile carpeting over the entire floor area and comfy fabric inserts, surrounded by grained vinyl, on the seats. And of course we already had the woodgrained upper door panels and dash accents that are also included with the option. We had added a custom center console/storage compartment both for convenience and to contain the Hy-Gain Hy-Range III CB radio. Consequently, over half of the interior battle was already handled.

So we began to contemplate ways in which we could further approach the finished look of Nancy's interior without spending a bundle of cash for one of his high-buck interiors. We decided to add our own low-buck carpeting accents to the kick panels, lower door panels, rear door panels and sides of the cargo area, to enhance the interior appointments while simultaneously providing scuff protection of these painted surfaces and to add sound deadening to the panels.

To actually cover these areas, we selected a high-quality indoor/outdoor carpet material with a corduroy-like ribbed surface. Since we intended to attach the carpeting to the panels with waterproof contact cement, we were careful to select a brand of indoor/outdoor carpeting that did not have a foam rubber backing, since that might tend to pull loose. Indoor/outdoor carpeting was selected also because it could be conveniently cut to shape with an X-Acto knife, leaving a clean edge that does not have to be bound or hemmed to prevent fraying. We purchased four square yards of the carpeting and three quarts of Wilhold contact cement to do the job. The total cost was about $35.

Applying the carpeting was time-consuming, but not difficult. The first step was to carefully cut a heavy paper pattern to fit exactly the area where the carpeting was to be applied. Then, using the pattern, the shape was marked on the back of the carpeting with a felt-tipped pen and cut out with an X-Acto knife.

After a check to ensure that the cut carpeting fit the panel correctly, a coat of contact cement was first applied to the back of the carpeting. This first coat tends to seal the carpeting as it soaks in. Another coat of cement was applied to the actual panel, carefully, to coat only the area actually to be covered by the carpeting, since excess contact cement is difficult to remove. Then a second coat of cement was applied to the back of the carpeting, just before the carpeting was actually placed on the panel. Care must be used to position the carpeting correctly as it is first brought into contact with the panel, since once the two cement-covered surfaces touch, adjustment is nearly impossible.

The result was an inexpensive yet professional-looking job. And the overall appearance of the interior was vastly improved. We even had enough scraps of carpeting left over to cut some simple floor mats to protect the stock carpeting from oil or grease that might accidentally get tracked in at the race track.

One word of caution, though. The contact cement can bears a warning that it should be used only in well-ventilated areas away from any fire or flame. The fumes given off as the cement dries can explode if they reach a sufficient concentration. And just breathing the fumes for a sustained time can give you a king-sized headache, as we learned to our regret.

at a time (limiting total timing changes to 8 degrees in either direction). The cam can be fine-tuned in 2-degree increments, once a direction is taken. As a guide, our Suburban responded most favorably when the cam was retarded 2 degrees and produced a .4-mpg gain over the stock setting. However, once we added headers, the best results were obtained with the cam advanced 4 degrees beyond the stock setting.

10. The addition of the Holley Street Dominator manifold and the 600-cfm R-6979 carb resulted in better gas mileage and driveability, although installation was a bit of a hassle (see text).

11. This aluminum air dam, shown with the bumper removed, deflects air around the sides of the vehicle rather than underneath it, resulting in a .4-mpg gain. Kills bugs too!

Selecting proper headers for this type of vehicle isn't easy, since such a big engine is happy operating at fairly low rpm. When empty, the engine virtually loafs along, but when heavily loaded, it labors, which produces excessive exhaust temperatures. Hooker installed a set of their No. 2812 headers with 1⅞-inch tubes. As anticipated, our Suburban showed no increase in gas mileage on our plotted test loop, since all tests were done with the vehicle empty without working the engine very hard. Based on past observations, we feel confident that mileage gains as high as 15 percent can be obtained with a trailer in tow, depending, of course, upon the weight of the load and the driving technique. If a Suburban was to be used primarily in city traffic with little or no load, then we'd recommend 1¾-inch headers to achieve a mileage gain in this mode of operation, although such headers would probably show little gain under heavily loaded conditions.

In further discussions with Chevy Engineering, they indicated that we might see a mileage gain with a flat front spoiler mounted just behind the front bumper. This type of spoiler pushes air around the sides of the vehicle rather than allowing it to come in contact with the aerodynamically dirty undercarriage. We fabricated one out of .060-inch aluminum. It is 63 inches wide and extends 6 inches below the bumper. The resultant mileage gain was another .4-mpg.

The final item we added was the then newest version of the Autotronics MSD-2 multiple-spark ignition system. We've proven this unit to be effective on previous mileage tests but were curious as to how this new unit would work in conjunction with the stock HEI breakerless system. The unit took less than 20 minutes to install, and the mileage gain was an impressive .9-mpg. The engine also starts quicker and idles more smoothly.

In the end we had increased our mileage to a respectable 15.2 mpg, a gain of over 24 percent, while simultaneously improving driveability. It was a fitting finish for a vehicle we've come to love.

—C.J. Baker

PRE-'76 MILEAGE MODS

It seems that every gas station operator in the world knows about the conspicuous gas consumption of the '75 and earlier 454 Suburbans. And you can see their faces beam with the anticipation of a big sale every time one of these vehicles rolls up to the pumps. But there are ways to greatly reduce that problem while simultaneously improving the driveability of the vehicle. Tony Nancy's 454 C-20 Suburban is an excellent example.

When Tony took possession of his truck, it got 8.4 mpg day-in and day-out. It didn't seem to matter if it was on the highway or in the city. And of course, when Tony had his Top Fueler rig in tow, the figure was usually in the 5-to-6-mpg range—not a thrilling fact with today's gasoline prices.

The first thing Tony did was replace the stock tires, which were just over 29 inches in diameter, with Formula Desert Dog PCV 12-16.5 tires that are 32 inches in diameter. Effectively this was about the same as dropping the 4.10:1 rearend ratio to the next lower ratio of 3.73:1. Then the exhaust system was freed up with a set of Cragar headers. The stock exhaust pipes and mufflers were retained, but tailpipe size was reduced from 2 inches to 1¾ inches and the resonators were eliminated.

In the engine compartment, things got a little more involved. The cam and lifters were discarded in favor of a Racer Brown EH-5 high-mileage hydraulic lifter cam and lifters. And since this truck would rarely experience engine speeds above 3000 rpm, the cam was installed 4 degrees (crankshaft) advanced to enhance low-speed torque and driveability. A Borg-Warner steel camshaft sprocket was also substituted, to prevent eventual retarding of the camshaft due to wear of the nylon covering on the stock sprocket teeth.

On top of the engine Tony replaced the stock cast-iron two-plane intake manifold with an Edelbrock Street Master manifold, just as we installed a Holley system on our '76 Project Suburban (see text), but unlike our approach, Tony retained the stock Quadrajet 4-barrel carb, complete with stock jetting.

For ignition, a Mallory photo-cell breakerless distributor was installed with the stock Mallory advance curve. Initial timing was set at 10 degrees BTC, with the appropriate vacuum lines temporarily disconnected as specified on the sticker affixed to the air cleaner.

The result was a phenomenal gain in efficiency that nearly doubled the stock gas mileage. Tony Nancy's Suburban now gets 14.9 mpg on the highway and 11.7 mpg while pulling his trailer. And the driveability and starting of the truck improved as well. The modifications were very straightforward and should produce equivalent results on similar '75 and earlier 454s.

Strip Show

A behind-the-panels peek at Jenkins Competition's Pro Stock Camaro.

We cried "show us your struts," and Team Jenkins complied. Encouraged by our shouts of "take it all off," Grumpy's crack crew pounced on their Pro Stock Camaro and peeled away nearly all its panels. Stripped of the outer garments that it normally wears only at public appearances, the racer's resemblance to Chevy iron ends almost entirely. So go ahead and grab a good long look. We're giving you a backstage pass to a "dragstriptease." It's a behind-the-scenes exhibit of engineering exotica, a top-billed acceleration act so thoroughly exposed it can only pass for a bare-bones Pro.

—Al Kirschenbaum

BILL JENKINS' NHRA CAMARO
GRUMPY'S TOY XV
DRIVER: LARRY LOMBARDO
CREW: JOE TRYSON, GEORGE AREFORD

ILLUSTRATION: STEVE AMOS

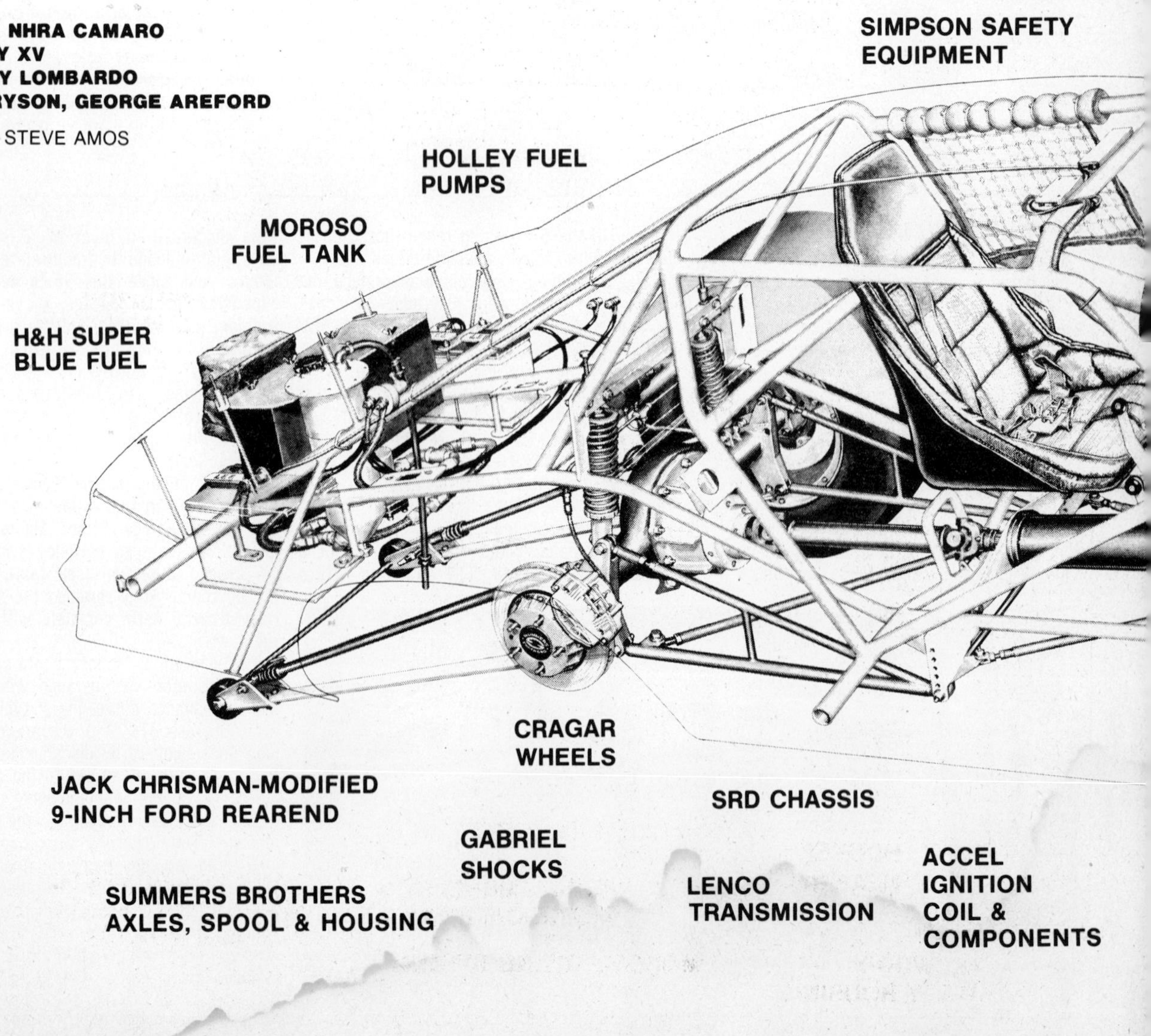

1

1. As the 15th model in a tall family tree of Grumpy's Toys, the Camaro is designed to meet a whole new set of quarter-miling circumstances. All of Jenkins' Chevys are constructed to cover the course in minimal time. With most modern race courses now equipped with concrete launch pads, the Camaro's primary function is to leave *hard!*

2. Its driveline functions as a loaded backbone member, and even the motor's mounting plates are stressed chassis components.

2

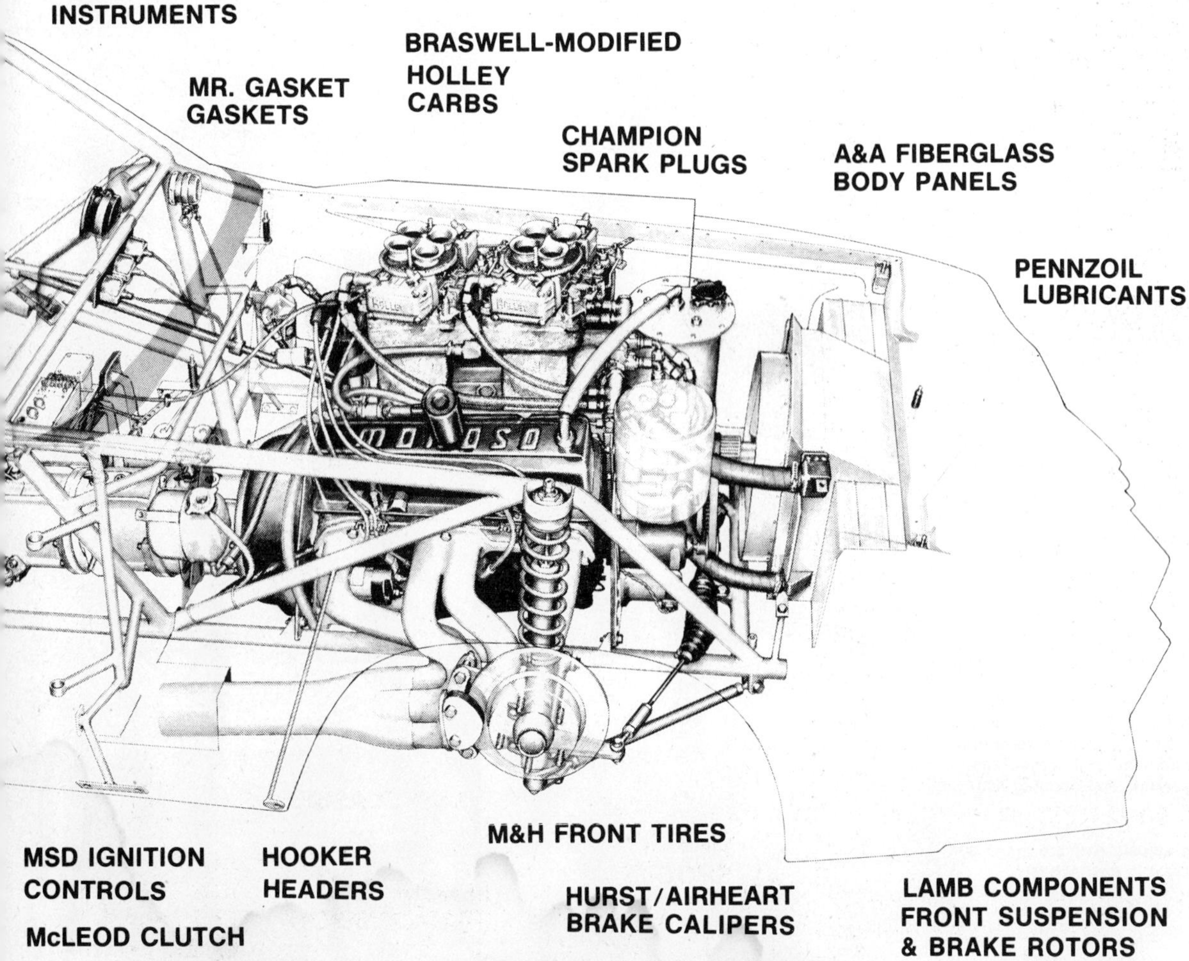

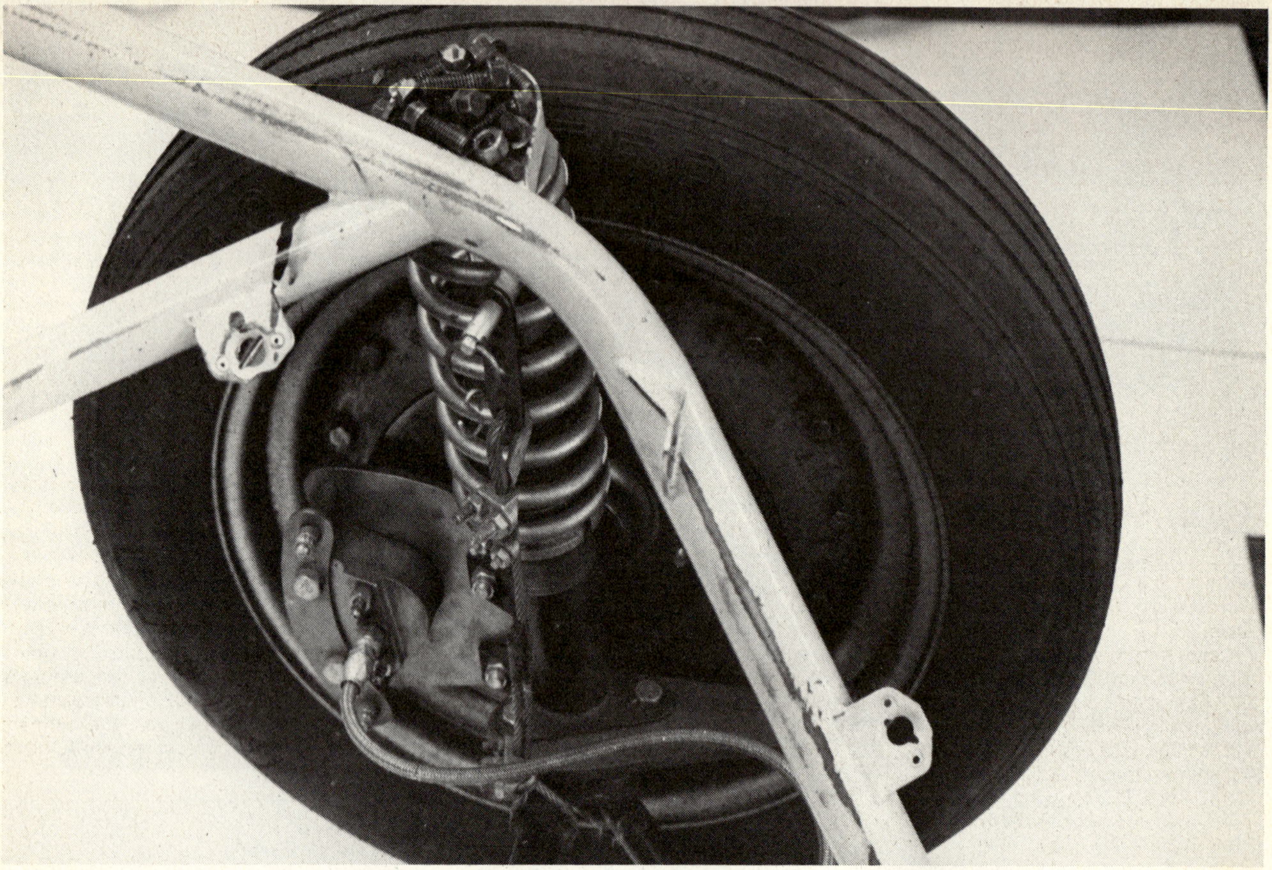

3

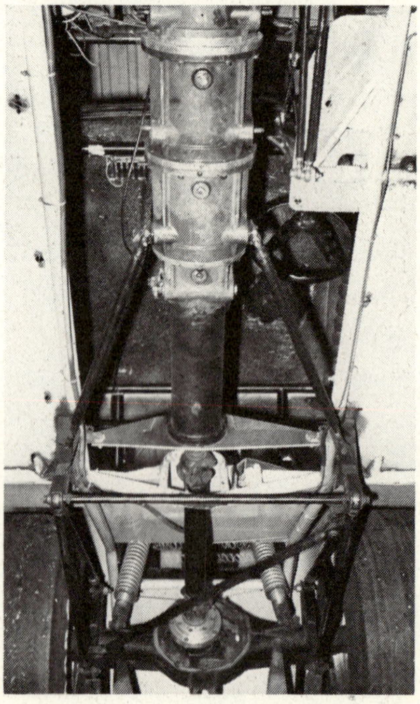

4

5

6

3. Once upon a time, Pro Stockers were built from real automobiles. But back in the early Seventies, one AMC racer rolled out a new Gremlin with completely custom-built underpinnings, and that radical departure in car construction prompted Jenkins, too, to turn to the tubes. Many of those primitive Pros pirated imported parts for their suspensions, and one of Jenkins' early Vegas rode on Datsun struts that actually protruded through the hood.

4. Grump now utilizes some very slick Lamb Components struts that work far better than the originals did, and even the three-link rear suspension system that Jenkins pioneered has been updated to a far simpler ladder bar design.

5. Jenkins Competition consistently leads the race to produce maximum power levels from the Chevrolet "little-block," as Grumpy refers to it. This 331-cubic-incher began life in much the same manner as many another racer's wedge, but attention to details and hard flogging at the leading edges of a number of sciences moves this mouse in a manner unmatched by any other gas-burning Stovebolt door car. Engine master Joe Tryson has worked with Team Jenkins since its Junior Stock days, and together they continue to develop such front-running systems as Toy XV's dry sump, its straight tube headers and its unique canted, annular-feed carburetors. Power output is currently approaching 700 horses.

6. Chassis loading adjustments are made by turning a single screw at each wheel.

Winning Wagon

Larry Severson's "grocery getter" Nova is a bracket racing bargain.

Trick doesn't always equal quick in bracket racing. Consistency and reliability are just as important as 8-second e.t.'s A car that is consistent, reliable, inexpensive to build and maintain, and still quick enough to run in Bracket I is all a bracket racer could hope for. Larry Severson, of West Fargo, North Dakota, is a lucky man, because his '66 Nova wagon does all of the above with ease.

Larry's highly visible red, white and blue wonder wagon runs in the high 10s at 120+ mph, yet the total investment is only $3850. The success formula is simple: a big-block Chevy and a Turbo 400 transmission in a lightweight body with a very consistent chassis—a combination that anyone could duplicate with very little trouble.

Rather than start from scratch, Larry saved time and money by purchasing a used Super Stocker (less engine and transmission) for $2300. Used class racers are excellent investments for bracket racers, because they are plentiful and priced at less than what they cost to build. A car that may be unfavorably factored for class racing can still be a great bracket car. A well-built class chassis mated to a less expensive (and more reliable) bracket racing drivetrain is a sure way to get a consistent bracket racer.

As a Super Stocker, the car ran a small-block; but Larry follows the low-rpm, big-inch theory of reliable power, so he installed a whopping 468-cubic-inch Chevy in the Nova's engine bay. Severson did the swap himself, using custom-made solid motor mounts made out of 3/16-inch mild steel bent to the desired shape. To clear the steering, the engine was mounted ½-inch higher and ½-inch to the right of the stock location. The spring towers were cut and rewelded to clear the headers. The headers are Hooker units designed for big-block Nova installations.

Suspension clearance is gained by using a stock '68 396 oil pan and modified tie rod. The center was cut out of the tie rod and the stub ends heated and bent to relocate the center section of the tie rod 4 inches forward of the original location. A piece of tubing was slipped over the bent ends and welded together in a locating fixture. For safety, a hole was drilled through the tubing and the tie rod at each end, and a spring pin was driven into the holes. Firewall clearance was obtained by cutting off the windshield wiper mount flush with the rest of the firewall.

The engine was built for long life and low rpm. It is a '73 454 bored .060-inch over for a total displacement of 468 cubic inches. The heads are '73 passenger car "smog heads" with oval ports. Jim Swedberg, at Valley Porting Service, in Grand Forks, North Dakota, opened the combustion chambers from 113cc to 117cc. Most of the opening was done around the exhaust valves. Valve size is 2 1/16-inch intake and 1 23/32-inch exhaust. They are used with General Kinetics rocker arms, solid lifters, stock pushrods, and a Competition Cams CB304-10 cam (.605-inch intake lift, .630-inch exhaust lift and 266/276 degrees duration, intake/exhaust, @ .050-inch valve lift). The pistons are 12.5:1 Arias open-chamber models used with Speed-Pro rings and steel connecting rods. The crank is a cross-drilled 454 LS-7 unit with .0025-inch clearance.

The engine gets its diet of high-test via a Holley Strip Dominator manifold and a single Holley 850-cfm double-pumper carburetor. To avoid fuel restriction, a dual parallel fuel pressure regulator system is used. The regulators are mounted on the right inner fender and flow through separate filters and then into a two-into-one connector at the carburetor. The gas is carried in a narrowed tank that fits neatly between the inboard-mounted rear springs necessitated by

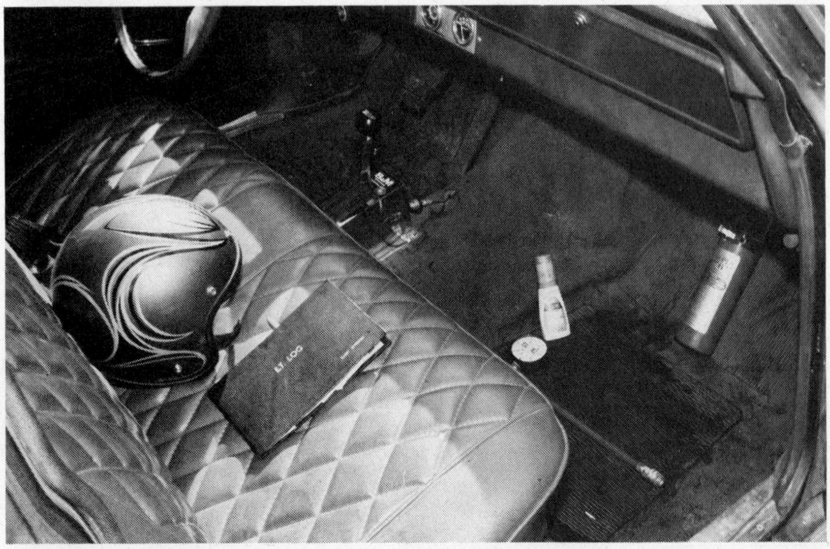

1

the narrowed rearend. The gas tank is a '78 Chevette unit narrowed 12½ inches. The tank was mounted upside down to take advantage of its natural shape as a sump. Half of a ⅜-inch pipe coupler was welded into the bung, and ½-inch line was connected to a Holley electric fuel pump.

The drivetrain consists of a manual-shift Turbo 400 with a 4500-rpm stall-speed torque converter made by Transmission Specialties, Inc., of Chicago. A B&M shifter and a Hayden trans cooler are used, along with a transmission blanket and a safety hoop. The rearend is a narrowed 12-bolt with 4.56 gears, Moroso Positraction and Summers Brothers axles. Drum brakes are used all around, with 15x13½-inch Cragar Super Trick wheels in front and 15x12 Super Tricks in back. The slicks are 13x30 M&H items and the front tires are 7.10x15 Morosos.

The narrowed rearend is suspended by mono-leaf springs and Monroe-Max 50/50 shocks. A rear subframe was fabricated from 1½x3x.120-inch mild steel tubing. The outside width of the frame rails is 30 inches. Two crossmembers are used, the one over the axle to mount the shocks. The original floor pan was removed and a new one fashioned from 20-gauge sheetmetal and riveted in place. The rear wheel tubs were made from 22-gauge metal. A 1⅞-inch, .120-inch-wall rollbar adds strength and safety.

To keep within his tight budget, Larry utilized much of the previous red, white and blue paint job. Earlier lettering and decals were removed and the "Grocery Getter" lettering was added by Gil "Butch" Schoener. The paint job is simple but highly visible, and the "Grocery Getter" logo on the innocuous-looking wagon draws lots of spectator support. The interior features red Naugahyde upholstery and black nylon carpeting over the rear subframe. Stewart-Warner gauges are used, along with a Hurst Roll Control.

While it isn't much on high-dollar trick items, there are plenty of good bracket

2

3

1. Just as important as any mechanical parts are Larry's logbook, tire gauge and shoe polish for dial-in changes.

2. The ladder bars are 36 inches long and made of 1¼-inch-o.d. x 3/16-inch-wall tubing. Binding is eliminated by using two rollers on either side of front of ladder bar.

3. The gas tank is a narrowed Chevette unit turned upside down to take advantage of its shape as a sump. The narrowed 12-bolt rearend uses an E-T finned cover.

4. The tie rod had to be altered to clear the 396 oil pan. The rod is 4 inches forward of its stock location. Roll pins through drilled holes in the tie rod are used for safety.

5. The stock shock towers had to be notched to clear the big-block Chevrolet engine. The headers are Hooker fenderwell-units.

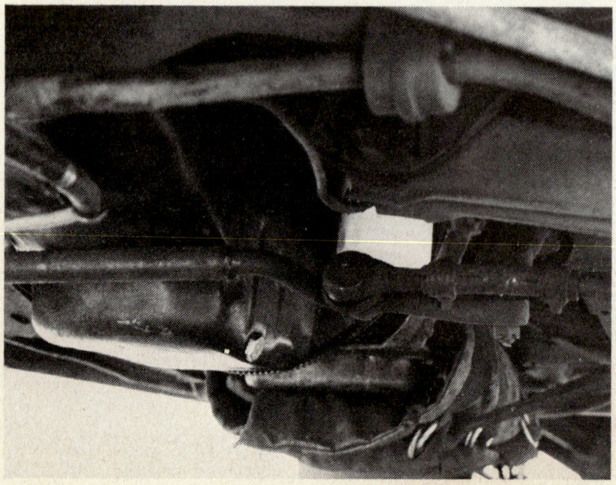

4

5

racing tips to be found on the car. Besides all the previously mentioned items, Severson's Nova features such things as a hinged fiberglass hood. Stock hinges with lightweight door springs from a hardware store don't weigh much, yet the hood is much easier for Larry to manage during quick between-round maintenance than a lift-off hood. For fast cool-downs, Larry installed a shower faucet on the front of the intake manifold. For more efficient transmission cooling, the cooler is mounted flat under the front bumper and an air scoop (made of aluminum) built around it to direct air to the cooler. To keep the battery covered and located where its weight will aid traction, Larry made a hatch in the right side of the rear subframe covering. The Sears battery rests right in front of the right rear well and is easily serviced by opening the rear door.

A nonmechanical item that Larry feels is very important to consistent bracket racing is his ever-present logbook. Track conditions and any changes made to the car are recorded, along with each run's performance. Larry also carries a dial tire gauge and a bottle of white shoe polish for dial-in changes.

Larry Severson is a member of the Bracket Racers Association of Fargo and races his colorful cargo carrier every week. Larry has raced a lot of cars in the past—everything from a '41 Studebaker to a Mustang Funny Car—but on a fun-per-dollar basis, the "Grocery Getter" will be a tough act to follow. A ten-second $3800 bracket bomber is a better bargain than two-dollar steaks.

—Bruce Caldwell

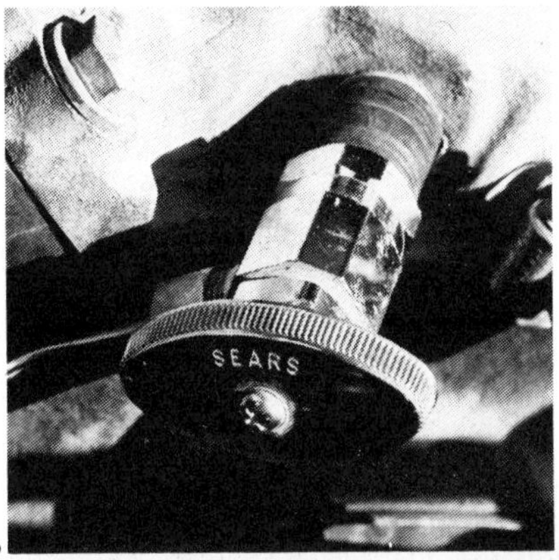

6

7

8

6. For quick cooldowns between rounds, Severson uses a bathroom faucet in the front of the intake manifold.

7. The Sears battery rests in a storage compartment under the rear subframe cover in front of the right rear wheelwell. Battery access is through right rear door.

8. To ease restriction problems, a dual-parallel fuel regulator and filter system is employed. The two lines join before reaching the 850-cfm Holley double-pumper carb.

Turbo Dies-El Camino

Turbocharging the GM diesel engine produces a powerful, responsive and economical powerplant in today's world of low-power cars and high-priced gas.

Sometimes we don't have to go looking for tech stories—they come looking for us. Such was the case with the turbocharged Olds diesel engine installation in the '64 Chevrolet El Camino shown here, which just happens to belong to Photo & Special Assignment Editor Eric "Rick" Rickman, in Petersen's Specialty Publications division. Longtime readers will remember Rick as being part of the HOT ROD Magazine staff from the early '50s until the end of 1970, so he's no newcomer to the world of high-performance and engine swaps.

With over 180,000 miles on the clock, the original 327-cubic-inch engine in the El Camino had gone through just about all the rebuilds and use that any engine can reasonably be expected to endure; but the truck was basically sound, so Rick began considering the engine swap alternatives. The goal was to find something reliable, responsive and, above all, economical to operate. The solution came in the form of a '78 350-cubic-inch Oldsmobile diesel V8 engine, which is offered as an option in Olds and Cadillac passenger cars and in GMC and Chevrolet pickup trucks. This engine would produce the reliability and economy, but the responsiveness left a bit to be desired, which is where Roto-Master (13402 Wyandotte St., North Hollywood, CA 91605) entered the picture. Roto-Master installed a prototype turbocharger system which was to serve as a basis for a turbo kit for the Olds diesel. That kit will soon be available to fit the stock Olds diesel V8 installations in both passenger cars and pickups.

The accompanying photos detail the turbo installation, exhaust system fabrication and the complete engine swap into the El Camino. To simplify the swap, an Olds 350 Turbo Hydra-Matic transmission was also used. Subsequent trips have proved the combination to be a respectable performer while virtually doubling the fuel mileage of the original 327. The turbo diesel added about 140 pounds to the total vehicle weight, including the dual batteries, larger radiator, etc. The majority of this weight increase (120 pounds) was on the front end, but air conditioning front spring spacers were all that was required to restore the original ride height to the El Camino. Boost is held to a maximum of five psi, which seems to be more than adequate. Rick reports that it takes some care to keep the El Camino from climbing above the double-nickel while cruising (see accompanying chart).

The only problems confronting anyone who contemplates a similar swap are: (1) the need for a heavy-duty dual-battery starting circuit to supply the required cranking power as used on the Olds diesel-equipped stockers; (2) a large and efficient radiator to dissipate the increased heat output of the diesel (Rick's El Camino uses a specially constructed 16x23.5-inch radiator with *five* rows of tubes); and (3) a thorough cleaning of the fuel tank and lines, plus the addition of a diesel fuel filter.

After studying the accompanying photos, we think you'll agree that ex-HOT ROD staffers don't just fade away; they get even sharper—even if they do need a little bit of a "boost" for climbing hills these days!

—C.J. Baker

1. We had planned to use a small-block Chevy engine, but there were plenty of 396s for under $100. Ours cost $75.

2. We installed a used battery from a fork-lift.

3. Spring towers were notched for header clearance.

4. It took a little looking, but we finally found a clean, straight '66 two-door, with a nice interior and a shot 6-cylinder engine, for $200.

2

3

BREAKDOWN OF MAJOR EXPENSES

Body/1966 Chevy II	$200
Engine/1967 Chevy 396 (325-hp)	75
Rear end (narrowing)	150
4.88 positraction (used)	100
Hooker headers (used)	-*
Holley intake manifold	110
Valve springs (used)	20
Valve job	35
Radiator/Corvette crossflow (used)	20
Rebuild Powerglide transmission (parts)	24
Motor mounts (used)	6
Camshaft/375-hp 396 (from a friend)	-
Transmission cooler	20
Recurve distributor	6
Distributor wires	12
Moroso oil pan	90
M&H slicks	75
Holley electric fuel pump	25
Gas line	6
Holley carburetor/800-cfm double-pumper (used)	50
Sun tachometer (used)	20
Moroso subframe	70
Wire and gauges	12
Wheels (rear)	18
Battery (used)	5
Battery cable	10
Hood scoop	15
Hood pins	6
Lakewood traction bars	—*
	$1180

*Chevy II engine traded for 396 headers
**Chevy II rear end traded for traction bars
Paint, lettering and Super Trick front wheels are optional—they're nice, but they don't make the car go any faster.

4

The Complete Chevy Parts Guide

Whether it's a complete aluminum block or a cowl induction hood, Chevrolet still has the parts. All you have to do is look for them. We did. . . .

We're happy to report that Chevy's performance parts program is alive and well. While certain low-demand parts (such as some oversize pistons and valves) have been discontinued, overall parts availability remains good. More parts than ever before are available for the timeless small-block. There is a new aluminum block with cast-iron liners similar to the old ZL1 big-block; 350 12.5:1 pistons featuring a revolutionary round wire pin retainer; mushroom cam and associated valvetrain parts; new 4-bolt-main cap with the outer bolt holes angled for increased strength; a Fel Pro head gasket with integral O-rings that should end small-block head sealing and cracking problems forever; and cylinder head and main bearing cap hold-down stud kits. Aluminum heads for the small-block should be available from the factory by fall, 1980; if you can't wait that long, the aftermarket has the situation well in hand (see sidebar). Also in the works is an aluminum water pump and new-design, super-duty, Carillo-style, small-block connecting rod that accepts 7/16-inch SPS capscrews. Not only is this the strongest connecting rod ever offered by the factory for the small-block, it's also much lighter (700-720 grams) than the current 7/16-inch small-block forging that's based on the big-block rod, and causes no cam interference problems in the 3.48-inch stroke 350 block.

The big-block parts list is no slouch, either. Aluminum blocks are still available, albeit from the aftermarket (see sidebar). All the good-guy cast-iron parts are still listed, along with the aluminum heads. In fact, a new design open-chambered aluminum head with a revised exhaust port that outflows all previous designs is now offered in two versions: with conventional water passages, and dry with no water passages. The LS6 11:1 short-block has been combined with cast-iron open-chamber heads to give an approximate compression ratio of 10.2:1, which should offer excellent performance on the street using today's poor gas. This assembly is offered complete, less exhaust manifolds, distributor and clutch. The ¾ LS7 12:1 454 assembly is also still available, as is the L88 427 short-block.

There are also heavy-duty parts for other engines. Most parts needed to build either a street or racing version of the 122-cubic-inch dual-overhead cam, four valve per cylinder Cosworth Vega motor are still offered. Preparing for the future, Chevrolet plans to introduce this fall aluminum heads for their 90 degree 223-cubic-inch V6 engine. This will probably be the 1980s small-block, as it is basically just a small V8 with the front two cylinders lopped off. And, that being the case, it will accept many V8 internal parts, such as connecting rods, pistons, oil pump, fuel pump and associated valves and valvetrain components.

Engines aren't the only Chevy heavy-duty parts. Most parts needed to build an M22 heavy-duty 4-speed are still obtainable, although there are no more complete Muncie transmission assemblies (standard or heavy-duty) to be had.

Most rearend parts are still carried; 12-bolt gears from 2.29-5.13 can be purchased. There are ratios up to 4.10 for late Camaros 10-bolts. For '63-'79 Vettes, gear selection ranges from 2.73-5.13. A new aluminum Dana carrier is in production for '80 Vettes. With the right (1980) halfshafts, it will backfit into earlier Vette IRS suspensions, offering both weight reduction and improved durability. Gear selection from the factory at press time was nearly nonexistent, as the new parts were not yet fully listed in the parts book. However, standard Dana 44 gears will fit.

For the Corvette and Camaro, there are special chassis components and body panels. These include F41, Daytona and Gymkhana suspension components utilized on the Vette and special Camaro Z28 pieces (although the Camaro '69 Trans Am suspension parts are long gone). There are special hoods, spoilers and, in some cases, fender flares for the two Chevy sports cars, as well as the Chevette (!) and Monza. Most of the parts needed for the old Cowl Induction and Camaro ducted hood assemblies, as well as the trick L88 hood and air cleaner parts, are still being carried. Also available is the recently introduced '80 Z28 ducted hood assembly.

HOW TO ORDER CHEVY PARTS

The trouble is: *Where* are they listed? Well, thanks to HOT ROD Magazine, you now have before you what we immodestly proclaim to be the most complete listing of factory performance parts yet. Because of the size of the effort, there may be a few mistakes in there somewhere, although every effort has been made to get these numbers right. If there's something

wrong—or something important left out—we'd like to know about it, so it can be corrected for future editions. Naturally, all numbers listed herein are subject to change or deletion by the "powers that be" at Chevrolet.

Many of the parts are heavy-duty off-road components that are not listed in the regular parts books. (GMPD) General Motors Parts Division plans to introduce a monthly bulletin for all dealers detailing the latest heavy-duty parts, but if your dealer for some reason doesn't yet have the bulletin or has misplaced it, just give him the *part number* and have him look it up in his Price Book, which lists in numerical order all the Chevrolet parts, standard and heavy-duty, with the current price. If your dealer is still unwilling to spend time ordering these parts, we suggest you deal with Rod, Stan or any other friendly counterman over at Baher Chevrolet (18600 Devonshire, Northridge, CA 91324, 213-875-2600), who have most of the parts in stock and are willing to ship anywhere. They graciously took time out of a busy day to provide many of the parts photographed for this article, and permitted the author to "move in" for several days and verify that all the parts presented herein were still in the Price Book.

But no matter what dealer you patronize, while you're there ordering the parts, pick up a copy of *Chevrolet Power.* At press time, the Third Edition was still on sale, but by the time you read this (or at least by late 1980), there ought to be a completely updated Fourth Edition available. This book tells you everything you ever wanted to know about building the Chevy big- and small-block V8 and Cosworth 122, setting up the Corvette for road-racing, designing a competition suspension and much, much more. At $5.95 (for the Third Edition), it's a great investment!

Read on and find out what it takes to join the Number One Team.

—Marlan Davis

BASIC ENGINES

SMALL-BLOCK V8: Built since 1955; displacements from 262-400 inches; most parts still stocked back to '55. Current HD parts program focuses on late 302-350 (except 305/307); many of these parts backfit to earlier motors. 302, 327, 350 have 4.00 (inch) bore, with strokes of 3.0, 3.25, 3.48, respectively. '67 and earlier came with 2.30 main journals, 2.0 rod journals ("small journal"); '68 and later 302/327 and all 350s have has 4⅛ bore x 3.75 stroke with 2.65 mains, 2.10 rods. These journal sizes are *not* interchangeable, but Chevy has raw and semi-finished cranks which can be finished to go in 400 block main bearings, available in stroke lengths from 3.0-3.625. Also, bearing makers like TRW have special bearings, allowing 2.45 journal crank to be used in 400. The 400 also has its own unique-length connecting rod, and requires special externally-balanced harmonic damper and flywheel.

90 DEGREE V6: Available since 1979 in 229-cubic-inch displacement only (3.736 bore x 3.48 stroke). Accepts many small-block V8 internal components, including 305 V8 pistons, front cover and timing gears, oil pump, etc. Will accept small V8 connecting rods if rods are narrowed. Will not accept heads, crank or cam. Heavy-duty aluminum heads and 4-inch-bore block currently under development.

BIG-BLOCK (MARK IV): Cast-iron: Built since 1965, installed in passenger cars through 1976, Corvette through 1973, pickups through 1979; still installed in larger trucks. Displacement ranges from 366 (truck)/396 (passenger) to 454 inches. 396, 402 (which replaces all 396s); 427 motors have bores of 4.094, 4.125, 4.250, respectively, all with 3.76 stroke. 454 has 427 4.250 bore with 4.0 stroke, requires special externally-balanced flywheel and harmonic damper. Aluminum: Displacement ranges from 427 to 510 inches. ZL1 427 shares most parts with cast-iron 427, including pistons, because it has cast-iron liners. 430s had 4.44 bore, 3.47 stroke, required special connecting rod no longer sold by factory. 465, 495 had 430 bore, but 427 and 454 strokes, respectively. 5.10 had 4.5 bore, with 454 stroke. Two blocks were offered for 430 and larger engines; one similar to ZL1 with cast-iron liners for displacements up to 495, the other with 390 alloy, with no liner, for displacements from 495 to 510; pistons are *not* interchangeable.

VEGA TWIN-CAM (COSWORTH) 122: Built from 1975-'76; displaces 122 cubic inches only (3.501 bore x 3.160 stroke). A modified Vega 140 motor, the "Cosworth" features linerless 390 alloy aluminum block and hemi heads, dual overhead camshafts actuating 4 valves per cylinder, beefed-up bottom-end, electronic fuel injection.

AFTERMARKET PIECES

CLOSING THE GAPS

If Chevy doesn't have it, you can bet the aftermarket does. We've assembled a listing of where to get those special parts that the factory doesn't carry.

ALUMINUM BIG-BLOCK

Yenko Chevrolet, Inc., 960 Washington, Route 19 S., McMurray, PA 15317: Linered block with bore sizes up to 4.44 (inches). Deck height up to .400 higher than stock available. Wet or dry decks on request; can build exact duplicate of ZL1 down to part number, if needed. Web Graphics, P.O. Box 2085, 17000 S. Vermont, Unit E., Gardena, CA 90247: Makers of the "Rodeck" block, a "dry" block for drag racing. Standard deck height is .400 higher than stock; can build up to .600 higher on request.

ALUMINUM SMALL-BLOCK

Milodon Engineering Inc., 7711 Ventura Canyon Ave., Van Nuys, CA 91402: Wet deck, cast-iron liners. Donovan Engineering Corp., 2305 S. Border Ave., Torrance, CA 90501: Wet deck, cast-iron liners.

ALUMINUM SMALL-BLOCK HEADS

Brodix Aluminum Products, P.O. Box D, Mesa, AR 71953 Brownfield Co., 3062 N. Lima St., Burbank, CA 91504; Speed Service, Inc., 3049 Irving Park Rd., Chicago, IL 60618

SPECIAL CONNECTING RODS

Carillo Industries, 33041 Calle Perfecto, San Juan Capistrano, CA 92675: Super-duty steel rods in special lengths.

SPECIAL CRANKSHAFTS

Hank the Crank, Inc., 7253 Lankershim Blvd., N. Hollywood, CA 91605: Blueprinted cranks, cranks in stroke lengths no longer offered by factory.

SPECIAL PISTONS AND BEARINGS

TRW Replacement Div., TRW Inc., 8001 E. Pleasant Valley Rd., Cleveland, OH 44131: High-compression forged pistons equal in quality to factory forged pistons (which are made by TRW), special high-compression state-of-the-art racing pistons (complete and unfinished), low-compression forged pistons for turbocharging and supercharging applications; special application pistons not covered by the factory (such as 400 small-block, stock and destroked). Also bearings, both stock replacement, racing and special applications, including thick main bearings to allow use of 350 crank in 400.

SUSPENSION, CHASSIS & BRAKE COMPONENTS

Dick Guldstand Enterprises, 11924 W. Jefferson Blvd., Culver City, CA 90230; Everything needed to set the Corvette chassis up for high-perf street, Gymkhana and/or road racing. Also brake parts and Camaro chassis parts Vette Brakes, Inc., 4540 21st Ave. N., St. Petersburg, FL 33713: High-performance Vette brake components, specializing in no-rust calipers lined with stainless-steel inserts and J56-style standard-calipers conversion.

CORVETTE FIBERGLASS

Eckler's Corvette Parts, P.O. Box 5637, Titusville, FL 32780: Complete line of stock replacement and competition body panels.

LIGHTING SYSTEMS, OFF-ROAD

Cibie Lights, 33195 Harper Ave., St. Clair Shores, MI 48082: Quartz-iodine headlights that fit in Corvette factory off-road headlight mounts.

REAREND PARTS

Tom's Differentials, 15551 Paramount Blvd., Paramount, CA 90723: Heavy-duty parts and rings and pinions for GM rearends; complete ring and pinion selection for new Corvette aluminum carrier that accepts Dana 44 parts.

ABBREVIATIONS USED IN PARTS LIST

AI=Air Induction ('80 Z28)
AR=Amount required
BB=Big-block (Mark IV) V8
CC=Closed-chamber (big-block)
cc=cubic centimeter
CI=Cowl Induction ('70-'72 Chevelle)
COS=Vega Twin-Cam (Cosworth) 122
CR=Compression ratio
DA=Ducted air hood ('69 Camaro)
DP=Double-pumper (Holley carb with accelerator pumps on both primary and secondary venturis)
FI=Fuel injection
HD=Heavy-duty
HEI=High Energy Ignition
IB=Inboard
IRS=Independent Rear Suspension (Corvette)
LH=Left-hand
LJ=Large-journal (small-block)
N/A=Part number not assigned by press time
OB=Outboard
OC=Open-chamber (big-block)
o.d.=Overall diameter
OS=Oversize
RH=Right-hand
SB=Small-block
SJ=Small-journal (small-block)
US=Under-size

GROUP 0.000 ENGINE—CLUTCH

0.000 ENGINE, Complete

All are 4-bolt main bearing engines, with forged crank, pistons & connecting rods, unless otherwise indicated.

3965774 454 HD, 12:1 CR, less intake & exhaust manifolds, distributor & clutch (LS7/cast-iron OC heads)
366250 454, 10.2:1 CR, less exhaust manifolds, distributor & clutch (1970 LS6 short-block assembly with 11:1 CR CC pistons and cast-iron OC heads, low-profile aluminum Holley 1x4 bbl. intake manifold)
3981903 402, 10.25:1 CR, less exhaust manifold, flywheel, water pump, & clutch ('65-'70 396/402 cid 325 hp standard passenger replacement: 2-bolt main cast bottom-end, Q-jet cast-iron spread-bore intake manifold)
14002400 350, 8.5:1 CR, less water pump, dampner, inlet/exhaust manifolds, distributor, flywheel, spark plugs, clutch ("Target Motor": '71-'80 universal replacement, 2-bolt-main, cast bottom end, small valve heads)

0.033 BLOCK ASSEMBLY, Cylinder

All are 4-bolt main bearing engines, with forged crank, pistons & connecting rods, unless otherwise indicated.

PARTIAL ENGINE (SHORT-BLOCK)

Includes block, crank, cam, timing gears, pistons & connecting rods, unless otherwise indicated.

3981820 454, 1970 LS6, 11:1 CR, for CC heads
3997548 454, 1971 LS6, 9.25:1 CR, for OC heads
3970699 427 HD, 1969 L88, 12.5:1 CR, for OC heads
3974228 427 HD, 1967-'68 L88, 12.5:1 CR, for CC heads
3970695 427, 1966-'69 L72, 11:1 CR, for CC heads
3981822 402, 1965-'70 L78, 11:1 CR, 2-bolt-main, for CC heads (replaces 396)
6259578 400, '71 & later, 8.5:1 CR, 2-bolt-main, cast bottom end
3966921 350, 1969-'70 LT1, 370 hp, 11:1 CR, mechanical cam
3970653 350, 1969-'70 L46, 350 hp, 11:1 CR, hydraulic cam
3997472 350, 1971-'72 LT1, 9.25:1 CR, mechanical cam
464152 350, 1973-'77 L82, 9.0:1 CR, hydraulic cam (includes block, pistons, connecting rods and crankshaft only)
474171 350, 1978-'79 L82, 9.0:1 CR, hydraulic cam (includes block, pistons, connecting rods and crankshaft only)
14019872 350, 1980 L82, 9.0:1 CR, hydraulic cam (includes block, pistons, connecting rods and crankshaft only)
465208 327, 1962-'65 L84 & FI (365/375 hp) and 1966-'67 L79 (350 hp), SJ 2-bolt-main (common replacement for partial assemblies 3970166 & 3970158). Does not include cam, cam/crank sprockets, front cover, timing chain, miscellaneous retaining hardware or engine oil support 1051858.
3970657 302, 1968-'69 Z28, 290 hp, 11:1 CR, mechanical cam, LJ
345010 122 COS, 8.5:1 CR, 2-bolt-main (includes block, crank, pistons & connecting rods only)

FITTED CYLINDER BLOCK

Includes pistons, pins, rings & bearings.

3981806 454, 1970 LS6, 11:1 CR, for CC heads
3997547 454, 1971 LS6, 9.25:1 CR, for OC heads
3970685 427, L72, 11:1 CR, for CC heads
3977676 400, 1970 LF6, 265 hp, 10:1 CR, cast crank & pistons
3966920 350, 1970 LT1, 11:1 CR
3970643 350, L46, 11:1 CR
3997428 350, 1971-'72 LT1/1973-'77 L82, 9.25:1 CR
474172 350, 1978-'79 L82, 9:1 CR
14019870 350, 1980 L82, 8.9:1 CR
3970136 327, 11:1 CR, SJ, 2-bolt-main
3970647 302, Z28, 11:1 CR, LJ
345009 122, COS, 8.5:1 CR, 2-bolt-main

CYLINDER & CASE ASSEMBLY (BARE BLOCK)

Includes main caps, bolts & cam bearings.

3963516 BB, 4.25-inch bore
458607 BB, 4.25-inch bore, .400-inch higher deck (427 truck block); use with aftermarket long rods to improve rod angularity.
330816 SB, 4⅛-inch bore (stock in 400 cid motor), 2-bolt main, siamesed cylinder walls (no water jacket between cylinder walls)
3970016 SB, 4-inch bore
366246 SB, 4-inch bore, high tin-content (increased strength)
366287 SB, 4-inch bore HD, siamesed bore with thick cylinder walls (block will accept up to ⅛-inch overbore), extra reinforcing in rear bulkhead area, blind head bolt holes (NASCAR "bowtie" block)
366300 SB, 4-inch bore, aluminum with cast-iron cylinder liners
370750 COS, 3.501-inch bore, 390 alloy aluminum (linerless)
N/A 90° V6, 4-inch-bore HD (available late 1980)

0.034 SLEEVE, Cylinder Liner Replacement, Aluminum Block

6263745 BB, 4.44 bore, standard
6263746 BB, 4.44 bore, .005 OS
6263747 BB, 4.44 bore, .010 OS
3992009 BB, 4.25 bore, standard
3992011 BB, 4.25 bore, .010 OS
14011043 SB, 4.0 bore, standard
14011044 SB, 4.0 bore, .005 OS
14011045 SB, 4.0 bore, .010 OS

0.056 BOLT—STUD, Crankshaft Bearing Cap

3909834 BB bolt, IB, (except aluminum block), 6 required (3 11/32-long x 1 3/16-thread length)
3859927 BB bolt, OB (except aluminum block), 10 required (2 39/64-long x 53/64-thread length)
3952321 BB bolt, IB (aluminum block), 6 required (½-13x4¼)
3952323 BB bolt, OB (aluminum block), 10 required (¼-13x3 1/16)
14011039 SB stud kit, replaces bolts
14011040 Washer for 14011039, 16 required
3942410 Nut for 14011039, 16 required
3932480 SB bolt, IB (7/16-14x3 11/32), 4-bolt cap, 10 required
3877669 SB bolt, OB (2 1/16-inch length), 6 required
3872781 SB bolt, through '68 (7/16-14x3 9/32), AR
3902885 BB stud, for mounting oil splash shield, except aluminum block (½-13x⅜-16x5 ⅛), 4 required
336719 BB stud, for mounting oil splash shield, with aluminum block, 4 required (replaces 3952322)

3960312 SB stud, for tray baffle mounting, '69 & later, (3/8-16x7/16-14x5 5/8), AR
3872718 SB stud, for tray baffle mounting, '62-'65, (7/16-14x3/8-16x 5¼), AR
3977137 COS bolt, (7/16-14x3 3/16)

0.093 PLASTIGAGE, Crankshaft Bearing
1390274 For measuring bearing clearance, green type, 12 pieces

0.095 CAP, Crankshaft Bearing
3877470 BB, 4-bolt-main replacement cap
14011072 SB, 4-bolt-main replacement cap (position 2, 3 & 4) for aluminum blocks made 1980 & later, features angled outer bolt holes for increased strength. Earlier aluminum & cast-iron blocks will accept this cap if block is remachined for it (large-journal blocks only) (available May 1980)
3970194 SB, '55-'69 2-bolt-main replacement cap (blue stripe)
3956688 SB, '70 & later 2-bolt-main replacement cap; and front replacement cap on 4-bolt-main motors
3932482 SB, replacement 2, 3 & 4 position on 4-bolt-main blocks & for upgrading 2-bolt-main large journal blocks to 4-bolt-main
3983483 COS, replacement 1-2-3 position

0.096 BEARING UNIT, Crankshaft No. 1-2-3 (V8/COS) & 4 (V8), (1 Upper & 1 Lower)

BIG-BLOCK
3965715 .001 OS 3885378 .002 US 347017 .020 US
347013 Standard 347016 .010 US
347014 .001 US

SMALL-BLOCK 400
3977522 Standard 3977524 .002 US 3977526 .010 US
3977523 .001 US 3977525 .009 US 3977527 .020 US

SMALL-BLOCK (LJ 302/327 & 350)
Large journal bearings of the same dimensions are interchangeable, although the parts book recommends the following applications:
3965718 .001 OS, upper & lower, #1-2-3-4
3912038 Standard, upper & lower: #1-2-3-4, '68-'69 327/'68-'75 350; #2-3-4, '76-'80 350
462661 Standard, upper, #1, '76-'80 350
462664 Standard, lower #1, '76-'80 350
3912039 .001 US, same application as 3912038
462673 .001 US, same application as 462661
462677 .001 US, same application as 462664
3912040 .002 US, same application as 3912038
472674 .002 US, same application as 462661
462678 .002 US, same application as 462664
3912037 .009 US, upper & lower: #1-2-3-4, '68-'69 327/'68-'74 350; #2-3-4, '75-'80 350
476011 .009 US, upper & lower, #1, '75-'80 350
3912041 .010 US, same application as 3912038
464047 .010 US, upper & lower, #1, '76-80 350
3912042 .030 US, same application as 3912038
464048 .030 US, same application as 464047

SMALL-BLOCK (SJ)
Bearings of same-size are interchangeable, but parts book recommends:
3876116 Standard, #1 3832931 .001 US, #2-3-4
3832930 Standard, #2-3-4 3876070 .009 US, #2-3-4
3876117 .001 US, #1 3876119 .010 US, #1
3832934 .020 US, #2-3-4

0.102 CAP, Crankshaft Main Bearing, COS No. 4
3983484 COS, replacement #4 position

0.103 BEARING UNIT, Crankshaft No. 4 (Cosworth DOHC 122 only)
6260600 Standard 6260802 .002 US 6260804 .020 US
6260801 .001 US 6260803 .010 US

0.105 CAP, Crankshaft Main Bearing, Rear
3877472 BB replacement cap, #5 position
3911007 SB '68-present replacement cap, #5 position
3983485 COS replacement cap, #5 position

0.106 BEARING UNIT, Crankshaft No. 5

BIG-BLOCK
3965717 .001 OS 347019 .001 US 347021 .010 US
347018 Standard 347020 .002 US 347022 .020 US

SMALL-BLOCK (400)
3977528 Standard 3977530 .002 US 3977532 .010 US
3977529 .001 US 3977531 .009 US 3977533 .020 US

SMALL-BLOCK (LJ 302/327 & 350)
3965719 .001 OS 3912033 .002 US 3912035 .010 US
3912030 Standard 3912034 .009 US 3912036 .020 US
3912032 .001 US

SMALL-BLOCK (SJ)
3779460 Standard 3779463 .010 US
3779461 .001 US

COSWORTH DOHC 122
6260981 Standard 6260983 .002 US 6260985 .020 US
6260982 .001 US 6260984 .010 US

0.137 SEAL ASSEMBLY—COMPOUND UNIT, Crankshaft Rear Bearing & Oil
3962863 BB rear oil seal, 1965-75, 2 required
473471 BB rear oil seal, 1976 & later, 2 required
9772826 SB rear oil seal, '55-'58, 1 required
473424 SB rear oil seal, '59-'80, 1 required
3983488 COS rear oil seal (rope type), 2 required
3997597 COS sealing compound for rear main bearing & oil (includes hardener, sealing compound, applicator, disc, rod, instruction sheet), use in place of 3983488

0.206 COVER, Crankcase Front End or Timing Belt
330850 BB
3991433 SB, with pointer, 1969-present
361925 SB, high-performance, 1962-'68 (replaces 3923290)
334029 COS, lower rear timing belt cover
334028 COS, lower front timing belt cover
352122 COS, upper rear riming belt cover

0.213 SEAL ASSEMBLY, Crankcase Front End
3860095 BB front oil seal
3980267 SB front oil seal, 1955-present
3973135 COS front crankshaft seal

0.219 POINTER—SNUBBER ASSEMBLY, Timing or Timing Belt
3992073 BB pointer
3991436 SB, high-performance pointer, 1969-present
352123 COS, timing belt snubber

0.269 HEAD, Cylinder

BIG-BLOCK
All are cast-iron with rectangular intake ports, 2.19-inch intake valves, square exhaust ports, and use 13/16 spark plugs, except as noted, Big-block heads are shipped bare.
14011076 Aluminum OC, finned exhaust port, 1.88-inch exhaust valve
14011004 Same as 14011076, but dry head (no water passages), for drag racing (available February 1980)
3946072 Aluminum OC, round exhaust port, 1.88-inch exhaust valve
6260482 OC, 1.88-inch exhaust valve, 5/8-inch spark plugs
3919838 Aluminum CC, 1.84-inch exhaust valve
3964287 CC, 1.88-inch exhaust valve, 5/8-inch spark plugs
3919839 CC, 1.72-inch exhaust valve

SMALL-BLOCK
All shipped assembled with guide plates and screw-in rocker arm studs. Small-block heads are machined for 2.02-inch intake and 1.60-inch exhaust valves, with 65cc combustion chambers; they accept 5/8-inch spark plugs, except as noted.
N/A Aluminum, angled spark plug, blocked heat riser (available late 1980)
3965784 Angled spark plug, blocked heat riser, casting number 340292. Features improved port flow design, additional cast iron around valve spring area, reduced valve guide diameter and increased valve spring pocket diameter, and (since 1976) tin alloy for improved hot strength.
336746 Angled spark plug (replaces 3365742), casting number 3991492. Features reduced valve guide diameter and increased valve spring pocket diameter for racing valve springs to accommodate 330585 spring.
3987376 Straight spark plug, 13/16 spark plugs, casting number 3991492. Service replacement for 1967-'69 302, 1964-'68 327 aand 1969-'70 350 high-performance engines.

90 DEGREE V6/229cid

N/A Aluminum cylinder head (available October, 1980)

COSWORTH DOHC 122:

The Aluminum heads are shipped with valve guides. and seats

6260845 Stock replacement head, use with head gasket.

366203 Off-road head, use with o-ring unit in place of head gasket.

0.289 GASKET, Cylinder Head

3976081 BB, 4.44 bore, HD composition, .045-inch compressed thickness

3969865 BB, L-88/ZL1, HO composition, .045-inch compressed thickness

3921940 BB, high performance, composition, .035-inch compressed thickness

330860 BB, steel shim, .022-inch thick

3974218 BB kit (includes head & manifold gaskets), L-88/ZL1

14011041 SB, HD service for 4- and 4¼-inch-bore engines, Fel-Pro-type composition Permatorque with built-in O-rings around cylinders & built-in shims around outer head bolt holes for stress relief, .040-inch compressed thickness; when used in 400 block holes may have to be drilled in gasket to match steam holes on block deck.

3965790 SB, HD service for 4- & 4 ¼-inch-bore engines, composition, .040-inch compressed thickness (reinstated March 1980)

3997790 SB, 4- & 4¼-inch-bore engines, composition, .046-inch compressed thickness

3830711 SB, 4-inch-bore, steel shim, .022-inch compressed thickness

3916336 SB, 4-inch bore, stainless steel/beaded type, .016-compressed thickness

462691 SB, '69-present, stock replacement

3931606 SB kit (includes head & manifold gaskets), 1957-'68

356805 SB kit (includes head & manifold gaskets), 1969-'71 (Some post-'71 smog engines take different gasket sets, so see your Chevy dealer if you need a stock replacement.)

352150 COS, stock replacement head gasket

347090 COS kit, stock replacement (includes head, cam cover & carrier and inlet & exhaust manifold gaskets)

366204 COS off-road, metallic O-ring unit, replaces gasket (see Power book for required modifications to head and block)

0.293 BOLT—STUD—PIN—WASHER, Cylinder Head

3965763 BB stud unit, cylinder head hold-down, replaces head bolts

3946079 BB stud, cylinder head hold-down, for aluminum heads on aluminum block, 4 required

3899696 BB & COS washer, cylinder head bolt/stud; 32 required on BB, 10 on COS

14014408 SB stud unit, cylinder head hold-down, replaces head bolts

14011040 SB washer, cylinder head stud, 34 required

8942410 Nut, 7/16-20 for all head studs; 32 required for BB, 34 for small-block

334011 COS cylinder head bolt, 10 required

452720 BB insert, replacement for aluminum heads, 32 required

3894327 BB pin, head locating, 4 required

585927 SB & COS pin, head locating, 4 required SB, 2 on COS

0.296 VALVE, Inlet

BIG-BLOCK

3879618 2.3-inch diameter, must rework head

340286 2.19-inch diameter, titanium alloy, light weight

3969815 2.19-inch diameter, hard-chrome stem

3969818 Same as 3969815, but .030 OS

3864808 2.19-inch diameter, standard-duty, standard-size

3876814 Same as 3864808, but .015 OS

3876815 Same as 3864808, but .030 OS

SMALL-BLOCK

366285 2.05-inch diameter, double vacuum melt stainless steel alloy with new flat face (no dish) head design with 10° angle on back of head; offers improved flow and durability over previous designs; must rework seat.

361976 2.05-inch diameter, swirl-polished, must rework seat.

3849814 2.02-inch diameter, swirl-polished

3849815 Same as 3849814, but .003 OS

3849816 Same as 3849814, but .015 OS

3849817 Same as 3849814, but .030 OS

COSWORTH DOHC 122

348637 Standard-size

355703 .003 OS

0.297 VALVE, Exhaust

BIG-BLOCK

340287 1.88-inch diameter, hollow stem, light weight

3946077 1.88-inch diameter, heavy-duty

3947872 Same as 3946077, but .003 OS

3879619 1.84-inch diameter

3860002 1.72-inch diameter, standard-size

6263735 Same as 3860002, but .015 OS

SMALL-BLOCK

3849818 1.60-inch diameter, swirl-polished, aluminized face, standard-size

3850040 Same as 3849818, but .003 OS

3850041 Same as 3849818, but .015 OS

COSWORTH DOHC 122

348639 Standard-size

355705 .003 OS

0.303 SPRING—SHIM, Valve

3970627 BB spring, street high-performance cams (2nd design spring)

3916164 BB spring, L-88/ZL1 design

3989354 BB spring, HD made from aircraft-quality steel wire, for cam 3994094. Offers durability improvement over 3916164 if rpm does not exceed 7000. Must use aftermärket seals

3911068 SB spring, for street use

3927142 SB spring, for cam 3927140 & to extend rpm range of street mechanical cams

330585 SB spring, dual spring for HD camshafts, except mushroom cam, use in castings 340292 (service assembly 3965784) and 3991492 (service assembly 336746) only

366282 SB spring, dual with damper (1.525 o.d.) for mushroom cam 366293; can only be installed in head casting 340292 (service assembly 3965784)

3891521 BB shim, .065-inch thick

3875916 BB shim, .015-inch thick

3731058 SB shim, 1/32-inch thick, use in place of valve stem oil shield

6260555 COS spring unit (inner & outer)

0.308 SEAL—SHIELD, Valve Spring Cap or Valve Stem

460527 BB seal, umbrella-style, used on standard '71-'76 454, also to replace discontinued seal 3866612, used on discontinued 1st design, street performance spring

1382076 Buick Motor Div. part number, stock Buick V8 valve seal, will fit on Chevy BB street and street/performance springs, offers better sealing than 460527. Order from Buick dealer or have local Chevy dealer punch up number on warehouse computer to obtain price and ordering information, since part number is not listed in any Chevy parts or price book. (Baher has these in stock.)

3835333 SB seal, "O"-ring, recommended for street cams only; will not work with mushroom cam & springs.

3836755 SB shield, cup shape

6260848 COS inlet valve stem seal unit, 1 required

0.309 CAP, Valve Spring

3859909 BB retainer, street performance springs, for discontinued 1st design spring

3989353 BB retainer, for spring 3989354

3964264 BB retainer & seal, for spring 3970627 (2nd design)

3879613 BB retainer & seal, for spring 3916164

3896934 SB retainer, steel, for springs 3911068 & 3927142, stamped "A"

330586 SB retainer, aluminum alloy, for spring 330585

366254 SB retainer, titanium, for spring 366282

334067 COS retainer

334065 COS seat & cup

0.310 KEY, Valve Stem

3947880 BB, hardened, green

3947770 SB, hardened, purple (replaces 838029)

334068 COS, split-lock design

0.333 ARM, Valve Rocker, with Adjusting Screw and Nut

3959182 BB, long slot for .600-lift camshafts, stamped "H"

3974290 SB, stamped "V" on inside of pad end, with ball

0.386 COVER—HOUSING, Valve Rocker Arm or Camshaft

325174 BB rocker arm cover, chrome, cast-in dippers, RH
325169 Same as 325174, but LH
3910298 SB, rocker arm cover, HD, no name/no holes, right or left, use to fabricate custom oil vent/breather systems, not stock on any engine
3923226 SB rocker arm cover, chrome, right or left, 1968 design
474207 SB rocker arm cover, ribbed aluminum, 1962-'67 design, LH
474208 Same as 474207, but RH
3965541 SB rocker arm cover, aluminum, with oil filler cap & cast-in dippers to improve rocker arm ball lubrication (1970-'77 high performance), LH
3965542 Same as 3965541, but RH (no oil filler cap)
464667 SB rocker arm cover, '78-'80 L82, black ribbed aluminum, RH
464668 Same as 464667, but LH with oil filter cap
348615 COS camshaft cover
348613 COS camshaft housing

0.413 RETAINER—BOLT—PIN, Rocker Cover or Camshaft Housing and Cover

3850874 SB spring bar, improves sealing, originally installed on Corvair sixes
180016 V8 ¼-20x9/16-inch rocker cover bolt
3877670 Reinforcement for 180016 bolt
334044 COS bolt, camshaft cover, 12 required
334081 COS bolt, camshaft housing to cylinder head, 17 required
334006 COS pin, cylinder head to camshaft housing, 2 required

0.426 ROD, Valve Push

3942416 BB 7/16-inch diameter, intake
3942415 BB 7/16-inch diameter, exhaust
3946067 BB ⅜-inch diameter, intake
3946069 BB ⅜-inch diameter, exhaust
3796243 SB 5/16-inch diameter, hardened tip, .060-inch-wall tubing, blue or black band on insert tip end
366277 SB 5/16-inch diameter, hardened tip, .075-inch-wall tubing, .100-inch longer than 3796243, black oxide-treated. Provides better valvetrain geometry with small-base circle cams, as well as greater strength.

0.429 BALL—NUT—STUD, Valve Pushrod Adjusting

3899622 BB rocker ball, heat-treated, grooved
3959142 SB rocker ball
3896648 BB rocker arm nut
3744341 SB rocker arm nut
3921912 BB rocker arm stud (If used in place of discontinued 3904396 on aluminum head, 452716 helicoil must be installed.)
3973416 SB rocker arm stud, screw-in type for 1970 & later high-performance heads

0.439 GUIDE PLATE, Push Rod

3879620 BB, for 7/16-inch pushrods
3860038 BB, for ⅜-inch pushrods
3973418 SB, for 1970 & later high-performance heads

0.459 TAPPET, Valve (or Lifter)

V8 TAPPETS

5232700 Hydraulic, BB
5232720 Hydraulic, SB
5231585 Mechanical, all V8, edge orifice metering for reduced oil flow, recommended for aftermarket roller rocker arms. Stock on '59-'65 high-performance 327, late Z28 302, 1970-'72 LT1
5232695 Mechanical, all V8, piddle valve design, recommended for production rocker arms. Stock on most Z28 302, all BB with mechanical lifters
366253 Mechanical, SB with mushroom cam 366293, has .960-inch-diameter foot. Bottom of lifter bore may have to be reworked.

COSWORTH DOHC 122

334071 Valve tappet
5233266 Shim, valve tappet, .111 thick (for adjusting valve lash)
5233267 Shim, .113 thick 5233275 Shim, .129 thick
5233268 Shim, .115 thick 5233276 Shim, .131 thick
5233269 Shim, .117 thick 5233398 Shim, .133 thick
5233270 Shim, .119 thick 5233399 Shim, .135 thick
5233271 Shim, .121 thick 5233401 Shim, .137 thick
5233272 Shim, .123 thick 5233402 Shim, .139 thick
5233273 Shim, .125 thick 5233403 Shim, .141 thick
5233274 Shim, .127 thick 5233404 Shim, .143 thick

0.519 CAMSHAFT

BIG-BLOCK All are non-grooved, for 1967 and later motors.

3883986 Hydraulic, .460/.480 lift, high-performance street or Marine, 1966-'72 396/427/454 (350-, 360-, 390-, 400-hp)
6272989 Hydraulic, .500 lift, high-performance street or Marine (never used on production cars)
3863143 MechanicalZ .520 lift, high-performance street or Marine, 1965-'71 396/427/454 (375-, 425-, 435-, 450-hp)
340282 Mechanical, .560 lift, NASCAR or short-track, 1x4-bbl. carb
3925535 Mechanical, .540/.560 lift, marginal street use or general competition, L-88 427, 1x4-bbl. carb
3925533 Same as cam 3925535, but reverse rotation for gear-drive applications
3959180 Mechanical, .560/.600 lift, general competition, ZL1/LS7, 1x4-bbl. carb/max power
3994094 Mechanical, .660-lift, drags & road-race on fuel-injected, multi-carbureted or large-displacement engines, used in Can Am series

SMALL-BLOCK

361995 Hydraulic, .372/.410 lift, high-torque & fuel economy, 1976-'79 305
3896929 Hydraulic, .390/.410 lift, good torque on street, used on 350
3863151 Hydraulic, .447 lift, good torque/excellent power, L-79 327 (350 hp) & L-46 350 (350 hp)
3896962 Hydraulic, .450/.460 lift, moderate torque/excellent power, 1973-'74 Z28, all L-82
3736097 Mechanical, .381/.380 lift, excellent torque & rpm in small-displacement motor, pre-1964 Corvette
3972178 Mechanical, .458/.485 lift, street high performance. Used on '70-'72 Z28, and to replace 3896292 "30-30" cam used in special high-performance & fuel injected 1964-'65 327, '67-'69 Z28 & '69-'72 LT1
3927140 Mechanical, .475/.495 lift, short track, 1st racing design
3965754 Mechanical, .495/.530 lift, short track & road racing, 2nd racing design
366293 Mechanical (mushroom lifter), .560/.575 lift, medium & long track, 3rd racing design

COSWORTH DOHC 122

348652 Mechanical, .340 lift, production cam, intake
348654 Mechanical, .340 lift, production cam, exhaust
366226 Mechanical, .400 lift, 289° duration, short track & rally, intake
366228 Mechanical, .400 lift, 284° duration, short track & rally, exhaust
366205 Mechanical, .400 lift, 334° duration, max power/racing, intake
366206 Mechanical, .398 lift, 324° duration, max power/racing, exhaust

0.529 GASKET—GASKET UNIT—SEAL, Camshaft Housing

345129 COS gasket unit, camshaft housing
347089 COS gasket unit, camshaft cover
334001 COS seal, camshaft front

0.533 BEARING, Camshaft Thrust

For BB gear-drive cam 3925533
3975949 Shim, camshaft sprocket, .015 thick
3887871 Plate, camshaft thrust
3817006 Bolt, thrust plate, 2 required

0.539 BEARING, Camshaft (Front or No. 1)

3967808 BB, stamped "1"
3723584 SB, stamped "1", '55-'77*
474005 SB, stamped "1", '78-'80*

*These two part numbers can be used interchangeably.

0.543 BEARING, Camshaft No. 2 (Front Intermediate)

3967810 BB, stamped "2"
3723585 SB, stamped "2", '55-'77*
474006 SB, stamped "2", '78-'80*

*These two part numbers can be used interchangeably.

0.546 BEARING, Camshaft (Center & Rear Intermediate)

3967809 BB, stamped "3 & 4"
3723586 SB, stamped "3 & 4", '55-'77*
474007 SB, stamped "3 & 4", '78-'80*

*These two part numbers can be used interchangeably.

0.549 BEARING, Camshaft Rear

3967810 BB, stamped "2" (0.543), except '65-'66
3876863 BB, stamped "2" (0.543), '65-'66 only
3723587 SB, stamped "5", '55-'63
3723585 SB, stamped "2" (0.543), '64-'75*
474008 SB, stamped "2" (0.543), '76-'80*
*These two part numbers can be used interchangeably.

0.553 PIN—PLUG, Camshaft Rear Bearing Hole

3952319 BB, 2 21/32 diameter, for aluminum block
3859926 BJ, 2 7/32 diameter, for cast-iron block, all through '73 & Corvette through '74
3999200 BB, '73-present, except Corvette
3704158 SB, 2⅛ diameter, through 1971
3997752, SB, '71-present
141344 COS pin, camshaft housing (dowel pin)
9427703 COS plug, Welch, rear cam housing

0.603 ROD ASSEMBLY, Connecting

BIG-BLOCK: All are 6.135 inches in length.
3856240 ⅜-inch bolt, Magnafluxed, pressed pin ('65-'69 396/375, 427/425-435), identified by rib on cap, coded blue
3909846 ⅜-inch bolt, Magnafluxed, floating pin ('67-'68 L88), coded pink
3963552 7/16-inch bolt, Magnafluxed, pressed pin (1970-'71 454 LS6)
3969804 7/16-inch bolt, Magnafluxed, shot-peened & heat-treated, floating pin (1969 L88/ZL1)

SMALL-BLOCK All are LJ, 5.7-inches in length, unless otherwise noted.
N/A 7/16-inch HD SPS capscrew, Magnafluxed, shot-peened & heat-treated, Carillo-style, 700-720 gram weight, offers increased strength over all other SB rods, clears high-lift cams on 350 (available late 1980).
N/A Same as above, but 6.0-inch length (better top-end power); requires special pistons (see pistons) (available late 1980)
3864881 11/32-inch bolt, 2.0-inch SJ
3973386 ⅜-inch bolt HD, Magnafluxed & shot-peened, pressed pin ('70-'72 LT1, '70-'74 Z28, all L82)
3946841 ⅜-inch bolt HD, Magnafluxed & shot-peened, floating pin ('68-'69 Z28, '69 LT1)
343710 7/16-inch HD BB forging with reduced journal diameter (2.1-inch large-journal), semi-finished/no pin holes
366230 Same as 343710, but fully finished, shot-peened & Magnafluxed, pressed pin
366232 Same as 343710, but fully finished, shot-peened & Magnafluxed, pressed pin, 6.0-inch length (better top-end power); requires special pistons (see pistons), pressed pin
361933 ⅜-inch bolt, 5.565-inch length for 400, pressed pin (replaces 346266)

COSWORTH DOHC 122
334020 Heat-treated & Magnafluxed, improved rod/cap surfaces over standard Vega rod

0.616 BEARING UNIT, Connecting Rod

BIG-BLOCK

3965716	.001 OS	347003	.001 US	347005	.010 US
347002	Standard	347004	.002 US	347006	.020 US

SMALL-BLOCK (400)

3910555	Standard	3910557	.002 US	3910559	.020 US
3910556	.001 US	3910558	.010 US		

SMALL-BLOCK (LJ 302/327 & 350)

3965720	.001 OS	3910556	.001 US	3910558	.010 US
3910555	Standard	3910557	.002 US	3910559	.020 US

SMALL-BLOCK (SJ)

3711828	Standard	3715046	.002 US	3715047	.020 US
3715045	.001 US	3715048	.010 US		

COSWORTH DOHC 122
3711828 Standard

0.623 BOLT, Connecting Rod

460405 11/32-24x1⅞, for rod 3864881 (replaces 3892670)
3951631 ⅜-24x1 23/32, for rod 361933
3916399 ⅜-24x1 59/64, for rod 3973386, 3946841
3862720 ⅜-24x2⅛, for rods 3856240, 3909846
N/A SPS Capscrew for Carillo-style SB rod (available late '80)
343713 Bolt, for all SB 7/16-inch bolt rods, except Carillo-style with capscrew
3981092 BB, 7/16-inch rod bolt, knurled shank type (stock on rod 3963552)
3969864 BB 7/16-inch HD rod bolt, boron type, stock on 3969804 rod (L-88/ZL1)
340285 BB 7/16-inch extra-HD rod bolt, SPS/H-11 alloy type
334021 COS rod bolt, 11/32-24x1⅞

0.626 NUT, Connecting Rod

3866766 ⅜-24 for all ⅜-inch rod bolts
3942410 7/16-20 for all 7/16-inch rod bolts
340289 7/16-20, 12-point aircraft type, for all 7/16-inch rod bolts, mandatory on 350 SB to clear cam; (except on SB Carillo-style rod, which has no nut); saves weight over standard nut
527599 11/32-24 for 460405 bolt, type EA, phosphate-coated & soaped

0.629 PISTON ASSEMBLY, with Pin

All are forged, made by Chevy by TRW, except as noted.

BIG-BLOCK
Note: piston assembly for 430/495 iron-liner aluminum engine has been discontinued.
3965764 4.44-bore aluminum bore, standard
3965766 4.44-bore aluminum bore, .010 OS
3965767 4.44-bore aluminum bore, .060 OS
3976014 454 OC, 12.5:1 CR standard
3976018 454 OC, 12.5:1 CR, .001 OS
3976022 454 OC, 12.5:1 CR, .020 OS
3976026 454 OC, 12.5:1 CR, .030 OS
3981075 454 OC, 12.5:1 CR, .060 OS
3976013 454 CC, 11.25:1 CR, standard
3976021 454 CC, 11.25:1 CR, .020 OS
3976025 454 CC, 11.25:1 CR, .030 OS
6262976 454 OC, 9.25:1 CR, standard
6262979 454 OC, 9.25:1 CR, .030 OS
3959105 427 OC, 12.5:1 CR, standard
3959106 427 OC, 12.5:1 CR, .001 OS
3981891 427 OC, 12.5:1 CR, .005 OS
3959107 427 OC, 12.5:1 CR, .030 OS
3909857 427 CC, 12.5:1 CR, standard
3909859 427 CC, 12.5:1 CR, .030 OS
3888342 427 CC, 11:1 CR, standard
3888344 427 CC, 11:1 CR, .030 OS
3969989 402 CC, 11:1 CR, standard
3969991 402 CC, 11:1 CR, .030 OS
3878231 396 CC, 11:1 CR, standard
3875906 396 CC, 11:1 CR, .030 OS

SMALL-BLOCK
3972160 400, 5.565-inch rod, 10:1 CR, cast, standard
3972166 400, 5.565-inch rod, 10:1 CR, cast, .030 OS
14011020 350, 12.5:1 CR, standard, RH
14011021 350, 12.5:1 CR, standard, LH
14011022 350, 12.5:1 CR, .030 OS, RH
14011023 350, 12.5:1 CR, .030 OS, LH
14011024 350, 12.5:1 CR, .060 OS, RH
14011025 350, 12.5:1 CR, .060 OS, LH
14011026 350, 12.5:1 CR, 6.0-inch rod, standard, RH
14011027 350, 12.5:1 CR, 6.0-inch rod, standard, LH
14011028 350, 12.5:1 CR, 6.0-inch rod, .030 OS, RH
14011029 350, 12.5:1 CR, 6.0-inch rod, .030 OS, LH
14011030 350, 12.5:1 CR, 6.0-inch rod, .060 OS, RH
14011031 350, 12.5:1 CR, 6.0-inch rod, .060 OS, LH
3942541 350, 11:1 CR, standard
3942542 350, 11:1 CR, .001 OS
3942543 350, 11:1 CR, .030 OS
336754 350, 9:1 CR (approximate; in late smog head), standard
336755 350, 9:1 CR (approximate; in late smog head), .001 OS
336757 350, 9:1 CR (approximate; in late smog head), .030 OS
3871208 327, 11:1 CR, standard
3850139 327, 11:1 CR, .030 OS
3946876 302, 11:1 CR, standard
3946882 302, 11:1 CR, .030 OS

COSWORTH DOHC 122

360171 8.5:1 CR, standard, Size S-4 (3.4975-3.4980)
360172 8.5:1 CR, standard, Size S-5 (3.4980-3.4985)
360173 8.5:1 CR, standard, Size S-6 (3.4985-3.4990)
360174 8.5:1 CR, standard, Size S-7 (3.4990-3.4995)
360175 8.5:1 CR, .020 OS (3.5173-3.5178)
366215 12.5:1 CR, standard
366217 12.5:1 CR, .010 OS

0.639 RETAINER, Piston Pin

3964238 BB, Spirolock design, .072-inch thick (454 replacement)
3942423 BB, Spirolock design, .050-inch thick (427 replacement)
366219 SB & COS, Spirolock design, .072-inch thick (HD use, must rework SB & 8.5:1 COS pistons)
3946848 SB, Spirolock design, .042-inch thick (stock 11:1 replacement when using full-floating pins)
14011033 SB, Posilock design, for 12.5:1 pistons only

0.643 RING UNIT, Piston

BIG-BLOCK

Sets for iron-liner engines are high-strength iron, moly-faced, radius-faced, 1/16-inch top and second with 3/16-inch three-piece oil rings. Rings for 390 alloy bores are high-strength special configuration, chromed, top, and plain cast-iron second compression, with low-tension 3/16 oil ring.

3965761 4.450 bore for 390 alloy engines
3965762 4.510 bore for 390 alloy engines
3992061 4.44 bore, iron-liner engines, standard
3992062 4.44 bore, iron-liner engines, .005 OS
3993828 4.25 bore, standard
3993829 4.25 bore, .005 OS
3993830 4.25 bore, .030 OS
3993831 4.25 bore, .060 OS
328526 402 BB/400 SB, standard & 396 BB, .030 OS; 1 cylinder only
328531 402 BB/400 SB, .030 OS & 396 BB, .060 OS; 1 cylinder only (Note: Most late 396 blocks WILL NOT accept a .060 overbore.)
328521 396, standard, 1 cylinder only

SMALL-BLOCK

Rings for production 11:1 pistons are low-tension, high-strength iron, moly-faced with a radius contact face, with 5/64 compression and 3/16 oil ring. A new 1/16-1/16-3/16 ring set for racing pistons (including the new 12.5:1 350 pistons) features low tension with moly facing and the latest design improvements. Rings for the 400 are the same as 402 rings; see big-block listing.

3995664 4.0 bore, for production pistons, standard
3995665 4.0 bore, for production pistons, .005 OS
3995666 4.0 bore, for production pistons, .020 OS
3995667 4.0 bore, for production pistons, .030 OS
366289 4.0 bore, for racing & factory 12.5:1 pistons, standard
366290 4.0 bore, for racing & factory 12.5:1 pistons, .020 OS
366291 4.0 bore, for racing & factory 12.5:1 pistons, .030 OS
366292 4.0 bore, for racing & factory 12.5:1 pistons, .060 OS

COSWORTH DOHC 122

Specially-designed chrome-plated rings are required for the 390 alloy silicon/aluminum bores. Production 8.5:1 pistons use 5/64 top and second compression rings with a 3/16 oil ring. For the 12.5:1 piston, a .040 thick top ring with a 1/16 second ring and 3/16 low tension oil ring is used.

6263197 For production 8.5:1 CR pistons, standard sizes
6263199 For production 8.5:1 CR pistons, .020 OS, 1 cylinder only
370753 For racing 12.5:1 CR pistons, .010 OS (file to fit)

0.646 CRANKSHAFT

COMPLETE BALANCED CRANKS

All are forged, Tufftrided and cross-drilled steel, unless otherwise noted. 454 & 400 cranks must be used with special externally-balanced flywheel and harmonic balancer.

3963524 BB, 4.0 stroke, 5140 steel, balanced for 12.5:1 454 (LS7)
3967811 BB, 3.76 stroke, 5140 steel, balanced for 12.5:1 396/427 with 7/16-inch-bolt rods
3963523 BB, 4.0 stroke, balanced for 9.25/11.25:1 454 LS6
3879621 BB, 3.76 stroke, balanced for 12.5:1 427 with ⅜-inch rods
3882842 BB, 3.76 stroke, balanced for 11:1 427
3882841 BB, 3.76 stroke, balanced for 11:1 396
3951527 SB, 3.75 stroke, cast nodular iron, for 400 motor
3941184 SB, 3.48 stroke (balanced for 350 with forged street pistons)
3838495 SB, 3.25 stroke, 1962-'67 SJ balanced for 11:1 327
3914681 SB, 3.25 stroke, 1968 & later LJ, balanced for 11:1 327
3917265 SB, 3.0 stroke, 1967-'68 SJ, balanced for 11:1 302
3941176 SB, 3.0 stroke, 1969 LJ, balanced for 11:1 302
326315 COS, balanced for production 8.5:1 pistons

SEMI-FINISHED CRANKS: Rough-turned on mains only.

3965753 BB, 4.0 stroke, 5140 steel
3997748 SB, 3.48 stroke

RAW FORGINGS

3963521 BB, 4.0 stroke, 5140 steel
3941182 SB, 3.4-3.625 stroke, 1053 steel
366280 SB, 3.25-3.625 stroke, 5140 steel
3941178 SB, 3.0 stroke (made from 5140 steel and with .0001-inch more material on main journals to allow machining for use in 400 after mid-1980)

0.659 BALANCER—PULLEY, Crankshaft (Harmonic Damper)

3963530 BB, 8-inch HD, for externally balanced 454
3879623 BB, 8-inch HD, for all internally balanced engines
3817173 SB, 8-inch HD, pre-1969 timing mark
3947708 SB, 8-inch HD, 1969-present timing mark
364709 SB, 8-inch extra-HD with malleable iron inertia ring
6272225 SB, externally-balanced standard balancer for 400
334012 COS damper, with belt groove
3955290 BB, crankshaft pulley, 1-groove
3955291 BB, crankshaft pulley, 2-groove high performance
3869978 BB, crankshaft pulley, 3-groove
3863108 BB, crankshaft pulley, 2-groove, L88/ZL1
3858533 SB, crankshaft pulley, 2-groove, deep groove

Many other pulleys available. See your Chevy dealer.

0.666 FLYWHEEL, with Ring Gear

MANUAL TRANSMISSION

3963537 BB, LW (lightweight) nodular iron, 12¾ diameter, externally balanced 454 (10½-inch clutch)
3991406 All V8 (except 454 & 400), LW nodular iron for 10½-inch clutch
3955151 All V8 (except 454 & 400), 14-inch diameter for dual-plate clutch
3992094 BB, 14-inch diameter for dual-plate clutch, externally-balanced 454
3998281 All V8 (except 454 & 400) with 11-inch clutch
3993827 BB with 11-inch clutch, externally-balanced 454
3986394 SB, standard clutch, for externally-balanced 400
334015 COS, LW high-strength iron

AUTOMATIC TRANSMISSION

336717 Externally-balanced 454, high performance
361950 All V8 (except 454 & 400), high performance
340296 All V8 (except 454 & 400), dual-bolt pattern
340298 Externally-balanced 400

0.724 BELT—BOLT—CHAIN—COVER, Camshaft Timing

3860036 BB link type, high-performance chain
3891519 BB double-roller type chain, HD as used in trucks
346261 SB, early link type chain
3735411 SB, double-row roller link type chain, HD as used in trucks
352133 COS, timing belt, camshaft & water pump drive
334051 COS, timing belt, distributor drive
334096 COS, timing belt bolt, upper cover, 2 required
333901 COS, timing belt front cover
352122 COS, timing belt cover assembly, upper rear

0.728 GEAR—SPROCKET, Crankshaft Timing

6272965 Gear, for gear-drive cam 3925533
3860035 Sprocket, for 3860036 chain
3891518 Sprocket, for 3891519 chain
3896959 Sprocket, for 346261 chain
3735413 Sprocket, for 3735411 chain
352132 Sprocket and hub for COS timing belt

0.736 GEAR—SPROCKET, Camshaft Timing

3860086 Gear, for gear-drive cam 3925533
330815 Sprocket, for 3860036 chain
3891517 Sprocket, for 3891519 chain
340235 Sprocket, for 346261 chain
3735412 Sprocket, for 3735411 chain
352130 COS, inlet camshaft sprocket
352131 COS, exhaust camshaft sprocket
352126 COS, distributor drive sprocket, mounts to inlet sprocket

0.738 KEY—LOCK—PIN, Camshaft Sprocket
3952320 BB, rear bearing plug lock, aluminum block
3836903 All V8, dowel pin (¼x⅝-inch), solid
334078 COS, camshaft retaining key, 2 required
334085 COS, lock for retaining cam sprocket bolts, 2 required

0.859 CLUTCH—PLATE, Clutch Pressure with Cover
6273958 V8 HD, 10.5-inch diameter, for 12 ¾-inch flywheel (L88)
231254 V8, for dual-plate clutch; consists of dual pressure plate and (1) 3993815 driven disc
3884598 V8, 11-inch diameter, high performance
3973142 COS

0.861 COVER, Clutch
3860361 V8, 11-inch diameter, high-performance
3973143 COS

0.863 PLATE, Clutch Pressure
3884630 V8, for 11-inch diameter clutch
3988716 COS

0.886 DISC—PLATE, with Facing, Clutch
3886059 V8, 10-spline for 10½-inch clutch
3959176 V8 10-spline, for converting 231254 dual-plate clutch for use with 10-spline imput shaft
3991428 V8, 26-spline for 10½-inch clutch
3993815 V8, 26-spline, replacement or additional plate required for 231254 dual-plate clutch
3908960 V8, 10-spline, 11-inch clutch
338124 COS

GROUP 1.000 COOLING—OIL SYSTEM—GRILLE

1.062 PULLEY—REINFORCEMENT, Water Pump
3995641 BB pulley (L88), 2-groove, deep-groove (6 9/32 OD), stamped 3848904
3770245 SB pulley, 2-groove, deep-groove (7 ⅛ OD)
3720616 SB water pump pulley reinforcement
Many other pulleys available. See your Chevy dealer.

1.0666 BELT, Accessory Drive
9433722 SB, HD fan & water pump, 34x38 captured belt
Other belts: See your Chevy dealer.

1.069 PUMP UNIT, Engine Coolant
3977630 BB, HD aluminum
N/A SB, HD aluminum (available late 1980)
3998207 SB, with short mounting legs (pre-'69 passenger & Corvette through 1970)
6258551 SB, with short mounting legs & large (1.50) bearing (1971 & later Corvette)
358175 COS

1.097 ELBOW, Thermostat Housing

442407 COS

1.153 OUTLET, Coolant
3877660 V8 water outlet & thermostat housing elbow, most cars
3841191 COS water outlet elbow

1.154 GASKET, Coolant Outlet
3701777 V8
334033 COS

1.219 RADIATOR—SUPPLY TANK—BRACKETS
3007436 Radiator, HD aluminum (L88 Corvette)
3016340 Tank, radiator supply (aluminum surge tank for 3007436)

1.272 DEFLECTOR, Radiator Lower Air
474475 Front spoiler, '80 Citation

1.287 EXTENSION, Radiator and Grille
3974013 '68-'72 Corvette front air dam
327695 '73-'78 Corvette front air dam, except '73-'74 454 with air conditioning
336176 '73-'74 Corvette front air dam, with air conditioned 454, will fit on all '68 and later, offers increased cooling
336177 Reinforcement for 336176
14016062 Cap, lower panel extension, '80 Corvette
14016063 Lower panel extension, left-hand, '80 Corvette
14016064 Lower panel extension, right-hand, '80 Corvette
Other air dams under Group 7.833.

1.426 PAN ASSEMBLY, Oil
3977591 BB, Corvette with trap door, 5-quart
3985999 BB, most passenger, 4-quart
359942 SB, Corvette with trap-door, 6-quart (replaces 3996670), stock on '63-'65 327 with mechanical lifters, '69-'71 LT1
360866 SB, Corvette 5-quart (replaces 3820000), stock on '69-'72 standard 350, '72 LT1 without HD M22 4-speed, all '73-'74 350
359940 SB, 1964-'67 Chevy II (front sump), 4-quart
465220 SB, '67-'74 Z28, '75-'79 L82, 4-quart (replaces 360009)
465221 SB, '75-'79 standard pan, most cars
3977122 COS, production
366211 COS, aluminum, cast for dry sump oiling
Other pans available. See your Chevy dealer.

1.429 GASKET, Oil Pan
3958614 BB kit, includes side gaskets, seals & packing
3931593 BB side gaskets
3909800 BB front seal
3955115 BB rear seal
3931597 SB kit, includes side gaskets, seals & packing, 1957-'74, with original pans
3887151 SB front seal, 1964-'74, original pan
3887152 SB rear seal, all through 1975, with original pans
458625 SB rear seal, all '76-'80 & pre '75 with replacement pans
359936 SB kit, includes side gaskets, seals & packing, 1975-'78, and pre-'75 with replacement pans
14002203 SB kit, includes side gaskets, seals & packing, 1979
357159 SB front seal, 1975 & later, and pre-'75 with replacement pans
3931599 SB side gaskets, through 1977
3977130 COS oil pan & baffle gasket

1.430 BAFFLE ASSEMBLY, Oil Pan
3967854 BB windage tray
3769768 SB, upper baffle, flat, Corvette long style, most often used with 359942 pan
3989075 SB, upper baffle, curved, Corvette short style, most often used with 360866 & 360009 pans in Corvette through 1975
3927136 SB, upper baffle, curved most often used with 465220 pan
348680 COS baffle
3992222 COS oil drain pipe assembly

1.453 PLUG, Oil Pan Drain
337185 Magnetic, for all pans

1.516 GAUGE ASSEMBLY—TUBE, Oil Level
3981088 BB dipstick, for 3977591 pan
3981089 BB tube, for 3981088 dipstick
3989391 BB dipstick, for 3985999 pan
3928901 BB tube, for 3989391 dipstick
3862730 SB dipstick, most pans, passenger '65-'67, Corvette '65-'74, Chevy II '65-'67; also fits COS
3865664 SB tube, for 3862730, upper (not for COS)
3981019 SB dipstick, '76-'79 Corvette (2nd design), 20 inches from cap to full mark & 21 from cap to add)
460147 SB tube, for 3981019, upper (not for COS)
3896912 SB dipstick, '67 Camaro
3951576 SB dipstick, '75 & 1st design '76-'78 Corvette (17 5/16 inches from cap to full mark & 18 3/16 from cap to add); also most Chevelle '65-present and passenger & Camaro '69-'79
14010286 SB dipstick, '80 Camaro
14010285 SB dipstick, '80 passenger
3923239 SB dipstick, most '68 passenger
3876870 SB tube, upper, for 3951576, 3896912, 3923239
3951600 SB tube, lower, for most pans except 359940 in '64 Chevy II
14010290 SB tube, for 14010286
14010289 SB tube, for 14010285
3927194 SB dipstick, for pan 359940 in '64 Chevy II
3860339 SB tube, for 3862730 in '65-'67 Chevy II
14010287 SB dipstick, '80 Corvette
348651 COS tube, for 3862730 when used in COS
Other gauges: See your Chevy dealer.

1.540 RADIATOR (COOLER), Oil
3157804 Cooler, oil (aluminum)

1.609 SPRING—VALVE—RETAINER—PLUG—PIN, Oil Pump
838839 All motors, pin, special groove
3876866 BB spring, pressure regulator, high performance & HD, green stripe

3848911 SB spring, pressure regulator, high performance, white stripe
3860377 BB valve, pressure regulating
3702366 SB valve, pressure regulating
3988701 COS valve, oil pressure regulator
3994264 COS spring, damping, oil pressure regulator valve
334026 COS spring, oil pressure regulator valve

1.639 SHAFT, Oil Pump Drive
3865886 BB, HD
3998287 SB, except '64-'67 Chevy II
3790872 SB, '64-'67 Chevy II

1.652 PUMP ASSEMBLY, Oil
475908 BB, street high-performance, with screen, passenger
475902 BB, street high-performance & L88, less screen, Corvette
3969870 BB, 1.3-inch gear (ZL1) HD, with screen
3848907 SB, high-pressure
3793310 SB, Chevy II (front sump), 1964-'67 (cast 3793558)
377178 COS

1.656 SCREEN, Oil Pump Intake
3860366 BB, 1965-'66
3955282 BB, 1967 & later, Corvette (except HD/L-88/ZL1)
3955281 BB, 1967 & later, passenger
3830080 SB, 1963-'75 with pump 3848907; also used on '80 Corvette
3790870 SB, Chevy II (front sump), 1964-'67
370793 COS, oil suction pipe
3977133 COS, oil suction pipe support

1.723 COVER, Oil Pump
3848909 SB, special high performance, part of pump 3848907
3836980 SB, Chevy II & most other standard applications

1.724 GASKET, Oil Pump Cover
3973114 COS oil pump gasket

1.837 VALVE ASSEMBLY, Oil Filter
5575416 Bypass valve assembly: BB, used on 4-bolt-main blocks to adapt for use of oil cooler; COS, used as oil filter bypass valve

1.840 ADAPTER ASSEMBLY, Oil Filter
326098 Adapter, oil cooler & filter, allows installation of oil cooler on any Chevrolet V8 engine, side outlets
340258 Same as 326098 but with forward outlets
326100 Gasket, oil cooler & filter
3737037 SB, seal, oil cooler & filter (BB discontinued)

1.855 CONNECTOR, Oil Filter Fitting
3853870 To enable screw-on filter to be used on aluminum block

GROUP 2.000 STARTER—IGNITION—LIGHTING

2.041 MOTOR ASSEMBLY, Starter
1108789 HD, for 12 ¾-inch flywheels
1108773 COS

2.042 BOLT—BRACE, Starter Motor
1366043 Bolt, starter motor long, special shoulder bolt, 1 required
1366042 Bolt, starter motor short, special shoulder bolt, 1 required
354353 Brace, front starter support, HD SB
3866768 Brace, front starter support, HD BB
343351 Brace, front starter support, COS

2.085 HOUSING, Starting Motor Drive, with Bushing
1893864 For large-diameter flywheels (11-inch clutch) & auto trans flex plates, offset bolt pattern (aluminum)
1968122 For small-diameter flywheels (10½-inch clutch), straight-across bolt pattern (aluminum)

2.170 COIL, Ignition
1115207 For transistorized ignition, part of unit 3997782 (embossed 176-12-V), all V8
1115444 For HEI, COS

2.171 WIRE, Ignition Coil
2985954 Coil-distributor wire, high tension, 16 39/64-inches OL, all V8 through '74 except Corvette
8912485 Coil-distributor wire, high tension, 5½-inch OL, BB Corvette through 1974, except 1972
12002014 COS, coil wire assembly

2.225 HARNESS ASSEMBLY, Electric Fuel Injection
8914616 COS

2.239 WIRE KIT, Spark Plug
8914473 Solid wire, silicone-jacketed, 90° boots, recommended for SB (includes 8 plug wires & 1 coil wire)
8914472 Same as 8914473, but straight boots, recommended for BB

2.240 WIRE, Spark Plug
12002015 COS wire, cylinder #1
12002016 COS wire, cylinder #2 & #3
12002017 COS wire, cylinder #4

2.251 LATCH, Coil & Distributor
8917297 COS distributor latch, holds plug wires to distributor
8917298 COS coil wire latch, holds coil wire to coil

2.270 PLUG, Spark
Following are recommended for racing use only. AC part numbers are used.

436TS, 438TS BB, cast-iron open chamber heads & SB angle plug head, track racing
441TS, 443TS, R42T BB, cast-iron open chamber heads & SB angle plug head, drag racing
436XLS, 438XLS BB, aluminum head, track racing
440XLS, 43XL BB, aluminum head, drag racing
R43LTSX COS, racing (Chevy #5613600)

2.274 FAN—PULLEY—GUIDE, Generator or Alternator
3829387 L-88 pulley, no fan included
3844100 COS pulley, no fan included
800594 Fan, 11-blade
460755 Guide, generator (alternator) belt, stock on '77 & later Corvette

2.275 GENERATOR (ALTERNATOR), less Pulley
1101040 70 amp Delcotron, with integral voltage regulator, for V8
1102849 63 amp Delcotron, with integral voltage regulator, for V8
1102848 42 amp Delcotron, with integral voltage regulator, for COS

2.277 BRACKET—BOLT—BUSHING—SPACER—SUPPORT, Generator (Alternator)*
*V8 components: Because of the wide variety of brackets, please consult your Chevy dealer.

357699 COS alternator bracket
360432 COS alternator bracket bushing, 2 required
3983422 COS alternator bracket spacer
360105 COS alternator mounting bolt

2.280 NUT—WASHER, Generator Pulley (All motors)
1915172 Alternator pulley nut
1941978 Alternator pulley washer

2.317 BRACE, Generator Bracket*
348691 COS alternator brace
355720 COS alternator brace bushing, 2 required

2.361 DISTRIBUTOR, Ignition
1111263 V8 transistorized magnetic impulse, tach drive, ball bearings, mechanical advance only, for gear-drive cam
1111267 Same as 1111263, but with plain bearings, mechanical & vacuum advance, for chain-driven cams
1110985 V8, dual points with tach drive
1110649 COS, HEI

2.362 HOUSING, Distributor
1851158 V8, '62-'74, for tach drive (Corvette) distributor
1966407 V8, '62-'74
360147 COS housing, distributor drive
1892189 COS housing, distributor base
334058 COS lock, distributor drive housing

2.363 CLAMP—GASKET—STUD—PIN, Distributor Mounting
3704912 V8 hold-down clamp
3980255 COS hold-down clamp
1950569 V8 housing oil seal
3932344 V8 distributor gasket
6256636 COS distributor gasket
3980253 COS stud, distributor hold-down clamp
334057 COS pin, distributor drive housing to cylinder head
360149 COS pin, locating, distributor cross shaft

2.364 LUBRICATOR UNIT—SEAL—PIPE—WASHER, Distributor
1852935 1962-'74 V8 lubricator unit, cam
1935621 '64-'73 passenger & '58-'70 Corvette felt wick retainer
1963858 '74 V8 passenger & '71-'74 Corvette felt wick retainer

1951099 V8 washer, housing oil seal, under pole piece
334001 COS seal, drive shaft, distributor drive housing
116341 COS elbow, distributor housing oil drain pipe at distributor housing, 1975
366838 Similar to 116341, but for 1976 COS
442411 COS elbow, distributor oil drain-back at oil pump
348626 COS seal, distributor drive housing to cylinder head O-ring
348627 COS seal, O-ring, drive housing to cylinder head

2.367 CAP, Distributor
1971244 V8, '58-'74
1880022 COS, HEI

2.372 SHAFT—POLE PIECE, Distributor
1961402 Shaft, dual-point distributor
1971037 Shaft, transistor ignition (plain-bearing), L88
830532 COS shaft
1964272 V8, rotating pole piece, replacement or to convert point-triggered system to magnetic impulse
1960779 Stationery pole piece, same application as 1964272
1891767 COS pole piece & plate
830446 COS retainer, pole piece & plate

2.374 GEAR, Distributor Drive
1958599 Drive gear & pin, to replace distributor drive gear or convert reverse-rotation distributor 1111263 to conventional rotation
352134 COS, sprocket, distributor drive unit

2.379 SHIM—WASHER, Distributor Shaft Thrust and Gear
1927529 V8, .005-inch thick shim
1837617 V8, .031-inch thick shim
817051 COS washer, under gear tang
811912 COS washer, under gear

2.382 ROTOR, Distributor
1852722 V8, '62-'74
333929 COS, HEI

2.383 AMPLIFIER—HARNESS—UNIT, Transistorized Ignition (V8) & HEI (Cosworth)
3997782 V8, transistor ignition unit
3955511 V8/COS, pulse amplifier (part of unit 3997782)
8901973 Wire assembly, part of unit 3997782
6297688 Pulse amplifier harness, part of unit 3997782
1880021 COS module harness
1875990 COS module

2.384 POINT KIT, Distributor (consists of Breaker Arm & Support)
1966294 V8, high performance

2.388 WEIGHT, Distributor Advance
1965843 L88 fast advance

2.389 SPRING, Distributor Weight
1965903 L88 fast advance

2.726 BRACKET—SHIELD, Headlamp
Following are parts required for Corvette racing off-road headlight system. The actual quartz-iodine lighting units must be purchased from an aftermarket source, such as Cibie.

3961463 Bracket, headlamp mounting (Corvette racing), replaces '68-'78 concealed headlamp assembly, LH
3961464 Same as 3961463, but RH
3961465 Shield, headlamp, plexiglass, replaces '68-'78 concealed headlamp assembly on Corvette, LH
3961466 Same as 3961465, but RH

GROUP 3.000 FUEL—EXHAUST—CARBURETION

3.265 MANIFOLD—SHIELD, Engine Intake
3947084 BB, aluminum intake manifold, 1x4-bb1 Holley, high-rise with divided plenum (396/375-425, 427/425) (L88/ZL1 open plenum intake discontinued)
3977609 BB, aluminum intake manifold, 1x4-bb1. Holley, low-profile (454 LS6)
3928514 BB, aluminum intake manifold, 3x2-bb1 Holleys, high-performance heads (rectangular intake ports), (427/435)
3928515 BB, aluminum intake manifold, 3x2-bb1 Holleys, low-performance heads (oval intake ports), (427/400)
3972114 SB, aluminum intake manifold, 1x4-bb1 Holley, high-rise with divided plenum (Z28/LT1)
14007377 SB, aluminum intake manifold, 1x4-bb1 Q-jet (spread-bore), high-rise with divided plenum ('79-'80 L-82)
361082 COS, aluminum intake manifold, electronic fuel injection
3931093 BB, intake manifold oil splash shield
3758369 SB, intake manifold oil splash guard, Z28/LT1
6271071 SB, intake manifold oil splash guard, L-82

3.270 GASKET, Intake Manifold & Exhaust Manifold
3955528 BB, intake, rectangular port, blocked heat passage (L-88/ZL1)
3957986 SB, intake, blocked heat passage
333903 COS intake
348618 COS header gasket

3.300 GASKET—GROMMET—MOUNT, Electronic Fuel Injection (COS)
334049 Gasket, distributor drive housing cover
336500 Grommet, injection control, 2 required
361096 Mount, M.A.P. valve

3.302 BRACKET—GROMMET—SUPPORT—VALVE, Fuel Injection (COS)
336497 Control bracket, LH
336498 Control bracket, RH
345143 Grommet, fuel rail bracket, 4 required
336499 Control support
345130 Valve
361085 Valve

3.304 O-RING—SEAL, Injection Valve to Intake Manifold (COS)
1606772 O-ring, 4 required
345141 Seal

3.308 WASHER, Fuel Injection (COS)
2454766 Washer, 2 required

3.320 CONTROL—PLATE—SENSOR—SHIELD, Electronic Fuel Injection (COS)
354819 Control
361091 Sensor mounting plate
334039 Speedometer switch sensor
345133 Inlet manifold air temperature sensor
361089 Coolant temperature sensor
361090 Inlet manifold air pressure sensor
354505 Shield, over heat switch

3.330 BRACKET—EXTENSION—FUEL RAIL—HOSE—MANIFOLD—PIPE, Fuel Injector (COS)
361088 Bracket
360188 Pulsair pipe extension
361087 Fuel rail
360124 Fast idle valve to throttle body, hose
360190 Pulsair manifold to valve hose
360189 Pulsair valve manifold
360186 Distributor drive housing oil drain pipe
360187 Pulsair pipe, 2 required

3.335 BODY—SWITCH, Fuel Injector Throttle (COS)
377369 Throttle body
345132 Throttle position switch

3.402 CLEANER, Carburetor Air
6423907 Complete assembly, open-element, low-restriction air cleaner (Z28 and "special high-performance" motors through 1972, except ducted air)
6423757 Complete assembly, BB 3x2 Holley air cleaner
8994368 COS, complete assembly
6485891 Assembly, '69 Camaro with DA hood & most 1970-'72 Chevelle CI
8997486 Assembly, '80 Z28 with AI (350 motor)
8997487 Assembly, '80 Z28 with AI (305 motor/California)

3.403 GASKET—STUD, Air Cleaner
3969843 4-bbl carb top to air cleaner base gasket
3899698 Holley 2300 2-bbl carb top to air cleaner base gasket, for 3x2 setup, 3 required
3785256 Air cleaner hold-down stud, cut to required length
559573 Wing nut

3.406 BASE—SEAL, Air Cleaner
6422188 Base, open-element & L-88 air cleaner
3934165 Seal, air cleaner to hood, L88-style fresh-air hood

3.407 PLATE—SUPPORT, Air Cleaner
14010482 Plate, hood inner panel seal, '80 Z28 AI
14013425 Support, air cleaner seal, '80 Z28 AI

3.410 COVER—DIFFUSER—ELEMENT—RETAINER—SCREEN—FILTER—BREATHER, Air Cleaner

6421746 Element, open-element/cowl induction/'69 DA Camaro
6422544 Element, '67-'68 Camaro with plenum air, tall-style, low-drag, can be used with open element & cowl induction air cleaners, if hood clearance permits
6423688 Element, 3x2 air cleaner
6484235 Element, '80 Z28 with AI, paper (A348C)
6421832 Chrome cover for open-element air cleaner
6423906 Diffuser, L88 air cleaner, fits inside of cold air box (equivalent to filter element)
3902394 Retainer, carb air intake screen, L88 air cleaner
3902396 Screen, carb air intake, L88 air cleaner
6488289 COS element
6484966 Breather, '80 Z28 AI, air cleaner & silencer

3.415 ACTUATOR—RETAINER—FRAME—SOLENOID—VALVE—WIRE ASSEMBLY—BRACKET—MOTOR—ROD—SENSOR—DOOR, Air Cleaner

'69 CAMARO DUCTED AIR HOOD/'70-'72 CHEVELLE COWL INDUCTION

Parts listed below fit both systems, unless otherwise indicated.
1998970 Actuator, hood panel door (CI)
3972676 Bracket, carb air control valve solenoid switch ('72 CI)
3955225 Frame assembly, carb air control valve (Camaro DA)
3975984 Frame assembly, carb air control valve (CI)
3922580 Grommet, carb air control valve vacuum ('70-'71 CI)
7729602 Relay assembly
3955227 Retainer, carb air control valve, 3 required
3972919 Screen assembly, hood mounting (CI)
3969803 Seal assembly, carb air control valve frame
6484242 Sensor unit, air cleaner thermal
1114427 Solenoid assembly, carb air control
3959115 Switch, carb air control solenoid (DA Camaro/'70 CI)
1242101 Switch, carb air control valve solenoid ('72 CI)
3955226 Valve, carb air control
6440776 Valve, hood panel door, use with 20-inch length of 2941942 hose (CI)

'80 Z28 CAMARO AIR INDUCTION

14010484 Bracket, air intake door solenoid
14010483 Door, air intake
8994262 Motor, air cleaner diaphragm
14010491 Rod, air intake door actuator
8997787 Sensor, air cleaner
14016262 Solenoid, hood air intake louver
12005715 Wire, hood louver
12005723 Wire, hood louver solenoid

3.417 DOOR—SCREW—SEAL—SPACER—SUPPORT—HOSE—STOVE, Air Cleaner

'67-'68 CAMARO WITH PLENUM AIR

3916621 Duct assembly, air cleaner to dash

'69 CAMARO DUCTED AIR HOOD/'70-'72 CHEVELLE COWL INDUCTION

Parts listed below fit both systems, unless otherwise indicated.
3945359 Bolt, actuator to door (CI)
3968560 Bumper, hood panel door stop, 2 required ('70 CI)
3968539 Door, hood panel (CI)
3975977 Extension assembly, air cleaner
3975986 Flange seal
3968558 Screw, hood panel door pivot, 2 required (CI)
3968559 Screw, hood panel door stop, 1 required (CI)
3955231 Seal, air cleaner to hood
3968556 Spacer, hood panel door actuator, support (CI)
3968557 Spacer, hood panel door pivot screw, 2 required (CI)
3968554 Support, hood panel door actuator (CI)

'80 Z28 CAMARO AIR INDUCTION

14009023 Hose, carb air inlet
3942604 Rivet, truss head, blind, amount required
338473 Stove, carb air heat, lower
472862 Stove, carb air heat, upper
3959544 Stud, carb air heat stove (#8x.375)

Other ducted air/Cowl Induction/air induction parts under Group 8.000.

3.601 MANIFOLD, Engine Exhaust

3960709 Header, exhaust, BB, Corvette side-mount, LH
3960710 Header, exhaust, BB, Corvette side-mount, RH
3977971 Header, exhaust, SB, Corvette side-mount, LH
3977972 Header, exhaust, SB, Corvette side-mount, RH
360193 Header, exhaust, COS

3.609 PIPE, Exhaust

3960713 Collector, side-mount headers, LH (RH discontinued)

3.611 GASKET, Exhaust Pipe

3960712 BB header gasket
3977970 SB header gasket

3.613 BOLT, Exhaust Pipe

3960711 Special exhaust bolt for headers

3.701 MUFFLER, Engine Exhaust

3977899 Slip-in style, for Corvette side-mount header collector
3869877 Corvair "Turbo" muffler, low-restriction, round "trash-can" style, for all high-performance exhaust
3858442 Same as 3869877, but oval-shaped for increased clearance

3.724 GASKET KIT, Carburetor

3930825 Gasket repair unit, 3x2 Holley carb setup, secondary carburetors (2 required)

3.725 CARBURETOR—REPAIR KIT

CARBURETOR ASSEMBLY

3957859 600-CFM Holley DP
3972121 780-CFM Holley, vacuum secondaries, stock on '70 4-speed LT1/Z28 (R-4555)
3965736 830-CFM Holley DP, 1-11/16 throttle bores (for NASCAR)
3955205 850-CFM Holley DP, stagger-jetted & modified for open-plenum manifold (L-88/ZL1) (R-4296)
3902353 Secondary carburetor, Holley BB 3x2 setup, 2 required (R-3659)
3940930 Primary carburetor, Holley BB 3x2 setup, automatic transmission (R-4056-1)
3940929 Primary carburetor, Holley BB 3x2 setup, 4-speed transmission (R-4055-1)

MINOR REPAIR KIT

Consists of pump, needle & seat unit, air horn gasket, thermostat housing cover gasket, clip & gauge (if applicable).
3964571 For 850-CFM DP

MAJOR REPAIR KIT

Consists of all parts necessary to overhaul (1) carburetor, including jets, nozzles, intake needle & seat, pump, lever, etc.
3964570 For 850-CFM DP
3979474 For 780 CFM 3972121
3914110 For 3x2 setup, front & rear carbs (2 required)
3914113 For 3x2 setup, center carb

3.726 GASKET—INSULATOR, Carburetor to Manifold

682914 Gasket, Holley 4-bbl
3921922 Gasket, Holley 2-bbl (for 3x2 setup), 3 required
3969835 Insulator, heat shield, Holley 4-bbl
3969837 Insulator, heat shield, Q-jet (spread-bore)

3.890 FILTER, Carburetor

5650030 High capacity, with screw-on element
854619 Large-capacity, in-line (AC-GF-61P)
854509 Element for 5650030

3.900 PUMP ASSEMBLY, Fuel

6415748 BB, high-capacity
6415325 SB, high-capacity
1608783 COS, fuel pump, on motor, with nuts
6470894 COS, fuel pump (electric), in fuel tank, with filter unit

3.901 MOUNT, Fuel Pump

345137 COS, mount, fuel pump, to prevent vibration

GROUP 4.000 TRANSMISSION—BRAKES

4.003 TRANSMISSION ASSEMBLY

All Muncie 4-speed transmission, '63-'72, have been discontinued as complete assemblies. The following are the most common complete replacement transmissions:
14016305 Warner, Super T10, wide ratio, universal
14019945 Turbo 350 automatic trans, universal
342451 Borg-Warner M75 5-speed, used on COS (Model HB)

4.005 GASKET KIT, Transmission

3840376 For '57-'63 Warner 4-speed
3917763 For '63-'72 Muncie 4-speed

3960199 For late Warner 4-speed
3886338 For '70-'75 Saginaw & '76-'78 Muncie 4-speeds
3958616 For '73-'74 Muncie 4-speed
8625905 M40 (Turbo 400) gasket kit, with seals
8626984 M40 (Turbo 400) overhaul kit
347126 M38 (Turbo 350) gasket kit, with seals
358045 M38 (Turbo 350) overhaul kit

BELOW ARE LISTED THE PARTS STILL AVAILABLE EXCLUSIVELY FOR THE M22 HD MUNCIE 4-SPEED. ALL OTHER TRANSMISSION PARTS: SEE YOUR CHEVY DEALER.

4.103 CASE, Transmission
3925659 M22 HD 1965 and all Muncie '66-'74

4.105 COVER—PLUG, Transmission Case
3968034 Magnetic drain plug, all '65 & later 4-speed
3977618 Cover, Muncie

4.311 SPRING, Shifter Lever
3831718 Muncie, high-load, helps prevent hop-out from all gears

4.351 GEAR, Transmission Main Drive (Clutch)
3978761 Muncie, HD 26-spline, to convert 10-spline to late-model clutches, or '71-'72 standard HD replacement
3925691 M22 gear, imput shaft, through '70

4.380 SYNCHRONIZER, Transmission
344241 Muncie, 1st gear spline lock, prevents hopup 1st gear
357238 Synchronizer, 1st-2nd gear (Muncie)
357239 Synchronizer, 3rd-4th gear (Muncie)

4.383 RING, Transmission Synchronizing
344243 Slotted for quicker shifting (Muncie)

4.395 GEAR, Transmission 2nd Speed
3879999 M22 HD

4.403 SHAFT, Transmission Main
3915087 '64-'70, all Muncie
3978759 '71-'74, all Muncie

4.415 GEAR, Transmission 3rd Speed
3880845 M22 HD

4.417 GEAR—KIT—BEARING Transmission 1st & Reverse Sliding
3924796 M22 HD 1st gear
357228 Muncie reverse gear
3965752 Unit, roller bearing 1st gear, M22 HD Muncie severe use

4.419 SLEEVE—WASHER—BEARING, 1st gear
326578 Sleeve, 1st gear for needle roller M22
326579 Washer, thrust, for needle roller M22
9433516 Needle roller bearing for 1st gear kit, M22

4.421 GEAR, Transmission Countershaft
3905466 M22 cluster gear

4.428 SHAFT, Transmission Counter
357223 Muncie

4.430 GEAR, Transmission Reverse Idler (with Bushings)
3879997 M22 HD front reverse idler gear
357231 Muncie (all) rear reverse idler gear

4.649 REPAIR, Hydraulic Brake Main Cylinder
18001018 Corvette, HD brakes

4.650 CYLINDER, Hydraulic Brake Main
18001004 Corvette, '67-'69 HD brakes

4.657 VALVE, Main Cylinder
3878944 Corvette, rear brake pressure regulator, adjustable, use to achieve optimum braking with different size front & rear tires

4.665A HOUSING, Wheel Brake Front Caliper
2621211 Corvette, HD brakes, outer, 2 required
2621209 Corvette, HD brakes, inner, LH
2621210 Corvette, HD brakes, inner RH

4.665B HOUSING, Wheel Brake Rear Caliper
5462016 Corvette, HD brakes, outer, LH (2nd design)
2621215 Corvette, HD brakes, outer, RH (2nd design)
5461991 Corvette, HD brakes, inner, LH (2nd design)
2621213 Corvette, HD brakes, inner, RH (2nd design)
5469499 Corvette, HD brakes, caliper assembly

4.667 REPAIR KIT, Brake Caliper
5467286 Corvette, all disc brakes, front; includes 1 ⅞-inch seal, boot & O-ring, repairs 1 piston bore
5467288 Corvette, all disc brakes, rear; includes 1 ⅜-inch seal, boot & O-ring, repairs 1 piston bore

4.668 PISTON ASSEMBLY—INSULATOR, Brake Caliper
5463823 Corvette, HD brakes (2nd design), front piston, 8 in kit
5469592 Corvette, HD brakes (2nd design)8 rear piston, 2 in kit
5463822 Corvette, HD brakes, front piston insulator, 8 in kit
5469593 Corvette, HD brakes, rear piston insulator (2nd design), 8 in kit
5472040 Corvette HD brakes, rear piston insulator, (1st design)
5471861 Corvette HD brakes, front piston (1st design)
5472039 Corvette HD brakes, rear piston (2nd design)

4.681 BRACKET, Brake Hose or Pipe
3947283 1967-'69 Camaro HD Disc brake LH
3947284 1967-'69 Camaro HD Disc brake RH

GROUP 5.000 BRAKES—REAR AXLE—PROPELLER SHAFT

5.001 BRACKET, Brake Caliper Rear
3948880 Bracket, caliper adapter bracket, Corvette HD brakes (1968 & later), RH
3948879 Same as 3948880, but LH

5.017 SHOE UNIT (PADS), Brake, with Facing (Corvette)
5468882 Unit, brake pad HD flanged backing plate (4 pieces per unit)
5470991 Unit, brake pad HD flanged thick backing plate (4 pieces per unit)

5.387 HOUSING, Axle
379846 Late Camaro 10-bolt (8.5-inch ring gear), with spacer
1254818 Late Camaro 10-bolt (8.5-inch ring gear)
553804 COS, 7.5-inch ring gear
3981667 Passenger 12-bolt, '65-'70
3981669 Chevelle 12-bolt, '68-'72
3981668 Chevelle 10-bolt, '68-'72
All other 10-12 bolt housings discontinued.

5.400 PLUG, Axle Filler & Drain
337185 12-bolt, magnetic, ½-inch-20
2309473 Corvette IRS, magnetic

5.420 SHAFT, Axle
360912 Corvette IRS half-shaft, '63-'79
14009998 Corvette IRS half-shaft, '80 (Dana center-section), automatic trans
14009999 Same as 14009998, but for manual trans
1249142 Late Camaro with 10-bolt (8.5-inch ring gear)
551566 COS, 7.5-inch ring gear (Vega/Monza '76-present)
3969348 '65-'67 Chevelle/'67-'69 Camaro/'68-'69 Chevy II, with 10-bolt carrier
3863726 '65-'67 Chevy II, with 10-bolt carrier
3868992 '65-'67 Chevy II, with 12-bolt carrier
3969349 '65-'67 Chevelle/'67-'69 Camaro/'68-'71 Chevy II, with 12-bolt carrier
3985720 '65-'70 Passenger, with 12-bolt carrier
3914059 '65 Passenger, with 10-bolt carrier
3985719 '66-'70 Passenger, with 10-bolt carrier
3969284 '68-'72 Chevelle/'70-'74 Camaro, with 10-bolt carrier
3969285 '68-'72 Chevelle/'70 Camaro, with 12-bolt carrier

5.505 CARRIER, Differential, with Cap

CARRIER, WITHOUT RING & PINION
3789870 '56-'62 Corvette
3917861 '63-'79 Corvette (IRS)

DIFFERENTIAL (POSITRACTION) (COMPLETE WITH RING & PINION)
3743970 '57-'62 Corvette (4.56 ratio)
3899137 '63-'79 Corvette IRS (3.70 ratio)
464907 '63-'79 Corvette IRS (3.08 ratio)
14024326 '80 Corvette IRS, Dana carrier, aluminum (3.07 ratio)

5.422 LOCK, Axle Shaft
3833322 .150 thick, '64-present
3959067 .160 thick, '64-present
3959068 .155 thick, '64-present

5.510 CASE, Differential

3982240 Conversion unit, HD Positraction Corvette (discs, plates, gears & springs)
3916234 Assembly, HD 12-bolt, 3.9-5.13 ratios, 18-plate Positraction, with internal gears
3993517 Similar to 3916234, but standard production
3993516 Similar to 3916234, but standard production, 3.07-3.73 ratios
3936444 Similar to 3916234, but standard production, 2.29-2.73 ratios
3918837 Bare case, used in 3916234 assembly (less internal gears)
3890129 Bare case, used in 3936444 assembly
3995877 Assembly, extra-HD 12-bolt, 3.90-5.13 ratios, 22-plate Posi
3947281 Bare case, 12-bolt, 4.10-5.13 ratios, for '68-'69 Camaro with 4-wheel-disc brakes
3993502 Assembly, 10-bolt, 1965-'72 with 8.20-inch ring gear, 3.08-3.70 ratios & Posi
3993504 Similar to 3993502, but for 2.56-2.73 ratios
3869342 Bare case, used in 3993502 assembly
407891 Assembly, 10-bolt HD ('73 & later with 8½-inch ring gear), 2.73-4.10 ratios with Posi
407890 Similar to 407891, but for 2.41-2.56 ratio
14024328 Assembly, '80 Corvette (with Dana carrier)
3997925 Assembly, '63-'79 Corvette, 2.73-3.70 ratio, with Posi
3997927 Similar to 3997925, but for 3.90-5.13 ratio
3997928 Bare case, used in 3997927 assembly
3997926 Bare case, used in 3997925 assembly
3853552 Assembly, '57-'62 Corvette, 4.11-4.56 ratio, with Posi
345938 Assembly, Vega/Monza with 7½-inch ring gear & Posi, 3.23-4.10 ratio
561091 Similar to 345938, but for 2.29-3.08 ratio
558648 Bare case, Vega/Monza, used in 561091, assembly

5.511 PLATE—SPRING, Differential Clutch (Positraction)

3998514 Plate, differential clutch, for locking rear axle differential (1.0693 ±.0035 thick)
3957939 Spring, differential, Positraction plate for 22-plate Posi
3957941 Guide plate, differential clutch pressure, for 22-plate Posi
3918834 Plate, Positraction, replacement for 3916234 case
3982240 Differential repair kit, 22 plates, springs & gears

Component parts for all Production differentials (early, 10 & 12 bolt and Corvette): See your Chevy dealer.

5.526 GEAR, Differential, Pinion

3957938 For differential case 3995876

5.528 GEAR, Differential, Side

3957937 For differential case 3995876

5.529 GEAR ASSEMBLY, Ring & Pinion Unit

EARLY: 1955-'64 Pqssenger & 1955-'62 Corvette

3758549 3.08
3748437 3.36
3795479 3.55
3748438 3.70
3748434 4.11
3748447 4.56

10-BOLT: 1965-'72. 8.20-inch ring gear

3961426	2.56	3961413	3.08	3961411	3.55
3961425	2.73	3961412	3.36	3961414	3.70

10-BOLT: 1973-present. 8.50-inch ring gear

Note: These part numbers have been subject to frequent changes.

1258715	2.41	1258709	3.08	1259441	3.73
1394766	2.56	1258713	3.23	3988515	4.10
1258708	2.73	1258710	3.42		

CORVETTE: 1963-'64. original carrier

3822990	3.55		
3822989	3.36	3822993	3.70

CORVETTE: 1965-'79 & 1963-'64 with replacement carrier

3961431	2.73	3961420	3.55	3961421	4.11
3961418	3.08	3961419	3.70	3961424	4.56
3961423	3.36	3970551	3.90	3963840	4.88
3970552	5.13, hardened				

CORVETTE: 1980 (Dana)

14024327 3.07
Accepts aftermarket Dana 44 gears and cases.

VEGA/MONZA: 1974-present with 7½-inch ring gear

560196	2.29	560203	2.93	560209	3.42
560200	2.56	560206	3.08	560211	3.73
560202	2.73	560208	3.23	560212	4.10

12-BOLT

3961433	2.29	3961429	3.90, hardened
3961434	2.56	3961409	4.10
3961435	2.73	3917971	4.10, hardened
3961407	3.07	3961430	4.33, hardened
3961405	3.31	3961410	4.56
3961406	3.55	3917973	4.56, hardened
3961408	3.73	3961404	4.88
3917972	4.88, hardened		
3961192	4.88, high-impact, drag-racing		
3961422	5.13, hardened		
3961195	5.13, high-impact, drag-racing		

5.542 WASHER. Differential Side Gear Thrust

3957940 For 22-plate Posi

5.560 SEAL. Propeller Shaft Universal Trunion

3982239 Seal assembly, Corvette (Prop shaft pinion flange oil)

5.812 BOLT. Hub or Drum

3910340 HD rear axle, 7/16-20x1.75 long
3819780 HD rear axle, ½-20x2⅞ long
3849110 HD rear axle, ½-20x1.75 long

5.822 SEAL. Rear Axle Shaft Bearing

3958078 For axle 3945184

GROUP 6.000 FRONT AXLE—STEERING

6.020 KNUCKLE ASSEMBLY. Steering

3954875 Steering knuckle arm LH, 5.312-inches long, '67-'69 Camaro
3954876 Same as 3954875, but RH

6.307 HUB AND DISC ASSEMBLY. Front

3991041 11 ¾ inch disc, 1967-'69 Camaro HD brakes

6.508 GEAR ASSEMBLY. Steering

7806396 20:1 ratio, '67-'75 Camaro
7819935 16:1 ratio, Monza/Vega

6.859 ARM. Pitman

3953225 5.75-inches long

GROUP 7.000 SPRINGS—STABILIZER SHAFT—SHOCKS—BUMPER

7.241 SHAFT. Stabilizer

356534 Late Z28 ('77-'78), 1¼-inch diameter, front, backfits to '70
3983084 Late Z28, rear, backfits to '70
3955782 '69 Z28, backfits to '67, front
3984557 '70-'74 Z28, 1-inch diameter, front
3983085 '70-'75 Z28, rear
334930 Corvette, front, standard, 13/16-inch
3831971 Corvette, front, standard, ¾-inch
3871318 Corvette, front, HD, ⅞-inch
351596 Corvette, front, '75-'78 FE7, 1 ⅛ inch
3967713 Corvette, rear, HD, .562-inch
351597 Corvette, rear, '75-'78 FE7, .44-inch

7.243 BUSHING. Stabilizer Shaft

388294 Late Z28, front, 2 required (1970-present)
480913 Late Z28, rear, 2 required (insulator) (1970-present)
486960 Late Z28, rear, 2 required (bushing)
3760439 '67-'69 Camaro, front, 2 required
3983072 '70-'74 Z28, rear support clamp, 2 required
3986427 Corvette, for ¾-inch bar, 2 required
3923674 Corvette, for 13/16 bar, ⅞ bar, 2 required
351595 Corvette, for 1 ⅛ bar, 2 required
481432 For .562-inch Corvette rear bar, 2 required
351600 For .44-inch Corvette rear bar, 2 required

7.345 ABSORBER. Shock

5535623 Late Camaro, front, 2 required (1978-'79)
4993544 '76-'77 Camaro, front, HD, 2 required
3188143 '76-'77 Camaro, rear, 2 required
4993545 Late Camaro, rear, 2 required (1978-'79)

3196957 Corvette, HD front, 2 required
4940459 Corvette, HD rear, 2 required
4993538 Corvette, HD FE7 (1978-'80), 2 required

7.412 SPRING, Front Coil

All Chevy springs, including late Corvette & Z28, are computer-matched to the weight of the vehicle and its options. Following are Corvette springs recommended for competition. For other cars or standard Corvette replacement springs, see your Chevy dealer.

3832518 HD suspension, RPO F41, 550 lb./inch rate, lowers vehicle approximately 1 inch
3986032 HD suspension, Corvette "Daytona", 860 lb./inch rate, lowers vehicle approximately 1-inch
3955727 '67-'69 Camaro, front coil, 400 lb./inch rate

7.425 SPACER, Front Spring

3789664 Spacer, coil spring, '67-'69 Camaro, ⅜-inch thick
3789665 Spacer, coil spring, '67-'69 Camaro, 9/16-inch thick

7.503 SPRING ASSEMBLY, Rear

370350 HD suspension, Corvette RPO F41, 1 required 305 lb./inch load
6258056 HD suspension, Corvette "Daytona", 1 required, 450 lb./inch load
14022104 Corvette '80 FE7, 1 required

7.516 BOLT, Rear Spring

3889964 U-bolt, extra-long for thick rear leaf springs

7.832 COVER—MOLDING, Front Bumper & Grille

3965793 Front air dam & grille, 1975-'80 Monza, racing fiberglass replacement for production unit

7.833 ATTACHMENT UNIT—DAM UNIT, Bumper Valance Panel

358377 '75-'79 Monza, production front air dam, LH
358378 Same as 358377, but RH
366210 Front air dam, lower extension, 1975-'80 Monza, for racing, must be used with 3965793 (Group 7.832)
366275 Chevette front air dam, offroad
3965771 '71-'73 Vega front air dam
3965776 '74-'76 Vega front air dam
3943249 '67-'68 Camaro front air dam
3938689 '69 Camaro front air dam
3990469 '70-'73 Camaro front air dam, except RS
3990470 '70-'73 Camaro front air dam, RS
339496 '74-'77 Camaro front air dam, except '77 Z28
339495 Attachment unit for 339496
3943251 Brace, outer, for 3943249, 2 required
3943253 Brace, center, for 3943249, 1 required
3990477 Brace, for 3990469 & 3990470, 2 required
339497 Brace, for 339496, 2 required
459137 '77 Z28 front air dam, LH
459138 '77 Z28 front air dam, RH
14000413 '78-'80 Z28 front air dam, LH
14000414 '78-'80 Z28 front air dam, RH
14000484 '78-'80 Z28 front air dam, center
14005993 '78-'80 Z28 front spoiler attachment kit
476241 '78 Corvette (Indy Pace Car) front spoiler, LH (silver)
476242 Similar to 476241, but RH
476243 Similar to 476241, but center
14009716 '79 Corvette front spoiler, center (similar to 476241, but unpainted)
14009717 Similar to 14009716, but LH
14009718 Similar to 14009716, but RH
3063100 Corvette, nut, front spoiler attach
14016186 Corvette, retainer, front spoiler attach (1980)

Pre-'78 Corvette front air dam & 1980 front spoiler under Group 1.287.

GROUP 8.000 FRONT END SHEETMETAL (including Fiberglass)

8.000 PANEL—SCOOP, Hood

3966673 '68-'69 Corvette L88/ZL1 hood with built-in fresh air box, use through 1972, if retaining pop-up windshield wiper door
339173 '73-present Corvette hood, L88-style (also fits '68-'72 if windshield wiper pop-up door is deleted)
3965797 Hood panel, Monza offrad/racing (IMSA)
3965799 Hood scoop, Monza offroad/racing (IMSA)
3949708 '69 Camaro, special ducted hood
3987028 '70-'72 Chevelle, Cowl Induction hood
14013421 '80 Z28 Camaro, air induction hood

8.036 CABLE ASSEMBLY—CLIP—GASKET—PIN—PLATE—STUD, Ducted Hood ('70-'72 Chevelle, with Cowl Induction)

3975342 Hood hold-down cable assembly, 2 required
3975343 Clip, cable, 2 required
3975340 Gasket, plate, 2 required
3975341 Pin, 2 required
3975339 Plate, 2 required
3975338 Stud, 2 required

8.054 LOUVER, Hood Air Intake

14010481 '80 Z28 air induction

8.055 PLATE, Hood

3968567 Plate (ornament): "COWL"
3968568 Plate (ornament): "INDUCTION"
3961496 '69 Camaro ducted hood, plate (ornament): "302"
3963196 Corvette, ornament: "L88"
3962692 Corvette, ornament (decal): LT1

'80 Z28 ducted hood ornamentation is available in many different decal colors. See your Chevy dealer.

8.131 EXTENSION, Front Fender (except Corvette)

3965779 Monza, front fender extension, IMSA-style, RH
3965780 Same as 3965779, but LH
366271 Chevette front flare unit, LH
366272 Same as 366271, but RH

8.153 EXTENSION, Front Fender (Corvette only)

3958681 '68-'78 L88/ZL1/SCCA-legal style, for tire clearance, LH
3958682 Same as 3958681, but RH
3965781 '74-'80 IMSA-style, for racing (will backfit to '68 with modification), LH
3965782 Same as 3965781, but RH

GROUP 12.000 REAR BODY MOLDINGS & SPOILERS

12.181 LID- EXTENSION, Rear Compartment

3965787 '74-'80 Corvette rear spoiler end cap (IMSA), will backfit to '68 if late-type rear is fitted, LH
3965788 Same as 3965787, but RH
3965786 '74-'78 Corvette IMSA rear spoiler center section
3949798 '69 Camaro rear spoiler
3916633 '67-'68 Camaro rear spoiler
3974538 '70-'72 Camaro rear spoiler, 1st design (1-piece unit)
3990475 '70-'73 Camaro rear spoiler, 2nd design (3-piece unit), left quarter panel spoiler
3990476 Same as 3990475, but for right quarter panel
480161 '70-'80 Camaro rear spoiler ('70-'73 2nd design), part of
339434 3-piece unit, center section

'74-'80 Camaro rear spoiler, part of 3-piece unit, right quarter panel

339433 Same as 339434, but for left quarter panel
356675 Monza rear spoiler, center section (production)
356677 Monza rear spoiler, left quarter panel section (production)
356678 Same as 356677, but for right quarter panel
3965794 Monza IMSA spoiler end cap, RH
3965795 Same as 3965794, but LH
3965796 Monza IMSA spoiler center section
3965772 Vega rear spoiler
3965798 Monza IMSA rear deck lid
474493 '80 Citation 2-door sport coupe LH spoiler extension
474494 '80 Citation 2-door sport coupe RH spoiler extension
474490 '80 Citation 2-door sport coupe spoiler center section
474491 Similar to 474493, but for 2-door hatchback
474492 Similar to 474494, but for 2-door hatchback
474489 Similar to 474490, but for 2-door hatchback

12.182 EXTENSION, Rear End Panel, Corvette (Production)

14012968 '79 Vette rear spoiler (paint to match)
476244 '78 Vette Indy Pace Car rear spoiler, pre-painted

12.940 EXTENSION, Rear Quarter Side Outer

3958683 '68-'69 L88/ZL1 Corvette fender flare, fits all '68 & later (SCCA), LH
3958684 Same as 3958683, but RH
3965777 '74-'80 Corvette rear fender & door extension (IMSA), will backfit to '68 if late rear is installed, LH
3965778 Same as 3965778, but RH
3965791 Monza IMSA rear fender & door extension, RH
3965792 Same as 3965791, but LH
366273 Chevette rear flare (2-door only), LH
366274 Same as 366273, but RH